Super-Intense Laser-Atom Physics

NATO Science Series

A Series presenting the results of scientific meetings supported under the NATO Science Programme.

The Series is published by IOS Press, Amsterdam, and Kluwer Academic Publishers in conjunction with the NATO Scientific Affairs Division

Sub-Series

I. Life and Behavioural Sciences	IOS Press
II. Mathematics, Physics and Chemistry	Kluwer Academic Publishers
III. Computer and Systems Science	IOS Press
IV. Earth and Environmental Sciences	Kluwer Academic Publishers

The NATO Science Series continues the series of books published formerly as the NATO ASI Series.

The NATO Science Programme offers support for collaboration in civil science between scientists of countries of the Euro-Atlantic Partnership Council. The types of scientific meeting generally supported are "Advanced Study Institutes" and "Advanced Research Workshops", and the NATO Science Series collects together the results of these meetings. The meetings are co-organized bij scientists from NATO countries and scientists from NATO's Partner countries – countries of the CIS and Central and Eastern Europe.

Advanced Study Institutes are high-level tutorial courses offering in-depth study of latest advances in a field.
Advanced Research Workshops are expert meetings aimed at critical assessment of a field, and identification of directions for future action.

As a consequence of the restructuring of the NATO Science Programme in 1999, the NATO Science Series was re-organized to the four sub-series noted above. Please consult the following web sites for information on previous volumes published in the Series.

http://www.nato.int/science
http://www.wkap.nl
http://www.iospress.nl
http://www.wtv-books.de/nato-pco.htm

Series II: Mathematics, Physics and Chemistry – Vol. 12

Super-Intense Laser-Atom Physics

edited by

Bernard Piraux

Université catholique de Louvain,
Louvain-la-Neuve, Belgium

and

Kazimierz Rzążewski

Center for Theoretical Physics,
Polish Academy of Sciences,
Warsaw, Poland

Kluwer Academic Publishers

Dordrecht / Boston / London

Published in cooperation with NATO Scientific Affairs Division

Proceedings of the NATO Advanced Research Workshop on
Super-Intense Laser-Atom Physics
Han-sur-Lesse, Belgium
24–30 September 2000

A C.I.P. Catalogue record for this book is available from the Library of Congress.

ISBN 0-7923-6863-0 (HB)
ISBN 0-7923-6864-9 (PB)

Published by Kluwer Academic Publishers,
P.O. Box 17, 3300 AA Dordrecht, The Netherlands.

Sold and distributed in North, Central and South America
by Kluwer Academic Publishers,
101 Philip Drive, Norwell, MA 02061, U.S.A.

In all other countries, sold and distributed
by Kluwer Academic Publishers,
P.O. Box 322, 3300 AH Dordrecht, The Netherlands.

Printed on acid-free paper

Printed in the Netherlands.

Contents

PART III : INTERACTION OF CLUSTERS WITH VERY INTENSE FEMTOSECOND LASER PULSES

PART IV : PRODUCTION OF VERY INTENSE FEMTOSECOND LASER PULSES

PART V : STABILIZATION AND RELATIVISTIC EFFECTS IN SUPER STRONG FIELDS

PREFACE

The study of atomic systems exposed to super-intense laser fields defines an important area in atomic, molecular and optical physics. Although the concept of super-intense field has no absolute meaning, it is now usual to call an electromagnetic field super-intense when it exceeds the atomic binding field. In the case of the simplest atomic system, hydrogen in its ground state, this occurs above an intensity of 3.5×10^{16} Watt/cm^2 which is the atomic unit of intensity. Presently at the laboratory scale and in extremely short and tightly focussed laser pulses, the electric field strength reaches peak values which are of the order of $10^{16} - 10^{18}$ Watt/cm^2 in the infrared frequency regime, the prospect being that such peak intensities may be reached within a few years in a regime of much higher frequencies (XUV or even X).

The interaction of such electromagnetic fields with an atomic system has a highly non-linear character which has led to the observation of totally unexpected phenomena. There are three fundamental processes which have marked the beginning of an intensive research in the field of super-intense laser-atom physics (SILAP). These processes which only involve one atomic electron are (i) the so-called above-threshold ionisation i.e. the absorption of additional photons in excess of the minimal number necessary to overcome the ionisation potential, (ii) the generation of (very) high order harmonics of the driving field which opened the route to the development of (high brightness) coherent sources of radiation in the XUV and X-ray range, and finally (iii), the stabilization of an atom exposed to an intense high frequency field which manifests itself by a decrease of the ionisation probability with increasing field intensity. These processes which have been the main topics of the first SILAP meetings are now well understood. However, there have been new developments recently due to important technological advances in the production of intense ultrashort (a few optical cycles) laser pulses.

Presently, the main research effort is essentially moving in the two following directions: the study of the interaction of complex systems (multi-electron atoms, molecules, clusters, ...) with ultrashort laser pulses and the relativistic effects in super strong fields. This volume which contains the reports of oral communications and some posters presented at the SILAP

2000 workshop (the 7th SILAP meeting) reflects these trends and is divided into the following sections:

1. the double ionisation of complex atoms;
2. the ionisation and dissociation dynamics of molecules;
3. the interaction of clusters with very intense femtosecond laser pulses;
4. the production of very intense coherent radiation of short wavelength;
5. stabilization and relativistic effects in super strong fields;
6. Rydberg atoms and strong field coherent control.

In the case of multi-electron atoms, the central question which is addressed is the role of the electron-electron correlation in processes involving two electrons such as strong field direct double ionisation. Recent experiments have produced extremely precise data. Although various model calculations give results which are in qualitative agreement with these data, their accuracy calls for ab initio calculations which are essential for a clear and non-ambiguous identification of the actual mechanisms behind the direct double ejection. Similarly, ab initio calculations are now becoming an essential step in the study of fundamental processes in molecules such as tunneling ionisation which according to recent experimental data is not yet well understood even for the simplest molecular systems. By contrast, studies of the interaction of clusters with ultrashort and intense laser pulses are mainly experimental. They indicate that ions with substantial kinetic energy are ejected as the result of the Coulomb explosion of these clusters. Application to deuterium-deuterium fusion is under development.

The production of intense coherent sources of radiation of short wavelengths proceeds in different directions. The radiation resulting from high order harmonic generation is in general relatively weak. However, there is some hope that harmonic generation from inner-shell electrons driven by very intense laser pulses could improve this situation. Besides, there are other important schemes under investigation. Amongst them, free electron lasers and X-ray generation via stimulated recombination were discussed during the meeting.

In the super strong field regime, relativistic effects such as for instance mass-shift, retardation and spin effects occur. At the present stage, it is only in the infrared frequency regime where experiments have been performed that relativistic laser intensities are available. In the high frequency regime where important questions such as the existence of atomic stabilization are presently addressed, the discussion rests on model and ab initio calculations.

The manipulation of the dynamics of quantum systems using strong external fields is what is meant by strong field coherent control. Based on the shaping of ultrafast high intensity laser pulses, new techniques have been developed to learn about and control the evolution of Rydberg wavepackets, chemical dynamics in molecular liquids and lattice dynamics in crystalline solids.

Silap 2000 was organized as a NATO Advanced Reasearch Workshop which was held the last week of September 2000 in Han-sur-Lesse, Belgium. As both editors of this volume and organizers of the conference, we gratefully acknowledge the financial support of NATO Scientific Affairs Division. We also benefitted from the financial support of the 'Fonds National de la Recherche Scientifique' of the French community of Belgium and the 'Université catholique de Louvain'. We also want to express our thanks to all whose efforts made this workshop a real success. Particular thanks go to Armelle de Bohan, Nicole Coisman, Mohand Brouri, Hamid Nebdi and the staff of the 'Domaine des Masures'.

B. Piraux
K. Rzazewski
Louvain-la-Neuve, November 2000

SILAP 2000: AN OVERVIEW

SILAP workshop was aimed at presenting, discussing and assembling information on physical processes which result from the interaction of a quantum system (atom, molecule, cluster...) with a super-intense laser field. Here, we give a brief account of the recent developments in that field in the light of the lectures delivered during the meeting.

1. The double ionisation of complex atoms

Over the last few years, many experimental and theoretical investigations concentrated on strong low-frequency field double (and triple) ionisation of atoms. Extremely precise measurements of the dependence of the *integral* yield of multicharged ions upon de laser intensity has revealed that a non-sequential ejection of electrons, a process entirely due to electron-electron correlation, has to be assumed to understand the experimental data. Several mechanisms such as shake-off, electron rescattering and double tunneling have been proposed but a clear identification of the actual process behind the direct double ejection was so far missing. During the workshop, accurate *differential* experimental data have been discussed. Electron and ion momentum distributions have been measured with COLTRIMS (Cold Target Recoil Ion Momentum Spectroscopy) and very accurate data about the energy distribution of the two electrons have been obtained by means of coincidence measurements. It turns out that the simple models which incorporate electron rescattering with possible collisional excitation of the parent ion and subsequent laser induced ionisation give results which are in good *qualitative* agreement with the experimental data.

Nevertheless, a clear and non-ambiguous interpretation of the mechanism behind the direct double ionisation calls for ab initio calculations based on the numerical solution of the corresponding time-dependent Schroedinger equation in its full dimensionality. Grid as well as spectral methods have been presented and discussed in detail. However, the numerical simulation of the double ionisation of a complex atom in the long wave length regime stays an enormous challenge. In addition to the difficulty of disentangling single and double ionisation contributions, the amplitude of the quiver motion in that case is such that the number of angular momenta to be included is very high leading to an untractable numerical calculation at the present stage.

2. The ionisation and dissociation dynamics of molecules

Molecules exposed to strong light fields exhibit several properties in addition to those found in the ionization of atoms. Most importantly, a molecule stripped-off some electrons have an unbalanced Coulomb repulsion between the nuclei, which leads to an explosion. Study of the dynamics of this process remains a central problem both for the theory and for the experiment. A significant progress on the theory side comes from the improvement of numerical techniques for solving the time dependent Schroedinger equation and from increasing computational power. For the first time three-dimensional calculations become available for the simplest possible molecule, which is a singly ionized hydrogen molecule. A new aspect, covered in SILAP2000, is the manipulation of the molecule as a whole. Cooling and trapping of neutral atoms with the help of light and magnetic fields proved to be perhaps the most fruitful aspect of atom/laser physics of last two decades. Most of the atom manipulation methods, such as the Doppler cooling method do not work since molecular electronic spectrum is dressed with large number of vibrational and rotational levels. For this reason it is very difficult to isolate a suitable duty cycle of transitions. There are, however, new intrinsically molecular manipulation methods just proposed and tested. Large polarizability of many molecules makes them susceptible to manipulation by the electric field of the laser light or even dc field. There are also proposed and experimentally demonstrated methods of controlling the angular motion of molecules. One can not only align molecular axis but also make them rotate and accelerate this rotation up to a break-up angular velocity.

3. The interaction of clusters with very intense femtosecond laser pulses

Going one step further in complexity we come to the interaction of super-strong femtosecond laser pulses with atomic clusters. The most important new feature found here is a very effective heating of ions by the ionized and accelerated electrons. The result is that ions with substantial kinetic energy are ejected when the heated cluster explodes. Recently performed experiments with molecular deuterium clusters, highly energetic ions are capable to undergo a DD fusion. It obviously opens a new, not much traveled, route to controlled thermonuclear reaction. On the theory side only crude models are available. They are mostly based either on classical physics simulations or on statistical models such as the Thomas-Fermi model. This area of strong field physics has strong overlap with plasma physics and increasingly with nuclear physics.

4. The production of very intense coherent radiation of short wavelength

Physics of superstrong laser-atom interaction is about to acquire a new dimension. Until now, available sources of superstrong light pulses all operated in the visible and infrared regions. A quest for strong coherent source in far ultraviolet and X-ray regime is on for a long time. An emerging winner is the Free Electron Laser. This year at the Hamburg based DESY institute the first laser action at 80-180 nanometers wavelength was demonstrated. In not-to-distant future high power pulses at this range are expected. The next stage will be even bigger free electron laser which would deliver X-ray pulses. The theoretical study of the interaction of VUV pulses with atoms is just starting. At these frequencies the outer electrons of a neutral atom are too slow to respond. What happens instead is that the inner shell electrons become active. There are early predictions of rapid inner shell ionization that might lead to the production of hollow atoms with one and even two vacancies in the inner shells. The ionization process is so rapid that its speed exceeds that of the Auger reconfiguration processes. The inner shell electrons responding in a nonlinear way to the high frequency driving should also effectively produce high harmonics of the driving frequency in excess of the conventional cut-off law.

5. stabilization and relativistic effects in super strong fields

Most studies of strong field laser-atom interaction are based on nonrelativistic dynamics. Atomic electron attains relativistic velocities in an infrared field of the intensity close to one atomic unit. The most relevant lowest order relativistic corrections include the magnetic field through the velocity dependent Lorentz force and the relativistic mass correction. In fact the magnetic force manifests its importance at lower fields in a subtle effect of stabilization. The stabilization of atoms in strong laser field is a somewhat paradoxical decrease of the ionization probability with increasing intensity of the pulse. It turns out that this effect is reduced due to the magnetic interactions. Significant amount of work has been devoted to the numerical studies of the wave-packet dynamics of the particle described by the Dirac equation. The results are hard to interpret if sizeable population of the negative energy states is produced. At this stage Dirac equation as an equation for a single charged particle looses is meaning and full QED is necessary. Nevertheless interesting new strong field relativistic phenomena were found in relativistic classical simulations. Good example is a cycloatom a bound motion of a charged particle in superstrong static magnetic field and in the circularly polarized light.

6. Rydberg atoms and strong field coherent control

With the continuing progress in shaping laser pulses in the femtosecond range new ways of controlling quantum dynamics become available. Systems under study include electron wave packets in Rydberg atoms, chemical dynamics in molecular liquids and lattice dynamics in crystalline solids. Quantum control aims to go beyond observations of atomic and molecular properties, to manipulate the quantum system in a predefined way. To this end the attempt is to build a machine with a feedback. The machine would analyze the efficiency of a given pulse and try to reshape the pulse to get closer to the desired state. This direction of research may have some relevance to the important and exploding field of quantum information. The methods developed may be used, for instance, to store and read-off information from the Rydberg wave-packets.

PART I

DOUBLE IONIZATION OF COMPLEX ATOMS

ENERGY DISTRIBUTION OF TWO-ELECTRON IONIZATION OF HELIUM IN AN INTENSE LASER FIELD

R. LAFON[1], J. L. CHALOUPKA[1], B. SHEEHY[1] AND L. F. DIMAURO[1]
Brookhaven National Laboratory
Upton, NY 11973 USA
P.M. PAUL[2], P. AGOSTINI[2]
SPAM, Centre d'Etudes de Saclay
91191 Gif Sur Yvette, France
AND
K. C. KULANDER[3]
Lawrence Livermore National Laboratory
Livermore, CA 94551 USA

1. Introduction

It is well known that a neutral atom interacting with a strong laser field will ionize at sufficiently high intensity even for photon energies well below the ionization threshold. When the required number of photons becomes very large, this process is best described by the suppression of the Coulomb barrier by the laser's oscillating electric field, allowing the electron to tunnel into the continuum. As the laser intensity is increased, more tightly bound electrons may be successively liberated by this mechanism. Such a sequential multiple ionization, long accepted as a reasonable approach to the formidable problem of a multielectron atom interacting nonperturbatively with an intense electromagnetic field, provides fair estimates of the various charge state appearance intensities[1] while the tunneling rates are in excellent agreement with single ionization yields[2]. However, more accurate measurements revealed systematic and very large deviations from the tunneling rates[2, 3, 4, 5]: near appearance intensity under standard experimental conditions, the observed double ion yield is several orders of magnitude larger than predicted by the sequential rate. It soon became clear that electrons could not be considered as independent and that electron-electron correlation had to be taken into account. Dynamic correlations have been

B. Piraux and K. Rzążewski (eds.), Super-Intense Laser-Atom Physics, 3–13.

considered in several theories. First qualitatively in the shakeoff model[4]; then empirically through the e-2e cross-section in the quantum/classical three-step model[6] (tunnel ionization, acceleration by the oscillating electric field and e-2e recollision with the ion); recently through the so-called intense field many-body-S-matrix theory[7] and a purely empirical model of collective tunnel ionization[8]. The validity of these ideas has been examined using numerical models[9, 10, 11, 12, 13]. The measurement of total ion yields over a dynamic range exceeding ten orders of magnitude[2], a major breakthrough made possible by the availability of high-repetition rate lasers at the beginning of the 90's, was for a long time the only quantitative data to confront theory.

Recently, experiments have succeeded in collecting new information. Almost simultaneously, Cold-Target-Recoil-Ion-Momentum (COLTRIM) [14, 15] and electron-ion coincidence time-of-flight measurements [16, 17] were applied to the process of double-ionization in a strong field. The ion recoil momentum distribution, obtained in neon[14] and helium[15], have been interpreted as a signature of the interelectron correlation but also of the dressing of the two-electron final state[18]. Electron-ion momentum distributions taken in argon[17] have clearly revealed the tendency for the two electrons to be ejected on the same side of the nucleus with a similar momentum. Similarly, Witzel et al.[16] measured a "hot" electron energy distribution from xenon in coincidence with the doubly charged ion.

Obviously, our understanding of the strong-field double ionization of helium would benefit from a coincidence measurement. However, this measurement has been hindered by a unique set of experimental difficulties specific to helium. In this paper, we present time-of-flight energy spectra of electrons generated in the strong-field double ionization of helium and detected in coincidence with the corresponding charge state of the ion. The behavior of the two-electron distributions are examined as a function of intensity for 0.78 μm, 100 fs pulses. Helium is the ideal candidate for the study of non-sequential (NS) double ionization, its simple structure affords theoretically tractable and greatly reduces the ambiguities present in more complex atoms due to higher-order "resonant" processes. Furthermore, strong-field ionization of helium has been the subject of intensive experimental and theoretical investigations, resulting in the most comprehensive understanding in all of intense laser-atom interaction. Previous studies[2, 19, 20] have shown quantitative agreement with quasi-classical models which validated the strong field limit of single electron rescattering dynamics. However, the two electron dynamics have remained unclear. In our experiment, the electrons detected in coincidence with a doubly ionized helium atom show a remarkable enhancement at high energies compared to the electrons involved in the single ionization process. We will show that,

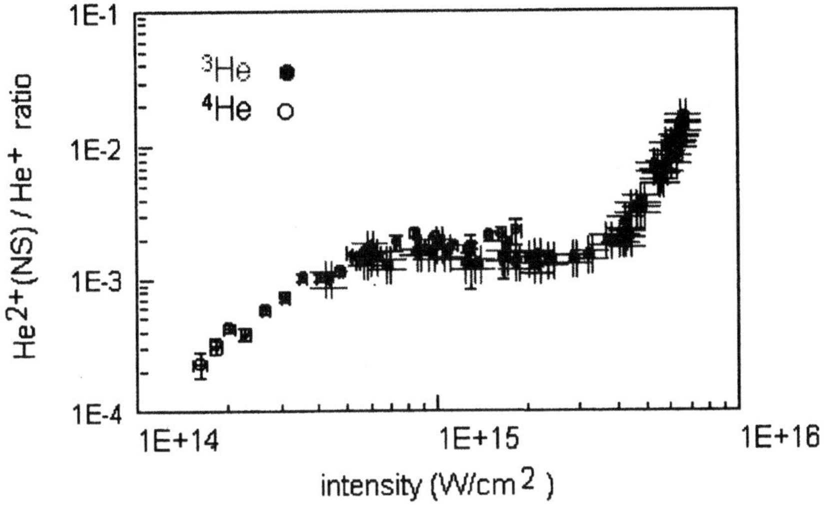

Figure 1. Plot of the ratio of He^{2+}/He^{+} for ^{4}He (squares) and ^{3}He (circles) as a function of intensity for 0.78 μm excitation. For helium, the ratio in the NS region is never larger than 0.002.

classically, this can only be due to rescattering in the backward direction.

2. Experimental Issues

A measure of the two-electron energy distribution for helium has been elusive since it poses unique experimental challenges. The most obvious is illustrated in Fig. 1. A plot of the intensity dependence of the He^{2+}/He^{+} ratio not only provides a sensitive measure of the NS dynamics but a immediate look at the problem of recording the two-electron distribution. The ratio of double-to-single ionization events is never larger than 0.002 or 1:500. Although this is also a problem for other atoms, the NS ratio in helium is at least an order of magnitude smaller. Thus, even in an ideal experimental system, e.g. perfect particle detectors and zero contamination, the problem remains the detection of the two electrons of interest from the plethora of single ionization.

In order to generate an electron distribution that correlates to a specific ion species and charge state, an ion charge-to-mass spectrum must be generated along with an electron time-of-flight spectrum. By detecting only one ion and one or more electrons in coincidence, it can be determined with some level of certainty that the detected ion and electrons were involved in the same ionization process. This level of certainty manifests itself in the number of "true" counts that are accumulated (where the electrons and ion are positively correlated) versus the number of "false", or accidental counts (where the electrons and ions are not correlated)[21]. This true-to-false ratio can be primarily improved in two ways. First, the collection and detection efficiencies of the ions and electrons must be made as close to unity as possible. In fact, if the ion efficiency is made to be perfect, then there will never be an opportunity for accidental coincidence counts. Second, the overall count rate can be lowered, making the likelihood of two ionization events per single laser shot as small as reasonable. For a certain set of experimental conditions, the true-to-false ratio can be calculated and used in the interpretation of the data.

In doing the coincidence experiment with helium, the general requirement to keep the overall count rate low is complicated by two factors. First, the ion yield of the second charge state with respect to the first is extremely low in the non-sequential region (<1/500). Therefore, if the detection rate of He^+ is kept relatively low (e.g., at an average of one detection per five laser shots), then the detection rate of He^{2+} will drop below one hit per 2500 shots. Depending on the electron detection efficiency, the rate of coincidence counts will be much lower still. In order to accumulate a significant amount of data, it is important to use a laser system at a high repetition rate and in a stable fashion over the course of many hours or even days. Second, the ionization threshold for the first charge state of helium is rather high (24 eV, corresponding to a saturation intensity for tunnel ionization of 8×10^{14} W/cm^2 at 0.78 μm and 100 fs). For many laser systems, this does not present a problem in terms of generating the peak intensity. Rather, it has the effect of enhancing the contribution of unwanted signal from the contaminants in the vacuum chamber. Since all the likely contaminants are ionized at much lower intensities, their contribution at the helium saturation intensity will come from a much larger focal volume. This effectively worsens the quality of the vacuum (e.g., the contribution from water is magnified by a factor of 250 at the helium saturation intensity). Clearly, the vacuum base pressure must be very low, and the purity of the helium introduced into the chamber must be as high as possible.

The laser used in this study was a titanium:sapphire chirped-pulse amplification system and has been described elsewhere[2]. It was operated at 1-2 kHz repetition rate and produced 100-fs pulses at a wavelength of 0.78 μm. The light was tightly focused to generate peak intensities as high as 10^{16} W/cm^2. The focal region was aligned to the center of a two-sided, pulsed-plate, ion-electron spectrometer (see Fig. 2). Electrons liberated in the ionization process are allowed to drift in a field-free region down the gold-plated flight tube towards the electron micro-channel plate (MCP) assembly. After a delay of 200 ns, plates 5 and 7 are pulsed from ground to -75V and -150V, respectively. Plates 6 and 8 remain grounded. The volt-

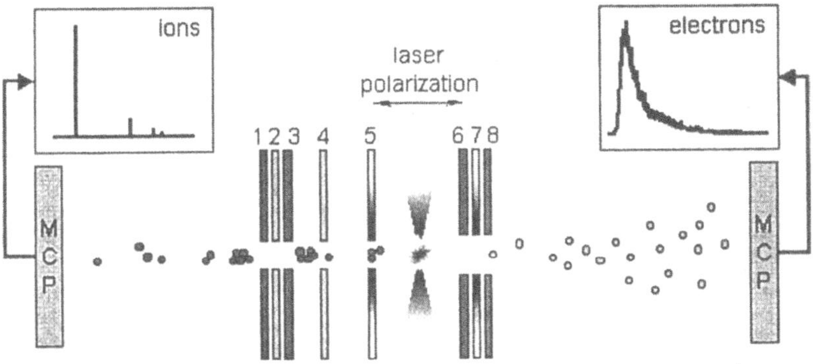

Figure 2. The two-sided electron-ion spectrometer. Electrons are allowed to drift in a static-field-free region towards the electron detector. Ions are extracted by pulsing the extraction plate to a negative voltage. Simultaneous detection of electrons and ions are performed for each laser shot.

age on plate 5 extracts the ions from the focal region, while the voltage on plate 7 blocks any secondary electrons generated by ion-plate impact from entering the electron flight tube. The delay of 200 ns is long enough to allow nearly all of the liberated electrons to drift into the protected, field-free flight tube before the turn-on of the plate voltages. The delay is also short enough that the ion displacement due to thermal motion does not affect their extraction towards the ion MCP. Both of these issues were verified in grounded and static-field tests. The electron detector subtends a solid angle of 10° and has an absolute collection and detection efficiency of roughly 1%. The voltages on plates 1 through 5 are tuned to optimize the extraction efficiency and resolution of the ion detector, giving it an absolute collection and detection efficiency of roughly 30%. The spectrometer sat within an ultra-high vacuum chamber with a nominal base pressure of 10^{-10} torr. The gas to be studied was delivered via a leak valve into the entire chamber, while continually pumping with turbo-molecular pumps. In our experiment, ^{3}He is used instead of ^{4}He to enhance the measurement sensitivity by removing the nearly degenerate m/q ratio between $^4He^{2+}$ and H_2^+, a major contaminant. To aid in the reduction of water contamination through the gas line, a liquid-nitrogen-cooled coil was used at times.

3. Coincidence Measurement

The spectrometer was first tested in non-coincidence experiments. Previously studied electron spectra and ion ratio curves were reproduced in both static field and pulsed plate modes. Then, to test the apparatus in a coincidence-style experiment, a controlled gas mix of krypton and xenon

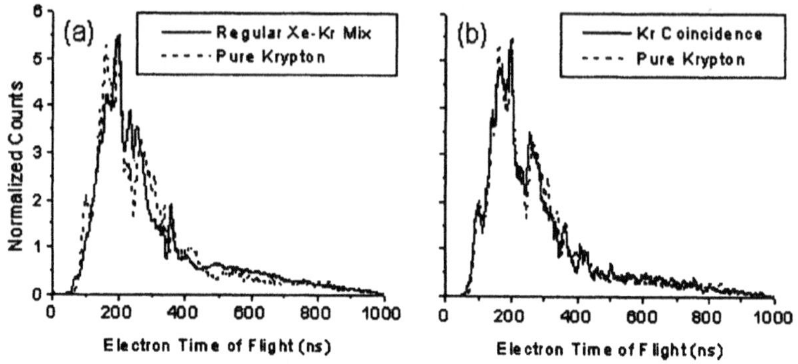

Figure 3. The Xe/Kr gas mix test. a) The ordinary electron spectrum generated with the gas mix (solid line) and the pure krypton spectrum (dashed line). b) The krypton coincidence spectrum taken with the gas mix (solid line) and the pure krypton spectrum (dashed line). All curves are normalized by their integrated values, smoothed from 1-nsec to 10-nsec resolution and uncorrected.

was used. First, the ordinary electron energy spectra of pure krypton and pure xenon were taken at an intensity of 4×10^{13} W/cm^2. Next, a coincident measurement using the gas mix was performed, giving a xenon:krypton ion count rate of 8:1 and an overall xenon detection rate of one hit per ten laser shots. The data from 65.9 million laser shots was accumulated at a repetition rate of 1 kHz over a total run time of nearly 20 hours. It consists of the total, non-coincident electron and ion spectra, as well as the ion and electron arrival times when only one ion is detected. With this information, it is straightforward to compile the coincident electron spectrum for any of the detected ions. Figure 3(a) shows the total electron time-of-flight spectrum (solid line) generated with thc gas mix, as well as the krypton spectrum (dashed line) generated with the pure gas. Both curves are normalized by their integrated values and are smoothed from 1-nsec to 10-nsec resolution. The curves are dissimilar since only 1/10 of the total spectrum is made up of electrons from krypton, their features are hidden by the more abundant xenon electrons. Figure 3(b) shows the krypton coincidence spectrum (solid line), which has been extracted from the total spectrum, and the pure krypton spectrum (dashed line). The much improved agreement is unmistakable. The calculated true:false ratio for this test was roughly 3:1, and the curve shown is uncorrected and is made up of a total of 7375 coincident electron hits. This test shows, in a clear and controlled fashion, the effectiveness of our coincidence technique.

Coincidence data using helium was taken at two intensities: the single ionization saturation intensity ($I_{sat}3D8 \times 10^{14}$ W/cm^2) and half of this value. Figure 4 shows the energy spectra for electrons detected in coincidence with the single and double ionization of ^{3}He at these two intensities. The curves are normalized by their integrated value, and the horizontal axis shows the electron energy as well as the ponderomotive potential (U_p) associated with the peak laser intensity. All of the curves are uncorrected for false counts and include representative error bars that reflect counting statistics and the true:false ratio. Figure 4(a) shows the energy spectra for electrons correlated to single (solid line) and double ionization (circles) at I_{sat}. A total of 205 million laser shots were taken. The detection rate of $^3He^+$ was roughly one in five, and the rate of background ions was one in three. The ratio of double to single ionization was 0.0013. These high count rates were used to improve the counting statistics on the double ionization electron spectrum, resulting in an overall true:false ratio that was calculated to be roughly 1:1. However, for the high energy electrons, where the abundance of "false" electrons is very small, the true:false ratio is much greater. For example, at 100 eV, the effective true:false ratio is 100:1. This is simply due to the paucity of high-energy electrons that can contribute to accidental counts. As a result, correcting the double ionization spectrum by subtracting away false counts has very little effect on the electron distribution beyond 50 eV. Figure 4(b) shows the energy spectra generated at 1/2 I_{sat}. A total of 190 million laser shots were taken. The detection rate of $^3He^+$ was roughly one in four, and the rate of background ions was one in five. The ratio of double to single ionization was 0.00059. Again, the overall true:false ratio was calculated to be roughly 1:1, but, as before, the effective ratio at high energies is much improved. The dramatic enhancement of these spectra at high energies for electrons involved in the double ionization process is readily apparent at both intensities. In fact, when scaled to U_p, it appears that the NS distributions are nearly identical.

4. Discussion

The double ionization electron spectra in Fig. 4 show some remarkable differences compared to the single ionization spectra. The single ionization features are well understood. The rapidly decreasing distribution for energies smaller than $2U_p$ is a result of the tunneling rate and the drift velocity corresponding to the phase at which the electron is set free in the field[6]. The long plateau extending to $10U_p$ is due to the electrons that are born after the peak of the field that return to the core where they backscatter elastically and are subsequently accelerated by the field to high energies. In contrast, the double ionization spectra of Fig. 4 show a much slower rate of decrease. They extend at least to $4U_p$ with profiles similar to that of the rescattering plateau of single ionization. The fact that the coinci-

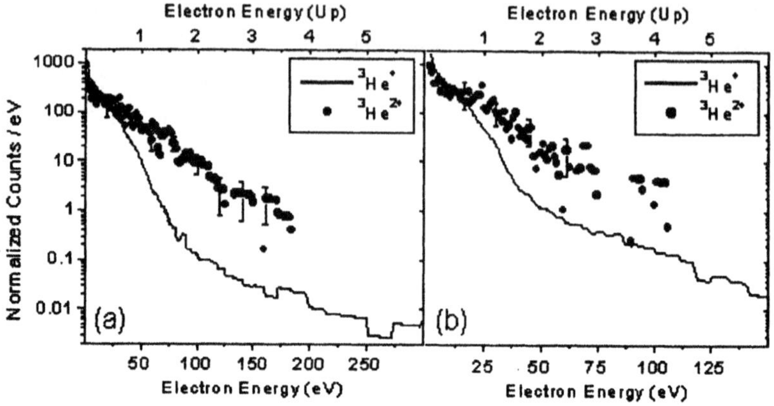

Figure 4. The energy spectra for electrons detected in coincidence with $^3He^+$ (dashed line) and $^3He^{2+}$ (solid line) at a) the single ionization saturation intensity and b) one-half of the saturation intensity. The enhancement of the high-energy electrons coming from the double ionization process is dramatic at both intensities, and appears to scale with the ponderomotive potential (U_p) associated with the peak intensity. Representative error bars reflect the counting statistics and the true:false ratio. All curves are uncorrected for false counts. Correcting the double ionization curves has a negligible effect on electrons above one U_p.

dence electrons can have energies beyond $2U_p$ clearly implies the release of electrons into the field with an initial non-zero backward velocity. In the rescattering model[6], an electron returning to the core with a kinetic energy larger than the binding energy of the second electron can liberate the second electron, giving rise to two electrons sharing the excess energy. The final electron energies depend on the amplitude and direction of the electron initial velocities, with the largest kinetic energies being reached for backward initial velocities. Measurements on e-2e triply differential cross sections (TDCS) for hydrogen and helium show that a substantial fraction of the electrons are scattered into the backward direction, particularly near threshold[22, 23]. Even at high collision energies, it is most likely that when one electron carries off most of the energy in the forward direction that the slower electron will emerge in the backward direction.

To illustrate the consequence of backscattering on the final electron energies, a simple 1D classical calculation was performed. Electron trajectories were initiated at all phases of the field, and were followed after the e-2e ionizing collision. Collision energies larger than the separation energy (40 eV) between the ground and first excited state in singly ionize helium are assumed to satisfy the condition for ionizing the second electron via an e-2e process. Those initial phases for an intensity of 0.8 PW/cm^2 are shown by the shaded region in Fig. 5. The excess energy is assumed to be

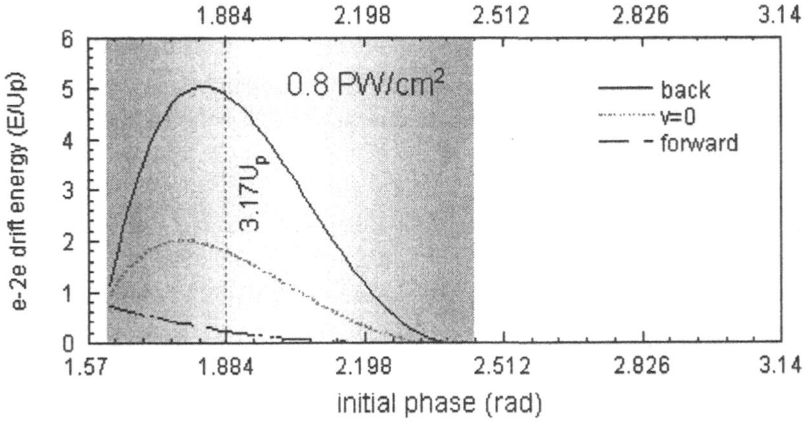

Figure 5. Plot of initial phase versus final drift energy after an inelastic e-2e collision. The shaded area indicates the initial phases having return energies equal to or greater than the He^+ binding energy of 40 eV. The results are calculated for an intensity of 0.8 PW/cm^2. The dotted vertical line indicates the initial phase which produces the maximum return energy of $3.17U_p$.

taken by one electron either in the forward (sign of velocity unchanged by the collision) or backward (sign of velocity changed by the collision) direction and the final drift energy is calculated. The forward (dashed line) and back (solid line) scattering curves are plotted in Fig. 5, along with a curve (dotted line) assuming zero velocity after the e-2e process. The zero velocity curve has an expected upper bound of $2U_p$ energy. However, any initial velocity along the forward direction will result in a final drift energy less than $2U_p$. In fact at the saturation intensity, the final energy is substantially reduced with a maximum around $0.75U_p$. Clearly, the measured double ionization electron distributions in Fig. 4 have energies well above $2U_p$ and the result in Fig. 5 shows that back scattering will produce energies up to $5U_p$. The calculation over-simplifies the problem but it unequivocally shows the essential role that backscattering must play in any rescattering e-2e process.

We conclude from these calculations, that the observed relatively flat electron energy distributions that extend out to $4U_p$ reflect the high probability of the scattered or ionized electron emerging initially in the backward direction. Since this is known to be the case in field-free e-2e measurements, the dramatic differences between the single and double ionization electron spectra are not surprising. Our results, in agreement with recent measurements [14, 15, 16, 17] are therefore compatible only with a rescattering mechanism of NSDI. However, our measurements clearly show electron energies larger than $2U_p$, implying drift motions with backward non-zero initial velocities. This is very different from that reported in argon[17].

In conclusion, our electron-ion coincidence measurements in helium, presented here for the first time, provide new support for the rescattering model of NSDI. They also provide new information about the maximum energy of the photoelectrons produced in the strong field NSDI of helium. This serves to emphasize the role of backward electron emission during the e-2e event in agreement with known TDCS in hydrogen and helium. It is expected that this work will stimulate full quantum calculations of the two-electron energy distribution.

5. Acknowledgments

The experiments were carried out at Brookhaven National Laboratory under contract No. DE-AC02-98CH10886 with the U.S. Department of Energy and supported by its Division of Chemical Sciences, Office of Basic Energy Sciences. P. A. acknowledge travel support from NATO under Contract No. SA.5-2-05(RG910678). This work was also carried out in part under the auspices of the U. S. Department of Energy at the Lawrence Livermore National Laboratory under contract No. W-7405-ENG-48.

References

1. Gibson, G., Luk, T.S. and Rhodes, C.K. (1990) Tunneling ionization in the multiphoton regime, *Phys. Rev. A* **41**, pp. 5049–5052.
2. Walker, B. *et al.* (1994) Precision measurement of strong field double ionization of helium, *Phys. Rev. Lett.* **73**, pp. 1227–1230.
3. L'Huillier, A., Lompre, L.A., Mainfray, G. and Manus, C. (1983) Multiply charged ions induced by multiphoton ansorption in rare gases at 0.53 μm, *Phys. Rev. A* **27**, pp. 2503–2512.
4. Fittinghoff, D.N., Bolton, P. R., Chang, B. and Kulander, K. C. (1992) Observation of nonsequential double ionization of helium with optical tunneling, *Phys. Rev. Lett.* **69**, pp. 2642–2645.
5. Larochelle, S., Talebpour, A. and Chin, S.L. (1998) Non-sequential multiple ionization of rare gas atoms in a Ti:Sapphire laser field, *J. Phys. B* **31**, pp. 1201–1214.
6. Corkum, P.B. (1993) Plasma perspective on strong field multiphoton ionization, *Phys. Rev. Lett.* **71**, pp. 1994–1997.
7. Becker, A. and Faisal, F.H.M. (1999) Interplay of electron correlation and intense field dynamics in the double ionization of helium, *Phys. Rev. A* **59**, pp. R1742–R1745.
8. Eichmann, U. *et al.* (2000) Collective multielectron tunneling ionization in strong fields, *Phys. Rev. Lett.* **84**, pp. 3550–3553.
9. Watson, J.B. *et al.* (1997) Nonsequential double ionization of helium, *Phys. Rev. Lett.* **78**, pp. 1884–1887.
10. Parker, J.S., Smyth, E.S. and Taylor, K.T. (1998) Intense-field multiphoton ionization of helium, *J. Phys. B* **31**, pp. L571–L578.
11. Kornberg, M.A. and Lambropoulos, P. (1999) Photoelectron energy spectrum in "direct" two-photon double ionization of helium, *J. Phys. B* **32**, pp. L603–L613.

12. Panfili, R., Szymanowski, C., Liu, W.-C. and Eberly, J.H. (2000) Spectroscopy in the neighborhood of the double ionization knee, in L.F. DiMauro, R.R. Freeman, K.C. Kulander (eds.), *Multiphoton Processes*, American Institute of Physics, New York, pp. 265–272.
13. Yudin, G. and Ivanov, M. (2000) Physics of correlated double ionization of atoms in intense laser fields: quasistatic tunneling limit, *Phys. Rev. A*, accepted.
14. Moshammer, R. *et al.* (2000) Momentum distributions of Ne^{n+} ions created by an intense ultrashort laser pulse, *Phys. Rev. Lett.* **84**, pp. 447–450.
15. Weber, Th. *et al.* (2000) Recoil-ion momentum distributions for single and double ionization of helium in strong laser fields, *Phys. Rev. Lett.* **84**, pp. 443–446.
16. Witzel, B., Papadogiannis, N.A. and Charalambidis, D. (2000) Charge-state resolved above threshold ionization, *Phys. Rev. Lett.* **85**, pp. 2268–2271.
17. Weber, Th., *et al.* (2000) Correlated electron emission in multiphoton double ionization, *Nature* **405**, pp. 658–661.
18. Becker, A. and Faisal, F.H.M. (2000) Interpretation of momentum distribution of recoil ions from laser induced nonsequential double ionization, *Phys. Rev. Lett.* **84**, pp. 3546–3549.
19. Walker, B., Sheehy, B., Kulander, K.C. and DiMauro, L.F. (1996) Elastic rescattering in the strong field tunneling limit, *Phys. Rev. Lett.* **77**, pp. 5031–5034.
20. Sheehy, B. *et al.* (1998) Single- and multiple-electron dynamics in the strong-field tunneling limit, *Phys. Rev. A* **58**, pp. 3942–3952.
21. Stert, V., Radloff, W., Schulz, C.P. and Hertel, I.V. (1999) Ultrafast photoelectron spectroscopy: femtosecond pump-probe coincidence detection of ammonia cluster ions and electrons, *Eur. Phys. J. D* **5**, pp. 97–106.
22. Roder, J., Ehrhardt, H., Bray, I. and Fursa, D.V. (1997) Absolute double differential cross sections for electron-impact ionization of helium, *J. Phys. B* **30**, pp. 1309–1322.
23. Rioual, S. *et al.* (1998) Absolute (e,2e) cross sections for the electron-impact ionization of helium in energy sharing kinematics at 44.6 eV, *J. Phys. B* **31**, pp. 3117–3127.

DOUBLE IONIZATION IN STRONG FIELDS: ION MOMENTA AND CORRELATED ELECTRON MOMENTA

A. STAUDTE AND T. WEBER
Institut für Kernphysik,
Universität Frankfurt,
Germany

H. GIESSEN, G. URBASCH AND M. VOLLMER
Fachbereich Physik,
Universität Marburg,
Germany

M. WECKENBROCK, O. JAGUTZKI AND L. SPIELBERGER
Institut für Kernphysik,
Universität Frankfurt,
Germany

AND

R. DÖRNER
Fakultät für Physik,
Universität Freiburg,
Germany

What happens when an atom is subjected to very intense light of a wavelength long compared to the ionization potential? We review recent experiments [1, 2, 3] in which this question is addressed experimentally by measuring the momenta of doubly charged Helium and Argon and the momentum correlation between the two emitted electrons from double ionization in the focus of a Ti:Sapphire laser at intensities of several $10^{14} W/cm^2$.

Already soon after short pulses with sufficient power became available it was discovered that double ionization occurs orders of magnitude more likely than what one would expect from a process in which each of the electrons couples to the field independently (see e.g. [4, 5] and [6] for a recent review). It became clear that electron-electron correlation was important to enable this process [7, 8, 9, 10, 11, 12, 13, 14]. The situation is comparable

B. Piraux and K. Rzążewski (eds.), Super-Intense Laser-Atom Physics, 15–23.

to ionization by synchrotron radiation. Here only one photon is absorbed and without electron-electron correlation the probability of ejecting the second electron would be zero. For synchrotron radiation a ratio of double to single ionization of up to about 3% (depending on the photon energy) is found. With synchrotron radiation today many very detailed experimental and theoretical studies of energy and angular distribution of both emitted electrons and the recoiling doubly charged ion are available (see eg. [15] for the pioneering experiment, [16, 17, 18] for studies using the technique used in the present study and [19] for a recent review). For the case of double ionization by femto second laser impact however, the experiments are still in their infancy. Until 1999 only total ion yield measurements were available. The reason is, that coincidence experiments in strong laser fields are extremely difficult. To assure that two fragments (two electrons or an electron and an ion) which are detected in coincidence originate from the same atom, the total single ionization rate in the focus (not only the number of particles detected in the detector) has to be far below unity. Given that the ratio of double to single ionization is only 10^{-4} for Helium and a few % for Argon extremely low counts rates are achievable with 1kHz repetition rates lasers. To facilitate such studies it is imperative to use spectrometers which have 4π solid angle for all particles to be detected.

In the present study we have used COLTRIMS (Cold Target Recoil Ion Momentum Spectroscopy) for momentum space imaging of ions and electrons. This technique has been developed originally for the study of ion impact ionization and has than be successfully used for experiment with synchrotron radiation, antiproton impact and electron impact (see [20] for a recent review). The laser is focussed onto a very dilute supersonic beam (background pressure 10^{-10} mbar, local jet gas pressure between 10^{-8} and 10^{-5} mBar. The gas pressure is adjusted such that only 0.1-0.3 ions are created per laser pulse. The ion and the electrons are guided by weak electric fields towards two position sensitive channel plate detectors. From the time of flight and the position of impact of each particle the charge state and the three dimensional momentum vector is obtained.

For single ionization the momentum of the recoiling ion is equal and opposite to the ejected electron momentum, if the pulse is short enough, that the electron does not leave the focus. For the coincident detection of electron and ion this allows an online monitoring of the events where more than one ion per pulse was produced and the ion and electron detected result from different atoms. Such events are located outside the diagonal of momentum conservation of figure 2. The width of the diagonal also shows the overall resolution of the system. It is about 0.1 a.u. for helium and

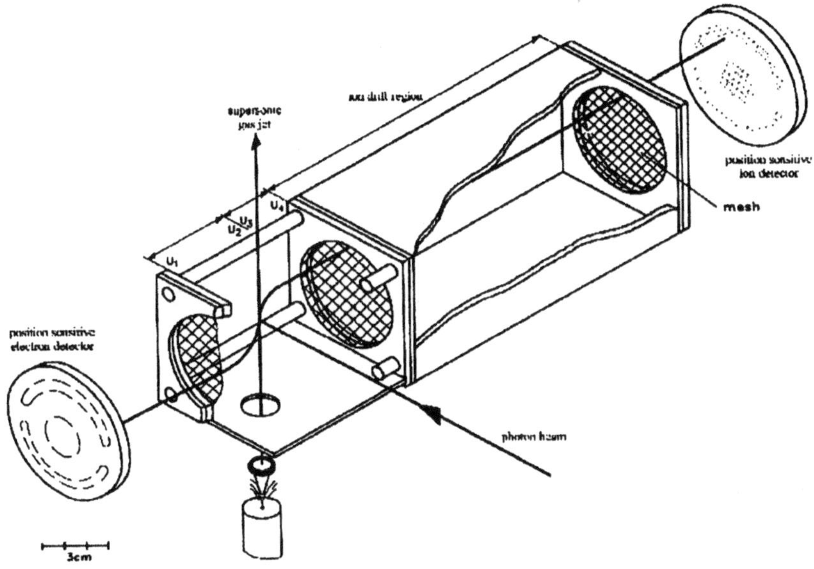

Figure 1. Typical COLTRIMS setup. The gas nozzle is cooled to 15-30 K, The laser is focussed into the supersonic gas jet. The electron detector is located on the left side of the spectrometer and the ion detector on the right side. (from [21]). For each particle the position of impact on the detector and the time of flight with respect to the laser pulse is measured.

about 0.4 a.u. for argon, since the gas jet yields much more efficient cooling for helium than for argon.

The ion momentum distribution for doubly charged He ions is shown in figure 3 for different intensities. At the highest intensity a double peak structure is found.

How does the ion receive its momentum? In the direction transverse to the electric field, the ionic momentum is solely a result of electron-ion interaction during the ionization process. This could be for example a scattering of the electrons at the core or a reminiscence of the initial state momentum distribution. Figures 3 d-f show that this transverse momentum distribution becomes wider with increasing peak power. In the rescattering model [22, 9], this transverse momentum would be mainly a result of the momentum transfer in the (e,2e) collision. However, the distributions shown in figure 3d-f are about a factor 2-4 narrower than the transverse recoil-ion momentum distribution for double ionization of helium by 270-2000 eV electron impact measured by Jagutzki et al [23]. The component of the momenta parallel to the polarization is mainly a result of the acceleration

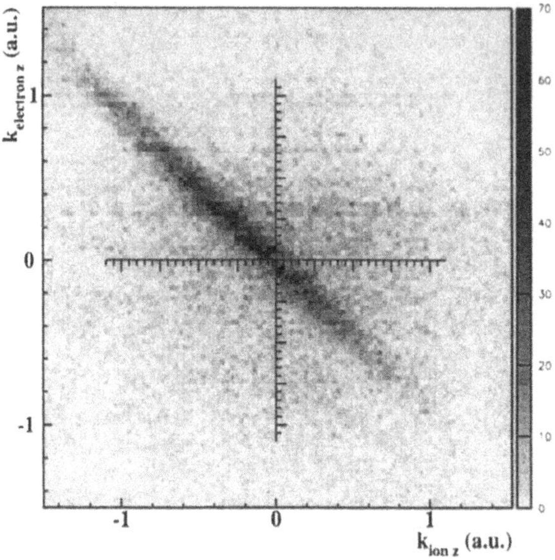

Figure 2. Momentum correlation between He^{1+} ion and electrons created in the focus of a 220 fsec, 800nm laser pulse at peak a intensities of $4 \cdot 10^{14}$ W/cm^2. The horizontal axis shows the momentum component of the He^{1+} ion along the polarization of the laser field, the vertical axis the momentum of the detected electron.

of the ion in the field. For an estimate, we assume that the first electron is removed at time t_1 and the ion switches its charge from 1^+ to 2^+ at time t_{12}. If both electrons are removed without momentum transfer to the ion, the ion momentum at the time t_∞ at the end of the laser pulse with envelope of the electric field strength E(t) is given by (in atomic units)

$$p_z^{He^{2+}}(t_\infty) = \int_{t_1}^{t_{12}} E(t)\sin\omega t dt + 2\int_{t_{12}}^{t_\infty} E(t)\sin\omega t dt. \qquad (1)$$

For the TS1/shake process, which is dominant for double ionization by synchrotron radiation, one might assume $\omega(t_1 - t_{12}) \ll \pi$, i.e. if the double ionization takes place in a time interval short compared to the optical cycle. Then equation 1 describes the motion of a particle with charge 2^+ in the oscillating field. The final momentum depends on the phase ωt_1 at the instant of creation of the ion. The maximum value of $2\sqrt{4U_p}$ (shown by the full arrows in figure 3(a-c)) corresponds to the creation of the ion when the optical field strength is zero. The data show clearly that the He^{2+} ion yield is strongly suppressed in the region approaching $2\sqrt{4U_p}$. The width of the peak in figure 3c corresponds to $\pm$ 1/4 field period centered around

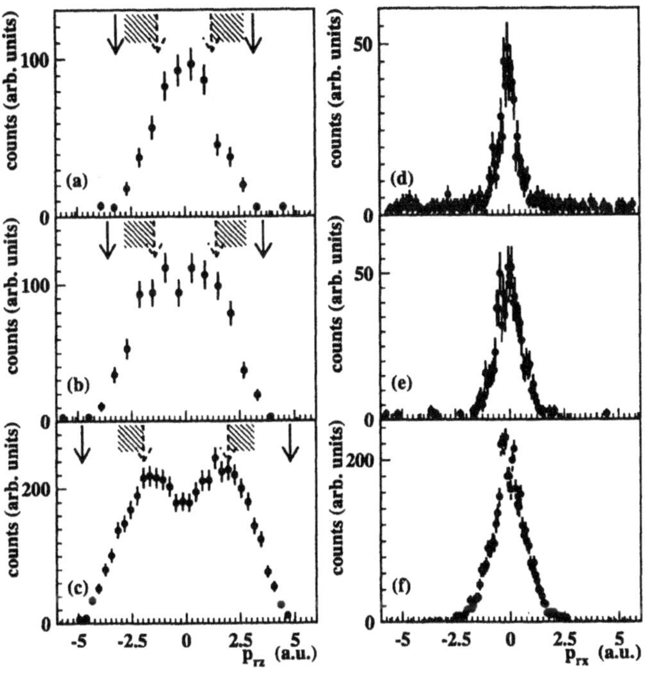

Figure 3. (a-c) distribution of He^{2+} ion momenta in the direction of the polarization integrated of the two momentum components perpendicular to the polarization, (d-f) momentum distribution in the direction of the photon propagation. The full arrows indicate a momentum of $2\sqrt{(4U_p)}$ which is the upper bound if the two electrons are liberated in a time interval short compared to the optical cycle, the dashed arrows indicate the momenta in a rescattering model without momentum transfer in the (e,2e) collision the hatched area is an estimate in the rescattering model with momentum transfer (see text). The peak intensities are $2.9\cdot 10^{14}$ W/cm^2 (a and d), $3.8\cdot10^{14}$ W/cm2 (b and e) and $6.6\cdot 10^{14}$ W/cm^2 (c and f) (from [1]).

the maximum field strength. This is consistent with prediction of a time dependent calculation of the He^{2+} yield of Parker and coworkers [24].

Contrary to the shake-off/TS1 process, in the rescattering model there is a significant time delay between the liberation of the first electron and its re-encounter with the ion which leads to double ionization. An estimate of the integral for the rescattering process is shown by the dashed arrow. This neglects a possible momentum transfer in the recollision process. The hatched area show an estimate including this momentum transfer. These findings are in agreement with results of a similar experiment performed by Moshammer and coworkers for Ne double and triple ionization [25]. Recently Becker and Faisal succeeded in the first calculations of recoil ion momentum distributions and could reproduce the observed double peak

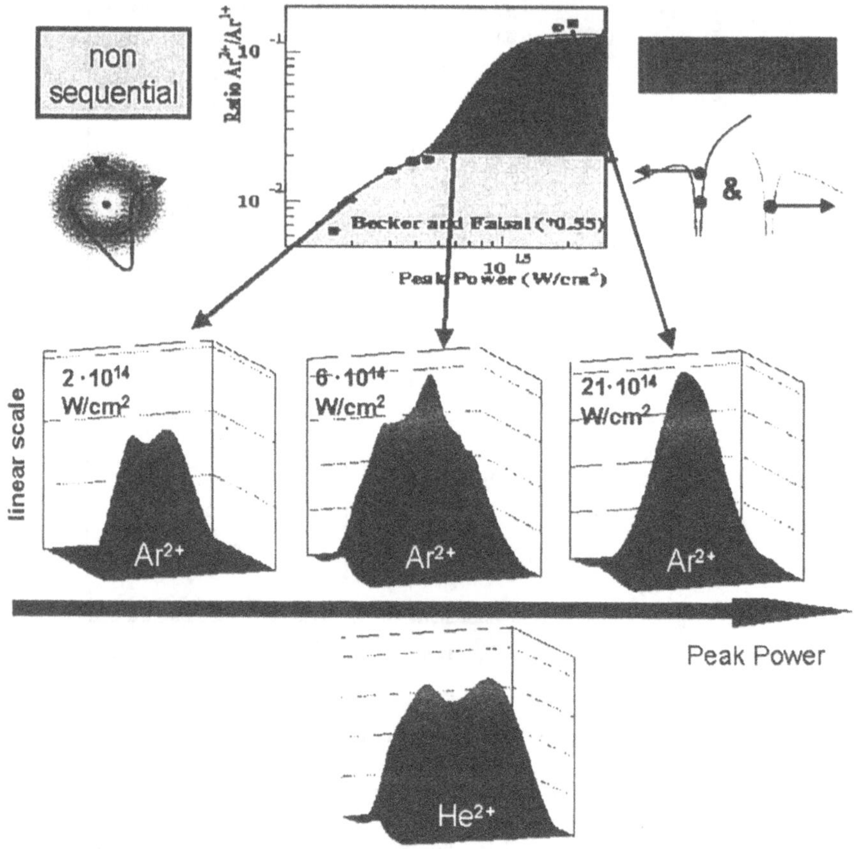

Figure 4. Evolution of the ion momentum distribution with increasing light intensity. Light intensities are pointed out in a distribution that shows the Ar^{2+}/Ar^{+} yield vs. intensity. A drastic change in the argon ion momentum distribution can be observed when entering the sequential ionization regime. Meanwhile at $6 \cdot 10^{14}$ W/cm^2 the helium distribution still shows the double peak. Measurements at intensities high enough to turn on sequential double ionization in helium are difficult due to the strong increase in signal from the residual gas.

structure [26]. They could show that it is a result of the final state interaction of the laser field with the electrons (and the ion).

For argon atoms we succeeded in measuring the ion momentum distribution also at intensities where double ionization proceeds sequentially [2]. In figure 4 the evolution of the argon ion spectrum with increasing laser intensity is shown. With the onset of sequential ionization the momentum distribution becomes narrower and peaks at zero momentum. It resembles

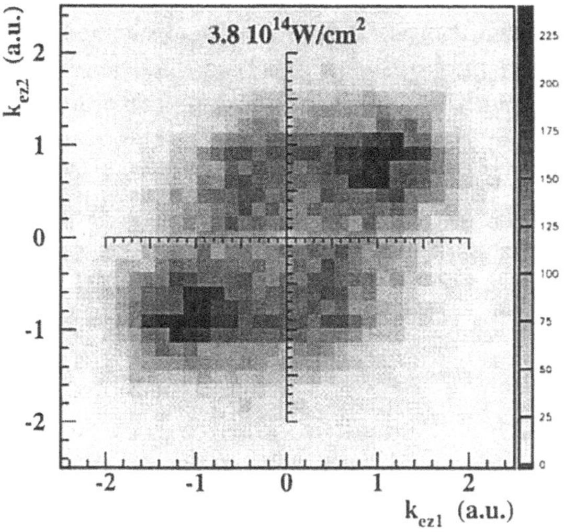

Figure 5. Momentum correlation between the two emitted electrons when a Ar^{2+} ion is produced in the focus of a 220 fsec, 800nm laser pulse at peak intensities of $3.8 \cdot 10^{14}$ W/cm^2. The horizontal axis shows the momentum components of one electron along the polarization of the laser field, the vertical axis the same momentum component of the corresponding second electron. Same sign of the momenta for both electrons means emission to the same half sphere. The data are integrated over the momentum components in the direction perpendicular to the polarization. The grey shading shows the differential rate in arbitrary units (adapted from [3]).

a distribution of singly charged ions convoluted with itself. Figure 5 shows the momentum correlation between the two electrons emitted from an argon atom at an intensity where nonsequential ionization dominates. We find a distinct maximum for both electrons traveling to the same side with similar momentum. While electron repulsion tries to drive the electrons to opposite sides (as observed with synchrotron radiation) here obviously the action of the laser field on the electrons determines their final momenta. It is the laser field which drives both electrons to the same side. Again this indicates that both electrons are set free simultaneously with a phase shift of about 35 deg with respect to the maximum of the field. This is in agreement with the expectation within the rescattering model if one takes excitation as an intermediate step into account. Recently Sacha and Eckhardt [27, 28] proposed a classical model for double ionization, which divides the process into two stages. First, a rescattering process creates a highly excited complex of two electrons. Then, the external laser field allows this compound state to decay in such a way that the electrons escape preferredly to the same side and therefore is in good agreement with our observations. In the second and fourth quadrant in figure 5 a weak accumulation of electron momenta is distinguishable. Events in these quadrants are corresponding to double ionization with both electrons escaping to opposite sides. The nature of these events will have to be examined in the future. At the moment the technique of COLTRIMS is being widely applied in ultrashort, intense laser physics not only to atoms but to molecules as well. We expect more light will be shed on light-matter interaction in the near future.

Acknowledgments

We are indebted to Horst Schmidt-Böcking for enthusiastic support of this project and R. Moshammer and J. Ullrich for many helpful discussions. Our analysis of the influence of the laser field on the final state momenta emerged from many fruitful discussion with A. Becker, F.H.M. Faisal. This work is supported by DFG, BMBF, GSI and DAAD. R.D. acknowledges supported by the Heisenberg-Programm of the DFG. The Marburg group thanks the Land Hessen and the DFG for support through their SFB383 and their Graduiertenkolleg 'Optoelektronik mesoskopischer Halbleiter'. We are grateful to W.W. Rühle for continuous support. We thank Roentdek GmbH for providing the position sensitive detectors.

References

1. Th. Weber, M. Weckenbrock, A. Staudte, L. Spielberger, O. Jagutzki, V. Mergel, G. Urbasch, M. Vollmer, H. Giessen, , and R. Dörner (2000) *Phys. Rev. Lett.*, **84**, 443.
2. Th. Weber, M. Weckenbrock, A. Staudte, L. Spielberger, O. Jagutzki, V. Mergel, G. Urbasch, M. Vollmer, H. Giessen, , and R. Dörner (2000) *J. Phys*, **B33**, L127.
3. Th. Weber, H. Giessen, M. Weckenbrock, A. Staudte, L. Spielberger, O. Jagutzki, V. Mergel, G. Urbasch, M. Vollmer, , and R. Dörner (2000) *Nature*, **404**, 608.
4. D.N. Fittinghoff, P.R. Bolton, B. Chang, and K.D. Kulander (1992) *Phys. Rev. Lett.*, **69**, 2642.
5. B. Walker, B. Sheehy, L.F. DiMauro, P. Agostini, K.H. Schafer, and K.C. Kulander (1994) *Phys. Rev. Lett.* **73**, 1227.
6. P. Lambropoulos, P. Maragakis, and J. Zhang (1998) *Physics Reports* **305**, 203.
7. A. Becker and F.H.M. Faisal (1996) *J. Phys*, **B29**, L197.
8. A. Becker and F.H.M. Faisal (1999) *J. Phys*, **B32**, L335.
9. P.B. Corkum (1993) *Phys. Rev. Lett.* **71**, 1994.
10. K.C. Kulander, J. Cooper, and K.J. Schafer (1995) *Phys. Rev.* **A51**, 561.
11. W.C. Liu, J.H. Eberly, S.L. Haan, and R. Grobe (1999) *Phys. Rev. Lett.* **83**, 520.
12. J.B. Watson, A. Sanpera, D.G. Lappas, P.L. Knight, and K. Burnett (1997) *Phys. Rev. Lett.* **78**, 1884.
13. D.G. Lappas and R. van Leeuwen (1998) *J. Phys* **B31**, L249.
14. K.J. LaGattuta and J.S. Cohen (1998) *J. Phys* **B31**, 5281.
15. O. Schwarzkopf, B. Krässig, V. Schmidt, F. Maulbetsch, and J. Briggs (1994) *J. Phys.*, L347–50.

16. R. Dörner, J. Feagin, C.L. Cocke, H. Bräuning, O. Jagutzki, M. Jung, E.P. Kanter, H. Khemliche, S. Kravis, V. Mergel, M.H. Prior, H. Schmidt-Böcking, L. Spielberger, J. Ullrich, M. Unverzagt, and T. Vogt (1996) *Phys. Rev. Lett.* **77**, 1024; see also erratum (1997) **78**, 2031.
17. R. Dörner, H. Bräuning, J.M. Feagin, V. Mergel, O. Jagutzki, L. Spielberger, T. Vogt, H. Khemliche, M.H. Prior, J. Ullrich, C.L. Cocke, and H. Schmidt-Böcking (1998) *Phys. Rev.* **A57**, 1074.
18. H.P Bräuning, R. Dörner, C.L. Cocke, M.H. Prior, B. Krässig, A. Kheifets, I. Bray, A. Bräuning-Demian, K. Carnes, S. Dreuil, V. Mergel, P. Richard, J. Ullrich, and H. Schmidt-Böcking (1998) *J. Phys.* **31**, 5149.
19. J. Briggs and V. Schmidt (2000) *J. Phys.* **33**, R1.
20. R. Dörner, V. Mergel, O. Jagutzki, L. Spielberger, J. Ullrich, R. Moshammer, and H. Schmidt-Böcking (2000) *Physics Reports* 330, 96–192.
21. V. Mergel (1996) PhD thesis, *Universität Frankfurt* ISBN 3–8265–2067–X. Shaker Verlag.
22. K.J. Schafer, Baorui Yang, L.F. DiMauro, and K.C. Kulander (1993) *Phys. Rev. Lett.* **70**, 1599.
23. O. Jagutzki, L. Spielberger, R. Dörner, S. Nüttgens, V. Mergel, H. Schmidt-Böcking, J. Ullrich, R.E. Olson, and U. Buck (1996) *Zeitschrift für Physik***D36**,5.
24. J. Parker, K.T. Taylor, C.W. Clark, and S. Blodgett-Ford (1996) *J. Phys* **B29**, L33.
25. R. Moshammer, B. Feuerstein, W. Schmitt, A. Dorn, C.T. Schöter, J. Ullrich, H. Rottke, C. Trump, M. Wittmann, G. Korn, K. Hoffmann, and W. Sandner (2000) *Phys. Rev. Lett.* **84**, 447.
26. A. Becker and F.H.M. Faisal (2000) *Phys. Rev. Lett.* **84**, 3546.
27. K. Sacha and B. Eckhardt (2000) *Phys. Rev.*, submitted.
28. K. Sacha and B. Eckhardt (2000) *Los Alamos National Laboratory e-print* physics/0010031.

MECHANISM OF THE NONSEQUENTIAL DOUBLE IONIZATION OF HELIUM

D.G. LAPPAS AND P.L. KNIGHT
Optics Section, The Blackett Laboratory,
Imperial College of Science, Technology and Medicine,
London, SW7 2BW, United Kingdom

P.P. CORSO
Istituto Nazionale di Fisica della Materia and Dipartimento di Scienze
Fisiche ed Astronomiche dell'Università, via Archirafi 36, 90123,
Palermo, Italy

1. Introduction

Model one-dimensional calculations show that there is a significant signature of nonsequential double ionization of He for all wavelengths, in the range from 248 to 1064 nm, provided that the laser pulse is sufficiently short. In this work, we show how the pulse duration can modify the "knee" structure of the double ionization yield curve. By using half-cycle pulses or by artificially decoupling the "inner" electron from the laser, we show how one can test the validity of the proposed shake-off or recollision mechanisms for the double ionization. The flexibility of our model calculations allows us to probe the basic mechanism of the double ionization.

The importance of the role of nonsequential (NS) ionization in multi-electron atomic systems subjected to an external intense laser field has been revealed in the last two decades in a series of experiments. The first observation of NS ionization in noble gases was made by L'Huillier *et al.* [1]. Fittinghoff *et al.* and Walker *et al.* [2] have made so far the most accurate measurements of the ion yields of Helium at infrared wavelengths. In both cases, the double-ionization signal is up to six orders of magnitude higher than expected from sequential ionization, and it saturates at the single ionization saturation intensity. For higher intensities it resumes its increase, but this time following the sequential electron ejection curve. This feature leads to the formation of a typical structure on the double ionization yield curve, known as a

B. Piraux and K. Rzążewski (eds.), Super-Intense Laser-Atom Physics, 25–39.

"knee", and it is evidence of the role played by the electron-electron interaction. More recent experiments [3] have revealed that NS ionization is also present in diatomic molecules and in rare gas atoms at various laser wavelengths.

The theoretical calculations that have been developed in the last few years in order to explain the very high degree of NS double ionization, are based on very different approaches. S-matrix theory by Becker and Faisal [4], "exact" fully correlated one-dimensional models [5,6,7] and numerical models with partial correlation [8,9,10] have all successfully reproduced the knee structure. Other attempts involving the full three-dimensional dynamics of the two correlated electrons have been made with supercomputers [11], but with only a very restricted range of intensities. In this work, we use one-dimensional models that have either full electron correlation [7], or treat the two electrons in a way that involves their mean fields in a partially correlated simple model that is also known as "crapola" [8,9,10]. For a review of some of the recent work regarding double ionization, see reference [12].

2. One-dimensional fully correlated 2-electron model

The numerical calculations that have been widely used in the last decade (see, e.g., Grobe and Eberly, Liu *et al.* [5], Lappas *et al.* [6,7]) approach the problem by reducing the dimensional complexity of the system and solving the two-electron Schrödinger equation exactly in one space dimension.

The basic scheme that is used to solve the two-electron system interacting with an external classical electric field starts with the total Hamiltonian in the minimal coupling form, in atomic units (a.u.) [5,6,7]:

$$H = \frac{(p_1 - A(t))^2}{2} + \frac{(p_2 - A(t))^2}{2} + 2V(x_1) + 2V(x_2) - V(x_1 - x_2) \qquad (1)$$

where x_1 and x_2 are the spatial coordinates of the two electrons, and p_1 and p_2 are their momenta. $A(t)$ is the vector potential of the laser field. We have chosen to work in the $p.A$ gauge, because in this gauge the time-dependent dipole potential due to the laser is included in the kinetic part of the Hamiltonian, thus keeping the potential term small. The time-evolution is performed with a standard numerical split-operator technique [5]. $V(x)$ is the Coulombic potential that describes the interaction of each electron with the core and with each other. Because of the fact that we are using a one-dimensional model, we employ the following soft-core potential, in order to avoid overestimating these interactions around the origin [5]:

$$V(x) = -\frac{1}{\sqrt{a^2 + x^2}} \tag{2}$$

For a^2=0.55 the ground state of the system matches exactly the ground state energy of the real He atom at –2.904 a.u. In what follows, the electric field is linearly polarized and given by: $E(t)=E_0(t)\sin(\omega_L t)=-\partial A(t)/\partial t$. The envelope function $E_0(t)$ takes into account the turn on and turn off effects of the laser pulse.

The double ionization probability P_2 is defined here (see also [11]) as the total probability that both electrons are at least a distance of R=5 a.u away from the origin at the end of the laser pulse. Accordingly, the single-ionization probability P_1 is defined as the probability that one and only one of the two electrons is within the distance R from the origin, i.e.:

$$P_1 = 2\int_{|x_1|<R} dx_1 \int_{|x_2|>R} dx_2 |\psi(x_1, x_2; t_{end})|^2 \tag{3a}$$

$$P_2 = \int_{|x_1|>R} dx_1 \int_{|x_2|>R} dx_2 |\psi(x_1, x_2; t_{end})|^2 \tag{3b}$$

where $\psi(x_1,x_2;t_{end})$ is the full two-electron wave function at the end of the laser pulse, and the factor of 2 is due to the symmetry under exchange of the two electrons. In such a model the two electrons are completely indistinguishable, and full correlation is considered.

3. An approximate model – "Crapola"

Although the exact model outlined above describes the experimental features in a very good qualitative manner, it is rather computationally demanding, and cannot be used easily for very long pulses, very high electron energies or for large regions of space. In order to avoid such problems, and also in order to probe the physical processes in more detail, an approximation to the previous model has been introduced by Watson *et al.* [8]. This is based on the fact that, in the intensity regime of the relevant experiments, one of the two He electrons ionizes rapidly, while there is only a small relative probability that the second electron also ionizes. It is then possible to assume that we can distinguish between the two electrons as if they were identifiable as an *external* and an *internal* one [13]. The external electron, because of fast tunnelling ionization, can be considered as bound only by a time-independent soft-core effective potential, the result of the nuclear attraction and screening. This single-active electron (SAE) approach has been quite successful in the past in describing single ionization [6]. The internal electron is subjected to the nuclear Coulombic potential and to the Coulombic mean field interaction with the external electron. It has been shown [7,14] that if both electrons are described by the same orbital (as in a time-dependent Hartree-Fock or density functional theory approach), the knee structure cannot be reproduced.

Neglecting the exchange term, we simply have for the wave function of two distinguishable electrons [8,9]: $\psi(x_E,x_I;t)=\psi_E(x_E;t).\psi_I(x_I;t)$, where x_E and x_I refer to the external and internal electron coordinates respectively. There are two different one-particle Schrödinger equations, one for the external electron and one for the internal one. For the external electron we have:

$$i\frac{\partial}{\partial t}\psi_E(x_E;t)=\left[\frac{(p_E-A(t))^2}{2}+V_E(x_E)\right]\psi_E(x_E;t) \qquad (4)$$

$$V_E(x_E)=-\frac{1}{\sqrt{a_E^{\,2}+x_E^{\,2}}} \qquad (5)$$

where V_E is the effective potential. We take the soft-core parameter to be a_E=0.697, so that the ground state corresponds to the true one-electron ionization potential of the real He atom. The equation for the internal electron contains an additional external field term due to the Coulombic interaction with the external electron:

$$i\frac{\partial}{\partial t}\psi_I(x_I;t)=\left[\frac{(p_I-A(t))^2}{2}+V_I(x_I;t)\right]\psi_I(x_I;t) \qquad (6)$$

$$V_I(x_I;t)=-\frac{2}{\sqrt{a_I^{\,2}+x_I^{\,2}}}+\int dx_E\frac{\psi_E^*(x_E;t)\psi_E(x_E;t)}{\sqrt{a^2{}_{E-I}+(x_E-x_I)^2}} \qquad (7)$$

In this case, a_I=0.7 and $a_{E\text{-}I}=\sqrt{2}$ yield the best match for the case of the real He atom.

The great advantage of this approach is that the two equations of motion are essentially decoupled, because the first equation (4) does not depend on x_I. Therefore, the two-dimensional problem is reduced to two separate one-dimensional Schrödinger equations, which can be easily solved numerically on any personal computer. We have solved eqs. (4) and (6) using a spatial and temporal grid with sufficient resolution for the electron energy even at intensities beyond the saturation of the double ionization. The length of the spatial grid is at least 1000 a.u.. The initial states for both electrons are their ground states.

We have considered a definition of ionization probability that is as close as possible to the definition in the case of the fully correlated two-electron model, given by eqs. (3a) and (3b) in the previous section. This definition requires the use of a region in space of radius R around the atom, inside which the electrons are not considered ionized. We define $P_E^{Out}(t)$ to be the probability that the *external* electron is outside this region of space. Similarly, $P_I^{Out}(t)$ is the probability that the *internal* electron is outside. In this case, the single ionization probability $P_1(t)$ is given by the probability that the internal electron is inside, times the probability that the external electron is outside, plus the probability that the external electron is inside, times the probability that the internal electron is outside. The double ionization probability $P_2(t)$ is given by the probability that the internal electron is outside times the probability that the external electron is outside too, i.e.

$$P_1(t) = (1 - P_I^{Out}(t)) \cdot P_E^{Out}(t) + P_I^{Out}(t) \cdot (1 - P_E^{Out}(t)) \qquad \text{(8a)}$$

$$P_2(t) = P_I^{Out}(t) \cdot P_E^{Out}(t) \qquad \text{(8b)}$$

Obviously, the above definition introduces the indistinguishability of the two electrons with respect to their experimental detection. One has the freedom of describing the two-electron system by any set of electronic orbitals that reproduce the two-electron dynamics, as long as at the end the two electrons are indistinguishable with respect to the only process that matters to the observer of ionization: measurement of the spatial joint probability distribution.

4. Wavelength dependence

We have analyzed the case of a one-dimensional He atom subjected to an external laser pulse using the model described in the previous section. The laser is a classical electric field whose pulse shape is given by a $\sin^2$ function in a reasonably realistic way: $E(t)=E_0(t)\sin^2(\pi t/n_c T_L)\sin(\omega_L t)$, where E_0 is the field strength, n_c is the total number of optical cycles (o.c.) in the pulse, and T_L is the laser optical period.

We have considered five different wavelengths: 1064 nm, 780 nm, 532 nm, 390 nm and 248 nm, that correspond to the Nd:Yag and Ti:Sapphire lasers, their second harmonics, and the KrF excimer laser, most commonly used today in strong-field physics experiments. For each wavelength, we have varied the field strengths in the range 0.05 a.u. - 0.6 a.u. (laser intensity 10^{14}-10^{16} W/cm^2). We have also looked at the He^+ ionization yield, in order to distinguish between the sequential and the NS contributions to the total double ionization yield of He.

In the case that is close to the experimental conditions, for a pulse duration of 32 o.c., the single ionization yields are in good agreement with those that are obtained from the SAE model for He, at least before the complete saturation region. Our results for all wavelengths, obtained from the approximate model, are shown in Fig. 1. We can

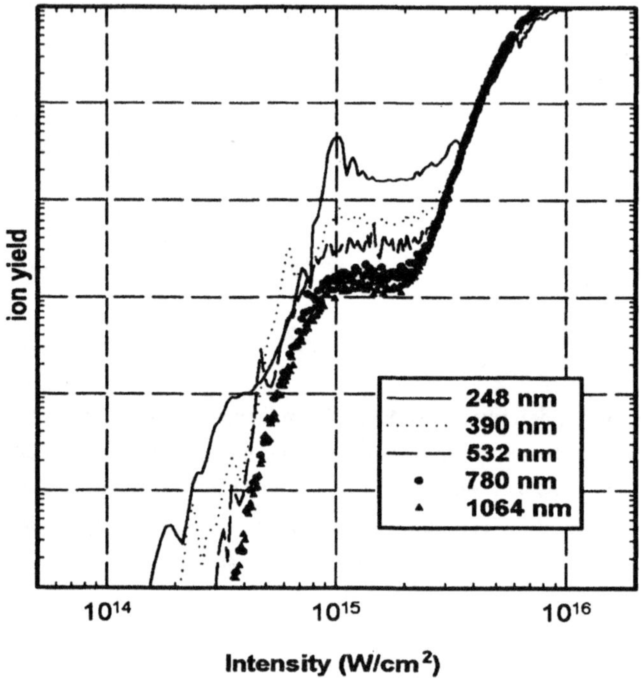

Figure 1. Double ionization yields of He as functions of intensity, for various wavelengths. The pulse has a $\sin^2$ shape with 32 optical cycles of total duration, for each wavelength.

observe that the relative strength of the double ionization signal changes as one varies the wavelength of the laser for a fixed pulse duration (measured in optical cycles, not in real time). For a fixed laser intensity, the double ionization signal increases as wavelength decreases. We can also see from Fig. 1 that the position of the knee in terms of laser intensity remains almost fixed as we change the wavelength.

We have also found a rather high NS ionization signal at 248 nm, in good agreement with ref. [7], but in disagreement with Becker and Faisal [4]. We believe that the very long laser pulses (longer than 600 o.c.) that were used in [4] caused saturation of the single ionization at low intensities, thus making the detection of the NS signal essentially too difficult to discern. In Fig. 2 we show our calculations of the ion yields using the exact one-dimensional model that was introduced earlier. The pulse here is only 8 o.c. long and has a trapezoidal shape with two cycles for turn on, four cycles of constant field strength and two cycles for turn off. We see that the NS

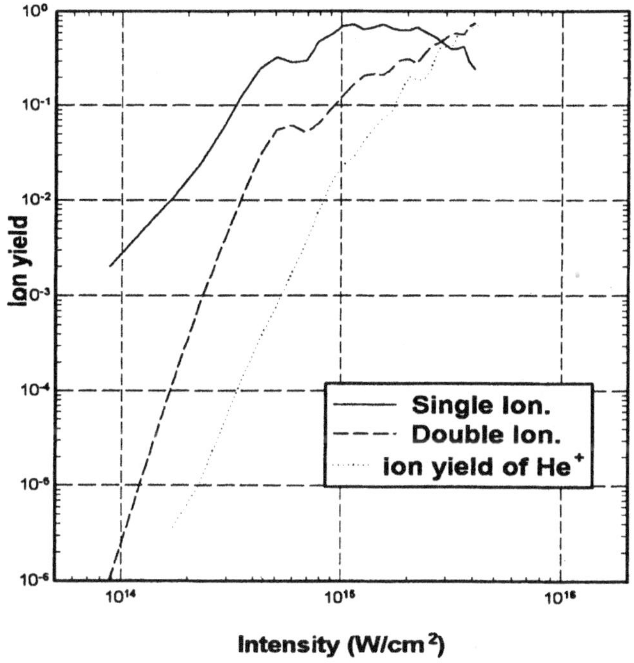

Figure 2. Single (solid line) and double (dashed line) ionization yield of He as a function of intensity, together with the ionization yield of the He^+ ion (dotted line) at 248 nm, with the "exact" fully correlated two-electron model. The pulse is trapezoidal with a total duration of 8 optical cycles and 2 cycles for turn on and turn off.

process can add more than an order of magnitude to the double ionization signal. The peak structure that is observable at 10^{15}W/cm^2 in the case of λ=248 nm in Fig. 1 (with the approximate model) is less pronounced in Fig. 2. We believe this peak is due to a resonantly enhanced ionization channel for the inner electron, and thus depends strongly on the detailed atomic energy level structure of each model.

When the wavelength is 390 nm or 248 nm, the raising part of the single ion yield curve is not as smooth as in the other cases, but has a rich structure characterized by the presence of small jumps (see Fig. 1). These jumps must presumably be connected with resonance atomic effects which appear at wavelengths such that multi-photon (or resonant) ionization is more important than tunnelling ionization.

5. Pulse duration effects

We have studied three different cases corresponding to pulse durations of 32, 8 and 4 optical cycles. The main results are shown in Fig. 3, for a 780 nm laser. Again, it is possible to observe some general features. At first, as we decrease the total pulse duration, passing from 32 to 8 and 4 o.c., we observe that all curves move towards higher laser intensity values. Such behavior can be simply explained, because, as the pulse duration decreases, the same effect is approximately obtainable with a higher intensity.

The same general features can also be observed with the 390 nm laser, and we can therefore consider the pulse duration effects as being in general independent of wavelength, in the regime of frequencies that we have studied.

An interesting result is connected with the shape of the knee. As we decrease the pulse duration, the knee structure tends to become smoother and eventually disappears. At the same time the relative magnitude of the NS ionization decreases. This phenomenon can be explained by the fact that when the pulse becomes very short, the time-limited interaction between the two electrons reduces the total amount of energy that the outer electron can transfer to the inner one during the laser pulse. In the following section we will discuss this process in more detail.

It is also interesting to note that although the NS ionization has been discovered experimentally because of the availability of gradually shorter laser pulses [1], it seems that in the limit of pulses that are shorter than a few optical cycles, the NS ionization is reduced. The pulse duration effects for pulses sufficiently long and weak that perturbation theory can be applied, have been shown by L'Huillier et al. [1] to predict a more easily observable NS ionization signal. That is simply a result of the rate equations for the ionization yields of different multiphoton processes that require very different number of photons. Therefore, the presence of a knee structure, corresponding to the saturation of the NS double ionization, should have been expected.

Given the sensitivity of the knee structure to the pulse duration, one cannot make easy conclusions regarding the existence, or non-existence, and the shape of the knee at various wavelengths, especially when the laser pulse is very long. New experiments at shorter wavelengths and with shorter pulses will be needed to verify these conclusions.

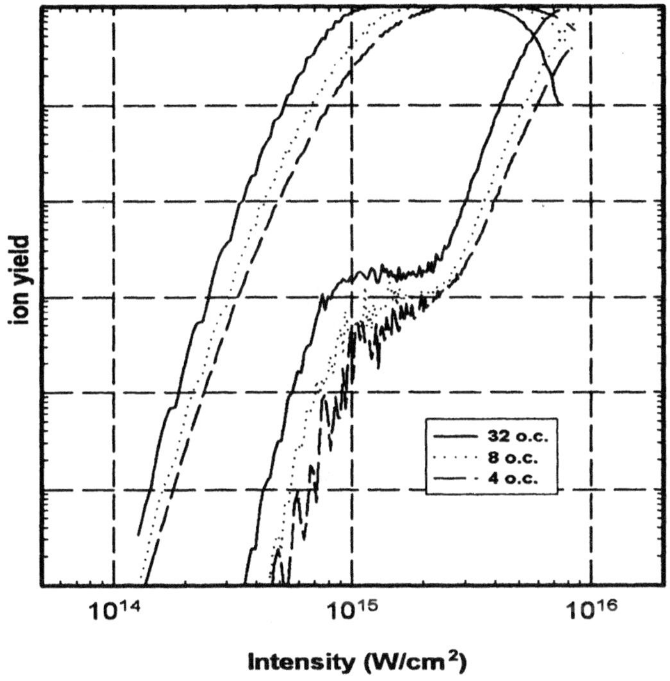

Figure 3. Single and double ionization yields of He as functions of intensity, as calculated from the approximate (Crapola) model, for various laser pulse durations, at 780 nm. In each case the pulse duration in optical cycles (o.c.) is shown in the box.

6. Discussion

In the last five years, two main mechanisms for the double electron ejection have been proposed. One of them, the "recollision" mechanism, is based on the strong-field quasi-classical approximation that has been introduced in order to describe high-harmonic generation in atomic gases irradiated by infrared laser pulses. According to this mechanism, proposed by Corkum [16], the external electron of He propagates freely in the laser field, after ionization. When some phase conditions are met, there is substantial probability that this electron returns to the atomic core within one optical period of the laser. The rescattering event is considered to be responsible for the emission of high-energy photons and for the enhanced ionization of the internal electron through collisional ionization. The second proposed mechanism [15] assumes that the fast ejection of the external electron leaves the internal one in a superposition of He^+ bound states (shake-off); this excited state contribution then leads to a fast double ionization due to the strong laser field.

We have performed model calculations that can distinguish between the two above processes. First, in order to exclude any possibility for rescattering we use a laser pulse with a duration of only one-half of the optical cycle. In this case, there is no time for the ejected external electron to interact with the internal one after its propagation in the laser field. Therefore, the only process that could then create a NS signal should be the shake-off effect. In Fig. 4, for the 780 nm laser, we show the single and double ionization yields of He, together with the ion yield of He^+, after irradiation with a half-cycle pulse. We see that the NS contribution is barely observable and still insufficient to create any knee structure. In particular, the one order of magnitude of double ionization yield above the sequential value is still very small (e.g., in comparison with Fig. 1), and one can state that the knee has disappeared. We conclude that the shake-off effect plays no role here.

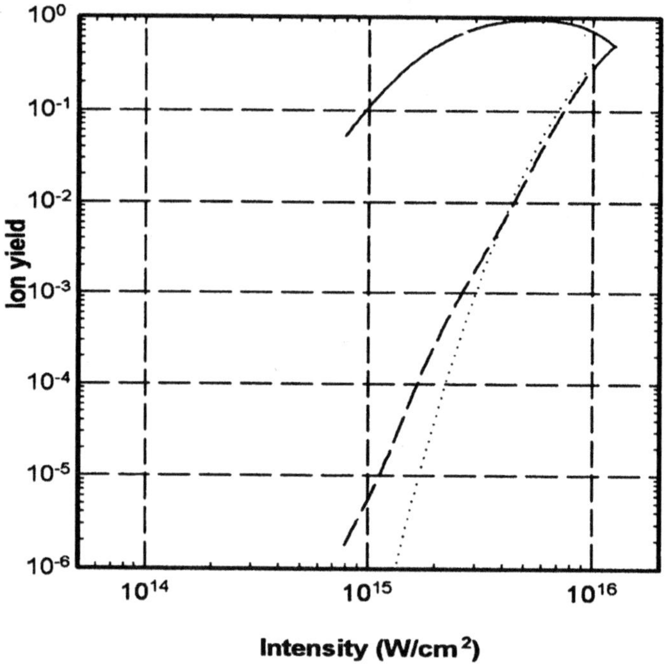

Figure 4. Single (solid line) and double (dashed line) ionization yield of He as function of intensity, together with the ionization yield of the He^+ ion (dotted line), at 780 nm, for a laser pulse that is only half an optical cycle long. The approximate model (Crapola) has been used.

Furthermore, in order to estimate the effect of the electron rescattering, in accordance with the recollision mechanism, we can artificially switch off the laser coupling with the internal electron in eq. (6), thus leaving only the electron-electron interaction term as an external field that can cause double ionization. As we can see from Fig. 5, this term alone can reproduce the very high NS signal of double electron ejection with excellent agreement, up to the end of the knee region, where the NS

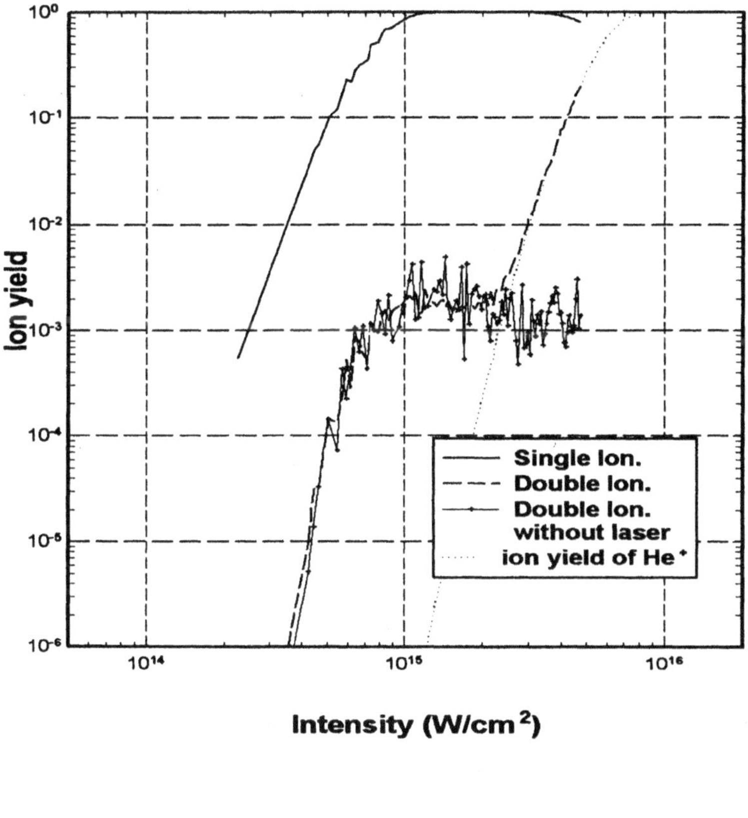

Figure 5. The double ionization yield of He (grey line connecting the crosses) when the internal electron is not coupled to the laser field, in the approximate (Crapola) model. The laser pulse has a total duration of 32 o.c., at 780 nm. The dashed line represents the double ionization of the full calculation.

process saturates due to the complete ionization of the external electron. For higher intensities there is no external electron packet density remaining to have an effect on the dynamics of the internal one, and the electron-electron interaction vanishes. This remarkable result confirms a previous study by Sanpera *et al.* [10], where the internal electron was considered to interact only with the dipole radiation field (harmonics) created by the motion of the external electron in the laser field. It is also reminiscent of the "atomic antenna" effect that was suggested by Kuchiev [17].

Finally, we note that the suggested interaction of the core electron with the outer electron is essentially different from the standard process of single-impact ionization. It is an electron-electron scattering process that occurs periodically, over many laser cycles, in a time interval of the order of the pulse duration. The energy that is transferred to the internal electron is provided by the kinetic energy of the returning

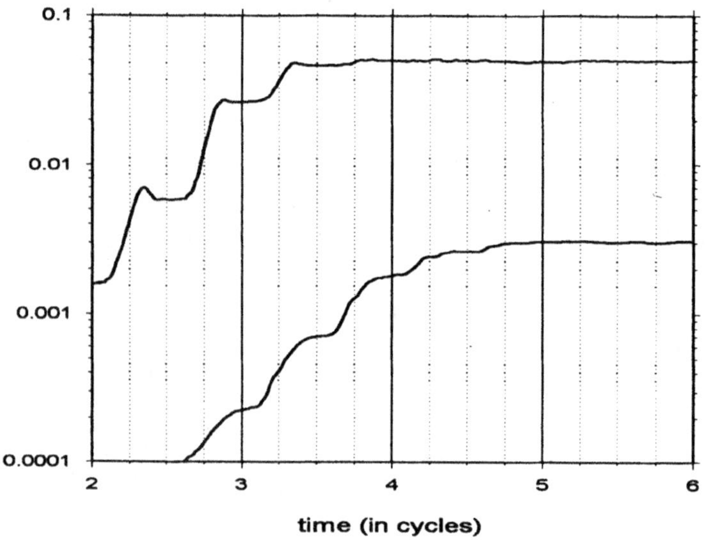

Figure 6. Single (upper line) and double (lower line) ionization probabilities as functions of time during a 6-cycle laser pulse at 780 nm for intensity 7.9×10^{14} W cm^{-2}. The pulse shape is $\sin^2$.

external electron, through periodic recollision events. This energy, in an amount that is restricted only by the finite duration of the external laser field, is provided by the laser pulse itself, through the multiphoton absorption by the outer active electron. In a purely classical treatment, multiple returns of the recolliding electron are needed in order for the total double ionization to match the experimentally observed NS yield [18].

From all the above, it has become evident that the basic mechanism of the NS double ionization involves a rescattering mechanism of the external electron with the internal. Furthermore, recent experiments seem to confirm the basic process of double electron ejection by measuring the momentum distributions of the recoil-ions [19]. However, the detailed mechanism of this process, e.g. the degree of excitation of the inner electron due to rescattering, is still under investigation. Kopold *et al.* [20] have demonstrated, that in a theoretical treatment based on the strong-field approximation (SFA) theory, inclusion of more excitation channels during rescattering is essential in understanding the experimental results.

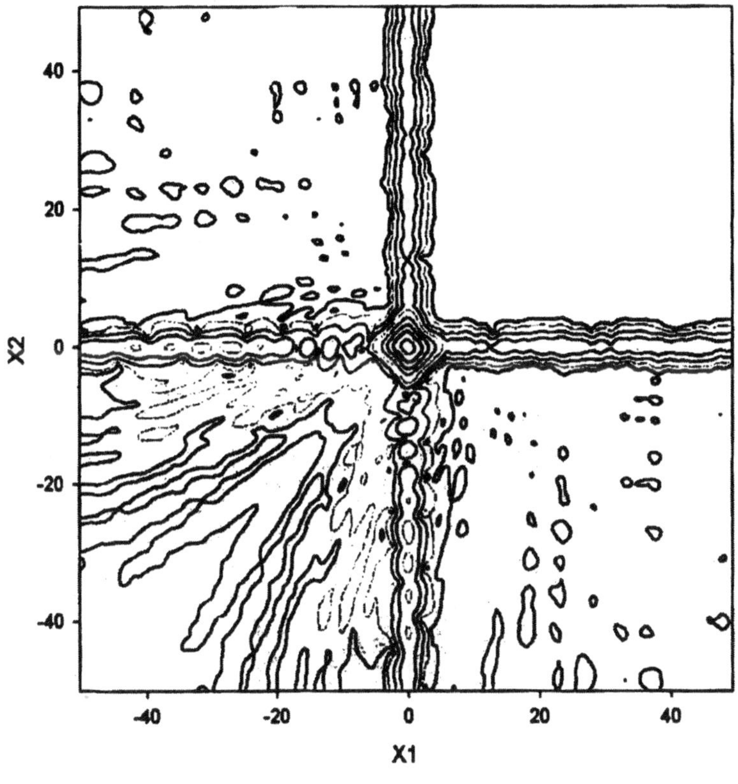

Figure 7. Contour plot of the 2-electron spatial joint probability distribution. This snapshot is taken at time 3.5 optical cycles in a laser pulse of sin(ωt) dependence, as in Fig. 6. Here the intensity is 4.2×10^{14} W cm^{-2}. The contours are spaced by one order of magnitude from each other, on a logarithmic scale.

In order to understand the time-dependent dynamics of double ionization, it is also useful to look into the explicit time-dependence of single and double ionization probabilities. For this purpose we have used the 1D fully correlated model of section 2. In Fig. 6 we show the time-dependent quantities of eqs. (3), i.e. single and double ionization, in the case of a pulse of total duration of 6 o.c., with sin(ωt) dependence. The maximum intensity is 7.9×10^{14} W cm^{-2}, for a wavelength of 790 nm. It is interesting to observe that the temporal peaks of single and double ionization are at intervals of the order of half a laser cycle, a fact that is consistent with the process of the outer electron periodically returning to the atomic core, twice per laser cycle, in order to eject the inner electron. A snapshot of the joint two-electron probability distribution in Fig. 7, taken for a slightly different intensity (4.2×10^{14} W cm^{-2}) indicates that there is a high probability that both electrons are ejected in the same direction at a time when a recollision is expected to occur. Similar calculations by other authors [21] have confirmed this process.

7. Acknowledgements

We thank the Imperial College Parallel Computing Centre for the allocation of its computer facilities and for technical support. This work was supported by the UK Engineering and Physical Sciences Research Council.

8. References

1. L'Huillier, A., Lompré, L.A., Mainfray, G., Manus, C. (1983), J. Physique **44**, 1247.
2. Fittinghoff, D.N., Bolton, P.R., Chang, C., and Kulander, K.C. (1992), Phys. Rev. Lett. **69**, 2642; Walker, B., Sheehy, B., DiMauro, L.F., Agostini, P., Schafer, K.J., and Kulander, K.C. (1994), Phys. Rev. Lett. **71**, 1227.
3. Talebpour, A., Larochelle, S., and Chin, S.L. (1997), J. Phys. B: At. Mol. Opt. Phys. **30**, L245; Larochelle, S., Talebpour, A., and Chin, S.L. (1998), J. Phys. B: At. Mol. Opt. Phys. **31**, 1201.
4. Becker, A., and Faisal, F.H.M. (1999), Phys. Rev. A **59**, R1742; Faisal, F.H.M, Becker, A., and Muth-Böhm, J. (1999), Laser Phys. **9**, 115.
5. Grobe, R., and Eberly, J.H. (1992), Phys. Rev. Lett. **68**, 2905; Liu, W.-C., Eberly, J.H., Haan, S.L., and Grobe, R. (1999), Phys. Rev. Lett. **83**, 520.
6. Lappas, D.G., Sanpera, A., Watson, J.B., Burnett, K., Knight, P.L., Grobe, R., and Eberly, J.H. (1996), J. Phys. B: At. Mol. Opt. Phys. **29**, L619.
7. Lappas, D.G., and van Leeuwen, R. (1998), J. Phys. B: At. Mol. Opt. Phys. **31**, L249.
8. Watson, J.B., Sanpera, A., Lappas, D.G., Knight, P.L., and Burnett, K. (1997), Phys. Rev. Lett. **78**, 1884.
9. Burnett, K., Watson, J.B., Sanpera, A., and Knight, P.L. (1998), Phil. Trans. R. Soc. Lond. A **356**, 317.

10. Sanpera, A., Watson, J.B., Shaw, S.E.J., Knight, P.L., Burnett, K., and Lewenstein, M. (1998), J. Phys. B: At. Mol. Opt. Phys. **31**, L841.
11. Dundas, D., Taylor, K.T., Parker, J.S., and Smyth, E.S. (1999), J. Phys. B: At. Mol. Opt. Phys. **32**, L231; Parker, J.S., Moore, L.R., Dundas, D., and Taylor, K.T. (2000), J. Phys. B: At. Mol. Opt. Phys. **33**, L691.
12. Lambropoulos, P., Maragakis, P., and Zhang, J. (1998), Phys. Rep. **305**, 203.
13. Tolley, A.J. (1999), J. Phys. B: At. Mol. Opt. Phys. **32**, 3449.
14. Petersilka, M., and Gross, E.K.U. (1999), Laser Phys. **9**, 105.
15. Sheehy, B., Lafon, R., Widmer, M., DiMauro, L.F., Agostini, P.A., and Kulander, K.C. (1998), Phys. Rev. A **58**, 3942.
16. Corkum, P.B. (1993), Phys. Rev. Lett. **71**, 1994.
17. Kuchiev, M.Yu. (1987), JETP Lett. **45**, 404.
18. Ivanov, M. (2000), private communication.
19. Weber *et al.* (2000), Phys. Rev. Lett. **84**, 443; Moshammer *et al.* (2000), Phys. Rev. Lett. **84**, 447; Weber *et al.* (2000), J. Phys. B: At. Mol. Opt. Phys. **33**, L127.
20. Kopold, R., Becker, W., Rottke, H., and Sandner, W. (2000), Phys. Rev. Lett. **85**, 3781.
21. Lein, M., Gross, E.K.U., and Engel, V. (2000), J. Phys. B: At. Mol. Opt. Phys. **33**, 433; Dörr, M. (2000), Opt. Express **6**, 111.

ELECTRON MOMENTUM DISTRIBUTIONS FOR DOUBLE IONIZATION IN THE STRONG FIELD LIMIT

S.P. GORESLAVSKI and S.V. POPRUZHENKO
Moscow State Engineering Physics Institute,
31 Kashirskoe shosse, Moscow, 115409, Russia

1. Introduction

The interaction between electrons ejected to the continuum by a strong laser field and their parent ions have been extensively investigated during the last decade. Different channels of that interaction, i.e. recombination, elastic and inelastic scattering give rise to the phenomena which are now known as high harmonic generation, high order above-threshold ionization and nonsequential double ionization. A Keldysh-type theories have been proposed for the latter process [1,2]. The intense-field many-body S-matrix theory has been successfully applied to calculate the total ion yield [2] and the momentum distribution for He^{2+} ions [3] measured in the experiment [4]. In that approach, the transition amplitude is given by a multifold integral and its calculation is not an easy task. Below we report on a simplified form of the electron distributions for nonsequential double ionization in strong laser fields which is derived from the S-matrix amplitude in the semiclassical limit, i.e. by applying a saddle-point method. The approximate results are compared with the recent theoretical [3] and experimental data [4, 5] on the momentum distribution of doubly charged ions and are used to provide qualitative estimates characterizing its shape.

2. Transition amplitude

The amplitude of nonsequential double ionization appearing in the quantum approach based on the Volkov states [1, 2] is of the form

$$M^{(2e)}\left(\vec{P},\vec{q}\right)=i\sum_{k}\int_{-\infty}^{+\infty}dt_1\left\langle\Psi_{\vec{P}\vec{q}}(t_1)\right|\frac{1}{r_{12}}\left|\Psi_{\vec{k}0+}(t_1)\right\rangle M^{(e)}\left(\vec{k},t_1\right) \tag{1}$$

where

$$M^{(e)}\left(\vec{k},t_1\right)=i\int_{-\infty}^{t_1}dt\left\langle\Psi_{\vec{k}0+}(t)\right|V_F(t)\left|\Psi_0(t)\right\rangle \tag{2}$$

B. Piraux and K. Rzążewski (eds.), Super-Intense Laser-Atom Physics, 41–50.

The quantity $M^{(e)}(\vec{k},+\infty)$ is the amplitude of direct one-electron ionization In equations (1), (2), $\Psi_0(t)$ is the initial bound two-electron state with the ionization potential I_1; $V_F(t)$ is the electron – laser interaction in the dipole approximation; the intermediate states are taken as $\Psi_{\vec{k}0+}(t)=\Psi_{0+}(t)\Psi_{\vec{k}}(t)$ where $\Psi_{0+}(t)$ is the ground state of a singly charged ion with the ionization potential I_2 and $\Psi_{\vec{k}}(t)$ is the Volkov state with the drift (asymptotic) momentum $\vec{k}$. Further on, $1/r_{12}$ is the electron-electron interaction; the final state, $\Psi_{\vec{P},\vec{q}}(t)$, is a product of two Volkov functions with the drift momenta $\vec{p}_1$ and $\vec{p}_2$ and is labeled by the total, $\vec{P}=\vec{p}_1+\vec{p}_2$, and relative, $\vec{q}=(\vec{p}_1-\vec{p}_2)/2$, momenta, of the ionized electrons. In actual calculations all two-electron wave functions in equations (1), (2) are symmetrized with respect to permutation of the electron coordinates. We use the length gauge and write down the Volkov state as

$$\Psi_{\vec{k}}(\vec{r},t)=\exp\left(i\vec{\Pi}_{\vec{k}}(t)\vec{r}-\frac{i}{2}\int_{-\infty}^{t}\vec{\Pi}_{\vec{k}}^{2}(\tau)d\tau\right) \tag{3}$$

where $\vec{\Pi}_{\vec{k}}(t)=\vec{k}-\vec{k}(t)$ is the kinetic momentum of an electron moving in an oscillating electric field with the drift momentum $\vec{k}$. Notice that the center-of-mass kinetic momentum is given by $\vec{\Pi}_{\vec{P}}(t)=\vec{P}-2\vec{k}(t)$. The field is linearly polarized along x-axis and both the electric force and $\vec{k}(t)$ have only x-projection equal to $F(t)=F\sin\omega t$ and $k(t)=p_F\cos\omega t$, respectively, where $p_F=F/\omega$.

We assume the tunneling regime, i.e. $F<F_a=(2I_1)^{3/2}$, $\omega<I_1$ and $\gamma^2=I_1/2U_p<1$, where $U_p=F^2/4\omega^2$ is the ponderomotive energy. That enables one to evaluate the integrals (1), (2) by a saddle-point method. Calculations proceed very much similar to the case of elastic rescattering [6-8] and include three stages: (1) evaluation of $M^{(e)}(\vec{k},t_1)$ by a saddle – point method; (2) summation over the intermediate momenta and reduction of the problem to calculation of the contribution from a single optical period and, finally, (3) evaluation of the latter by a saddle – point method. After the first two stages the rate of transition to the state $\Psi_{\vec{P},\vec{q}}(t)$ is a sum where each term contains $\delta(I_1+I_2+\vec{P}^2/4+\vec{q}^2+2U_p-m\omega)$. The sum over m is replaced by an integral and the latter is evaluated by using the δ-functions. The resulting ionization rate describes an envelope of the δ-function peaks and is given by

$$dW=2\pi\omega^{-1}\left|B(\vec{P},\vec{q})\right|^2 d^3P d^3q/(2\pi)^6 \tag{4}$$

where the one-cycle amplitude is

$$B(\vec{P},\vec{q})=\frac{\Lambda\omega}{(2\pi)^2}\int_0^{2\pi/\omega}dt_0F^{1/2}(t_0)\int_{t_0}^{+\infty}dt_1A^{(e\to 2e)}\frac{\exp(-F_a/3F(t_0)+iS)}{F_a/(2I_1F(t_0))+i(t_1-t_0)} \tag{5}$$

Here the phase S equals to

$$S=S(\vec{P},\vec{q},t_0,t_1)=I_1t_0+I_2t_1+\vec{q}^2t_1+\frac{1}{4}\int_{-\infty}^{t_1}dt\vec{\Pi}_{\vec{P}}^2(t)-\frac{1}{2}\int_{t_0}^{t_1}dt\vec{\Pi}_{\vec{k}(t_0)}^2(t) \tag{6}$$

and the amplitude in the Born approximation for impact ionization [9] produced by the incident electron with the momentum $\vec{k}_* = \vec{\Pi}_{\vec{k}(t_0)}(t_1)$, so that the emitted pair has the total, $\vec{P}_* = \vec{\Pi}_P(t_1)$, and relative momenta, $\vec{q}_* = \vec{q}$, is defined as

$$A^{(e\to 2e)}(\vec{P}_*,\vec{q}_*;\vec{k}_*)=\frac{32\pi^2(2I_2)^{5/4}}{\left(2I_2+(P_*-k_*)^2\right)^2}\left[\frac{1}{\left(\vec{k}_*+\vec{q}_*-\vec{P}_*/2\right)^2}+\frac{1}{\left(\vec{k}_*-\vec{q}_*-\vec{P}_*/2\right)^2}\right] \tag{7}$$

The cross-section of the process is related to the amplitude (7) by the equation

$$d\sigma^{(e\to 2e)}(\vec{P}_*,\vec{q}_*,\vec{k}_*)=P_*\left|A^{(e\to 2e)}(\vec{P}_*,\vec{q}_*,\vec{k}_*)\right|^2 d\Omega_n d^3q_*/(16\pi^5k_*) \tag{8}$$

where $d\Omega_{\vec{n}}$ is the solid angle along the unit vector $\vec{n}=\vec{P}_*/|\vec{P}_*|$. The factor Λ in equation (5) does not depend on momenta and has no effect on the shape of the momentum distributions. It arises from calculation of the direct ionization amplitude and includes, in particular, the overlap integral $\langle\Psi_{0+}|\Psi_0\rangle$.

3. Momentum distribution

Evaluation of the double integral (5) by a saddle – point method leads to two equations for stationary points

$$\frac{I_1}{F\sin\omega t_0}+\frac{F}{\omega^2}\left[\omega(t_1-t_0)\cos\omega t_0-\sin\omega t_1+\sin\omega t_0\right]=0 \tag{9}$$

$$\frac{1}{2}\vec{\Pi}_{\vec{k}(t_0)}^2(t_1)=I_2+\vec{q}^2+\frac{1}{4}\vec{\Pi}_{\vec{P}}^2(t_1) \tag{10}$$

Solutions (t_0,t_1) depend on the final momenta (actually only on three parameters $(P_x,P_\perp,q=|\vec{q}|)$) and may be real or complex. According to that criteria, the momentum space is classified under two domains, classical and quantum, respectively. For momenta in the quantum domain the integral (5) contains additional small exponential factor in comparison to its value in the classical domain and, hence, the resulting momentum distribution has a well–defined cut-off at the boundary between the domains.

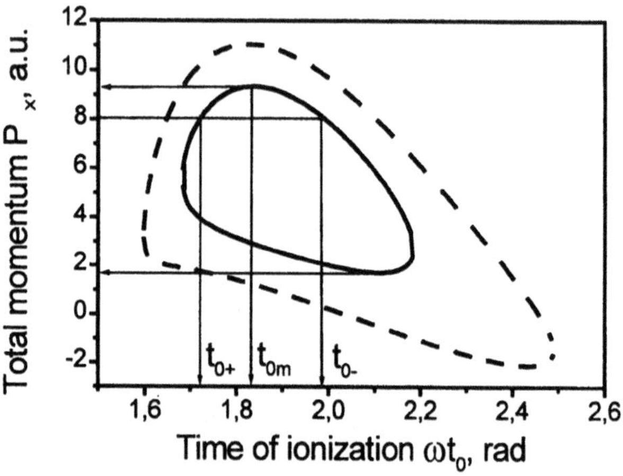

Figure 1. Total momentum along the laser field as a function of the time of ionization, $P_x = P_x(t_0, P_\perp, q)$, *calculated from equations (9), (10) for ionization of Ne by Ti : Sa laser at intensity* $1.310^{15}\ W\,sm^{-2}$. *The curves are for* $P_\perp = 0$, $q = 2.32\,a.u.$ *(solid) and* $P_\perp = q = 0$ *(broken).*

In the classical domain, equations (9), (10) are identical to the kinematics relations of the simple-man model for double ionization via rescattering [10]. Equation (9) states that the electron starting in the laser field with zero velocity at the time t_0 at the turning point of the oscillating potential barrier returns to the ion at the time t_1. Equation (10) is the energy conservation law for inelastic $e \to 2e$ collision occurring at the time t_1. The inverse function to $t_0 = t_0(P_x, P_\perp, q)$ is shown in figure 1. The maximum (minimum) value of that function (marked by the arrow in the figure 1 and denoted below as $P_x(t_{0m})$) is an upper (lower) cut-off of the distribution in P_x for given $(P_\perp, q)$. The closed curves are shrinking in sizes with increasing q and/or $P_\perp$ and their largest size is presented by the broken line.

The function $P_x = P_x(t_0, P_\perp, q)$ takes a particular value between the cut-offs (i.e. inside the classical domain) at two different times $t_{0\pm}$ (see figure 1). When only the first return to the ion is accounted for then the double integral (5) has two stationary points, (t_{0-}, t_{1-}) and (t_{0+}, t_{1+}). To follow the standard saddle-point practice one has to expand the phase (6) in deviations from one of the points and calculate a two-fold Gaussian integral, then repeat calculations for the second point and sum the results. Contributions from both stationary points interfere in the amplitude (5) and the distribution (4) takes the form

$$dW^{(2e)}(\vec{P},\vec{q}) = \left(w_- + w_+ - 2(w_- w_+)^{1/2}\sin S_{+-}\right)d^3Pd^3q \tag{11}$$

where $S_{+-} = S(\vec{P},\vec{q},t_{0+},t_{1+}) - S(\vec{P},\vec{q},t_{0-},t_{1-})$ and the contribution from an individual stationary point (subscripts $\pm$ are omitted) is given by:

$$w(\vec{P},\vec{q}) = \frac{A^2 I_1}{64\pi^7}\frac{\omega^3}{F^2 F_a}\left|A^{(e\to 2e)}\left(\vec{\Pi}_{\vec{P}}(t_1),\vec{q},\vec{\Pi}_{\vec{k}(t_0)}(t_1)\right)\right|^2 \frac{\sin^2\omega t_0}{\Delta_\perp^2(t_1,t_0)|D|}\exp\left(-\frac{2F_a}{3F(t_0)}\right) \tag{12}$$

Here, the transverse width of the photoelectron wave packet at the times of ionization and return equals to $\Delta_0 = \left(F_a/2IF(t_0)\right)^{1/2}$ and $\Delta_\perp(t_1,t_0) = \left(\Delta_0^2 + \Delta_0^{-2}(t_1 - t_0)^2\right)^{1/2}$, respectively, and $D = s_{00}s_{11} - (s_{01})^2$ is a determinant of the second derivative matrix. In equation (12) and below we use the scaled phase $s = S/z_F$ and denote its derivatives with respect to dimensionless variables ωt_0 and ωt_1 as s_1, s_{01} and so on.

The assumption of isolated stationary points fails for P_x near $P_x(t_{0m})$ when the points t_{0-} and t_{0+} are close to each other and to the position of the cut-off, t_{0m}, (see figure 1). The inapplicability of the standard approach manifests in a singularity of the distribution (12) so as $D \to 0$ when $P_x \to P_x(t_{0m})$. The reason is that $\partial P_x/\partial t_0 \propto D$ and the equality $D = 0$ is the extremum condition for the function $P_x = P_x(t_0, P_\perp, q)$. For final momenta near the cut-offs the calculational procedure has been modified by taking into consideration the third order terms in the expansion of the phase S [7, 8]. That yields the distribution of the form

$$dW^{(2e)} = w_b d^3Pd^3q \tag{13}$$

where the density w_b is derived from equation (12) by the replacement

$$|D|^{-1} \to 2\pi z_F^{1/3}|s_{00}|^{-1}|s_{111}/2|^{-2/3} Ai^2\{z_F^{2/3}Y\} \tag{14}$$

and, in the argument of the Airy function, $Y = \left(2/|s_{111}|\right)^{1/3}\left[s_1 - D^2/\left(2|s_{111}|s_{00}^2\right)\right]$. In equations (13), (14), the derivatives of s are taken at the point (t_{0m}, t_{1m}) whereas P_x is treated as a free parameter which might be either smaller or larger than $P_x(t_{0m})$. The function Y turns to zero for P_x exactly equal to the cut-off, $P_x(t_{0m})$, and according to the properties of $Ai(x)$ the distribution w_b is finite at the boundary of the classical and quantum domains.

4. Discussion

The formulas (11) and (13) jointly provide a comprehensive description for the two-electron distribution over the 6D momentum space. They match each other inside the classical domain at some distance from the boundary. It is seen from these formulas that the effect of the $e-e$ interaction *ab initio* accounted for in the S-matrix amplitude (1) is expressed in terms of the Born cross-section for impact ionization.

4.1. SIMPLE-MAN MODEL

The momentum distribution of the simple-man model may be deduced from the equation in the classical domain (11). To do that it is necessary to disregard the interference term in equation (11) and to introduce new variables defining the state of motion of the ionized electron pair, namely, $\left(t_0, \vec{n} = \vec{\Pi}_P(t_1)/\left|\vec{\Pi}_{\vec{P}}(t_1)\right|, \vec{q}\right)$ instead of $\left(\vec{P}, \vec{q}\right)$. The two sets of variables are related by equations (9), (10) and direct calculations show that

$$p_F^{-3}|D|^{-1} \sin \omega t_0 d^3 P d^3 q = 2\omega \left|\vec{\Pi}_{\vec{k}(t_0)}(t_1)\right|^{-1} \left|\vec{\Pi}_{\vec{P}}(t_1)\right| d\Omega_n dt_0 d^3 q \qquad (15)$$

Considering that $W_{st}(F) = I_1 \Lambda^2 (F/F_a) \exp(-2F_a/3F)$ is the rate of ionization by a dc field F and in view of equations (7) and (15) the distribution (11) takes the form

$$dW^{(2e)} = d\sigma^{(e\to 2e)}\left(\vec{\Pi}_{\vec{P}}(t_1), q, \vec{\Pi}_{\vec{k}(t_0)}(t_1)\right) \frac{W_{st}(F(t_0))}{\Delta_\perp(t_0, t_1)} \frac{\omega\, dt_0}{2\pi} \qquad (16)$$

which is just the basic equation postulated by the simple-man theory.

4.2. 6D-DISTRIBUTION

Figure 2 serves to illustrate the impact of the cross-section on the shape of the distribution. A wide solid curve shows the normalized distribution in P_x for fixed q and $P_\perp$ calculated with a ,"flat" inelastic cross section, i.e. assuming the factor $\left|A^{(e\to 2e)}\right|^2$ to be constant. The inner part of that curve exhibiting the interference pattern is described by the equation (11), while vicinities of the side maxima are calculated from the equation with the Airy function (13). The Coulomb cross section of impact ionization in the Born approximation (7) is shown separately by a broken line. The actual distribution in P_x (not shown in figure 2) is the product of these two curves and, obviously, is confined to a region of the narrow maximum in the inelastic cross section. For the parameters used in figure (2), (i) the interval of P_x where equation (11) is applicable do exist and (ii) the maximum of the cross section falls into it. In such situation the simple-man version (16) could be used to obtain the 6D momentum distribution with disregarded interference term. Nevertheless, it should be kept in mind that the interval of applicability for equation (11) decreases and even may disappear for larger q and/or $P_\perp$ and lower intensity in which case the momentum distribution is to be calculated with the equation (13).

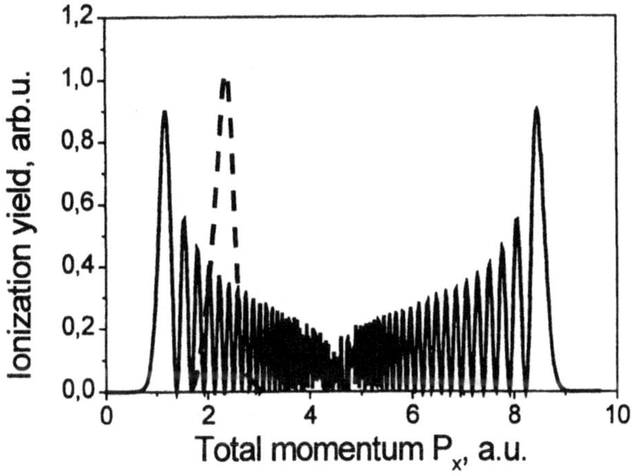

Figure 2. Distribution in P_x for the "flat" cross section (solid line) and the Coulomb cross-section for impact ionization (broken line) calculated with the parameters of the solid curve in figure 1.

4.3. ION MOMENTUM DISTRIBUTION

In the dipole approximation for the laser field and for short laser pulses preventing a ponderomotive acceleration in the laser focus, momentum conservation relates the momentum of the doubly charged ion, $\vec{P}_{ion}$, and the total momentum of the two ionized electrons so that $\vec{P}_{ion} = -\vec{P}$. Therefore, theoretical electron distributions in $\vec{P}$ could be immediately compared with the measured ion momentum distributions [4, 5]. We derive the former by integrating the 6D momentum distribution over the relative momentum analytically in directions of $\vec{q}$ and numerically in $q = |\vec{q}|$. It is seen in the course of calculations that for intensities of interest the contribution to the integral in dq from region near the cut-offs is small and so the curves shown in figures 3, 4 have been computed by integrating the distribution (11) between the cut-offs. The calculation of one curve requires about a minute with an average PC.

A comparison of the distribution in P_x resulting from the present calculations with the one obtained by using Monte Carlo simulations to evaluate multiple integral in the amplitude (1) [3] is exhibited in figure 3. A close correlation between the curves is indicative of a sufficiently good accuracy provided by our approximate calculational scheme. Both theoretical distributions qualitatively agree with the measurements at that intensity in ref. [4].

Figure 4 exhibits the distribution in P_x for the case of Ne^{2+} and $Ti:Sa$ laser at intensity 1.3 $10^{15} W\,sm^{-2}$. The symmetric maxima of the experimental [5] curve (bold squares) are at $\pm 4\,a.u$. Two theoretical curves have been derived in the above described

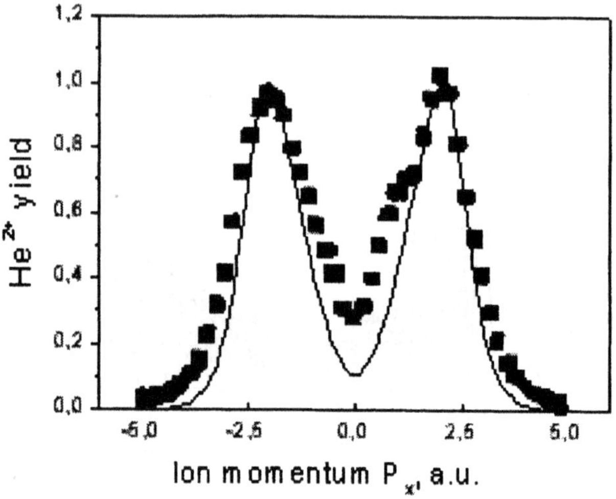

Figure 3. Distribution of the total momentum component in the direction parallel to the laser field for the case of He^{2+} *and* $Ti:Sa$ *laser at intensity* 6.6 $10^{14}\,W\,sm^{-2}$. *The result of numerical calculations in ref. [4] is shown by squares and the one from the present work by solid line.*

way by using the Born (solid) and „flat' (broken) cross section for the $e \to 2e$ process (actually, the broken curve is calculated in the framework of the „dominant phase" model [11] which accounts for the phase factors and disregards the preexponential factors in the transition amplitude (1)). Their maxima are at $\pm 2.7\,a.u$ and $\pm 6\,a.u$, respectively. It is seen that though the two-humped structure of the distribution in P_x owes its origin to the kinematics of the rescattering process real position and width of the maxima are acutely sensitive to the shape of the inelastic cross section. A better fit of the experimental data may be reached by a heuristic replacement of the Born cross section by a more smooth function of the electron momenta. That may be the case if for that purpose one uses the inelastic cross section beyond the Born approximation more suitable for the electron energy range under consideration.

The analytic formulas provide simple estimates for characteristic features of the momentum distribution. A dramatic effect of the cross section on the shape of the distribution is mainly due to the factor $\left(2I_2 + \left(\vec{\Pi}_{\vec{p}}(t_1) - \vec{\Pi}_{\vec{k}(t_0)}(t_1)\right)^2\right)^{-4}$. In the tunneling regime, it has a narrow maximum at $\vec{\Pi}_{\vec{p}}(t_1) = \vec{\Pi}_{\vec{k}(t_0)}(t_1)$, i.e. when all of the kinetic momentum is transferred from the incident electron to the ionized electron pair. That condition reeds explicitly $P_{\perp} = 0$ and $P_x = p_F(\cos\omega t_1 + \cos\omega t_0)$. On the half-period $0 < \omega t_0 < \pi$ the direct ionization is effective at the maximum field, $\omega t_0 \approx \pi/2$, and the corresponding time of the first return ωt_1 is not far away from 2π.

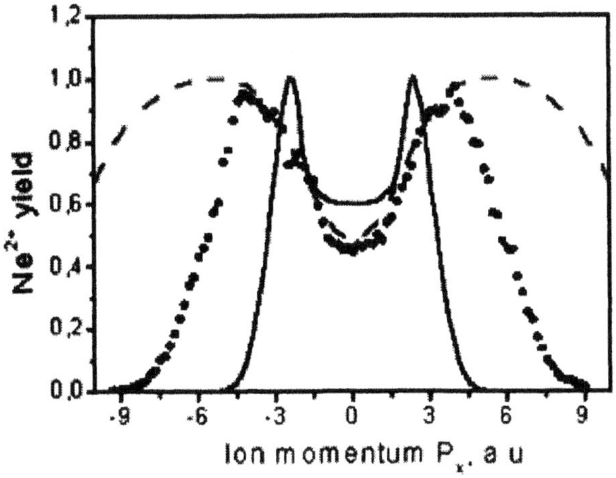

Figure 4. Distribution of the total momentum component P_x for the case of Ne^{2+} and $Ti:Sa$ laser at intensity 1.3 10^{15} $W\,sm^{-2}$. Curves are for the experimental data [5] (circles) and present calculation with the Born (solid) and "flat" (broken) cross sections.

Hence, a crude estimate for the position of the maximum is $P_x^{(\max)} \approx p_F$. The symmetric maximum at $P_x^{(\max)} \approx -p_F$ arises on an adjacent half-period of ωt_0 with the opposite direction of the laser field. The solid curve in figure 4 is maximum at $P_x^{(\max)} \approx \pm 0{,}8\, p_F$. The width of either peak estimated from the dominant factor in the cross section is of order of $\Delta P_x \approx (2I_2)^{1/2}$. A one-humped distribution in $P_{\perp}$ centered at $P_{\perp} = 0$ (it is not shown in figures) has the same width $\Delta P_{\perp} \approx (2I_2)^{1/2}$. Consequently, the widths under consideration are expected to be of comparable value and to have a very weak intensity dependence. These widths decrease with increasing laser intensity if the distributions are presented as a function of momenta in units of $p_F = F/\omega = (4U_p)^{1/2}$. The total width of the distribution in P_x is basically determined by the distance between the symmetric maxima and could be estimated as $\Delta P_{tot} = 2p_F + (2I_2)^{1/2}$.

5. Conclusions

In conclusion, we have developed an approximate, semiclassical and largely analytical theory of the two-electron distributions arising from nonsequential double ionization of atoms by strong laser fields and applied it to derive qualitative estimates for the shape of the momentum distribution of doubly charged ions. It has been shown that the effect of electron – electron correlation *ab initio* accounted for by the Keldysh-type theories of the nonsequential double ionization in is expressed, in the tunneling regime, in terms of the cross-section for impact ionization in the Born approximation. As well it is has been shown that though the two-humped structure of the distribution in the ion momentum along the direction of the laser field owes its origin to the kinematics of the rescattering process real position and width of the maxima are acutely sensitive to the shape of the inelastic cross section.

6. Acknowledgements

The authors thank W. Becker and R. Kopold for valuable discussions. The support from the RFBR (projects 99-02-17810 and 00-02-17078) and from the INTAS (project 99-450) is acknowledged.

7. References

1. Kuchiev, M. Yu. (1995) ATI as a source for multiply charged ion production in a laser field, *Journal Physics B: Atomic, Molecular and Optical Physics* **28**, 5093-5115.
2. Faisal, F.H.M .and Becker, A. (1996) Effect of rescattering on ATI and $e-e$ correlation on double ionization in intense laser fields, in P Lombropoulos and H Walther (eds.), *Multiphoton Processes 1996,* The Institute of Physics Conference Series **154**, Bristol, pp.118- 131.
3. Becker, A. and Faisal, F.H.M. (2000) Interpretation of momentum distribution of recoil ions from laser induced nonsequential double ionization, *Physical Review Letters* **84**, 3546-3549.
4. Weber, Th., Weckenbrock, M., Staudte, A., Spielberger, L., Jagutzki, O., Mergel, V., Afanch, F., Urbasch, G., Volmer, M., Giessen, H., and Dorner, R. (2000) Recoil-Ion Momentum Distribution for Single and Double Ionization of Helium in Strong Laser Fields, *Physical Review Letters* **84**, 443-446.
5. Moshammer, R., Feuerstein, B., Schmitt, W., Dorn, A., Schroder C.D., Ulrich, J., Rottke, H., Tramp, C., Wittmann, M., Korn, G., Hoffmann, K., Sandner W. (2000) Momentum Distribution of Ne^{++} Ions Created by Intense Ultrashort Laser Pulse, *Physical Review Letters.* **84**, 447-450.
6. Goreslavskii, S.P., and Popruzhenko, S.V. (1998) Simple quantum theory of the high energy above - threshhold ionization in the tunneling regime, *Physics Letters A* **299**, 477 – 482.
7. Goreslavskii, S.P., and Popruzhenko, S.V. (1999) Rescattering and quantum interference near the classical cutoffs, *Atomic, Molecular and Optical Physics* **32**, L531-L538.
8. Goreslavskii, S.P., and Popruzhenko, S.V. (2000) Tunneling limit in the theory of photoelectron rescattering by the parent ion, *Zhurnal Experimental'noi i Teoreticheskoi Fiziki* **117**, 895-905 (Transation: *JETP* **90**, 778-787).
9. Landau, L.D., and Lifshiz E.M. (1977) *Quantum Mechanics,* Pergamon Press, Ch. 18, sec. 145.
10. Corkum, P.B. (1993) Plasma Perspective on Strong – Field Multiphoton Ionization, *Physical Review Letters.* **71,** 1994-1997.
11. Kopold, R., Becker, W., Rottke, H., Sandner, W. (2000) Routes to nonsequential double ionization, *Physical Review Letters.*, (in press).

S-MATRIX THEORY OF 'RECOIL-ION' MOMENTUM DISRIBUTION FOR DOUBLE IONIZATION IN FEMTOSECOND LASER FIELDS

F.H.M. FAISAL AND A. BECKER
Fakultät für Physik, Universität Bielefeld, D-33615 Bielefeld, Germany

1. Introduction

Double ionization of atoms in intense femtosecond laser pulses is currently one of the most vigorously investigated problems in strong-field atomic physics. Use of sophisticated coincidence techniques has recently permitted measurements of the 'recoil-ion' momentum distributions during double ionization of He [1] and Ne [2] by intense Ti:sapphire laser pulses. Such data for triple ionization for Ne has also been obtained [2]. These results provide important new tests of the mechanism of double ionization process in intense ultrashort laser pulses.

Here we shall briefly discuss the theoretical challenge posed by these observations and the progress made using the recently developed so-called 'intense-field many-body S-matrix theory (IMST)' to this end [3, 4, 5, 6, 7]. The theoretical difficulties for intense-field laser-atom interaction arise from a combination of three factors:
(a) the highly non-perturbative light interaction,
(b) the quantum many-body problem, and
(c) the non-separable Coulomb correlation interaction, $\sum_{i\neq j} 1/r_{ij}$.
One approach of attacking such problems is to try to integrate the Schrödinger equation of the system (atom coupled to the laser field) numerically. This approach is essentially exact but it is limited by the computing power of the state-of-the-art computers to solve high dimensional partial differential equations over realistically large space-time grids. Much progress have been made in treating one-electron systems (hydrogenic systems, or single-active-electron models) in 3-dimensions. It is also expected to provide valuable information to the double ionization problem at hand, for two-electron systems involving 6-D partial differential equations that reaches the limit

B. Piraux and K. Rzążewski (eds.), Super-Intense Laser-Atom Physics, 51–66.

of simulation techniques using today's supercomputers (e.g. [8] and contributions in this volume).

An alternative approach to the numerical simulation, is a systematic approximation method. The recently introduced 'intense-field many-body S-matrix theory' (IMST) is designed to this end [3]. In this theory the expansion of the S-matrix (or the probability amplitude) of the ionization process is rearranged in such a way that the dominant features of the process can appear in the leading terms of the asymptotic series of the S-matrix. Such a rearrangement has been shown to be possible [3] by introducing three partitionings of the total Hamiltonian of the interacting system, namely, an initial, a final *and* an intermediate partitioning. At the same time the laser interaction with the free electron (whether in the final or in the virtual intermediate states) is taken into account to all orders *non*-perturbatively by using the Volkov solution (i.e. the exact solution of a free electron in a laser field, e.g. [9], p. 11) and the Volkov propagator [10, 11]. As will be seen below, introduction of the intermediate partitioning permits a remarkable flexibility to the S-matrix expansion that is not present in other versions, such as the usual 'prior' or 'post' [12, 13] form of the S-matrix (or evolution operator) series. This novel expansion allows one, in particular, to take into account, already in the leading terms of the theory, the virtual fragment-states of a *many*-body system, through which the reaction of interest may proceed predominantly. This can be done in IMST *without* affecting the orthogonal projections to the desired initial and final states of the transition of interest. In contrast, the usual 'prior' or 'post' expansions would have to be carried out to large (if not to infinite) orders, in order to account for such intermediate fragment-states. This is because the 'prior' or the 'post' expansion is constrained to use the *same* reference propagator that is appropriate for either the initial or the final state of the process.

2. Intense-field many-body S-matrix theory (IMST)

We briefly outline the present S-matrix theory (IMST) and discuss the recently obtained results of analysis with its help [7] of the observed recoil-ion momentum distributions for the double ionization in an intense femtosecond laser pulse. The Schrödinger equation of the interacting system (laser-field coupled to the two electrons of He atom) is a time-dependent generalization of the Lippmann-Schwinger equation satisfying the initial condition, namely, the atom is in the ground state in an initial time t_i. Thus, at first, we partition the total Hamiltonian of the system $H(t)$, as usual

$$H(t) = H_i^0 + V_i(t). \tag{1}$$

i.e., as a sum of the initial reference Hamiltonian H_i^0 (unperturbed He atoms) and rewrite

$$\left[\left(i\hbar\frac{d}{dt} - H_i^0\right) - V_i(t)\right] |\Psi(t)>= 0 \tag{2}$$

as an integral equation

$$|\Psi(t)>= |\phi_i(t)> + \int_{t_i}^{t} dt_1 G(t,t_1)V_i(t_1)|\phi_i(t_1)>, \tag{3}$$

where we have introduced the total Green's function (or propagator) $G(t,t')$ by the definition

$$\left[i\hbar\frac{d}{dt} - H(t)\right] G(t,t') = \delta(t-t'). \tag{4}$$

And, $\phi_i(t)$ is the solution of the initial unperturbed Hamiltonian, H_i^0 (e.g. He atom),

$$\left[i\hbar\frac{d}{dt} - H_i^0\right] |\phi_i(t)>= 0 \tag{5}$$

that is chosen to satisfy the initial condition. The initial interaction is $V_i(t)$. For the double ionization problem this is given by the interaction Hamiltonian of the two electrons coupled to the vector potential of the laser field,

$$V_i(t) = \sum_{j=1,2}\left[-\frac{e}{mc}\mathbf{p}_j\cdot\mathbf{A}(t) + \frac{e^2}{2mc^2}A^2(t)\right]. \tag{6}$$

Next, we introduce the final state partitioning of the total Hamiltonian

$$H(t) = H_f^0(t) + V_f(t), \tag{7}$$

where, $H_f^0(t)$ is the final reference Hamiltonian (e.g., the two-electron Volkov Hamiltonian [10, 11]), and $V_f(t)$ is the final state interaction (e.g., the electron-electron correlation, and also any residual interaction with the nucleus). The associated Green's function satisfies

$$\left[i\hbar\frac{d}{dt} - H_f^0(t)\right] G_f^0(t,t') = \delta(t-t'). \tag{8}$$

The total Green's function is now expanded in terms of the final Green's function in the form

$$G(t,t') = G_f^0(t,t') + \int_{t_1}^{\infty} dt_1 G_f^0(t,t_1)V_f(t_1)G(t_1,t') \tag{9}$$

and is inserted in the expression for the total wavefunction $|\Psi(t)>$ above, Eq. (3), to get

$$\begin{aligned} |\Psi(t)> &= |\phi_i(t)> + \int_{t_i}^{t} dt_1 G_f^0(t,t_1) V_i(t_1)|\phi_i(t_1)> \\ &+ \int_{t_i}^{t}\int_{t_i}^{t_2} dt_2 dt_1 G_f^0(t,t_2) V_f(t_2) G(t_2,t_1) V_i(t_1)|\phi_i(t_1)> \end{aligned} \tag{10}$$

In this way we have arrived at a useful form for the wavefunction that not only satisfies the initial state condition but also is well arranged for computing the transition amplitude by projection to any state of the final reference Hamiltonian, even when the latter is *not* identical to the initial reference Hamiltonian. This is because the projection is orthogonal and hence extracts only *one* term from the final reference Green's function (given in the proper state representation). Such a formulation is, therefore, useful not only in the present context but also for many other problems, e.g., charge exchange and chemical reactions (which invariably involve unequal reference Hamiltonians for the input and output channels). Thus, projecting onto the final state $|\phi_f(t)>$, belonging to $H_f^0(t)$, we write the transition amplitude for the $i \to f$ transition as:

$$\begin{aligned} (S-1)_{fi}(t) &= -\frac{i}{\hbar}[\int_{t_i}^{t} dt_1 < \phi_f(t_1)|V_i(t_1)|\phi_i(t_1)> \\ &+ \int_{t_i}^{t}\int_{t_i}^{t_2} dt_2 dt_1 < \phi_f(t_2)|V_f(t_2) G(t_2,t_1) \times \\ &\times V_i(t_1)|\phi_i(t_1)>]. \end{aligned} \tag{11}$$

This is an exact *master equation* for the S-matrix, whose most important features, as indicated above, are (a) that the orthogonal projections for the initial and the final states are carried out independently of equal *or* unequal reference Hamiltonians, and (b) that the total Green's function appears *between* the initial and the final interactions. This is unlike the appearence of the Green's function at one end of the usual 'prior (or direct time)' or 'post (or time reversed)' (see e.g.,[12, 13])) expansion of the S-matrix theory (or the Dyson series of the evolution operator). As mentioned, it is this flexibility of Eq. (11) that permits the introduction of any *virtual* fragments propagator of interest, already in the leading terms of the IMST. Note that, inclusion of the effects of such virtual fragments-channels in the usual 'prior' and 'post' expansions may only be obtained indirectly, if at all, by summing them to very high orders, if not to infinite orders. It is this difference with the other forms of S-matrix (or evolution operator) expansions that makes IMST more useful for many-body reaction problems. It allows us to introduce a third intermediate partitioning of the total Hamiltonian:

$$H(t) = H_0(t) + V_0(t) \tag{12}$$

which corresponds to the virtual fragments-Hamiltonian $H_0(t)$; the corresponding propagator $G_0(t,t')$, satisfies

$$\left[i\hbar\frac{d}{dt} - H_0(t)\right] G_0(t,t') = \delta(t-t'). \tag{13}$$

For the double ionization problem of present interest, the virtual intermediate fragments can consist of one electron in the Volkov states of all virtual momenta $\{\mathbf{k}\}$, and a singly charged residual ion in its virtual eigenstates; the corresponding fragments-Hamiltonian is, therefore,

$$H_0(t) = [(\mathbf{p}_1 - e\mathbf{A}(t)/c]^2/2m + [\mathbf{p}_2^2/2m - Ze^2/r_2]; \qquad (Z=2, \text{ for He}). \tag{14}$$

The corresponding two-particle fragments-Green's function, satisfying Eq. (13), is [3, 7]

$$\begin{aligned} G_0(t,t') &= -\frac{i}{\hbar}\theta(t-t')\sum_j \frac{1}{(2\pi)^3}\int d\mathbf{k}|\mathbf{k}> |\phi_j^+(2)> \\ &\times e^{-\frac{i}{\hbar}\int_{t'}^{t}\{[(\mathbf{k}-e\mathbf{A}(\tau)/c]^2/2m+E_j\}d\tau} <\mathbf{k}| <\phi_j^+(2)| \end{aligned} \tag{15}$$

where, $\theta(t-t')$ is the Heaviside theta function, $\{|\mathbf{k}>\}$ is the complete set of plane wave states with wavevectors $\mathbf{k}$, and $\{|\phi_j^+(2)>\}$ is the complete set of residual ionic states.

Once $G_0(t,t')$ is available, we can expand $G(t,t')$ as

$$G(t,t') = G_0(t,t') + \int_{t_i}^{t} dt_1 G_0(t,t_1)V_0(t_1)G_0(t_1,t') + \cdots \tag{16}$$

and substitute it in the expression for the total wavefunction, Eq. (10) above, to obtain the wavefunction expansion,

$$\begin{aligned} |\Psi(t)> &= |\phi_i(t)> + \int_{t_i}^{t} dt_1 G_f^0(t,t_1)V_i(t_1)|\phi_i(t_1)> \\ &+ \int_{t_i}^{t}\int_{t_i}^{t_2} dt_2 dt_1 G_f^0(t,t_2) \times \\ &\times V_f(t_2)G_0(t_2,t_1)V_i(t_1)|\phi_i(t_1)> + \cdots \end{aligned} \tag{17}$$

We note that alternative physical mechanisms can be investigated by alternative and/or additional intermediate partitionings, that can be introduced exactly in the same way at this stage, and/or after one or more interactions, systematically.

Finally, to obtain the transition amplitude explicitly, we substitute the Green's function expansion, Eq. (16), in Eq. (11), and write the resulting amplitude as an S-matrix series (or the IMST),

$$(S-1)_{fi}(t) = \sum_{j=1}^{\infty} S_{fi}^{(j)}(t) \tag{18}$$

with

$$S_{fi}^{(1)}(t) = \int_{t_i}^{t} dt_1 < \phi_f(t_1)|V_i(t_1)|\phi_i(t_1) > \tag{19}$$

$$S_{fi}^{(2)}(t) = \int_{t_i}^{t}\int_{t_i}^{t_2} dt_2 dt_1 < \phi_f(t_2)|V_f(t_2)|G_0(t_2,t_1)V_i(t_1)|\phi_i(t_1) > \tag{20}$$

$$\begin{aligned} S_{fi}^{(3)}(t) &= \int_{t_i}^{t}\int_{t_i}^{t_3}\int_{t_i}^{t_2} dt_3 dt_2 dt_1 < \phi_f(t_3)|V_f(t_3) \times \\ &\times G_0(t_3,t_2)V_0(t_2)G_0(t_2,t_1)V_i(t_1)|\phi_i(t_1) > \end{aligned} \tag{21}$$

......

We note that the first term of IMST is the same as the so-called KFR theory (first term) [14, 15, 13]. This is because it is not difficult to show by partial integrations that the first term of the 'prior', 'post' and IMST series are generally *equivalent* for long interaction times (as shown in [16], footnote 11). In fact the 'prior' and 'post' series could be obtained as special cases of IMST, but the latter rearrangement (IMST) differs term by term *qualitatively*, as discussed above, from the next term onward.

3. Application to 'recoil-ion' momentum distribution for double ionization in intense fields

Using IMST, we investigate and discuss below the recently obtained momentum distributions of the doubly charged ions of He, both *parallel* as well as *perpendircular* to the polarization axis of the laser. The momentum of the doubly charged ions is related to that of the two outgoing electrons via the momentum conservation relation. Thus, neglecting the very small contribution from the momentum of the (long wavelength) laser photons, the overall momentum conservation requires that the sum of the electron momenta parallel to the laser polarization axis, (z-axis), satisfies, (i) $\mathbf{P}_{par.} = -[(\mathbf{k}_a)_{par.} + (\mathbf{k}_b)_{par.}]$ and that perpendicular to it satisfies, (ii)

$\mathbf{P}_{perp.} \approx -[(\mathbf{k}_a)_{perp.} + (\mathbf{k}_b)_{perp.}]$ where, $\mathbf{P}$ is the momentum of the doubly charged ion, and $\mathbf{k}_a$ and $\mathbf{k}_b$ are the momenta of the two outgoing electrons.

Before presenting the results of calculations, let us summarise the most prominent features of the experimentally observed distributions as reported recently [1, 2]. They are:

(a) the component of the recoil momentum *parallel* to the laser polarization direction shows a prominent *double*-hump distribution with a central minimum,
(b) the perpendicular component of the recoil momentum shows a single-hump distribution,
(c) the parallel component distribution is very broad,
(d) the perpendicular distribution is narrow,
(e) the maximum momentum transfer of the parallel component shows a cut-off at a value $\mathbf{P}_{par.} \approx Re(4\sqrt{U_p})$ for He [1] and a still larger value for Ne [2],
(f) the distributions are qualitatively different from the corresponding distributions for the $\gamma + 2e$ reaction by (weak field) synchrotron photons, which favors the Warnier process of back to back emission near zero momentum.

Previously we had extensively investigated the intensity dependence of the *total* double ionization yields using the so-called 'correlation mediated energy sharing diagram' (see Fig. 1 below) derived from the present theory [3, 4], as well as using other much simplified models [5, 6, 16] suggested by it. In fact, these investigations for the first time permitted us to identify the above mentioned quantum mechanical diagram that automatically incorporates, in appropriate parameter domains, the well-known classical 'rescattering model' proposed by Corkum [17] (see discussion below and in [3],[6](a),[7]).

For the double ionization problem, the leading two terms of the IMST series, Eqs. (19) and (20), provide two first rank, and six second rank diagrams (these are drawn in [3]). For the near infrared Ti:sapphire laser wavelength, $\lambda = 800$ nm, and the intensity, $I = 6.6 \times 10^{14}$ W/cm^2 used for the He experiment [1], the contributions from the first rank diagrams (corresponding to $S_{fi}^{(1)}$) as well as five other second rank diagrams (belonging to $S_{fi}^{(2)}$) are found to be much smaller than the remaining correlation mediated energy sharing diagram. (The 'rank' of a diagram corresponds to the number of times the generalized interactions (vertices) appear in a diagram.) To avoid any confusion let us point out that the relative importance of the eight leading IMST diagrams are in general quite different for different domains of the laser parameters, in particular, in different wavelength domains. Thus, for example, one of the first rank diagrams was found to be dominant at $\lambda = 248$ nm, in the UV wavelength domain, over the entire range of the intensity in the UV experiment of interest (see, experiment [18], and its analysis [6](a),[16]). This is in contrast to the dominant contribution of the correlation mediated energy sharing diagram in the present case of Ti:sapphire laser wavelength. Thus, we see that IMST can be helpful in determining the main *mechanism* involved in different parameter domains of the non-linear field interaction.

The quantum-physical picture behind the double ionization process for near infrared wavelengths and for the so-called non-sequential (NS) domain of intensity (e.g. the He double ionization yields experiment by Walker et al. [19] and the recoil momentum distributions experiments by Weber et al. [1] and Moshammer et al. [2]) can in fact be visualized directly by reading the correlation mediated energy sharing diagram (Fig. 1) from the bottom to the top, in the indicated direction of the flow of time. Thus, at an initial time t_i the two electrons (say, 1 and 2) are in the ground state of the unperturbed He atom. At a time t_1 (when the field phase is $\phi_1 = \omega t_1$) one of the bound electrons (say, 1) absorbs, via the generalized ATI-like interaction (V_{ATI}), a large amount of field energy by a virtual above-threshold ionization (ATI) process, while the other electron propagates in the virtual states of the residual ion. This intermediate ensemble of virtual states corresponds to the two-particle propagator or Green's function $G_0(t, t')$ Then at a time t_2 (field phase $\phi_2 = \omega t_2$) the two electrons interact or 'internally collide' via the electron-electron correlation, V_{corr} (that may, in general, include the interaction with the nucleus as well), and share the energy until both of them may have enough energy to escape together from the binding force of the atom, into the final two-electron Volkov state (c.f., [10, 11]), and proceed with the final momenta $\mathbf{k}_a$ and $\mathbf{k}_b$ to the detector(s), at t_f.

Note that quantum mechanically speaking, since the electrons are not observed in the above intermediate states in the experiment, contributions to the double ionization amplitude, in principle, must be integrated for all values of $\Delta t = t_2 - t_1 = \tau$ (or the phase difference, $\Delta\phi = \phi_2 - \phi_1$), as it naturally occurs in the time integrations of the amplitude of the process (see below). Under appropriate limiting conditions, however, the time integrations over the interval τ may be estimated semiclassically. Under the latter conditions (e.g. $U_p \gg \hbar\omega, E_B \gg \hbar\omega$) a significant contribution can arise from time intervals of the order of the classical return time(s) of the first emitted electron, to the core region, with a corresponding return energy of the order of $3U_p$. This provides the theoretical connection between the present quantum theory and the classical 'rescattering model' of Corkum [17], mentioned above. In the actual calculation below we shall not make the semiclassical approximation of the time integrations but evaluate them *exactly* (see Eqs. (24),(25) below) after expanding all the Volkov waves in

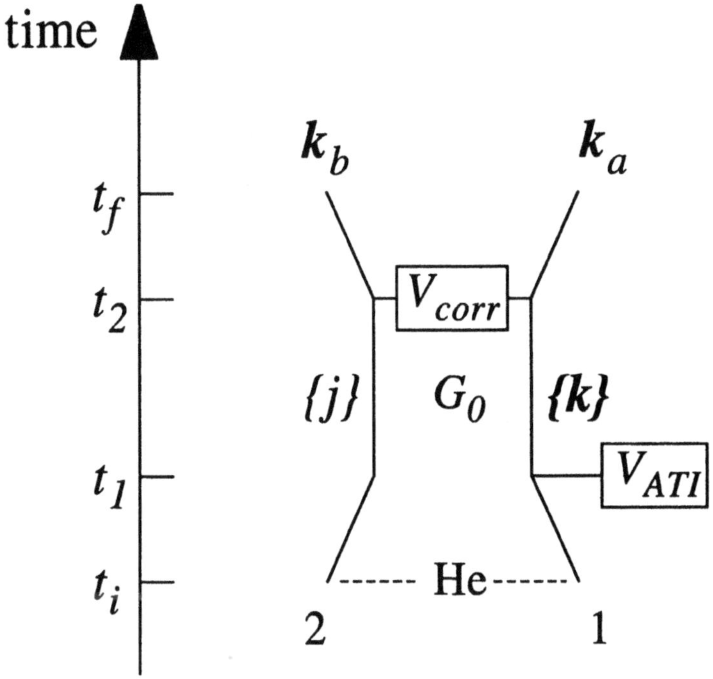

Figure 1. Correlation mediated energy-sharing diagram for laser induced non-sequential double ionization of He. For interpretation see text.

the amplitude in terms of the generalized Bessel functions.

For the purpose of actual calculations, we write down the non-sequential (NS) double ionization amplitude, following the diagram in Fig. 1, as,

$$S_{fi}^{(2)}(t_f, t_i)|_{NS} = -\frac{i}{\hbar}\int_{t_i}^{t_f} dt_2 < \phi_f^V(\mathbf{k}_a, \mathbf{k}_b; \mathbf{r}_1, \mathbf{r}_2; t_2)|V_{corr}(t_2)|\Psi_i(\mathbf{r}_1, \mathbf{r}_2; t_2) > \tag{22}$$

where, the angular brackets stand for the spatial integrations; $\phi_f^V(\mathbf{k}_a, \mathbf{k}_b; t_2)$ is the two-electron product Volkov final state in the field (c.f. [10]), $V_{corr}(t_2)$

stands for correlation operator and $\Psi_i(t_2)$ is the total wavefunction of the system, at time t_2. In the present approximation, $\Psi_i(t_2)$ is given by (c.f. time evolution up to t_2 in Fig. 1):

$$\begin{aligned}\Psi_i(\mathbf{r}_1,\mathbf{r}_2;t_2) &= \int_{t_i}^{t_2} dt_1 G_0(\mathbf{r}_1,\mathbf{r}_2,t_2;\mathbf{r'}_1,\mathbf{r'}_2,t_1)|V_{ATI}(t_1)|\phi_{1S}(\mathbf{r'}_1,\mathbf{r'}_2;t_1)> \\ &= -\frac{i}{\hbar}\int_{t_i}^{t_2} dt_1 \sum_j \frac{1}{(2\pi)^3}\int d\mathbf{k}|\phi^V(\mathbf{k};\mathbf{r}_1;t_2)\phi_j^+(\mathbf{r}_2;t_2)> \times \\ &\quad \times <\phi_j^+(\mathbf{r'}_2;t_1)\phi^V(\mathbf{k};\mathbf{r'}_1;t_1)|V_{ATI}(t_1)|\phi_{1S}(\mathbf{r'}_1,\mathbf{r'}_2;t_1)>\end{aligned} \quad (23)$$

where $G_0(t_2,t_1)$ is the intermediate two-electron Green's function, Eq. (15); $V_{ATI}(t_1)$ is the interaction operator for the virtual above-threshold ionization or ATI-like process at the time t_1, and $\phi_{1S}(\mathbf{r}_1,\mathbf{r}_2;t_1)$ is the ground state wavefunction of the He atom with binding energy $E_B = 2.904$ (a.u.). An exact evaluation of this amplitude, including all orders of correlation and ATI interaction, is practically an impossible task. We have, therefore, restricted ourselves in this work to the lowest significant terms of the theory by replacing $V_{corr}(t_2)$ by e^2/r_{12} and $V_{ATI}(t_1)$ by $(-\frac{e}{mc}\mathbf{A}(t_1)\cdot\mathbf{p}_1 + \frac{e^2}{2mc^2}A^2(t_1))$ This simplification still requires calculation of a formidable multiple integral which we have evaluated approximately as follows. The Jacobi-Anger formula ([20], p. 7) is used to expand the Volkov wavefunctions, $\phi^V(\mathbf{k},\mathbf{r};t)$, in terms of the corresponding Fourier components defined by the generalized Bessel functions of two arguments (e.g., [9]). This allows us to evaluate the two-fold time integrations, over the instants t_1 and t_2 of the two interactions, exactly; for pulses much longer than a laser period (as in the experiments of present interest) this gives,

$$\begin{aligned}S_{fi}^{(2)}(\infty,-\infty)|_{NS} &= -2\pi i \sum_N \delta\left(\frac{\hbar k_a^2}{2m} + \frac{\hbar k_b^2}{2m} + E_B + 2U_p - N\hbar\omega\right) \times \\ &\quad \times T^{(N)}(\mathbf{k}_a,\mathbf{k}_b),\end{aligned} \quad (24)$$

where,

$$\begin{aligned}T^{(N)}(\mathbf{k}_a,\mathbf{k}_b) &= \sum_{n,j}\int \frac{1}{(2\pi)^3} d\mathbf{k} <\phi^0(\mathbf{k}_a,1)\phi^0(\mathbf{k}_b,2)|\frac{e^2}{r_{12}}|\phi_j^+(2)\phi^0(\mathbf{k},1)> \\ &\quad \times \frac{J_{N-n}\left(\alpha_0\cdot(\mathbf{k}_a+\mathbf{k}_b-\mathbf{k});\frac{U_p}{2\hbar\omega}\right) J_n\left(\alpha_0\cdot\mathbf{k};\frac{U_p}{2\hbar\omega}\right)}{\frac{\hbar k^2}{2m} - E_j + E_B + U_p - n\hbar\omega + i0} \times \\ &\quad \times (E_j - E_B - \frac{\hbar k^2}{2m}) <\phi_j^+(2)\phi^0(\mathbf{k},1)|\phi_{1S}(1,2)>\end{aligned} \quad (25)$$

where $\phi^0(\mathbf{k}, i), i = 1, 2$ are plane wave states, and $\phi_{1S}(1, 2)$ is the ground state of the two-electron atom.

Next, the six-fold space integrations are carried out analytically, the radial integration in k is performed by pole approximation and the integrals over the angles of $\mathbf{k}$ and the sum over n are performed numerically. Contribution of the lowest term of the sum over j (corresponding to the ground state of the ionic state) is retained only since it is found to dominate over the contribution from any excited state (c.f. [4]). Finally, Monte-Carlo sampling method is used to evaluate the differential rate of double ionization, $\sum_N \frac{dW^{(N)}}{d\mathbf{k}_a d\mathbf{k}_b}$, as a function of the two outgoing momenta $\mathbf{k}_a$ and $\mathbf{k}_b$, from the formula,

$$\frac{dW^{(N)}}{d\mathbf{k}_a d\mathbf{k}_b} = \frac{2\pi}{\hbar}\delta\left(\frac{\hbar k_a^2}{2m} + \frac{\hbar k_b^2}{2m} + E_B + 2U_p - N\hbar\omega\right)|T^{(N)}(\mathbf{k}_a, \mathbf{k}_b)|^2 . \quad (26)$$

We may note that earlier we have developed useful approximate models starting from the amplitude, Eq. (24,25), and making further simplifying assumptions, such as neglect of the non-diagonal terms after modulo-squaring the amplitude (incoherent n-sum), use of the maximum classical energy [5] or the back-scattering energy [6] as the 'incident' energy in conjuction with the semi-empirical 'e−ne' rate formulas, that were found to provide excellent double and multiple ionization yields when compared with the available experimental data in a wide variety of cases [5, 6]. In contrast, here, as also in [3, 4, 7], we have *not* made these 'model' simplifications. We also note that here as in [7] we are interested as in the experiments [1, 2], in the distributions in a relative scale. Estimate of the absolute scale of the distributions may be obtained on applying the WKB estimate of the Coulomb correction [6, 21].

Computations are carried out for He atom, and the distributions of the parallel and the perpendicular components of the *sum* of the two momenta, $(\hbar\mathbf{k}_a + \hbar\mathbf{k}_b)$, are determined, in each case by integrating over the remaining variables and summing the contributions from all significant Ns. We recall that under the condition of the experiment the He^{2+} recoil momentum $\mathbf{P} \approx -(\hbar\mathbf{k}_a + \hbar\mathbf{k}_b)$ [1, 2].

We show in Fig. 2 the experimental result (panel a) obained for the recoil momentum of He^{++}, *parallel* to the (linear) polarization axis by Weber et al. [1], in the field of a Ti:sapphire laser pulse of 200 fs, and a peak intensity of 6.6×10^{14} W/cm^2 and $\lambda = 800$ nm. They are compared with the corresponding results of the present theory [7] in panel b. We note that the experimental data are available only in arbitrary units. One data point is therefore fitted to the theory to determine the *relative* scale for the entire set. As it can be seen from the comparison, all the essential features of the experimental distribution of the parallel component of the recoil momentum are reproduced by the theoretical results. Thus, both the distributions show a double-hump structure with a central minimum. The positions and the heights of the two maxima are also well reproduced. We observe that the maximum size of the recoil ion momentum (cut-off momentum) of the experiment and the calculation of the sum electron-momenta also agree well with each other and in fact in both cases it is as large as ≈ 5 a.u.

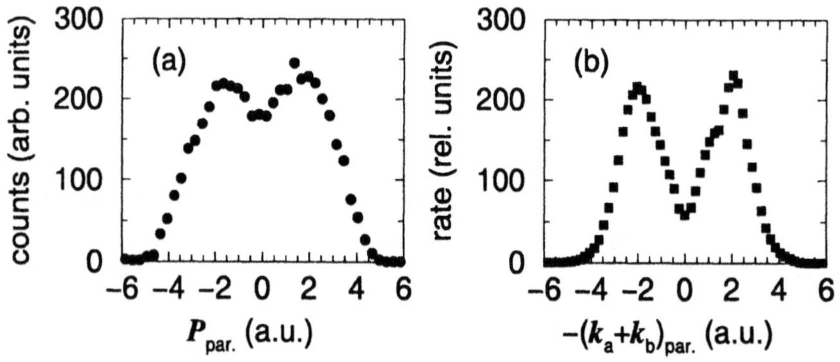

Figure 2. Recoil ion momentum distribution of He^{2+} parallel to the polarization direction, $P_{par.}$ (experimental data, panel a) and the sum-momentum of the two outgoing electrons in the opposite direction (present theory, panel b).

Finally, we note that there is a quantitative difference at the minimum of the distribution. This suggests that either the uncertainty in the intensity measurement and momentum resolution [22] and/or the higher order contributions in the theory might be involved here, which should be investigated in the future.

In Fig. 3 we show the distributions for the perpendicular component of the recoil momentum. Again the essential features of the experimental distribution, panel a, are reproduced by the theoretical calculation for the sum momentum of the electrons, panel b. Moreover, comparison of Fig. 2 with Fig. 3 shows that the theoretical calculations are consistent with the observation with respect to relative widths in the parallel and perpendicular directions, the latter being much narrower than the former.

What is the origin of the double-hump structure in the parallel case *and* its absence in the perpendicular case? To gain a greater insight into their origin we calculated the corresponding distributions by deliberately neglecting the *final*-state interaction of the two electrons with the laser pulse, i.e. dropping the 'Volkov dressing' of the two electron in the final state. This is easily done in the present theory by replacing the final state two-electron Volkov wavefunction, by two free (plane) waves of momenta $\mathbf{k}_a$ and $\mathbf{k}_b$, and keeping everything else the same.

The result so obtained for the parallel component is shown in Fig. 4a. A comparison of the theoretical calculation *with* (Fig. 2b) and *without* (Fig. 4a) the Volkov dressing in the final state, clearly shows that the double-hump character of the parallel component of the distribution collapses into a single-hump structure in the *absence* of the final state Volkov dressing. This unequivocally suggests that the final-state field interaction of the two electrons is primarily responsible for the double-hump structure of the observed distribution in this case.

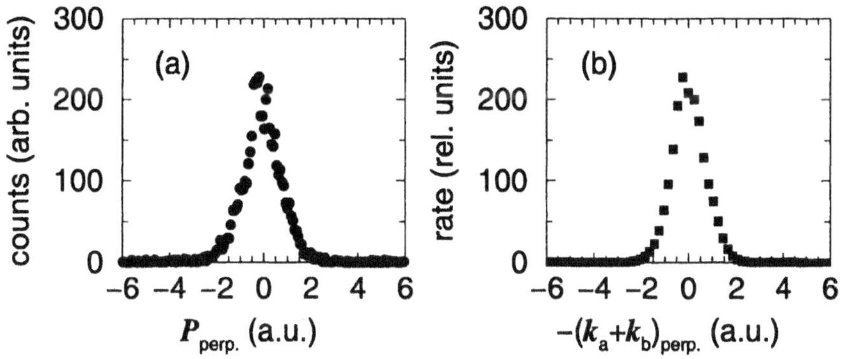

Figure 3. The same as in Fig. 2 but *perpendicular* to the polarization direction; panel a (experiment), panel b (theory)

In Fig. 4b we show the corresponding result of calculation for the perpendicular component in the absence of the final state laser interaction. Comparison of this distribution with that of Fig. 3b shows that in this (perpendicular) case the presence of the final state laser interaction does not play a significant role. This is as might be expected in the present case, since the force due to the electric field is negligible in the direction perpendicular to the polarization direction. This is also consistent with the observed narrow width of the perpendicular distribution due to the absence of significant laser coupling in the final state. Therefore, in this case we have the interesting situation of observing the double ionization in the laboratory as if the laser field is switched off as soon as the two electrons are freed from the bound atomic system.

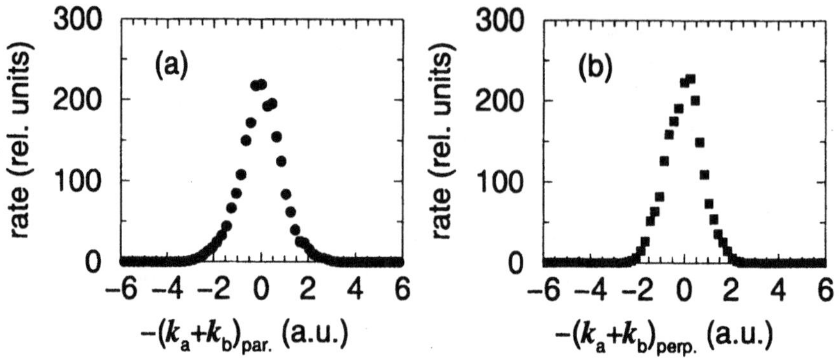

Figure 4. Sum-momentum distributions calculated ***without*** the final state 'Volkov dressing' of the two outgoing electrons: (a) parallel case (theory), c.f. Fig. 2 ; (b) perpendicular case (theory), c.f. Fig. 3.

We may finally ask, what determines the observed large width of the parallel distributions in Fig. 2? From the explicit expression of the transition matrix element, Eq. (25), and using the properties of the (generalised) Bessel functions involved, one could derive [7] the following expression for the cut-off momentum of the distribution,

$$|\mathbf{P}_{par.}|_{cut-off} \approx |(\mathbf{k}_a)_{par.} + (\mathbf{k}_b)_{par.}|_{cut-off} \approx Re(4\sqrt{U_p} + \sqrt{(8U_p - E_B)}), \tag{27}$$

where E_B is the binding energy. For the case of the distribution shown in Fig. 2 use of this formula predicts the cut-off momentum ≈ 5 a.u., in good agreement with both the experimental value, and the numerical result. Satisfactory agreement between the above cut-off formula and the available experimental values at other intensities for He [1] and Ne [2] has also been found [7].

4. Conclusion

Intense-field many-body S-matrix theory (IMST) is briefly reviewed. A dominant 'correlation mediated energy-sharing diagram' for non-sequential double ionization in intense femtosecond laser pulses, is used to describe physically the time-evolution of the process. Its connection, in the appropriate limit of large poderomotive energy and small frequency, with the classical 'rescattering model' [17] is pointed out. It is then applied to the current problem of 'recoil-ion' momentum distribution for double ionization in intense laser pulses. Calculations are performed for the double ionization of He at $800nm$, peak intensity $6.6 \times 10^{14} W/cm^2$ and pulse durartion $200fs$. The results for the sum-momentum of the two electrons in the parallel and perpendicular directions (measured with respect to the direction of polarization) are found to reproduce the characteristic features of the experimental distributions of the ion momentum. The observed double-hump distribution in the parallel case and its absence in the perpendicular case are shown to be due to the presence of the final-state *'Volkov dressing'* of the two outgoing electrons in the former (and its effective absence in the latter) case. The large (parallel) cut-off momenta observed at different intensities for He [1] and for Ne [2] are given by a simple formula derived theoretically.

5. Acknowledgement

It is a pleasure to thank R. Dörner, R. Moshammer and J. Ulrich for communicating their experimental results prior to their publications and for stimulating discussions. This work was partially supported under the project number, FA 160/18-2, SPP: "Wechselwirkung intensiver Laserfelder mit Materie", by Deutsche Forschungsgemeinschaft (Bonn).

References

1. Th. Weber et al., Phys. Rev. Lett. **84**, 443 (2000).
2. R. Moshammer et al., Phys. Rev. Lett. **84**, 447 (2000).
3. F.H.M. Faisal and A. Becker, (a) in *Selected Topics on Electron Physics*, editted by D.H. Campbell and H. Kleinpoppen, Plenum Press: N.Y. (1996), p. 317; (b) in *Multiphoton Processes 1996*, editted by P. Lambropoulos and H. Walther, Int. Nat. Conf. Ser. No. 154 (IOP: Bristol, 1997) p. 118.
4. A. Becker and F.H.M. Faisal, J.Phys. B **29**, L197 (1996).
5. F.H.M. Faisal and A. Becker, Laser Phys. **7**, 684 (1997).
6. A. Becker and F.H.M. Faisal, (a) Phys. Rev. **A59**, R1742 (1999); (b) *ibid* **A59**, R3182 (1999); (c) J. Phys. B **32**, L225 (1999).
7. A. Becker and F.H.M. Faisal, Phys. Rev. Lett. **84**, 3546 (2000).
8. J.S. Parker, K.T. Taylor, C.W. Clark and S. Blodgett-Ford, J. Phys. B **29**, L33 (1996); E.S. Smyth, J.S. Parker and K.T. Taylor, Comp. Phys. Comm. **144**, 1 (1998).
9. F.H.M. Faisal, *Theory of Multiphoton Processes*, Plenum Press: N.Y. (1987).
10. F.H.M. Faisal, Phys. Lett. A **187**, 180 (1994).
11. A. Becker and F.H.M. Faisal, Phys. Rev. **A50**, 3256 (1994).
12. C. Joachain, *Quantum Collision Theory*, 3rd edn., North-Holland, Amsterdam (1983).
13. H.R. Reiss, Phys. Rev. A **22**, 1786 (1980).
14. L.V. Keldysh, Zh. Eksp. Teor. Fiz. **47**, 1945 (1964) [Sov. Phys. JETP **20**, 1307 (1965)].
15. F.H.M. Faisal, J. Phys. B **6**, L89 (1973).
16. F.H.M. Faisal, A. Becker and J. Muth-Böhm, Laser Phys. **9**, 115 (1999).

17. P. Corkum, Phys. Rev. Lett. **71**, 1994 (1993).
18. D. Charalambidis et al. J. Phys. B **30**, 1467 (1997).
19. B. Walker et al., Phys. Rev. Lett. **73**, 1227 (1994).
20. A. Erdélyi (Ed.), *Higher Transcendental Functions*, Vol. 2, (New York: McGraw-Hill, 1953).
21. V.P. Krainov, J. Opt. Soc. Am. B **14**, 425 (1997).
22. The uncertainty in the intensity measurement is ≈ 15 - 30%, and that of momentum resolution ≈ 0.2 - 0.4 a.u. [R. Dörner, and H. Rottke (private communication)].

NONSEQUENTIAL DOUBLE IONIZATION: A MINIMAL-CORRELATION APPROACH

R. KOPOLD AND W. BECKER
Max-Born-Institut
Max-Born-Str. 2a, 12489 Berlin, Germany

Introduction

Nonsequential multiple ionization, where two or more electrons are ripped off the atom by the laser field in one coherent process, has attracted much interest for over a decade. Representative references include [1-6] for experimental data and [7-16] for theoretical work; the not-too-recent work is covered in the review articles [17, 18]. Unlike above-threshold ionization (ATI) or high-harmonic generation (HHG), this process cannot proceed in the absence of electron-electron correlation. Moreover, this correlation affords a sizeable premium for collaboration between the two electrons: for optimal laser intensities, the ratio of nonsequential over sequential double ionization, which materializes in the famous "knee", can be as high as six orders of magnitude [4]. Until about a year ago, experimental data were restricted to the total yields. Recently, however, application of the COLTRIMS (cold-target recoil-ion-momentum spectroscopy) technique [19] has allowed for the determination of the (three-dimensional) momenta of the ion and the electrons so that now the completely differential ionization rate has become accessible [20-24].

Early on, rescattering was identified [7] as a promising candidate for the physical mechanism of nonsequential double ionization (NSDI), the same mechanism that is known to be responsible for ATI and HHG. However, serious quantitative problems with this scenario precluded its acceptance until recently [25]. This problem has now been revisited and appears to have been resolved [26]. More and more, rescattering is emerging as the dominant mechanism [27-30], but depending on the atomic species and the laser intensity, others may be required in addition.

Thus far, there are only few theoretical results for the momentum distributions in NSDI and even fewer explicit calculations. The intense-field many-particle S-matrix theory [10] which has successfully reproduced many data for the total yields has been extended to calculations of the differential rates [31]. A calculationally simpler version [32] of essentially the same model yields similar results

B. Piraux and K. Rzążewski (eds.), Super-Intense Laser-Atom Physics, 67–77.

in those parameter regimes where it is applicable. There is also a simulation on the basis of the time-dependent Schrödinger equation in one dimension for each electron [33]. Much can be learned already from purely classical kinematical considerations [34] along the lines of the so-called simple-man model of intense-laser atom processes.

In this paper, we present a simple fully quantum-mechanical model [35] that is able to predict the *momentum distributions* for any given scenario, such as sequential ionization, nonsequential ionization via rescattering, or nonsequential ionization via rescattering into an excited state followed by tunneling. The *absolute yield* for a given scenario cannot be predicted, but its intensity dependence can. Most importantly, different scenarios generate very different momentum distributions. Comparison of the calculated with the measured momentum distributions then can support or discredit the respective scenario.

The Model

The model is an extension of the successful Keldysh-Faisal-Reiss (KFR) model of ATI [36] or the Lewenstein model [37] of HHG to a many-particle situation. We make consistent use of the strong-field approximation (SFA), that is, we ignore the effect of the binding potential on electrons in the continuum as well as the effect of the laser on electronic bound states. All interactions are modeled by contact interactions, including the electron-electron interaction which is only taken into account at the instant of rescattering. The approach is in the spirit of the work of Landau and Dykhne (cf. [38]) which underlines the relative insignificance of the detailed form of the potentials.

In order to motivate our ansatz for the matrix element for double ionization, we first focus on single ionization. The single-active-electron KFR matrix element for direct ionization (not involving rescattering) into a state with asymptotic momentum $\mathbf{p}$ is

$$M_{\mathbf{p}} = -\frac{i}{\hbar}\int_{-\infty}^{\infty} dt \langle \psi_{\mathbf{p}}(t)|V|\psi_0(t)\rangle \sim \int_{-\infty}^{\infty} dt e^{iS_{\mathbf{p}}(t)/\hbar}, \tag{1}$$

where $\psi_{\mathbf{p}}(t)$ and $\psi_0(t)$ denote the Volkov state with momentum $\mathbf{p}$ and the initial atomic bound state with binding energy $E_0 < 0$, respectively, and $V(\mathbf{r})$ is the binding potential. To get the last expression of Eq. (1), a zero-range potential has been used which takes care of the spatial integration. The exponent displays the action of the electron along its path from the initial bound state to the final continuum state,

$$S_{\mathbf{p}}(t) = -\int_{t'}^{\infty} d\tau \varepsilon_{\mathbf{p}}(\tau) - \int_{-\infty}^{t'} d\tau E_0, \tag{2}$$

approximated according to the SFA: at time t', the electron tunnels out of the binding potential; for earlier times the laser field and for later times the binding potential is neglected. Hence, up to the time t', the accumulated action is just the binding energy of the electron integrated over time. Thereafter, the action is given by the negative of the kinetic energy $\varepsilon_{\mathbf{p}}(\tau) = [\mathbf{p} - e\mathbf{A}(\tau)]^2/2m$ of the electron in the laser field, integrated over time.

In the event of rescattering, the freed electron interacts one more time with the atomic potential. In this process, it can acquire (classical) drift energies of up to $10U_P$. This is described by the matrix element [39]

$$\begin{aligned} M_{\mathbf{p}} = & -\frac{1}{\hbar^2}\int_{-\infty}^{\infty} dt \int_{-\infty}^{t} dt' \int d^3\mathbf{k} \langle \psi_{\mathbf{p}}(t)|V|\psi_{\mathbf{k}}(t)\rangle\langle\psi_{\mathbf{k}}(t')|V|\psi_0(t')\rangle \\ & \sim \int_{-\infty}^{\infty} dt \int_{-\infty}^{t} dt' \int d^3\mathbf{k}\, e^{iS_{\mathbf{p}}(t,t',\mathbf{k})/\hbar}, \end{aligned} \tag{3}$$

which contains an additional integration over the time t where the electron rescatters. In between the rescattering time t and the ionization time t', the electron occupies any one of the Volkov states with drift momentum $\mathbf{k}$, which have the wave functions (in the length gauge)

$$\langle \mathbf{r}|\psi_{\mathbf{k}}(t)\rangle = \frac{1}{(2\pi\hbar)^{3/2}} \exp\frac{i}{\hbar}\left[(\mathbf{k} - e\mathbf{A}(t))\cdot\mathbf{r} - \int^{t} d\tau\varepsilon_{\mathbf{k}}(\tau)\right]. \tag{4}$$

Actually, the integral over $\mathbf{k}$ in Eq. (3) yields the Volkov propagator [39]. As above, after evaluation of the matrix elements with the assumption of a zero-range potential, the last expression of Eq. (3) results, which is dominated by the approximated action

$$S_{\mathbf{p}}(t,t',\mathbf{k}) = -\int_{t}^{\infty} d\tau\varepsilon_{\mathbf{p}}(\tau) - \int_{t'}^{t} d\tau\varepsilon_{\mathbf{k}}(\tau) - \int_{-\infty}^{t'} d\tau E_0 \tag{5}$$

of the electron. It differs from Eq. (2) only by the presence of the second term, which is the classical action of the electron traveling along a closed orbit from time t' to time t.

Equations (3) and (5) suggest how to model nonsequential double or N-fold ionization in the same vein on the assumption of a recollision scenario, where the first electron when it returns to its parent ion simultaneously kicks out the remaining $N-1$ into states with drift momenta $\mathbf{p}_i\,(i = 1,...,N)$. The analogous ansatz is

$$M_{\{\mathbf{p}_i\}} \sim \int_{-\infty}^{\infty} dt \int_{-\infty}^{t} dt' \int d^3\mathbf{k} \exp[iS_{\{\mathbf{p}_i\}}(t,t',\mathbf{k})/\hbar], \tag{6}$$

where now

$$S_{\{\mathbf{p}_i\}}(t,t',\mathbf{k}) = -\int_{t}^{\infty} d\tau \sum_{i=1}^{N} \varepsilon_{\mathbf{p}_i}(\tau) - \int_{t'}^{t} d\tau\, [\varepsilon_{\mathbf{k}}(\tau) + E_{0_{N-1}}] - \int_{-\infty}^{t'} d\tau E_{0_N} \tag{7}$$

approximates the action of N electrons in the manner of the SFA: Bound electrons are approximated by unperturbed bound states of the atom or the ion, whereas free electrons are approximated by non-interacting Volkov-states. Up to the time t' the contribution to the action of all N bound electrons is their total binding energy E_{0_N} integrated over time. From time t' to time t, one electron makes an excursion driven by the laser field. In the SFA, its contribution to the action is — as in the generalized KFR matrix element (3) — its oscillating kinetic energy $\varepsilon_{\mathbf{k}}(\tau)$ integrated over time. During this period of time, the remaining $N-1$ electrons are still bound to the ionic core with the total binding energy $E_{0_{N-1}}$. Finally at time t the returning electron kicks out the still bound electrons. Neglecting their interaction with the ionic core and with each other, the contribution to the action is given by the integral over the sum of their kinetic energies in the laser field. The first ionization at time t' was out of a zero-range potential, and the recollision at time t assumed one joint contact interaction.

In the following we will evaluate the matrix element (6) for double ionization ($N=2$) and a linearly polarized laser field with the vector potential $\mathbf{A}(t) = \mathbf{A}_0 \cos \omega t$. The generalization to arbitrary polarization is rather straightforward. All but one of the five integrations in the matrix element (6) can be carried out analytically. The explicit result is [40]

$$\begin{aligned} M_{\mathbf{p}_1,\mathbf{p}_2} &\propto \hbar\omega \sum_n \delta\left[2U_p - E_0 - E_1 + \frac{1}{4m}(\mathbf{P}^2+\mathbf{p}^2) - n\hbar\omega\right] \\ &\times \sum_{l=-\infty}^{\infty} J_{n+2l}\left(\frac{e\mathbf{P}\cdot\mathbf{A}_0}{m\omega}\right)\left(\int_0^\infty d\tau \left(\frac{m}{i\tau}\right)^{3/2} \exp i\left(\frac{E_0\tau}{\hbar\omega}\right)\right. \\ &\times \left\{\exp i\left[\frac{4\eta\sin^2\frac{\tau}{2}}{\tau} - \eta\tau - l\phi(\tau)\right] J_l[\eta y(\tau)] - J_l(\eta)\right\} \\ &\left. + J_l(\eta)\int_0^\infty d\tau \left(\frac{m}{i\tau}\right)^{3/2} \exp i\left(\frac{E_0\tau}{\hbar\omega}\right)\right), \end{aligned} \tag{8}$$

where the $J_l(x)$ are Bessel functions. The real functions $y(\tau)$ and $\phi(\tau)$ are given by

$$y(\tau)e^{i\phi(\tau)} = 1 - ie^{i\tau}\left(\frac{4\sin^2\frac{\tau}{2}}{\tau} - \sin\tau\right), \tag{9}$$

and τ is the dimensionless travel time $\omega(t-t')$. The binding energies of the first and the second electron are E_0 and E_1, respectively. The matrix element is analyzed in terms of the total momentum $\mathbf{P} \equiv \mathbf{p}_1 + \mathbf{p}_2$ and the relative momentum $\mathbf{p} \equiv \mathbf{p}_1 - \mathbf{p}_2$ of the two final electrons. Note that the latter only appears in the argument of the delta function in the first line on the right-hand side of Eq. (8).

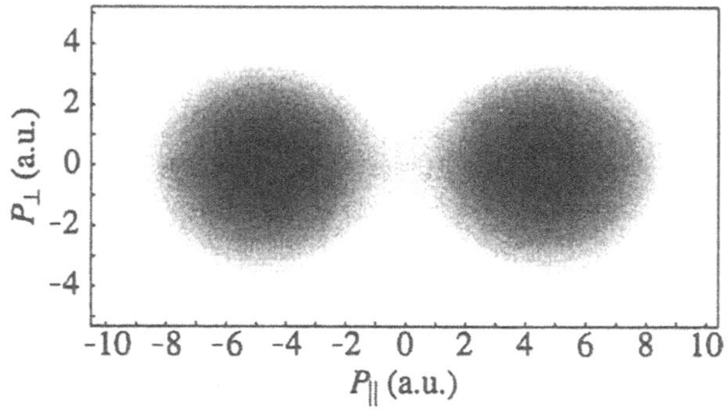

Figure 1. Linear-scale density plot of the distribution of the total electronic momentum $\mathbf{P} = (P_{||}, P_\perp)$ in nonsequential double ionization of neon at 8×10^{14} W/cm^2 and $\hbar\omega = 1.55$ eV.

Therefore, for a fixed total momentum **P** the relative momentum distribution is isotropic as a consequence of the assumptions of this model. The delta function allows for the net transfer of an arbitrary number n of photons from the laser field (with ponderomotive potential $U_p = \eta\hbar\omega$ and laser frequency ω) to the charged particles,

$$\frac{1}{4m}(\mathbf{P}^2 + \mathbf{p}^2) = n\hbar\omega - 2U_P - |E_1| - |E_2|. \tag{10}$$

However, photon transfers in excess of $n\omega \approx 3.17U_p$ will be more and more suppressed. On the technical side, please, note that the first integral over τ on the right-hand side of Eq. (8) converges absolutely owing to the function in the curly bracket (for $\tau = 0$, we have $\phi(\tau) = 0$ and $y(\tau) = 1$) while the second one can be calculated analytically by analytic continuation in the fractional power of τ [41].

Results

The first COLTRIMS experiments [20, 21, 22] detected the momentum of the ion. The short-pulse optical lasers that are being used in these experiments allow us, to a very good approximation, to disregard the momenta of the photons transferred to the atomic system. Momentum conservation then implies that the momentum of the ion is equal in magnitude and opposite in direction to the sum of the electronic momenta. We calculate the latter. Figure 1 exhibits results of calculations based on the S-matrix element (8) for the case of neon and an intensity of 8×10^{14} W/cm^2 such that the maximal energy of the returning electron in the rescattering picture is well above the binding energy E_2 of the second electron. The density plot gives the distribution of the total-momentum component $P_{||}$ parallel and an arbitrary component $P_\perp$ perpendicular to the linearly polarized laser field. The remaining third component of **P** (perpendicular to the field) as well as the relative momentum

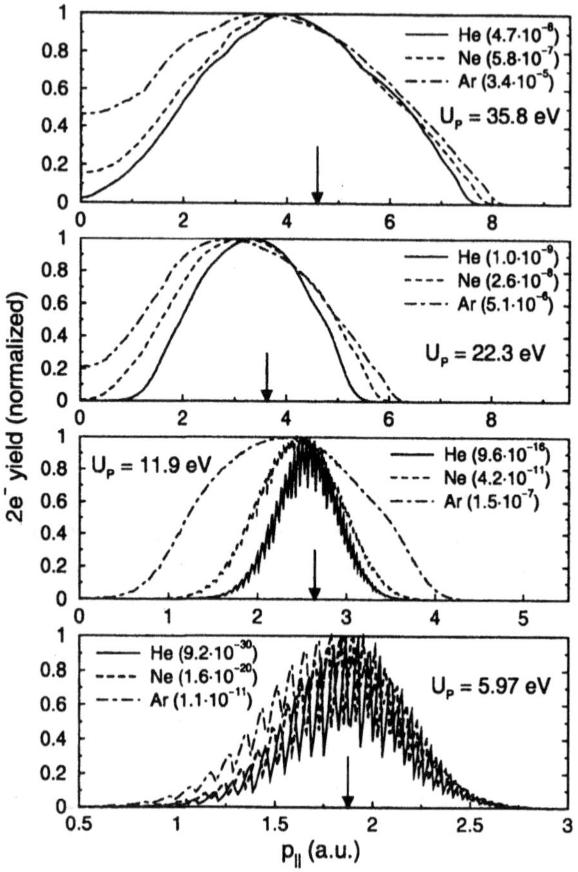

Figure 2. Distribution of $P_{||}$ for $\mathbf{P}_{\perp} = 0$ for He, Ne, and Ar, $\hbar\omega = 1.55$ eV, and various intensities characterized by the corresponding ponderomotive potential U_P. Because of symmetry, the distribution is only shown for $P_{||} > 0$. Each curve has been normalized to a maximum of unity; the number given in parentheses for each curve specifies the respective normalization factor. The vertical arrows mark the position of $4\sqrt{mU_P}$ whose significance is discussed in the text.

p have been integrated over. The figure clearly shows a two-peaked distribution of the shape that has been observed in the recent experiments.

In the most naive application of the classical rescattering model one might assume that the returning electron promotes the second electron into the continuum such that both find themselves at rest at about the time when the electric field of the laser goes through zero. Under these conditions, the total momentum of the two electrons will be $P_{||} = 4\sqrt{mU_p}$ (with the exact travel time of the simple-man model the number is 3.8 in place of 4). This is indeed the approximate position where the distribution peaks, even though in the case of this figure the energy of the returning electron is much higher than required for the second ionization.

In Fig. 2 we compare the distributions of the total momentum for helium, neon, and argon for various intensities. In the uppermost panel, the maximal classical excess energy of the returning electron over the second ionization potential, viz. $\Delta E = 3.17U_p - |E_1|$, is comfortably positive for all three rare gases considered ($E_1 = 54.4$ eV, 27.6 eV, and 41.0 eV for He, Ar, and Ne, respectively). In the bottom panel, it is negative for all three. The second ionization still occurs, though with greatly reduced yield, in close analogy with HHG and high-order ATI spectra extending beyond their respective classical cutoffs. The mechanism is tunneling of the second electron assisted by the returning first electron. From top to bottom, the distributions narrow, and total momenta near zero become more and more suppressed. The larger ΔE, the more is the peak of the distribution shifted below the naive classical value of $3.8\sqrt{mU_p}$. In the lowest panel, the distributions for all three rare gases peak almost exactly at this value. This is no surprise, as the assumptions that led to it are perfectly satisfied in this case. For sufficiently negative values of ΔE, the distributions display a very regular sawtooth structure which reflects the discrete energy transfer (10) from the laser field. The message of Fig. 2 is this: for a pure rescattering scenario, the larger the binding energy of the second electron and/or the lower the laser intensity, the narrower is the two-peak structure of the distribution of $P_{\|}$.

Very recently, experimental data have been obtained for the distribution of both electron momenta $\mathbf{p}_1$ and $\mathbf{p}_2$ [23, 24]. In Figs. 3 and 4 we present distributions of the components $p_{1\|}$ and $p_{2\|}$ parallel to the laser field, calculated from the matrix element (8). The remaining four momentum components have been integrated over. The figures reveal a strong correlation between the parallel components of the two momenta. There are virtually no events in the second and the fourth quadrant. Owing to the lower intensity underlying Fig. 4 (and in spite of the lower ionization potential of argon), the two-peak structure is more pronounced for argon than for neon, in agreement with the results shown in Fig. 2.

Comparison with the Experimental Data

The calculated distribution of the total momentum presented in Fig. 1 agrees well with the data [21] provided we assume for the calculation a lower intensity (8×10^{14} W/cm^2) than the peak intensity (1.3×10^{15} W/cm^2) given in the experimental paper [21]. In this regard, one should note that (i) a fixed-intensity calculation at the experimental peak intensity would ignore the volume effect and (ii) the experimental spectrum of singly ionized neon that was recorded under the same conditions was best reproduced for an intensity of 9×10^{14} W/cm^2, using the theoretical spectrum of Ref. [42]. In contrast to this good agreement, the data in helium [20] exhibit a weakly developed two-peak structure for 6.6×10^{14} W/cm^2 which disappears for lower intensities. This is in direct conflict with the prediction of the rescattering model, shown in Fig. 2, that the two-peak structure

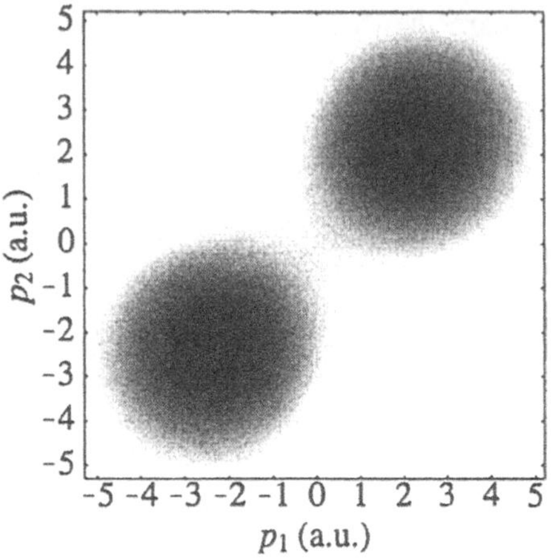

Figure 3. p_1-p_2 momentum distribution of neon for $\hbar\omega = 1.55$ eV and 8×10^{14} W/cm^2.

should become more distinct when the intensity decreases. For neon, unfortunately, data are only available for one intensity. The data in argon [22] do not show any two-peak structure at all, but rather a plateau within a small intensity interval. The intensities covered by Fig. 2 have been chosen such that they can be directly compared to the data [20, 22]. The rescattering scenario of this paper with the additional assumption of a zero-range electron-electron interaction yields the peaks of the two-peak distribution at or slightly below $P_{\|} = 4\sqrt{mU_P}$. Other theoretical approaches that exclusively model rescattering [32] or include it among possible other mechanisms [31], but describe the electron-electron interaction in terms of the Born approximation and the Coulomb potential, produce the position of these peaks at a lower value of $P_{\|}$. This latter description, however, ignores the interaction of the returning electron with the ion. It should be mentioned that the fact that we neglected the interaction between the two electrons in the final state has no effect on the distribution of the total momentum.

Comparing the model to the most recent measurements of the distribution of $p_{1,\|}$ and $p_{2,\|}$ we encounter much the same situation. For neon there is good agreement: the data [24] as well as the theoretical results presented in Fig. 3 show strongly correlated electron momenta and practically no events in the second and fourth quadrant. The theory makes qualitatively the same prediction for argon which is displayed in Fig. 4. Here, the correlation is even stronger than in Fig. 3 for neon, but this primarily due to the lower intensity used for argon. In contrast, in the data [23, 24] there is a sizeable number of events in the second and the fourth quadrant where the two electron-momentum components have opposite sign. The intensities are such that sequential double ionization is negligible.

Hence, in general, the results of this model with the assumption of a rescatter-

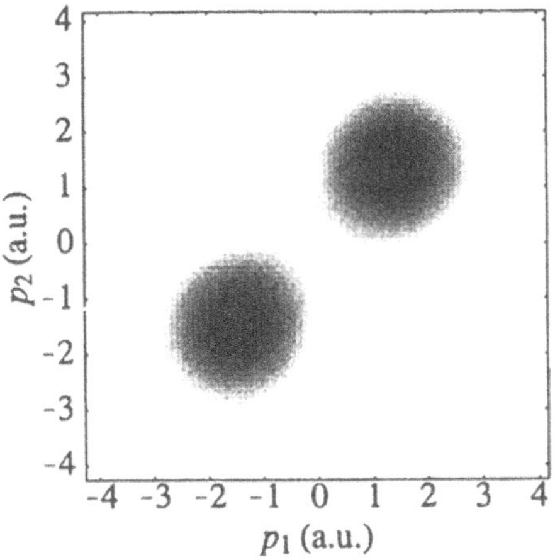

Figure 4. p_1-p_2 momentum distribution of argon for $\hbar\omega = 1.55$ eV and 3×10^{14} W/cm^2.

ing scenario agree well with the data in neon [21, 24] and not so well or not at all with the data in helium [20] and argon [22, 23, 24]. However, general conclusions from this observation are impeded by the fact that these data are taken at different intensities.

Discussion and Conclusions

The model laid out in this paper allows for a simple and straightforward exploration of the consequences for the S matrix of a given scenario of how double or multiple nonsequential ionization proceed. In this paper, we adopted the rescattering scenario. The model owes its conceptual simplicity and its convenience for explicit calculations to the use of contact interactions for all particles involved. How well is this approximation justified? A partial answer, regarding the electron-electron interaction, can be extracted from Ref. [32]: generally, it appears that the contact interaction yields results that are the closer to the results of a Coulomb potential, the lower the intensity is and the stronger the second electron is bound. Of course, the adequacy of the Born approximation for the returning electron as it is used in Refs. [31, 32] is itself an issue that is not resolved.

As discussed above, we obtained good agreement with the data for neon while there are major discrepancies with the data in argon and helium. A possible resolution should lie in the identification of a different scenario that reproduces the observed electron distributions for argon and helium. Their characteristic traits are a less well developed or totally absent peak structure in the total-momentum distribution and a less pronounced correlation between the two electron momenta. Qualitatively, these features emerge when we assume that the returning electron does not immediately kick out the other one, but first promotes it into an excited state from which it tunnels out at a later time. This scenario has been implemented by corresponding approximations to the action and the distribution of the total momentum has been evaluated, allowing each electron one dimension [35]. Indeed, the region in between the two peaks of the distribution of Figs. 1 and 2 is filled in as expected: the excited second electron preferentially tunnels out near the peaks of the laser field and, consequently, acquires only a small drift momentum.

Nonsequential double ionization cannot happen in the absence of an electron-electron interaction. The model presented in this paper allows for this interaction only at the instant of time when the recollison of the first electron with its parent ion occurs and approximates it by a contact interaction. The fact that on the basis of this assumption the data in neon can be satisfactorily reproduced underlines the dominance of the electron-field interaction over the inneratomic dynamics in intense-laser atom physics.

We benefitted from discussions with M. Dörr, S. P. Goreslavskii, S. V. Popruzhenko, H. Rottke, and W. Sandner. This work was supported in part by the Deutsche Forschungsgemeinschaft.

References

1. L'Huillier, A., Lompré, L. A., Mainfray, G., and Manus, C. (1983), *Phys. Rev. A* **27**, 2503-2512.
2. Fittinghoff, D. N., Bolton, P. R., Chang, B., and Kulander, K. C. (1992), *Phys. Rev. Lett.* **69**, 2642-2645.
3. Kondo, K., Sagisaka, A., Tamida, T., Nabekawa, Y., and Watanabe, S. (1993), *Phys. Rev. A* **48**, R2531-R2534.
4. Walker, B., Sheehy, B., DiMauro, L. F., Agostini, P., Schafer, K. J., and Kulander, K. C. (1994), *Phys. Rev. Lett.* **73**, 1227-1230.
5. Augst, S., Talebpour, A., Chin, S. L., Beaudoin, Y., and Chaker, M. (1995), *Phys. Rev. A* **52**, R917-R920.
6. Larochelle, S., Talebpour, A., and Chin, S. L. (1998), *J. Phys. B* **31**, 1201-1214.
7. Corkum, P. B. (1993), *Phys. Rev. Lett.* **71**, 1994-1997.
8. Lerner, P. B., LaGattuta, K. J., and Cohen, J. S. (1994), *Phys. Rev. A* **49**, R12-R15.
9. Kuchiev, M. Yu. (1995), *J. Phys. B* **28**, 5093-5115.
10. Becker, A. and Faisal, F. H. M. (1996), *J. Phys. B* **29**, L197-L202; (1999), *Phys. Rev. A* **59**, R1742-R1745.
11. Watson, J. B., Sanpera, A., Lappas, D. G., Knight, P. L., and Burnett, K. (1997), *Phys. Rev. Lett.* **78**, 1884-1887.
12. Bauer, D. (1997), *Phys. Rev. A* **56**, 3028-3039.
13. LaGattuta, K. J., and Cohen, J. S. (1998), *J. Phys. B* **31**, 5281-5291.
14. Liu, W.-C., Eberly, J. H., Haan, S. L., and Grobe, R. (1999), *Phys. Rev. Lett.* **83**, 520-523.
15. Dundas, D., Taylor, K. T., Parker, J. S., and Smyth, E. S. (1999), *J. Phys. B* **32**, L231-L238.
16. LaGattuta, K. J. (2000), *J. Phys. B* **33**, 2489-2494.
17. DiMauro, L. F. and Agostini, P. (1995), *Adv. At. Mol. Opt. Phys.* **35**, 79-120.
18. Lambropoulos, P., Maragakis, P., and Zhang, J. (1998), *Phys. Rep.* **305**, 203-293.
19. Ullrich, J., Moshammer, R., Dörner, R., Jagutzki, O., Mergel, V., Schmidt-Böcking, H., and Spielberger, L. (1997) *J. Phys. B* **30**, 2917-2974.
20. Weber, Th., Weckenbrock, M., Staudte, A., Spielberger, L., Jagutzki, O., Mergel, V., Afaneh, F., Urbasch, G., Vollmer, M., Giessen, H., and Dörner, R. (2000), *Phys. Rev. Lett.* **84**, 443-446.

21. Moshammer, R., Feuerstein, B., Schmitt, W., Dorn, A., Schröter, C. D., Ullrich, J., Rottke, H., Trump, C., Wittmann, M., Korn, G., Hoffmann, K., and Sandner, W. (2000), *Phys. Rev. Lett.* **84**, 447-450.
22. Weber, Th., Weckenbrock, M., Staudte, A., Spielberger, L., Jagutzki, O., Mergel, V., Afaneh, F., Urbasch, G., Vollmer, M., Giessen, H., and Dörner, R. (2000), *J. Phys. B* **33**, L127-L133.
23. Weber, Th., Giessen, H., Weckenbrock, M., Urbasch, G., Staudte, A., Spielberger, L., Jagutzki, O., Mergel, V., Vollmer, M., and Dörner, R. (2000), *Nature* **405**, 658-661.
24. Rottke, H., Trump, C., Wittmann, M., Korn, G., Sandner, W., Moshammer, R., Schröter, C. D., Fischer, D., Dorn, A., Crespo, J., Feuerstein, B., Neumeir, P., and Ullrich, J., to be published.
25. Kulander, K.C., Cooper, J., and Schafer, K. J. (1995), *Phys. Rev. A* **51**, 561-568.
26. Bharwaj, R., Yudin, G., Meheandele, M., Aseyev, S., Villeneuve, D., Corkum, P., and Ivanov, M., this volume.
27. Lein, M., Gross, E. K. U., and Engel, V. (2000), *J. Phys. B* **33**, 433-442 (2000).
28. Dörr, M. (2000), *Opt. Express* **6**, 111-116.
29. Haan, S. L., Hoekema, N., Poniatowski, S., Liu, W.-C., and Eberly, J. H. (2000), *Opt. Express* **7**, 29-38.
30. Watson, J. B., Burnett, K., and Knight, P. L. (2000), *J. Phys. B* **33**, L103-L109.
31. Becker, A. and Faisal, F. H. M. (2000), *Phys. Rev. Lett.* 84, 3546-3549.
32. Goreslavskii, S. P. and Popruzhenko, S. V., this volume.
33. Lein, M., Gross, E. K. U., and Engel, V. (2000), *Phys. Rev. Lett.* **85**, to be published.
34. Feuerstein, B., Moshammer, R., and Ullrich J. (2000), *J. Phys. B* **33**, L823-L830.
35. Kopold, R., Becker, W., Rottke, H., and Sandner, W. (2000), *Phys. Rev. Lett.* **85**, 3781-3784.
36. Keldysh, L. V. (1964), *Zh. Eksp. Teor. Fiz.* **47**, 1945-1957 [*Sov. Phys. JETP* **20**, 1307-1314]; Faisal, F. H. M. (1973), *J. Phys. B* **6**, L89-L94; Reiss, H. R. (1980), *Phys. Rev. A* **22**, 1786-1813.
37. Lewenstein, M., Balcou, Ph., Ivanov, M. Yu., L'Huillier, A., and Corkum, P. B. (1994), *Phys. Rev. A* **49**, 2117-2132.
38. Landau, L. D. and Lifshitz, E. M. (1977), *Quantum Mechanics (Nonrelativistic theory)*, 3rd edition (Pergamon Press) §53.
39. Lohr, A., Kleber, M., Kopold, R., and Becker, W. (1997), *Phys. Rev. A* **55**, R4003-R4006.
40. Kopold, R. (2000), Doctoral dissertation, Technical University of Munich.
41. Becker, W., MvIver, J. K., and Wódkiewicz, K. (1993), *Laser Phys.* **3**, 475-481.
42. Krainov, V. P. (1997), *J. Opt. Soc. Am.* **14**, 425-431.

NON-SEQUENTIAL DOUBLE IONIZATION OF ATOMS IN STRONG FIELDS

KRZYSZTOF SACHA
Fachbereich Physik, Philipps Universität Marburg, D-35032 Marburg, Germany;
Instytut Fizyki im. M. Smoluchowskiego, Uniwersytet Jagielloński, ul. Reymonta 4, PL-30-059 Kraków, Poland

AND

BRUNO ECKHARDT
Fachbereich Physik, Philipps Universität Marburg, D-35032 Marburg, Germany

A series of recent experiments by Weber *et al* [1, 2, 3] and Moshammer *et al* [4] have provided data on the correlation between outgoing electrons in multiple ionization in strong lasers: the total momentum of the electrons is aligned along the field axis and the joint distribution of parallel momenta for the two electrons has pronounced maxima along the diagonal, showing that the electrons typically come with the same momenta [3]. As a step towards a better understanding we discuss here the pathways to double ionization within a classical model for electrons in a combined Coulomb and external field.

According to the currently accepted models [5, 6, 7, 8, 9], the process of multiphoton multiple ionization can naturally be divided into two steps: in the first step a compound state of highly excited electrons close to the nucleus is formed and in the second step several electrons can escape from this compound state to produce the multiply ionized final state. The first step clearly requires a quantum description. The second one, because of the high excitation in the complex and the multiphoton nature of the whole process, should be accessible to a classical analysis. And it is this second step, the escape of two or more electrons from the highly excited compound state close to the nucleus, on which we focus. We restrict ourselves in the following to double ionization but the model can be easily extended to the multiple escape.

B. Piraux and K. Rzążewski (eds.), Super-Intense Laser-Atom Physics, 79–83.

There are various paths of decay of the excited complex: single ionization when one electron escapes and the other one is trapped to the ion, sequential double ionization corresponding to an escape of the second electron when the first electron has already been ionized and finally simultaneous non-sequential double ionization. If the complex decays when the field is on, both electrons are pulled towards the Stark saddle. Electron repulsion then singles out a symmetric motion, with both electrons at the same distance from the nucleus, as the most likely and preferred one: almost any deviation between the electrons will be amplified and the symmetry destroyed. If one electron is further out than the other electron repulsion will help it to ionize and will at the same time block the other. The symmetric subspace is an invariant subspace of the dynamics, that is, if initial positions and momenta are chosen symmetric with respect to the field polarization axis this will not change in the course of time and the trajectories will not leave the subspace. We therefore propose that non-sequential ionization is dominated by the dynamics near this subspace and that much of the behavior can be calculated within the subspace. This proposal is similar to Wanniers description of double ionization after electron impact [10]. The main difference is that in the present case the external field has to be taken into account.

With the polarization of the field along the z-axis the symmetric configuration (the C_v symmetry subspace) corresponds to coordinates (in the cylindrical frame) $z_1 = z_2 = Z$, $\rho_1 = \rho_2 = R$ and $\phi = \varphi_1 - \varphi_2$ with conjugate momenta $p_{z_1} = p_{z_2} = p_Z/2$, $p_{\rho_1} = p_{\rho_2} = p_R/2$ and p_ϕ, respectively. The ϕ-degree of freedom corresponds to a bending motion of the electrons against each other with respect to the field axis (i.e. for $\phi = \pi$ and $p_\phi = 0$ the electrons move in a plane). The Hamiltonian of the system for zero total angular momentum projection along the polarization axis reads

$$H = \frac{p_R^2 + p_Z^2}{4} + \frac{p_\phi^2}{R^2} + V(R, Z, \phi) \tag{1}$$

with potential energy

$$V = -\frac{4}{\sqrt{R^2 + Z^2}} + \frac{1}{R\sqrt{2(1 - \cos\phi)}} + 2ZFf(t)\cos(\omega t + \theta). \tag{2}$$

F, ω and θ stand for the amplitude, frequency and initial phase of the field respectively while

$$f(t) = \sin^2(\pi t/T_d) \tag{3}$$

is the pulse envelope with the duration T_d.

For the field frequencies used in the experiments [1, 2, 3, 4], $\omega = 0.057$ a.u. (800 nm), fast motion of the electrons close to the nucleus can be analyzed in the adiabatic approximation for fixed phase of the field. Then it is easy to show that in the (R, Z, ϕ) space the external field opens up a saddle at $R = r_S \sin\theta_S$, $Z = r_S \cos\theta_S$ and $\phi = \pi$ where $r_S^2 = \sqrt{3}/|Ff(t)\cos(\omega t + \theta)|$ and $\theta_S = \pi/6$ or $5\pi/6$ depending on the phase of the external field, see Fig. 1. The saddle allows the electrons to ionize even if the initial energy of the complex is negative. Once the electrons cross the saddle they are accelerated by the field and rapidly gain energy so that they can escape to infinity even when the field vanishes.

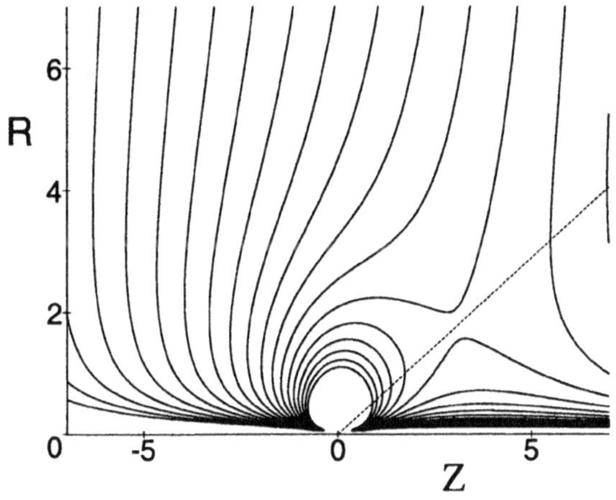

Figure 1. Section through equipotential surfaces of the adiabatic potential Eq. (2) for fixed time t and for $\phi = \pi$. The saddle moves along the dashed line when the electric field points in the positive Z-direction and along a second one obtained by reflection on $Z = 0$ during the other half of the field cycle.

To investigate signatures of non-sequential escape in double ionization experiments we calculate the distributions of final ion momenta within the C_v symmetry subspace. Our model possesses two parameters: the energy E of the complex and the time t_0 during the pulse when the complex is formed. It turns out that the latter parameter determines the width of the ion momentum distribution parallel to the field polarization axis. The complex ionizes usually during the first field cycle after it has been created and consequently the maximal ion momentum component parallel to the field can be estimated as

$$p_{par} = 2\sqrt{2U_p} = 2Ff(t_0)/\omega \tag{4}$$

where U_p denotes the ponderomotive energy of a free electron. Eq. (4) corresponds to parallel emission of two electrons with the maximal energy that can be acquired by a free electron in the field with the amplitude $Ff(t_0)$. Fitting the width of the calculated distribution to the experimental distribution allows one to estimate a moment in the pulse when the majority of doubly ionized ions are created.

The initial energy of the complex is the higher the more energy is brought in by the returning electron in the rescattering process which in turn depends on the intensity of the laser field. Our simulations show that for a broad range of the initial energy the final ion momentum distributions are essentially similar, with a double hump structure for the parallel momentum of ions. However for the initial energy close to the minimal saddle energy the parallel momentum distribution changes its character: the two peaks disappear and are replaced by a single maximum around zero momentum value. This observation is in agreement with the experimental results [1] where the double hump structure in the distribution appears only when the field intensity is sufficiently high.

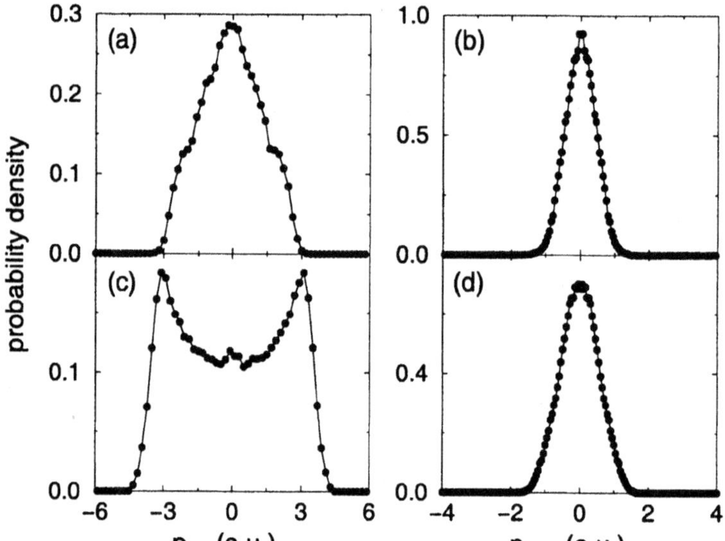

Figure 2. Final distributions of ion momenta parallel, p_{par} and transverse, p_{trans} to the field polarization axis calculated in the C_v symmetry subspace for field frequency $\omega = 0.057$ a.u. (800 nm) and initial time $t_0 = 0.33T_d$ where $T_d = 20 \times 2\pi/\omega$. Panels (a)-(b) correspond to field peak intensity 2.9×10^{14} W/cm^2 and initial energy $E = -0.6$ a.u. while panels (c)-(d) to 6.6×10^{14} W/cm^2 and $E = -0.4$ a.u..

In Fig. 2 we show the final ion momentum distributions both parallel and transverse to the field polarization axis for two different peak field intensities 2.9 and 6.6×10^{14} W/cm^2. The distributions have been calculated for the initial time $t_0 = 0.33T_d$ (where $T_d = 20 \times 2\pi/\omega$) with average over the phase of the field and over microcanonically chosen initial conditions. The calculated distributions are in a remarkable agreement with the experimental data [1] clearly indicating the dominance of the symmetric non-sequential electron escape in the double ionization process. Note that the parallel ion momentum distribution can reveal single maximum around zero momentum value even if one considers the non-sequential ionization process only. The reason is that for the low initial energy the electrons that cross the saddle are slow and the combined Coulomb and external field interactions shape the distribution. This is different from the situation at higher energy where the distribution is mostly shaped by the external field because after crossing the saddle the electrons quickly leave the region where the Coulomb interaction is significant.

According to the model and our calculations non-sequential double ionization can occur only in or near a symmetric invariant subspace where both electrons have the same distance from the nucleus. This symmetric motion is unstable and the bottleneck for double ionization is a saddle in phase space. The ion momentum distributions calculated in the subspace are in a very good agreement with the experimental data. Further details of the model and further numerical explorations are given in [11].

Acknowledgments

Financial support by the Alexander von Humboldt Foundation and by KBN under project 2P302B00915 are gratefully acknowledged.

References

1. Weber, Th., Weckenbrock, M., Staudte, A., Spielberger, L., Jagutzki, O., Mergel, V., Afaneh, F., Urbasch, G., Vollmer, M., Giessen, H. and Dörner, R.: 2000, *Phys. Rev. Lett.* **84**, pp. 443–447
2. Weber, Th., Weckenbrock, M., Staudte, A., Spielberger, L., Jagutzki, O., Mergel, V., Afaneh, F., Urbasch, G., Vollmer, M., Giessen, H. and Dörner, R.: 2000, *J. Phys. B* **33**, pp. L128–L133
3. Weber, Th., Giessen, H., Weckenbrock, M., Urbasch, G., Staudte, A., Spielberger, L., Jagutzki, O., Mergel, V., Vollmer, M. and Dörner, R.: 2000, *Nature* **405**, pp. 658–661
4. Moshammer, R., Feuerstein, B., Schmitt, W., Dorn, A., Schöter, C.D., Ullrich, J., Rottke, H., Trump, C., Wittmann, M., Korn, G., Hoffmann, K. and Sandner, W.: 2000, *Phys. Rev. Lett.* **84**, pp. 447–450
5. Corkum, P.B.: 1993, *Phys. Rev. Lett.* **71**, pp. 1994–1998
6. Kulander, K.C., Cooper, J. and Schafer, K.J.: 1995, *Phys. Rev. A* **51**, pp. 561–568
7. Becker, A. and Faisal, F.H.M.: 1996, *J. Phys. B* **29**, pp. L197–L202
8. Sheehy, B., Lafon, R., Widmer, M., Walker, B., DiMauro, L.F., Agostini, P.A. and Kulander, K.C.: 1998, *Phys. Rev. A* **58**, pp. 3942–3952
9. Becker, A. and Faisal, F.H.M.: 2000, *Phys. Rev. Lett.* **84**, pp. 3546–3549
10. Wannier, G.H.: 1953, *Phys. Rev.* **90**, pp. 817–825
11. Eckhardt, B. and Sacha, K.: *Physica Scripta* (in press)

DOUBLE-ELECTRON IONIZATION OF TWO-ELECTRON SYSTEM IN A STRONG LASER FIELD

O.V.TIKHONOVA
Institute of Nuclear Physics, Moscow State University,
119899 Moscow, Russia, e-mail: ovt@mics.msu.su

Introduction

The advance in the creation of the high-intensity laser sources allow for the investigation of the atomic and molecular dynamics in the strong field region, where the perturbation theory does not appear to be valid. A lot of new interesting features of the quantum systems interacting with a strong laser field were discovered: the phenomenon of atomic stabilization, the above threshold ionization and dissociation, high harmonic generation and other [1-3].

In a series of experiments [4-7] measured the ionization yields produced by a laser pulse in He, surprisingly large signal of double ionization has been observed. The data appeared to be much larger than that predicted by the single active electron model. The additional double ionization yield found can be attributed to a nonsequential process and is observed as a characteristic "knee" or "shoulder" structure on the curve of intensity dependence of the ion yield.

Many theoretical and experimental studies are aimed to find out the mechanisms of such nonsequential ionization in a strong laser field. Several distinct processes are supposed to contribute to the appearance of the doubly charged ions:

a) The "shake-off" model of double ionization, suggested by Fittinghoff [5]. According to this model, the change of the potential caused by escape of one of electrons is assumed to occur nonadiabatically, so the remaining electron does not stay in the ground state and can be ionized.

b) The classical rescattering or recollision process, proposed by P.Corkum [8]. According to this model, one electron is promoted into the continuum by the laser field and then oscillates like a classically free electron. This electron can return to the core with the maximal classical kinetic energy E_{max}=3.17U_p (equal to 3.17 of ponderomotoric potential) and kick out the second electron by a *e*-2*e* scattering-like process.

c) The mechanism due to *e-e* correlation, proposed by F.Faisal [9] and Kuchiev [10] separately in 1996. The double escape appears to be a quantum mechanical process of absorption of photon energy by one of the electrons, which is shared cooperatively with the other electron through the Coulomb correlations.

The goal of our work is to investigate the dynamics of two-electron system in a strong laser field and to analyze the contribution of different channels to a double ionization process.

In our simulations the 1D model of the two-electron system in electromagnetic field is studied. Although reduction of the dimension of the model from 3D to 1D does not allow for a quantitative comparison with the experiment, qualitatively, the physical behaviour of the system can be easy understood.

B. Piraux and K. Rzążewski (eds.), Super-Intense Laser-Atom Physics, 85–94.

The exact numerical simulations are performed by direct numerical integration of the nonstationary Schrödinger equation (NSE):

$$i\hbar\frac{\partial\psi}{\partial t} = [H_0 - e(x_1 + x_2)\varepsilon(t)\cos\omega t]\psi$$

Here H_0 is a field-free Hamiltonian of the system, the laser frequency ω corresponds to the photon energy of 2 *eV* and the laser pulse envelope $\varepsilon(t)$ is chosen of the smooth trapezoidal form with turn on time t_f=2*T* and plateau duration t_p=5*T*. The method of the numerical solution of the NSE is similar to that explained [11].

The field-free Hamiltonian can be presented in such a way:

$$H_0 = \frac{p_1^2}{2m} + \frac{p_2^2}{2m} - \frac{ze^2}{\sqrt{\alpha^2 + x_1^2}} - \frac{ze^2}{\sqrt{\alpha^2 + x_2^2}} + \frac{e^2}{\sqrt{\alpha^2 + (x_1 - x_2)^2}}.$$

Here α is a smoothing parameter and it is chosen to be of the order of 2 a.u.

At first, the stationary Schrödinger equation with the field-free Hamiltonian is solved:

$$H_0\psi = E\psi$$

If z = 2, this Hamiltonian describes the model of two-electron atom — a *Xe*-like atom with ionization potential I_i=12.95 eV. The electron-density distributions for the ground and a few of excited states of the *Xe*-like atom are presented at Fig. 1.

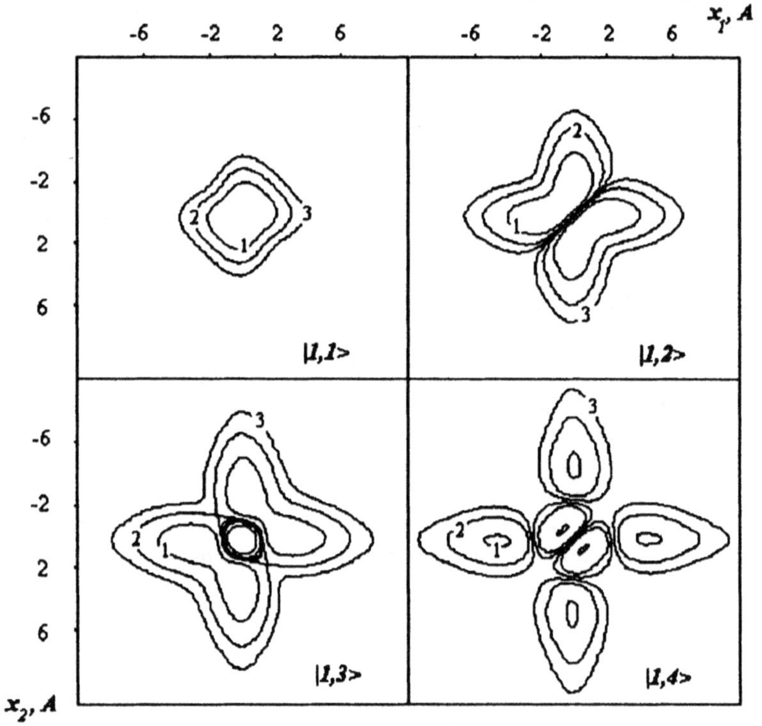

Fig. 1. 2D electron-density distributions for the ground and a few excited states of the *Xe*-like atom. The levels of equal probability are: 1 - 10^{-2}, 2 – 10^{-3}, 3 – 10^{-4}.

To perform the numerical calculations, it's more convenient to use new coordinates ξ and η:

$$\xi = \frac{x_1 + x_2}{\sqrt{2}}, \quad \eta = \frac{x_1 - x_2}{\sqrt{2}}.$$

The parameters of the spatial grid are the following:
$\xi \times \eta = 540 \times 810$, $\xi \in$ (-150 Å, 150 Å), $\eta \in$ (-300 Å, 300 Å).

Using the time-dependent wave function of the system found from the nonstationary Schrödinger equation the two-dimensional electron-density distribution can be obtained at any instant of time. At such pictures the single electron ionization is characterized by the so-called cross-distribution, while the wave-packets of double ionization are found to appear in I – IV quadrants. Using such space pictures, the probability of non-ionization W^b can be calculated by integration inside the square area *14*14* Å near the nucleus: $W_b = \int_{-a}^{a} dx_2 \int_{-a}^{a} dx_1 |\psi(x_1 x_2 t)|^2$. The probability of single ionization W_1 is calculated by integration over the cross area:

$$W_1 = \int_{-a}^{a} dx_1 \int_{-\infty}^{\infty} dx_2 |\psi(x_1 x_2 t)|^2 + \int_{-a}^{a} dx_2 \int_{-\infty}^{\infty} dx_1 |\psi(x_1 x_2 t)|^2 - W^b$$

And then the probability of double ionization is found as follows:

$$W_2 = 1 - W_1 - W^b$$

Results.

The main result of our simulations is the probability of a single and double ionization obtained for the *Xe*-like atom in dependence on laser intensity presented at Fig. 2.

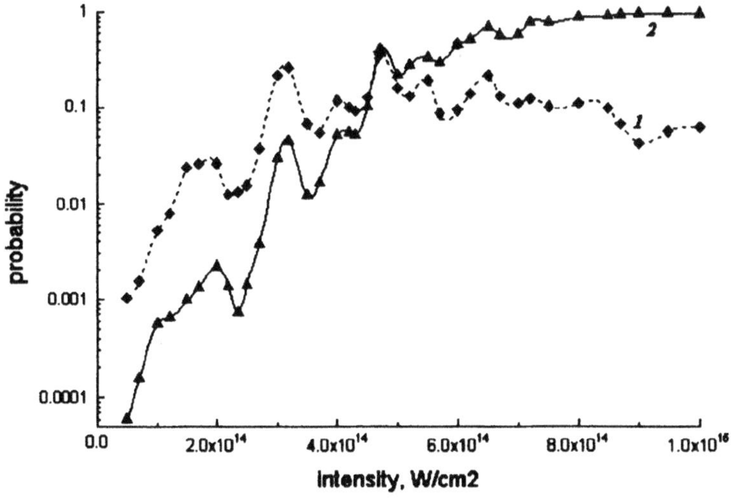

Fig. 2. The probability of single (1) and double ionization (2) of the *Xe*-like atom versus laser intensity.

In the low intensity region the doubly charged ion yield is shown to be characterized by a non-monotonous structure. Quite similar double ionization signal with a few peaks was obtained by Eberly and co-workers [12]. In our opinion such a behavior can be attributed to the appearance and the vanishing of the multiphoton resonances between the initial and one of the excited bound states of the system.

In experiments, the results are usually averaged over a focal spot of the laser beam. If we perform this procedure

$$\langle W(P_0)\rangle = \frac{1}{\pi r_0^2}\int_0^\infty W(P(r))\,2\pi r dr\,,$$

assuming the Gaussian distribution of the laser intensity

$$P = P_0 \exp\left\{-\left(\frac{r}{r_0}\right)^2\right\},$$

we obtain the averaged single and double ionization yields presented at Fig. 3.

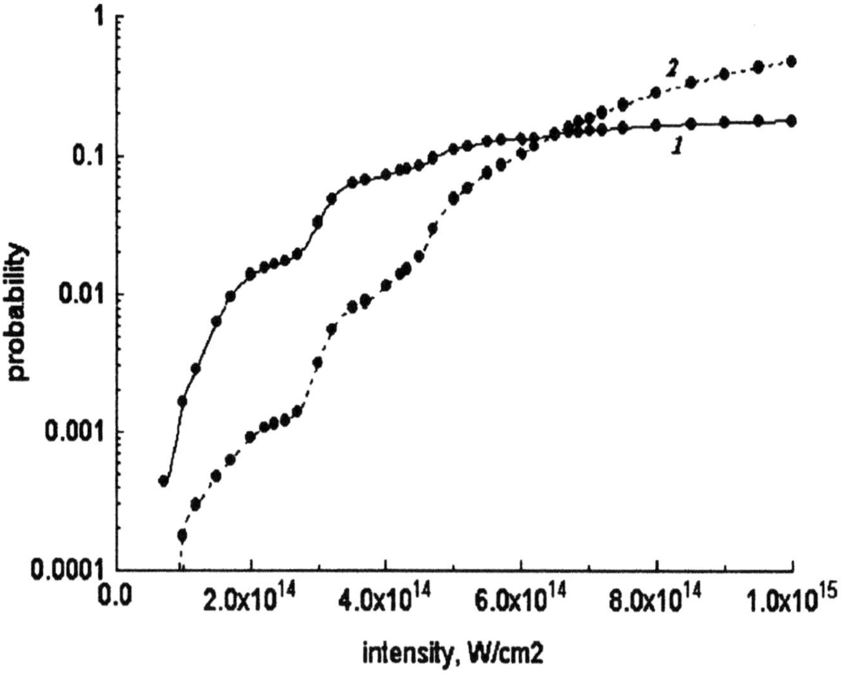

Fig. 3. The single (solid curve) and double ionization yield (dashed curve) averaged over a laser focal spot.

Similar results with the changing slope of the curves were obtained experimentally by the group of professor Chin and can be found for example in [13].

Thus, the multiphoton resonances are found to contribute to the ionization of the *Xe*-like system at rather low laser intensities. It should be noticed that such resonances seam to be more pronounced when the number of photons *n* needed for the single ionization is not very big (in our case $n = 7$) and the Keldish parameter γ is of the order or more than unity.

To examine the contribution of the rescattering process to the double ionization, the results of the two-electron calculations are compared with that obtained by the passive electron model. According to this model one of the electrons is assumed not to interact with the electromagnetic field. In this case the nonstationary Schrödinger equation can be written by following way:

$$i\hbar\frac{\partial\psi}{\partial t} = [H_0 - ex_1\varepsilon(t)\cos\omega t]\psi$$

So, the ejection of the second electron can occur because of the *e-e* interaction only.

At rather low laser intensities, the classical recollision process appears not to take place. But the energy exchange between electrons is found to occur. This fact is illustrated by the space-density distribution shown at Fig. 4 at different instants of time. The correlated energy exchange results in the formation of the cross. Although one of the electrons is passive, the picture becomes rather symmetrical. By the end of the pulse it is difficult to distinguish active electron from the passive one, because of the energy exchange. And this energy exchange is caused by *e-e* correlations.

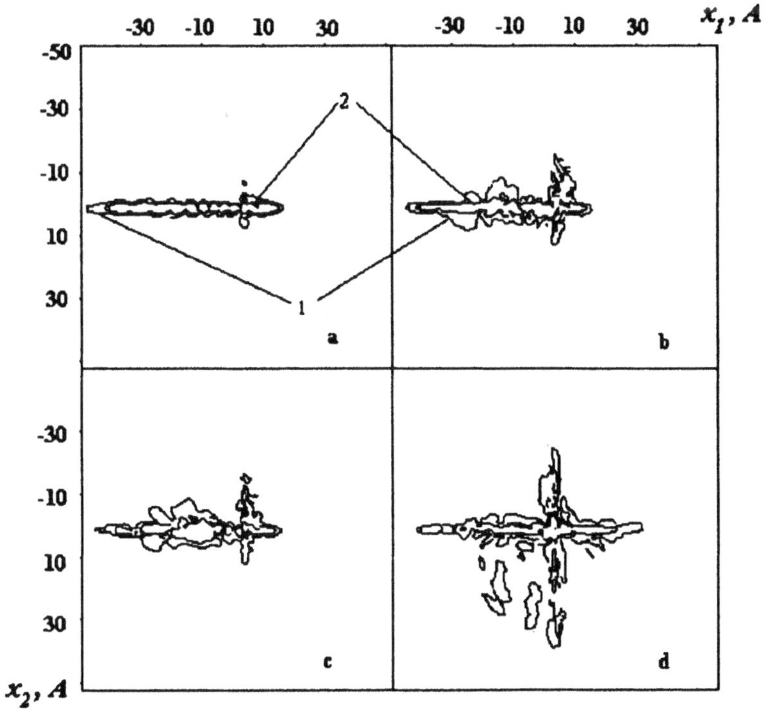

Fig. 4. The electron-density distributions obtained for the *Xe*-like atom in the passive electron model for laser intensity $P=5\cdot10^{14}$ W/cm^2 at different instants of time: a) *2T*, b) *4T*, c) *7T*, d) *9T*. The levels of equal probability are: 1 - 10^{-3}, 2 – 10^{-4}.

So, we can conclude that in the low intensity region in the case of the *Xe*-like atom both the multiphoton resonances and the energy exchange between electrons due to the *e-e* correlations appear to be very important for the double ionization process.

In the high intensity region the results of the passive electron model can be easily understood in terms of classical rescattering process, which is illustrated by dynamics of the mean coordinates of electrons calculated in such a way:

$$x_{1,2} = \int x_{1,2} |\psi(x_1,x_2)|^2 dx_2, dx_1 .$$

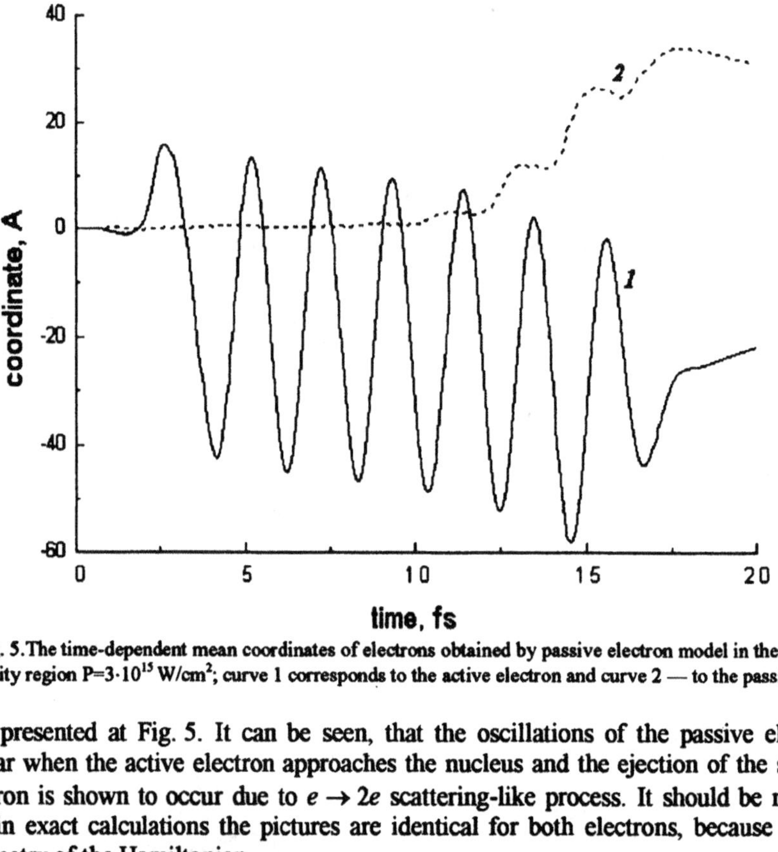

Fig. 5.The time-dependent mean coordinates of electrons obtained by passive electron model in the high intensity region P=3·10^{15} W/cm^2; curve 1 corresponds to the active electron and curve 2 — to the passive one.

It is presented at Fig. 5. It can be seen, that the oscillations of the passive electron appear when the active electron approaches the nucleus and the ejection of the second electron is shown to occur due to $e \rightarrow 2e$ scattering-like process. It should be noticed that in exact calculations the pictures are identical for both electrons, because of the symmetry of the Hamiltonian.

On the space distribution picture the rescattering process is characterized by the density wave-packets that are formed from the cross domain rather than from the zero point (Fig.6).

As for the exact calculations in this intensity range, the electron density distribution is characterized not only by the recollision wave-packets formed from the cross domain, but also by some jets that are coming from the zero-point (Fig.7). They are labeled by the arrows. Similar space distributions are obtained also by Eberly and co-workers [14]. But in our opinion, the formation of jets should be attributed to the simultaneous ejection of both electrons due to the direct action of the laser field on each electron. It can be seen that both electrons move in the same direction, but they remove each other because of the Coulomb repulsion. In dependence on the face of laser field the electron wave packets are formed on the one or another side of the nucleus. Thus, the collective ejection of two electrons is said to occur.

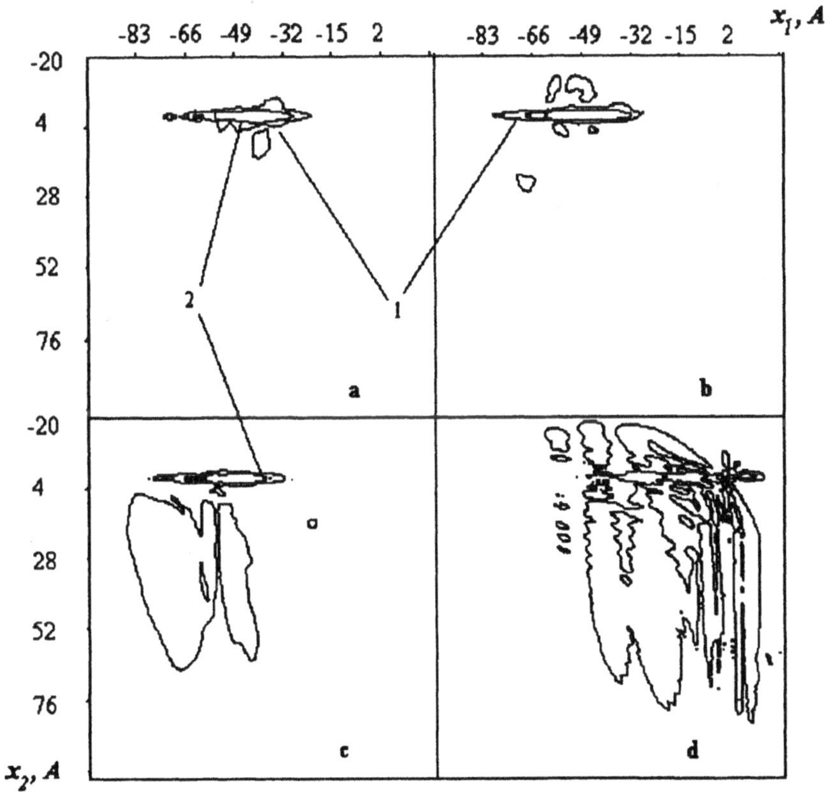

Fig.6. The electron-density distributions obtained for the *Xe*-like atom in the passive electron model for laser intensity P=3·10^{15} W/cm^2 at different instants of time: a) *2T*, b) *4T*, c) *7T*, d) *9T*. The levels of equal probability are: 1 - 10^{-3}, 2 – 10^{-4}.

Since at these laser intensities the double ionization yield is found to exceed considerably the results of passive electron model (Fig. 8), the direct action of the electromagnetic field on both electrons is confirmed to be very significant for the ionization process.

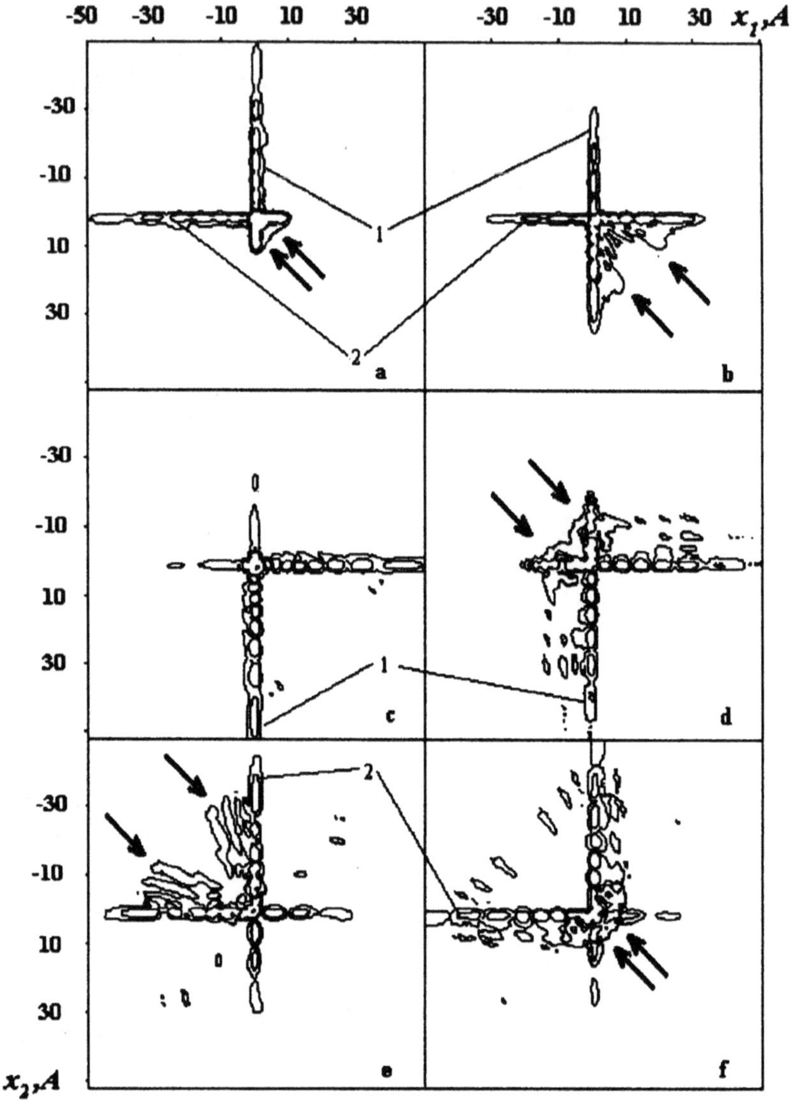

Fig. 7. The electron-density distribution obtained for the *Xe*-like atom in exact calculations for laser intensity P=10^{15} W/cm^2 at different instants of time: a) 3T, b) 3.2T, c) 3.4T, d) 3.6T, e) 3.8T, f) 4T. The levels of equal probability are: 1 - 10^{-3}, 2 – 10^{-3}

Thus, classical rescattering process is proved to take place, but another channel of double ionization — collective ejection of both electrons — appears to be very important.

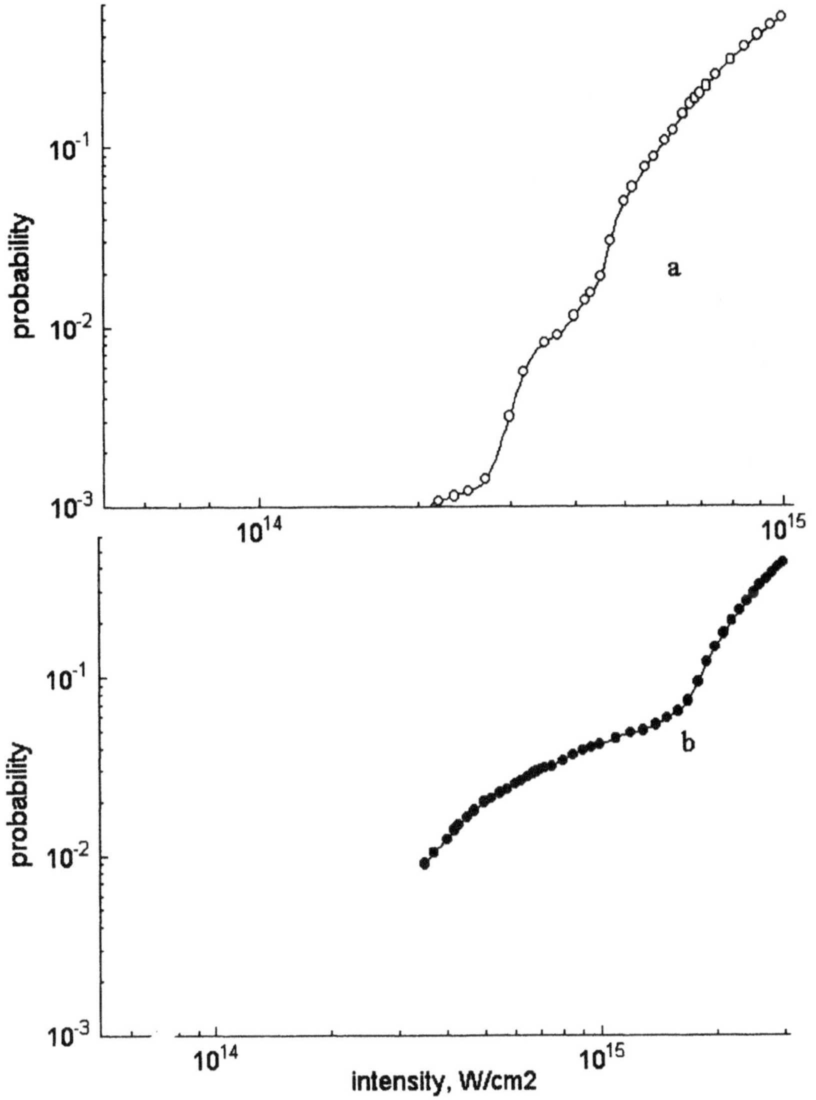

Fig. 8. ' he averaged double ionization yield calculated for the *Xe*-like atom in dependence on laser intensity; **a**: exact calculations; **b**: results of passive electron model.

Conclusions.

The laser-induced single and double ionization of a model two-electron *Xe*-like system is studied by means of direct numerical integration of nonstationary Schrödinger equation. The contributions of different channels to the double ionization are analyzed.

It is found that

1) At rather low laser intensities both a correlated energy exchange between electrons and possible multiphoton resonances are proved to play a role in ionization process.

2) The rescattering process is shown to take place at rather high intensities but does not appear to be the main mechanism of double ionization.

3) The direct action of the laser field on both electrons is found to be very significant and collective ejection of two electrons is confirmed to be a very important channel of double ionization.

4) The *e-e* correlations play dramatic role in the ionization. For this reason the system does not seem to be described correctly in terms of the single electron orbitales.

This study was supported by Russian Foundation for Basic Research, grant 00-02-16046 and INTAS, grant 99-1495

References.

1. Gavrila, M., (1992) Atoms in intense laser field, ed. by M.Gavrila, (New-York: Academic), p.435
2. Delone, N.B., and Krainov, V.P., (1994) Multiphoton processes in atoms, (Berlin: Springer)
3. Fedorov, M.V., (1997) Atomic and free electrons in a strong light field, (Singapore: World Scientific)
4. Augst, S., Strickland, D., Meyerhofer, D.D., Chin, S.L., and Eberly, J.H., (1989) Tunneling ionization of noble gases in a high-intensity laser field, *Phys.Rev.Lett.*, **63**, 2212-2215
5. Fittingoff, D.N., Bolton, P.R., Chang, B., and Kulander, K.C., (1992), *Phys.Rev.Lett.*, **69**, 2642-2645
6. Fittingoff, D.N., Bolton, P.R., Chang, B., and Kulander, K.C., (1994), *Phys.Rev.A*, **49**, 2174
7. Walker, B., Sheehy, B., DiMauro, L.F. et al, (1994), Precision measurment of strong-field double ionization of helium, *Phys.Rev.Lett.*, **73**, 1227-1230
8. Corkum, P.B., (1993) Plasma perspective on strong field multiphoton ionization, *Phys.Rev.Lett.*, **71**, 1994-1996
9. Becker, A., and Faisal, F.H.M., (1996) Mechanism of laser-induced double ionization of helium, *J.Phys.B*, **29**, L197-L202
10. Kuchiev, M.Y., (1996), *Phys.LettA.*, **212**, 77
11. Volkova, E.A., Popov, A.M., and Tikhonova, O.V., (2000), Double-electron ionization of a quantum system in a laser field: rescattering and interparticle correlations, *JETP*, **91**, 706-712
12. Liu, W.-C., Eberly, J.H, Haan, S.L., and Grobe R., (1999), Correlation effects in two-electron model atoms in intense laser fields, *Phys.Rev.Lett.*, **83**, 520-523
13. Talebpour, A., Chien, C-Y., and Chin, S.L., (1996), Population trapping in rare gases, *J.Phys.B.*, **29**, 5725-5733.
14. Eberly, J.H., (2000) Article in this issue

CALCULATION OF DOUBLE IONIZATION OF HELIUM

H.G. MULLER
FOM-Institute for Atomic and Molecular Physics
Kruislaan 407, 1098 SJ Amsterdam, The Netherlands

1. Introduction

One of the most interesting open problems in super-intense laser-atom physics is the exact mechanism of non-sequential double ionization (the famous 'knee' in the double-ionization yield curve [1]). The most fundamental instance of this process, in helium, is open to 'exact' numerical simulations, since for linear polarization it reduces to a problem in 'only' five dimensions. This paper presents the first preliminary results from a computer code designed to give a detailed answer to the question how exactly the non-sequential ionization occurs in this case.

2. Why Bother?

Solving the time-dependent Schrödinger equation in five dimensions is a very demanding computational task. Is it really worth it to embark on such a venture, now that many approximate methods and models have been presented? After all, some of these methods claim excellent agreement with experiments, not only for helium at 800 nm, but also for many other multiple ionization studies in other gases at other wavelengths. The main reason for answering this question affirmatively is that there seem to be very serious consistency problems in some of the competing methods. Furthermore, the high difficulty level of the problem is exactly what is making it attractive; it would not be very interesting working on a problem that would easily yield to brute force numerics. The 3-D double ionization problem fortunately is totally out of this class: so far it has not yielded to the most advanced numerical methods[2] that were especially designed for the purpose, even with the aid of a very powerful supercomputer. To make myself really impopular, I will first highlight some of the problems in competing methods.

B. Piraux and K. Rzążewski (eds.), Super-Intense Laser-Atom Physics, 95–106.

2.1. CRAPOLA

The Crapola model[3] is an interesting variant on the Hartree method (or Hartree Fock; in a two-electron singlet states these methods are identical), where one electron (the 'core' electron) feels the field of the laser, nucleus and the other electron averaged over the probability distributon for its position, while the second ('outer') electron feel the laser, nucleus and a screening due to the core electron frozen in its initial state. Some of the problems with this method are intrinsic too Hartree-like methods in general: the effect of the average field of one subsystem on another is not necessarily the same as the average of the effect of the actual field. The difference is especially embarrasing in highly non-linear situations such as tunneling or multiphoton ionization.

At fist glance it would seem that due to this problem non-sequential ionization should be totally absent from Crapola under realistic conditions. For instance, if only 1% of the population of the outer electron tunnels out in each optical half cycle, the field that the core electron feels from this packet when it recollides is a hundred times weaker than that of an actual electron following the same trajectory as the packet, and even if some of the frequency components in this weak time-dependent field would be capable of resonantly coupling the core-electron initial state to its continuum, one would expect a 10^4-fold (field-squared) suppression of the second-ionization yield, rather than the 100-fold suppression just expected from the low probability of the encounter.

In Crapola, however, two artifacts conspire to counteract this suppression. The first effect is caused by the 99% (or so) of the population of the outer electron that remained behind in the ground state. Although this part of the charge distribution does not generate any rapidly time-dependent field that can help to couple the core electron to the continuum, it does generate a static field that screens the nuclear charge. This field lowers the energy needed for dislodging the core electron to a value very close to the first ionization potential, lowering the threshold for an (e,2e) process that might occur on recollision, and thus strongly enhancing it. In a two-particle treatment the future evolution of a part of the wave function that is the product of an ionic ground state and a returning continuum wave packet is, (according to the superposition principle), completely independent from the presence of other contributions.

The second artifact is even worse. Because the returning 1% packet will at some point spatially overlap the 99% initial state, it will create a rapid beating at the frequency corresponding to the energy difference of the two. This oscillating overlap charge distribution has many multipole moments (the dipole moment usually being thought responsible for high-harmonic generation), and the corresponding fields can also couple directly with the core electron in order to eject it. Just like in the treatment of harmonic generation one predicts the correct probability for the generation of an XUV photon from the time-dependence of the dipole moment, the multipole couplings predict the correct probability for ejection of the core electron. However, to know what happened to the original wave packet, one would have to go one order higher in this coupling, and consider the back-action of the generated harmonic field or ejected core electron on this packet. Treating this back-action correctly will lead to the result that emission of a photon, or ejection of the core electron, will always be perfectly correlated with a return of the original wave packet to the ground state. This component of Crapola thus does not calculate double ionization, but rather the exchange contribution of the rescattering process to single ionization.

The conclusion is that the Crapola approximation almost totally supresses the mechanisms that can be responsible for double ionization in real life, replacing it by a 'knee' that is partly due to the effective appearence of an artificially low second ionization potential, but mostly due to mislabelling of a singly ionized final state as doubly ionized. Any similarity to the real process thus seems totally fortuitous.

2.2. S-MATRIX TS1

The S-matrix theory[4] for describing the non-sequential double ionization process is more solidly based. In addition it produces superb agreement with a wide range of experiments. Nevertheless some of its features are puzzling. Specifically, the simplified expression[5] contains the (e,2e) cross section at an unrealistically large recollision energy of (over) 8Up, while (semi-)classical models predict an upper limit to this energy of 3.17 Up. The corresponding Feynman diagram takes into account the elastic rescattering off the potential in the 'ATI-leg' of the diagram, the rescattering boosting the electron energy of the first ejected electron into the plateau region of the ATI spectrum, and then using this energy to kick out the second electron in an (e,2e) event. This is remarkable, since in classical approximations the high energy after rescattering is only reached when the electron is very far from the nucleus The rescattering event itself is completely elastic, and thus limits the energy immediately after backscattering also to 3.17Up. It is hard to imagine how an electron speeding away at a distance of 50 Bohr could transfer its energy to the core electron.

It is my personal suspicion that the appearence of 8Up is an artifact of the approximation in one of the integrations for evaluating the matrix element, in particular the neglect of the oscillating off-diagonal cross terms between the various intermediate Volkov states. This way the relative phase of the various Volkov states populated by the initial ATI step is completely removed from the calculation. This seems very dangerous, because the relative phases of a superposition of Volkov states (which at any given time are just plane waves) determine where the wave-packet is located. In reality the position of the wave packet of course is the all-determining factor for the possibility to interact with the second electron. This makes at least the simplified TS1 expression[5] suspect, and creates a need for exact ab-initio results to which it could be compared. In addition, it is important to test the approximations that go in the full evaluation [4] of the lowest-order S-matrix expression, such as the neglect of the Coulomb potential in the intermediate state.

3. Why TDSE?

Solving the time-dependent Schrödinger equation on a grid has proved an increasingly powerful method as computers improved. In particular the multiply split-operator method[6] seems extremely efficient, especially under conditions where high accuracy is desired, reducing integration times at least one order of magnitude compared to competing methods. Its main features can be summarized as folows:

- Spherical coordinate system
- Radial coordinate grid of constant spacing
- Angular-momentum basis
- Radial derivatives through fourth-order implicit finite-differences
- Hamiltonian split in atomic part, and a large number of interactions between neighboring angular-momentum subspaces
- Each factor in the propagator expressed as an (exactly unitary) Crank-Nicholson form

To calculate ionization yields with the aid of this code for 800-nm fields at several hundred TW/cm^2 takes only about 5 minutes. Equiping the model system with an extra core electron, described by 600 basis states, would thus drive up the required integration time to a still affordable 50 hours.

4. What to calculate?

Considering the large difference in difficulty level between calculation ion yields and electron spectra, calculating the latter at the moment seems too ambitious, and for now we settle for calculation of ion yields. Given this, the description of one of the two electrons can be limited to a fairly small space only. In this work a sphere of radius $R_{min} = 8$ Bohr is used. As soon as the electron leaves this sphere, and the other electron is either farther out *or* has an angular momentum larger than any partial wave contributing significantly to the (polarized) ground state, it is recorder as a double ionization event.

This procedure is a little different from the more commonly employed one, which records the population in the region where both electrons have a distance from the nucleus larger than R_{min}, and it can be considered a refinement of it. Apart from including situations where one electron happens to be close to the nucleus, but not in the ground state, in the double ionization count, the main difference with the other procedure is that it does not take account of any return of the wave packet from the doubly ionized region into the less-ionized regions. This is not only more convenient: one can argue that this is actually better. Although it is in theory possible that the action of the laser field would treat any returning current in such a way that it ends up back in the ground state, this will not happen to any significant extent in practice. This can be proved as follows: since the overall evolution of the wave function is unitary, and since any returning current initially was orthogonal to the ground state (which is vanishingly small beyond R_{min}), it must stay orthogonal at any later time to what evolves out of this ground state. A large amount of return to the ground state in the few cycles that the packet hangs around near the nucleus is thus only possible under conditions where the depletion of the ground state during the same period is at least as large, or the two could never stay orthogonal. In usual situations, the ground-state depletion stays well below 1% per optical cycle, however, providing a hard upper limit to the returning fraction.

Note that this procedure ignores the possibility for population of high excited states, and is bound to fail under conditions where such population is important. It would thus be more accurate to say that this procedure calculates the (double) *depletion of the ground state* rather than the ionization[7]. The usual assumption, though, is that any removal from the ground state will lead to rapid ionization, if necessary through a sequential process (emptying any excited states that might have been formed). In so far as excited states of the ion do play a role, we further remark that the $n = 2$ and $n = 3$ shells still fit within the R_{min}-sphere, so that their effect as intermediates on the route to double ionization is fully accounted for in an explicit way.

5. How?

The algorithm for propagating the wave function is a straightforward extension of that used for the two-dimensional (single-active-electron, SAE) case. The wave function is represented as

$$\Psi(\mathbf{r}_1,\mathbf{r}_2)=\sum_{ll'm}\frac{1}{r_1r_2}f_{ll'm}(r_1,r_2)(Y_l^m(\Omega_1)Y_{l'}^{-m}(\Omega_2)+Y_l^{-m}(\Omega_1)Y_{l'}^m(\Omega_2))/\sqrt{2}. \tag{1}$$

The choice to use an *uncoupled* angular basis (i.e. using the internal azimuthal quantum number m rather than the total angular momentum L as in [2]) was done to limit the basis size. From classical estimates the wave function is expected to peak strongly around $m = 0$, so that only very few m values have to be included[8]. This can not be exploited in an L basis, where a single m is described as a superposition of all L states with approximately equal weights.

The most efficient gauge is chosen for description of each electron seperately, i.e. the 'core' electron (i.e. the electron that is within R_{min} from the nucleus) is described in the length gauge, the other electron (which might or might not be within R_{min}) in the velocity gauge. Even at intensities around $10^{15} W/cm^2$, five to seven ℓ-components are sufficient for a length-gauge description of an electron that does not venture further than 8 Bohr[1]. The SAE propagator is applied twice, one time for each electron. The one-electron hamiltonians are chosen as

$$H_2^{at}=\frac{1}{2}p_2^2-2/r_2, \tag{2}$$

$$H_1^{at}=\frac{1}{2}p_1^2-1/r_1-V_{SR}(r_1).$$

The short-range screening potential $V_{SR}(r)=(1-r/2)^4$ present in H_1^{at} is only used when $r<2$, and causes a ground state energy of that hamiltonian very close to the first ionization potential of helium, as wel as the correct asymptotic behavior at $r\downarrow 0$. Note that both hamiltonians are dependent on operators for only one of the electrons, which makes them (and the individual factors of their respective Crank-Nicholson propagators) commute trivially.

In addition to the SAE operators, the two-electron code requires an implementation of the electron repulsion $1/r_{12}$. This is so far the weakest part of the code, implementing the multipole expansion of this operator only upto the dipole term,

$$1/r_{12}\approx\{1/r_2-1/r_1\}_{r_1<r_2}+V_{SR}(r_1)+\sum_{\ell}\frac{r_<^{\ell}}{r_>^{\ell+1}}P_\ell(\cos\Theta_{12}). \tag{3}$$

[1] In the velocity gauge the description of an unpolarized ground state requires about 20 ℓ-components.

The first three terms, together with the potentials included in the atomic hamiltonians, constitute the monopole term of the hamiltonian. They are nonzero only on a small part of the grid, where they cause simple multiplication with a phase factor. They are treated in a separate (cheap) split operator to make the atomic hamiltonians commute.

The dipole interaction is local in the radial basis, but couples grid points in the angular dimensions according to $\Delta\ell_1 = \pm 1$, $\Delta\ell_2 = \pm 1$, $\Delta m = 0, \pm 1$. This means coupling to 12 neighbors. Each of these couplings is implemented as a separate split operator (a 2×2 matrix) applied both before and after the other operators. To derive the matrix elements unfortunately requires a division which is unique to that particular grid link. Even in single precision this is very expensive, but the alternative of storing these time-independent coefficients (requiring a memory access) in even more expensive on a fast PC, and would require three times as much storage as the wave function itself!

The large number of couplings is the price we pay for the reduction of the basis set by using m: in an L-basis the corresponding selection rule would be $\Delta L = 0$, and we would have only coupling to 4 neighbors. This is partly made up for by the reduced complexity of the laser field operators, which obey $\Delta m = 0$ rather than $\Delta L = 0, \pm 1$. Unfortunately, the price becomes progressively higher for the higher multipoles, the angular part of which can be expressed as higher powers of the dipole. The quadrupole operator would already couple with 45 neighbors. Fortunately the higher multipoles also decay much faster with r, and their implemetation thus might be restricted to smaller fractions of the grid.

One way to ameliorate matters is by transforming the ℓ_2 dimension to an angular ϑ_2-representation before applying the $1/r_{12}$ propagators. This would diagonalize the interaction with respect to this dimension ($\Delta\vartheta_2 = 0$ in stead of $\Delta\ell_2 = \pm 1$), so that the electron repulsion only spreads out in a two-dimensional (ℓ_1, m) space. This would strongly limit the growth of the number of links (6 for the dipole, 15 for the quadrupole). The transformation can be done quite cheaply because of the low number of ℓ_2 needed in the basis[2], and due to step chaining only one forward and one backward transform are needed per step.

6. Absorbing boundary

Making a grid so large that the wave packet will never reach the boundary usually is prohibitively expensive for anything except one-dimensional problems. We thus have to deal with the grid boundary, preferably in such

[2]For five angular momenta, the transformation is a 5×5 matrix, which blocks out as a 3×3 plus 2×2 on the even and odd functions.

a way that it causes as little disturbance of the wave function on the grid as possible. Ignoring the problem, by cutting the hamiltonian matrix in a hermitian way, causes total reflection of any part of the packet that reaches the boundary, strongly disturbing the calculation on the inner part of the grid. More benign behavior occurs if we use an absorbing boundary condition, aimed at mimicking the behavior of a semi-infinite grid.

In order to describe the core electron with a sufficiently small number of degrees of freedom, it is essential to waste as few of them to the absorber at $r_2 = R_{min}$ Apart from the problem of returning current from beyond the absorber, which can not be helped, most absorbers also deform the wave function as it approaches the boundary, and there is no a priori guarantee that this deformation does not disturb the interactions that are going on in this region. It is thus preferable to faithfully represent the wave function on as large a number of grid points as possible.

For one selected outgoing momentum it is possible to get perfect absorption and perfect representation on every grid point, i.e. have a 'zero-point absorber'. In a three-point finite-difference approximation to the Laplacian only the last point of the grid feels the boundary, and the corresponding row of the hamiltonian matrix can be taylored to make the wave $\exp(ik_0r)$ an exact eigenfunction. This can be realized by giving the last diagonal element the proper (complex) value. It is important that the only deviation from hermiticity is on the diagonal: in this case the imaginary part can be treated as a separate (diagonal) split operator, which amounts to a multiplicative mask. This is a sure guarantee against instability, provided all imaginary parts have the proper sign.

Since the reflection coefficient is a continuous function of momentum k, momenta close to k_0 are absorbed as well, but onfortunately the range of efficient absorption is rather small. To do better requires that the wave function deviates from a pure outgoing wave in at least a single point: the amount of charge absorbed by multiplication with a fixed mask function depends only on the magnitude of the wave function, and outgoing waves of the same amplitudes but different k carry different amounts of current that have to be absorbed. We thus leave the last point of the wave function as an arbitrary (complex) dynamical variable. In the exact eigenfunctions it is fully and uniquely determined by the (given) value of the wave function in the forelast point and the last row of the hamiltonian. This last row contains one real off-diagonal and one complex diagonal adaptable parameter, so that one more real degree of freedom is needed for making two independently selected momenta exact eigenfunctions (one complex condition linking the last three points for each of the two k). The forelast diagonal element can be used to this end, leading to the following matrix for the

Laplacian:

$$\Delta_2 = \frac{1}{h^2}\begin{pmatrix} \cdot & \cdot & & & \cdot \\ \cdot & -2 & 1 & & \\ & 1 & -2 & 1 & \\ & & 1 & A & B \\ & & & B & C \end{pmatrix} \tag{4}$$

Under the assumption that the eigenvalues of this matrix for the wave with momentum k is $-k^2$, (which seems accurate enough for the low k that we want to absorb), A, B and C can be derived with the aid of some algebra from k_0 and k_1, the two momenta that we want perfectly absorbed. If we choose those momenta not too far apart, the reflection of all momenta lying in between will also be very low. For instance, choosing $k_0 = 0.6$ and $k_1 = 1.5$ will provide an absorption of better than 99% for all energies between 3eV and 30eV, and better than 90% from 1eV to 50eV. This seems good enough for the purpose of calculating the total amount of double ionization, since even the reflected part will simply be absorbed on a later encounter with the absorber.

The TDSE solver does not use the simple three-point finite-difference approximation to the Laplacian, but instead the fourth-order implicit (Numerov) version[6]

$$f" = (1 + \frac{h^2}{12}\Delta_2)^{-1}\Delta_2 f. \tag{5}$$

The easiest way to make an absorbing termination of this operator is to simply use as Δ_2 in this expression the real (hermitian) part of Eq.4, and implement the imaginary part as a mask multiplication (in the last point only!) before or after the time step[3]. This probably underutilizes the available degrees of freedom available in the quotient of two tridiagonal matrices, but is a sure way to avoid instabilities.

7. Results

The results presented here are preliminary, since the code has not yet been sufficiently tested to guarantee its correctness. Expanding the grid size in any of the angular dimensions had no visible effect on the results, changing the r_2-absorber parameters only very little. Changing the position of this absorber from 8 to 10 Bohr had the expected effect (the double ionization current arrived a little later). This kind of convergence is a necessary, but unfortunately not sufficient condition for correctness. Furthermore, beware that the electron repulsion is included only upto the dipole term.

[3]It does not really matter to which step we think this multiplication belongs, since the value in the termination point was arbitrary anyway.

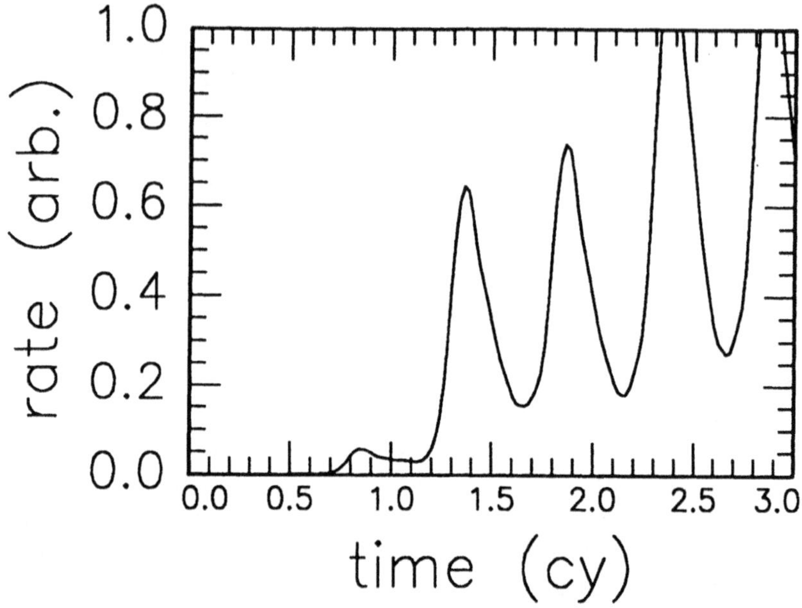

Figure 1. The double-ionization current at $R = 8$ Bohr during the flat part of a laser pulse at 390 nm with field amplitude 0.150 a.u.. A half-cycle turn-on ended at $t = 0$. The double ionization only switches on after the maximum of the third half-cycle of full intensity.

Fig. 1 shows the current hitting the absorber at $r_2 = R_{min}$ as a function of time, for a 10-cycle flat-top pulse of 390-nm light with half-cycle sine-squared turn-on. One salient feature is that the first cycle of the flat part (which starts at $t = 0$) does not seem to produce any double ionization at all, which strongly points to some kind of recollision mechanism. The first full-strength burst of double ionization appears slightly before the zero crossing at the end of the third half-cycle. Because of the travel time to $R = R_{min}$ this means the ionization itself took place near the field maximum of this half-cycle. This is unexpected from the viewpoint of (e,2e) on recollision, because recollisions at this time have very low impact energy.

The strength of the first burst of double ionization (in the second cycle) correlates well with the intensity. The behavior on later cycles is very sensitive for the exact intensity in a non-obvious way, and can stay low or grow to many times the magnitude of the first burst (Fig. 2). This points to involvement of very long trajectories, in a resonant way.

At 800 nm the behavior is slightly different. Although the current in Fig. 3 still peaks slightly after the field maxima, there now is also significant current in between these peaks. This current starts about one quarter cycle

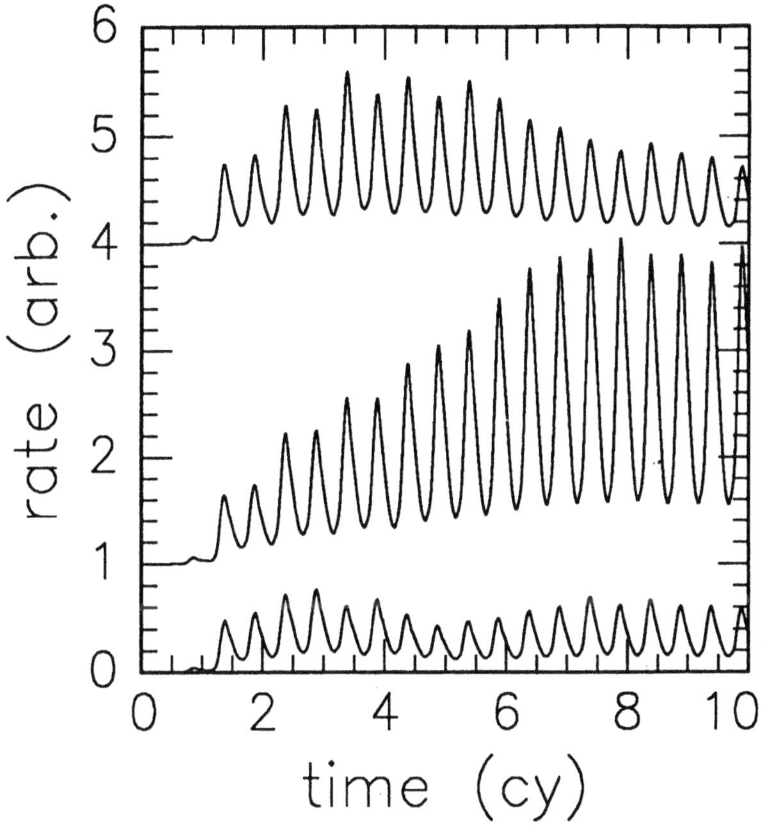

Figure 2. The double-ionization current at 390 nm during the flat part of laser pulses with field amplitude 0.152, 0.150 and 0.146 a.u. (top to bottom). A half-cycle turn-on ended at $t = 0$. The curves are offset vertically for ease of viewing.

before the first peak, and is compatible with maximum-energy impact of the single-ionization wave packet produced in the first half-cycle of the flat part. Also here, later cycles do in general not just repeat the second one, showing the importance of long trajectories.

In conclusion, it seems that accurate calculations on the double ionization problem are well feasible with the current method at 800 nm for intensities upto 1 PW/cm^2. Current integration times on a 500-MHz Pentium-II are under 2 μs per grid point per time step. Reprogramming the electron repulsion in a smarter way would allow inclusion of the quadrupole interaction without driving up this time much. Using the new SSE instructions of Pentium-III for this purpose would even speed up the code significantly (at the expense, however, of the number of computers that could run it!). Preliminary results suggest that double ionization depends on recollision. The involved trajectories seem longer than originally thought, though.

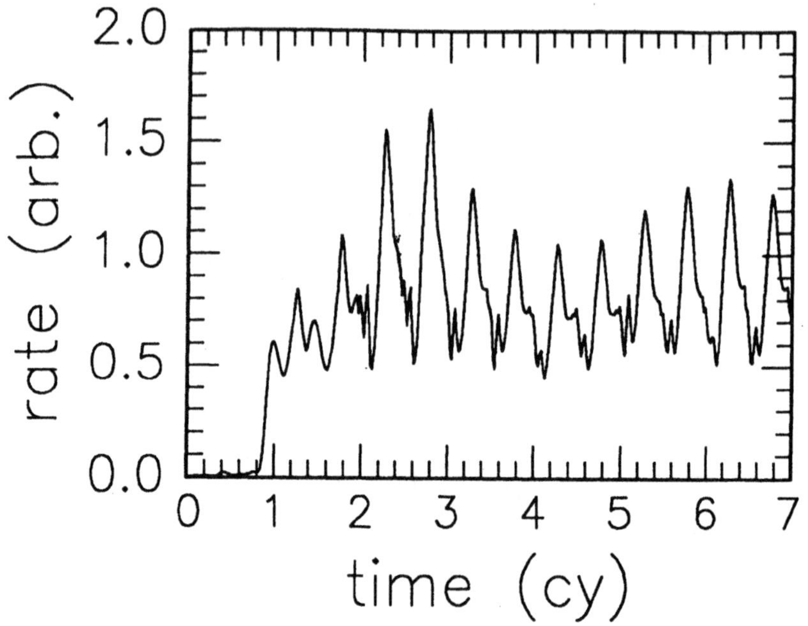

Figure 3. The long-term behavior of the double-ionization current at 800 nm during the flat part of laser pulses with field amplitude 0.105 a.u.: even 7 cycles after turn-on a steady state is not yet reached.

Acknowledgement - This work is part of the research program of Fundamental Research on Matter (FOM), which is supported by the Netherlands Organization for the Advancement of Research (NWO).

References

1. B. Yang, K.J. Schafer, B. Walker, K.C. Kulander, P. Agostini and L.F. DiMauro (1993) *Phys. Rev. Lett.* **71**, 3770.
2. E.S Smyth, J.S Parker and K.T. Taylor (1998) *Comp. Phys. Comm.* **114**, 1.
3. J.B. Watson, A. Sanpera, D.G. Lappas, P.L. Knight and K. Burnett (1997) *Phys. Rev. Lett.* **78**, 1884.
4. A. Becker and F.H.M. Faisal (1996) *J. Phys. B* **29**, L197.
5. A. Becker and F.H.M. Faisal (1999) *Phys. Rev. A* **59**, R1742.
6. H.G. Muller (1999) *Laser Phys.* **9**, 138.
7. J.H. Eberly, personal communication.
8. H.G. Muller (2000) in L.F. DiMauro, R.R. Freeman and K.C. Kulander (eds.), CP525 *Multiphoton Processes: ICOMP VIII, 8th International Conference*, American Institute of Physics, pp.257

THE TWO-ELECTRON RESPONSE IN LASER DRIVEN HELIUM

LAURA R MOORE, JONATHAN S PARKER, DANIEL DUNDAS, KAREN J MEHARG AND K T TAYLOR
Department of Applied Mathematics and Theoretical Physics, Queen's University of Belfast, Belfast, BT7 1NN, Northern Ireland.

1. Introduction

The four main goals of our full-dimensionality integration of the Time-Dependent Schrödinger Equation (TDSE) have been to:

1. Enable the calculation of intense-field (two-electron) phenomena that no low-dimensionality or ad hoc theory can adequately model [1] [2].
2. Support the design of simplified theoretical models of multi-electron, atom-laser interactions and assess their ranges of validity [3] [4].
3. Support, interpret and guide laboratory experiment [2].
4. Provide a stepping stone towards a full-dimensional treatment of the 3-body and 4-body problems presented by laser-driven H_2^+ and H_2 which are of fundamental importance to femto-second Chemistry [5].

In this paper, attention is focussed on points 1 and 3 above. A description of the method and associated computer algorithms is given in section 2. Section 3 concentrates on recent results obtained at a laser wavelength of 390 nm which take the form of ratios of double- to single-ionization yields. A direct comparison with experimental work is presented. Section 4 gives details of an investigation into ionization dynamics at a laser wavelength of 20.6 nm. Momentum distributions for doubly-ionizing wavepackets are given and preliminary results on two-electron above threshold ionization are reported.

2. The Method and the Code

In performing the integration of the TDSE, two assumptions are made. The first is that nuclear mass can be taken to be infinite and the second is that the laser field is linearly polarized. Beyond these constraints the problem is treated in its full generality.

B. Piraux and K. Rzążewski (eds.), Super-Intense Laser-Atom Physics, 107–116.

To solve the TDSE, a mixed basis set, finite-difference (FD) method is adopted. In this approach the angular variables $\theta_1, \theta_2, \phi_1, \phi_2$ are dealt with by expanding the wavefunction on a basis set of coupled spherical harmonics ('partial waves '), $|l_1 l_2 LM >$, and the radial variables r_1, r_2 are modelled on a FD grid. With the assumption of linearly polarized light, the z component of total angular momentum, M, remains constant in time, effectively reducing the problem to that of a 5D time-dependent partial differential equation. Taking the atom to be initially in its ground state ($M = 0$) the wavefunction is written as

$$\Psi(r_1, r_2, \theta_1, \theta_2, \phi_1, \phi_2) = A \sum_{L,l_1,l_2} f_{L,l_1,l_2}(r_1, r_2, t)|l_1 l_2 L >, \tag{1}$$

where A is the Pauli symmetrization operator. Electron spin S is conserved so with an initial singlet state ($S = 0$), A enforces even symmetry of the wavefunction under exchange of electron spatial coordinates.

The wavefunction is propagated on the FD grid through exponentiation of the Hamiltonian in the Krylov subspace [6]. To remove the possibility of reflections of the wavefunction at the edge of the grid, the wavefunction is split into a coherent superposition of an inner part and an outer part as it approaches the outer boundary. The outer part is zero at small radial distances from the nucleus but at the outer boundary it contains the total wavefunction so that the inner part goes smoothly to zero at the outer edge. At present the outer part of the wavefunction is discarded so that the net effect is equivalent to that of an absorbing boundary.

There are four parameters in the code which may be adjusted so as to make the numerical integration arbitrarily close to the Schrödinger equation from which it was derived. These four parameters are: the grid point separation δr, the maximum total angular momentum L_{max}, the radius of volume of integration R and the number of terms N retained in the expansion of the $\frac{1}{r_{12}}$ electron-electron interaction. The settings used as standard are $\delta r = 0.33$ au, $L_{\text{max}} = 24$ (so in total 819 partial waves come into play), $R = 90$ au and $N = 4$. Using these settings should lead to calculations of double-ionization rates being converged to within 10% of their true values.

The Hamiltonian H may be written as $H = H_{\text{at}} + H_{\text{ee}} + H_{\text{pA}}$ where H_{at} is the atomic Hamiltonian, H_{ee} is the $\frac{1}{r_{12}}$ electron-electron Hamiltonian and H_{pA} is the laser interaction Hamiltonian. The differential operators in H are approximated by their 5 point FD operators. The quality of the FD representation of H can be assessed by testing each of the above terms of H independently. Setting $H = H_{\text{at}}$ reduces the problem to that of two singly-ionized helium atoms and it is verified that in the $\delta r \to 0$ limit the eigenenergies match those of He^+. The next step is to check H_{pA} by setting $H = H_{\text{at}} + H_{\text{pA}}$ and verifying that the resulting ionization rates of He^+ are the same as those of an independently written single-electron code. Finally H_{ee} is tested by setting $H = H_{\text{at}} + H_{\text{ee}}$ and verifying that as $\delta r \to 0$ the eigenenergies for the helium atom approach their theoretical values.

At larger values of δr, the code employs three arbitrary parameters to ensure that bound state energies on the FD grid are correct to within 0.02%. These three parameters are chosen to ensure that the $^1S^e$ground state of He, the 1s2p $^1P^o$excited state and the ground state of He^+ are shifted to their correct energies. These three lowest energy states are ordinarily the most poorly represented by the FD lattice since they lie close to the nucleus and hence span relatively few grid points. Shifting them to their correct energies has the effect of improving all other states. Table 1 shows energies of the lowest energy eigenstates of the full-dimensional FD Hamiltonian at $\delta r = 0.33$ au, calculated using the Arnoldi method. Error is calculated by comparison with the most accurate values in the literature [7].

TABLE I. Eigenenergies of the two-electron finite-difference Hamiltonian at $\delta r = 0.33$ Bohr.

State energy (Hartrees)	Error (Hartrees)
L = 0	
-2.9037244	0.0000000
-2.1455872	0.0003868
-2.0611908	0.0000811
-2.033579	0.000007
L = 1	
-2.1238431	0.0000000
-2.0551509	0.0000045
-2.0310725	0.0000029
-2.0198931	0.0000128
L = 2	
-2.0556187	0.0000020
-2.0312786	0.0000012
-2.0200127	0.0000031

3. Double-ionization at 390 nm

We have carried out calculations at a laser wavelength of 390 nm in order to compare directly our results for double-ionization of helium with those obtained by DiMauro and co-workers at Brookhaven National Laboratory [8]. These results take the form of ratios of double- to single-ionization yields.

Support for laboratory experiment underscores one of the most important practical applications of our work. It is difficult experimentally to measure accurately absolute laser intensities, with uncertainties of the order of ±25% to ±50% being reported over the past decade [9]. The accurate calibration of laser intensities can be made possible by making available, as functions of intensity, reliable double- and single-ionization cross-sections for helium, the simplest of all multi-electron systems.

3.1. CONVERGENCE TESTING

In any investigation, to produce results which are meaningful, it is necessary to check the accuracy of the calculation being performed. In the two-electron integration, error estimations are made primarily through convergence tests which verify that results are insensitive to variations in R, N, L_{max} and δr. The values of these parameters which give converged results depend on the particular laser frequency and intensity under consideration. The following discussion on convergence relates specifically to the calculation carried out at 390 nm, but we implement similar convergence tests in all of our calculations.

Convergence in the radius of integration R and in the number of terms N retained in the expansion of the $\frac{1}{r_{12}}$ electron-electron interaction is monitored by setting $R = 90$ au and 125 au and by setting $N = 4$ and 5 and checking that any change in the ionization rate is less than 2 or 3%. For studying convergence in L_{max} it is verified that the truncation of the basis set of partial waves does not have much effect on the He^{++} yield when L_{max} is reduced from 24 to 18.

The sensitivity to δr is assessed with the help of a one-electron integration designed to model single-electron ionization of helium [10]. This one-electron model scales almost identically with δr as the two-electron model, thus allowing us to extrapolate the results to $\delta r \to 0$ using the less computationally demanding one-electron integration [3]. This one-electron integration has a further use. It can be used to calculate $He^+ \to He^{++}$ rates to a high degree of precision. In this way, it is found that the sequential double-ionization process contributes negligibly to the He^{++} yields at all intensities less than 4.0×10^{15} W cm^{-2}.

Upper bounds on error are estimated by comparing the highest accuracy results with those obtained using the standard settings of R, N, L_{max} and δr. Summing these contributions indicates that an upper bound on error may be conservatively set at 10%.

3.2. CALCULATION OF RATIO HE^{++}/HE$^+$

The helium code computes He, He^+ and He^{++} populations, and it is the rate of change of these populations with time that gives ionization rates. To calculate these populations, the FD grid is divided into three regions which may be identified as He ($r_1 < R_0$ and $r_2 < R_0$ so both electrons are simultaneously close to the nucleus), He^+ (where one electron is close to the nucleus and the other far away) and He^{++} ($r_1 > R_0$ and $r_2 > R_0$ so that both electrons are simultaneously far from the nucleus). Usually R_0 is set to 8 au, but it is verified that ionization rates are insensitive to the actual position of the boundary R_0.

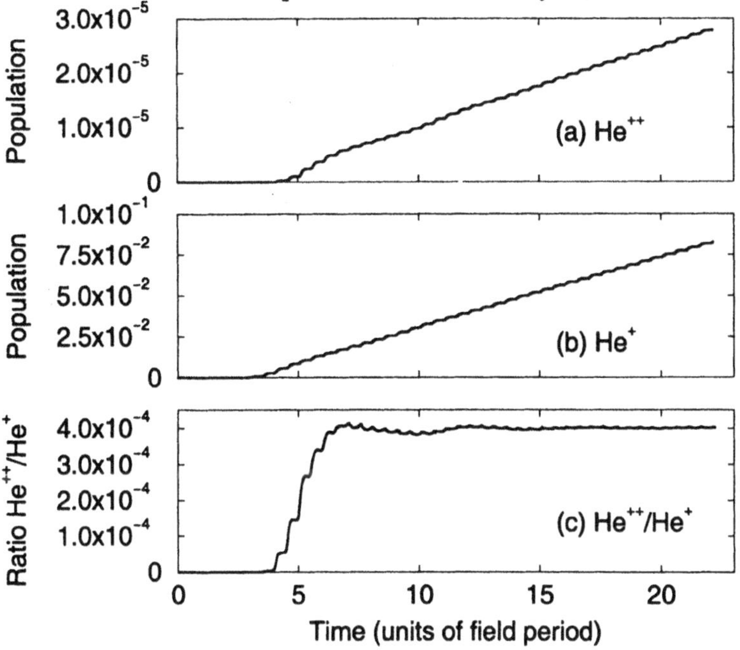

Figure 1. The calculation and time evolution of the value of the ratio of He^{++} to He^+ at $I = 6.5 \times 10^{14}$ W cm^{-2}. The ramp-on of the laser is complete after 4 field periods.

The value of the ratio of double- to single-ionization rates is determined at eight different intensities. Figure 1 shows how the ratio He^{++}/He^+ is calculated at the intensity 6.5 x 10^{14} W cm^{-2}. Plotted on graph (a) is He^{++} population, plotted on graph (b) is He^+ population, and on graph (c) is the resulting ratio obtained by dividing graph (a) by graph (b) at every point. The ratio He^{++}/He^+ maintains a steady value after any transients have died away. This behaviour is observed at all intensities studied.

3.3. RESULTS

The procedure described above yields ionization rates and ratios that are functions of pulse peak intensity only and that are independent of pulse profile. In order to be able to make a direct comparison with experimental data, the theoretically derived results must take into consideration the experimental pulse characteristics.

The experimental data points arise from ionization due to 120 fs (FWHM) laser pulses of hyperbolic secant temporal profile, $I(t) = I_0 \mathrm{sech}^2(t)$, and, assuming a TEM(0,0) mode, the pulses have gaussian intensity distributions transverse to propagation. Modelling this geometry of the experimental pulses involves a temporal and spatial averaging of the theoretical ionization rates. Full details of this procedure may be found in [2].

The resulting direct comparison between calculation and experiment is displayed in Figure 2. The experimental data points have been shifted in intensity by 50%, $I \to 1.5I$. As remarked earlier, a 50% error in the experimental determination of laser intensities is typical. Good agreement is now obtained, with our calculated values of the ratio of He^{++}/He^{+} lying within the error bars of the experimental data.

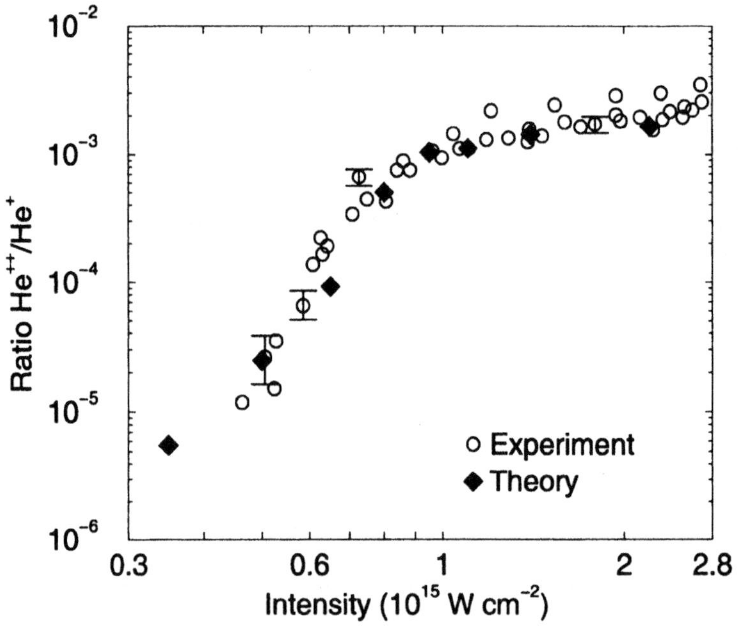

Figure 2. Ratio of He^{++} to He^{+} ion yields. Experimental data points are given by open circles and the theoretically derived ratios by solid diamonds.

4. Double-electron Above Threshold Ionization at High Frequency

Another area of interest to us is that of the high frequency domain. The free-electron laser source in Hamburg (to be made available to experimentalists in 2003) will be capable of producing pulses with wavelengths of the order of 10 nm. At these wavelengths the absorption of only one or two photons is sufficient to cause double-ionization of helium [11]. This suggests that the ionization mechanisms here should be substantially different to those at 390 nm where 25 photons are required for double-ionization.

In this section the laser frequency is set fixed to the frequency separation of the ground state and the 2s2p state (2.21 au, or 20.6 nm). A laser beam of this frequency will pump population into the 2s2p state. The subsequent auto-ionization of this state is described in section 4.1 as it provides a stringent test of the quality of the representation of doubly-excited states on the FD lattice. Section 4.2 analyses the momenta and energies of the ionizing electrons.

4.1. AUTO-IONIZATION OF THE 2S2P $^1P^o$STATE

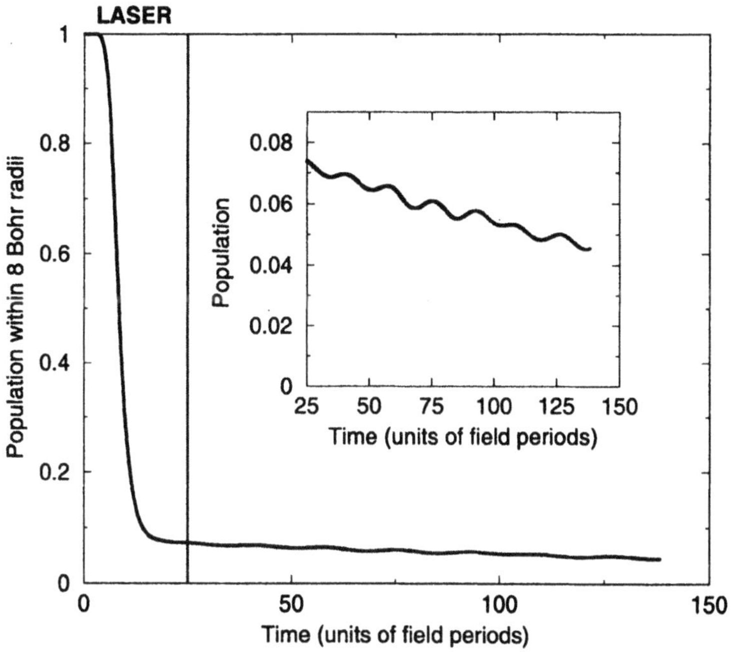

Figure 3. Population within 8 Bohr radii of the nucleus, during excitation with a laser pulse of wavelength 20.6 nm and of peak intensity 4.0×10^{16} W cm^{-2}. The laser pulse is ramped on over 8 field periods and ramped off similarly. The total duration of the laser pulse is 25 field periods, after which there is no laser acting on the system. This point in time is marked by a vertical line. The subsequent decay profile of the population within 8 Bohr radii of the nucleus is magnified in the inset.

Figure 3 shows population within 8 au of the nucleus during the excitation and auto-ionization process. The 2s2p state decays to a singly-ionizing 1skp state of positive momentum k and so the lifetime of the 2s2p state is found by measuring the rate of decay of this population after the field has been ramped to zero. The lifetime is found to be 720 au which is in good agreement with the theoretical value of 732 au.

Due to the rapid turn-on, the spectrum of the pulse is broad enough to populate the (2s3p+3s2p) $^1P^o$state. This state then decays to the same final state as

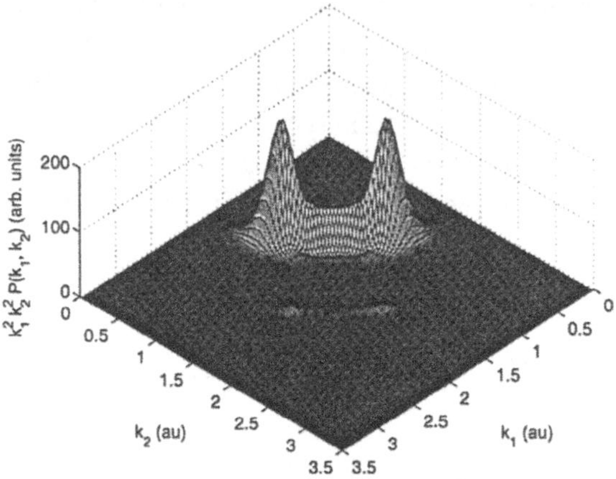

Figure 4. Probability density of ionizing electrons in momentum space after excitation with a 20 field period laser pulse of wavelength 20.6 nm and of peak intensity 2.0×10^{16} W cm^{-2}.

the 2s2p $^1P^o$, giving rise to the low amplitude oscillations seen in the inset of Figure 3. This interpretation of the oscillations is verified by confirming that the beat frequency of the decay curve exactly equals the frequency separation of the two doubly-excited states.

After ramping the field off, a Lanczos-Arnoldi eigendecomposition of the wavefunction calculates the energy of the 2s2p state, on a $\delta r = 0.33$ au grid, to be -0.694 au, which compares well with the theoretical value of -0.693 au [7].

4.2. MOMENTUM DISTRIBUTIONS OF DOUBLY-IONIZING WAVEPACKETS

The final state momenta and energies can be computed by performing the Fourier transform of the wavefunction at the end of the laser pulse. The angular variables of the resulting momentum distribution function are integrated away, leaving a distribution $P(k_1, k_2)$ in the two radial variables k_1 and k_2.

The calculated momentum probability distribution at the end of a 20 field period pulse is displayed in Figure 4. To translate the momentum k into energy E we use the approximation $E = \frac{1}{2}(k_1^2 + k_2^2)$. This verifies that the twin lobed structure in Figure 4 lies $2\hbar\omega$ in energy above the ground state so that it must correspond to the first ionization peak. The twin lobes arise from a sequential process in which one electron experiences an ionization potential of -0.9037 au and the other experiences the ionization potential of He^+ (-2.0 au) so that the two

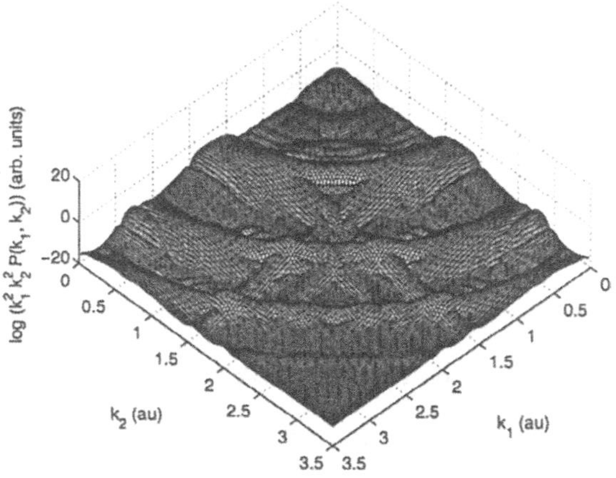

Figure 5. Log plot of the probability density of ionizing electrons in momentum space after excitation with a 20 field period laser pulse of wavelength 20.6 nm and of peak intensity 2.0×10^{16} W cm^{-2}.

electrons ionize with different energies. Higher energy peaks are visible and are made more clear on a logarithmic scale as presented in Figure 5.

Figure 5 shows a series of circular arcs along which $k_1^2 + k_2^2$ is constant, so that within an arc the total energy shared by both electrons is constant. Further examination leads to the discovery that the arcs are separated by exactly $\hbar\omega$ in energy. We call this double-electron above threshold ionization (DATI).

Non-sequential DATI can easily be distinguished from sequential DATI. Sequential DATI is just ordinary ATI of He^+ so that the momentum of one electron increases while that of the other remains fixed. In Figure 5 this process is therefore represented by the appearance of straight lines at constant k_1 or k_2. It should be noted that the straight line segments between successive circular arcs are many orders of magnitude less than the peaks they join together so that these segments are really just the background of the He^+ ATI spectrum. This indicates that the circular arcs in Figure 5 present us with evidence of non-sequential DATI, whereby both electrons absorb energy from the field and non-sequential double-ionization occurs.

Acknowledgements

JSP and DD acknowledge financial support from the UK Engineering and Physical Sciences Research Council (EPSRC). The EPSRC have also provided financial support towards computational resources used at Computer Services for Academic Research, Manchester. LRM and KJM acknowledge the receipt of research studentships provided by the Department of Higher and Further Education, Training and Employment.

References

1. Dundas D, Taylor K T, Parker J S and Smyth E S, *J. Phys. B: At. Mol. Opt. Phys.*, **32**, L231 (1999).
2. Parker J S, Moore L R, Dundas D D and Taylor K T, *J. Phys. B: At. Mol. Opt. Phys.*, **33**, L691 (2000).
3. Parker J S, Moore L R, Smyth E S and Taylor K T, *J. Phys. B: At. Mol. Opt. Phys.*, **33**, 1057 (2000).
4. Parker J S, Glass D H, Moore L R, Smyth E S, Taylor K T and Burke P G, *J. Phys. B: At. Mol. Opt. Phys.*, **33**, L239 (2000).
5. Dundas D, McCann J F, Parker J S and Taylor K T, *J. Phys. B: At. Mol. Opt. Phys.*, **33**, 3261 (2000).
6. Smyth E S, Parker J S and Taylor K T, *Comput. Phys. Commun.*, **114**, 1 (1998).
7. Scrinzi A and Piraux B, *Phys. Rev. A*, **58**, 1310 (1998).
8. Sheehy B, Lafon R, Widmer M, Walker B, DiMauro L F, Agostini P A and Kulander K C, *Phys. Rev. A*, **58**, 3942 (1998).
9. Augst S, Meyerhofer D D, Strickland D and Chin S L, *J. Opt. Soc. Am. B*, **8**, 858 (1991);
   Fittinghoff D N, Bolton P R, Chang B and Kulander K C, *Phys. Rev. Lett.*, **69**, 2642 (1992);
   L'Huillier, A Lewenstein M, Salières, Balcou Ph, Ivanov M Yu, Larsson J and Wahlström C G, *Phys. Rev. A*, **48**, R3433 (1993);
   Walker B, Sheehy B, DiMauro L F, Agostini P, Schafer K J and Kulander K C, *Phys. Rev. Lett.*, **73**, 1227 (1994);
   Zhou J, Peatross J, Murnane M M, Kapteyn H C and Christov I P, *Phys. Rev. Lett.*, **76**, 752 (1996).
10. Parker J S, Smyth E S and Taylor K T, *J. Phys. B: At. Mol. Opt. Phys.*, **31**, L571 (1998).
11. Kornberg M A and Lambropoulos P, *J. Phys. B: At. Mol. Opt. Phys.*, **32**, L603 (1999);
    Nikolopoulos L A A and Lambropoulos P, 'Multichannel theory of two-photon single and double ionization of Helium ', submitted to *J. Phys. B: At. Mol. Opt. Phys.*(2000).

THE HELIUM ATOM IN STRONG AND SHORT LASER PULSES: MULTIELECTRON EFFECTS

H. BACHAU AND R. HASBANI
Centre Lasers Intenses et Applications (UMR 5107), Université Bordeaux I, 351 Cours de la Libération, F-33405 Talence, France
http://www.celia.u-bordeaux.fr

1. INTRODUCTION

Two-electron ejection at low frequencies receives a constant attention since 15 years from both theoretical and experimental sides. We note in particular the recent measurements of Weber et al (2000) and Moshammer et al (2000), which provide informations on the total ion momentum distribution in the non-sequential regime. Numerically, simulating double electron ionization at 800 nm is an enormous challenge. The main difficulty is that the full resolution of this problem requires the non-perturbative solution of the time-dependent Schrödinger equation (TDSE) together with an accurate representation of electron correlations. These latter requirements lead to very demanding calculations. Much has been learned in the single active electron model (SAE) but very little is known beyond the approximations sustaining the time-dependent Hartree-Fock (TDHF) approach (see Kulander, 1987, 1988, Horbatsch et al, 1992, Maragakis and Lambropoulos, 1997 and Hasbani et al, 2000). Nevertheless, progress have been realized. In particular the resolution of the TDSE with approaches based on finite element developments have been shown to be particularly efficient in the XUV domain. For high frequency fields and femtosecond regime, theoretical investigations are stimulated by the perspectives offered by the harmonic generation and free electron laser sources (FEL). In this context the spectral method has been shown to be particularly efficient. It consists in developing the time-dependent solution of the Schrödinger equation on a truncated basis of antisymmetrised products of orbital functions (see Zhang and Lambropoulos, 1995). This type of method includes explicitly the characteristics of the pulse (e.g. the temporal dependence)

B. Piraux and K. Rzążewski (eds.), Super-Intense Laser-Atom Physics, 117–126.

and the atomic structure (bound and autoionizing states, continua..). Also, it is worth noticing that the various informations on the final product of the laser-atom interaction (bound-state populations, ionization yield and angular distribution of ejected electrons, branching ratio in electron spectra..) can be easily extracted from the wavefunction at the end of the pulse. Another approach consists in describing the radial wavefunction on a grid. This method is capable of modelling intense-field two-electron effects such as double photoionization (see Dundas et al, 1999) or autoionization effects (see Parker et al, 2000). Due to memory limitation on available computers, rather severe space-time limitation constrains exist. Nevertheless the grid method has been adapted for parallel processing (see Smyth et al, 1998) and it takes advantage of the rapid development of massively parallel machines. Details and references on the various approaches are reviewed in Lambropoulos et al, 1998. Most of the more recent developments are presented in the proceedings of the SILAP2000 Nato Advanced Research Workshop.

In this paper we present a method which incorporates the electron correlations at a high degree of accuracy. The method consists in expanding the radial part of the total wavefunction of the system on antisymmetrised products of B-spline functions (de Boor, 1978). The efficiency of such an approach has been demonstrated in atomic structure calculations by Decleva et al, (1995) who calculated the ground state position of He with an error of 7.8 10^{-8} a.u.. Accurate calculations of bound and autoionizing states of two-active electron systems have been performed by van der Hart and Hansen (1992) using large B-spline basis sets. In these calculations, the basis of two-electron configurations is built with products of the hydrogenic functions, expanded on B-splines. Provided that there is no truncation of the two-electron basis set, the approach of van der Hart and Hansen is equivalent to our method. The relevance of the direct B-spline expansion has been also demonstrated in the particular case of the ionization of H^- in strong fields just above the lower ionization threshold (see Lagmago Kamta et al, 2000). Details on structure and TDSE calculations in H^- and He can be found in Hasbani, (2000). Calculations have been performed, in the case of He, with photon energies ranging from 5.4 eV to 34. eV (see Hasbani et al, 2000). In this latter publication, emphasis was put on the problem of the excitation of autoionizing states. In the present contribution we analyse results obtained for three-photon ionization above the double ionization threshold of helium with 30. eV photons.

A linear polarization of the field is assumed and atomic units are used throughout unless otherwise stated.

2. THEORETICAL APPROACH

2.1. ATOMIC STRUCTURE CALCULATIONS

2.1.1. *Bound and continuum states.*

Let us first briefly describe the method. The electronic structure of the helium atom satisfies the equation

$$(H - E_n)\Psi_n = 0 \tag{1}$$

H is the non-relativistic Hamiltonian which reads

$$H = H_0(1) + H_0(2) + \frac{1}{r_{12}} = -\frac{1}{2}\Delta_1 - \frac{1}{2}\Delta_2 - \frac{Z}{r_1} - \frac{Z}{r_2} + \frac{1}{r_{12}} \tag{2}$$

where $Z = 2$ and the labels 1 and 2 refer to the electron 1 and 2 respectively. For a given total angular momentum L and projection M, the solution of equation (1) is expanded on a basis of two-electron configurations which are products of one-electron functions as follows

$$\Psi_n^{L,M}(\mathbf{r}_1, \mathbf{r}_2) = \mathcal{A} \sum_{\alpha \equiv i,j,l_1,l_2} c_{n,\alpha}^{L,M} \frac{B_i^k(r_1)}{r_1} \times \frac{B_j^k(r_2)}{r_2} \times \mathcal{Y}_{l_1,l_2}^{L,M}(1,2) \tag{3}$$

where $\mathcal{A}$ is the antisymmetrisation operator, l_i is the angular momentum of electron i. The radial function $B_i^k(r)$ is a B-spline of order k, i.e. a piecewise polynomial of degree $k-1$. The N_b B-spline functions are spanned, along the radial axis, in a "box" defined from $r = 0$ to R_{max}. As usually, we choose the B-spline sequence so that $B_1^k(0) = 0$ and $B_{N_b}^k(R_{max}) = 0$ in order to fulfil the boundary conditions. $\mathcal{Y}_{l_1,l_2}^{L,M}(1,2)$ is a bipolar spherical harmonic. A direct diagonalization of the equation (1) gives the eigenenergies and the expansion coefficients $c_{n,\alpha}^{L,M}$ of the bound and continuum states. Since the basis has a finite dimension, we obtain a discretised representation of the whole atomic spectrum.

2.1.2. *Doubly excited states.*

Doubly excited states may play a significant role in multiphoton ionization processes. It is therefore of particular interest to calculate the population of the resonant part of the autoionizing states (R-AIS) at the end of the pulse. Fano (1961) defined the R-AIS as a zero-approximation bound state, embedded in a continuum of states. The concept of autoionization is introduced through the configuration interaction which "mixes" the R-AIS and the adjacent continuum. In the Feshbach-O'Malley approach (Feshbach, 1962, and O'Malley and Geltman, 1965), the R-AIS located below the $He^+(2l)$ threshold are solutions of the equation:

$$(QHQ - E_n^r)\Phi_n^r = 0 \tag{4}$$

where E_n^r is the unshifted energy of the resonance and Q is a projection operator given by

$$Q = 1 - P \tag{5}$$

with

$$P = P_1 + P_2 - P_1 P_2 \tag{6}$$

The labels 1 and 2 refer to the electrons 1 and 2 respectively and

$$P_i = |\phi_{1s}(\mathbf{r_i}) >< \phi_{1s}(\mathbf{r_i})| \tag{7}$$

$\phi_{1s}(\mathbf{r})$ is the hydrogenic orbital calculated for a nuclear charge $Z = 2$. The projection onto the operator Q in equation (4) ensures that any eigenstate Φ_n^r is devoid of $1s$ hydrogenic component. The states located below the $He^+(2l)$ threshold ($E_n^r < -0.5$ a.u.) represent the R-AIS associated to lowest autoionizing state series.

2.2. TIME-DEPENDENT CALCULATIONS

2.2.1. *The time-dependent configuration interaction approach.*
Let us now consider the interaction between the atom and the laser field. Within the dipole approximation, the time-dependent Schrödinger equation to be solved is given in the velocity gauge, by

$$i\frac{\partial}{\partial t}\Psi_v(\mathbf{r}_1, \mathbf{r}_2, t) = [H + \sum_{i=1,2} \mathbf{A}(t) \cdot \mathbf{p}_i]\Psi_v(\mathbf{r}_1, \mathbf{r}_2, t) \tag{8}$$

The vector potential, polarized along the z–axis, is defined as

$$\mathbf{A}(t) = A_0(cos\frac{\pi}{T}t)^2 sin(\omega t)\mathbf{e}_z \tag{9}$$

where T is the total pulse duration and ω the photon energy. Similarly, in length gauge, the time-dependent Schrödinger equation reads

$$i\frac{\partial}{\partial t}\Psi_l(\mathbf{r}_1, \mathbf{r}_2, t) = [H - \sum_{i=1,2} \mathbf{E}(t) \cdot \mathbf{r}_i]\Psi_l(\mathbf{r}_1, \mathbf{r}_2, t) \tag{10}$$

with the electric field defined as

$$\mathbf{E}(t) = -\frac{\partial}{\partial t}\mathbf{A}(t) \tag{11}$$

The time-dependent total wavefunction is expanded on the basis of the field-free atomic eigenstates (see the equations (1), (2) and (3))

$$\Psi_{v,l}(\mathbf{r}_1, \mathbf{r}_2, t) = \sum_{\alpha \equiv n,L}^{N_s, L_{max}} C_\alpha^{v,l}(t)\Psi_n^{L,M}(\mathbf{r}_1, \mathbf{r}_2) \tag{12}$$

Equations (8) and/or (10) are integrated over the total pulse duration T. The initial conditions are given by

$$\Psi_l(\mathbf{r}_1, \mathbf{r}_2, t = -T/2) = \Psi_v(\mathbf{r}_1, \mathbf{r}_2, t = -T/2) = \Psi_{in}^{L,M=0}(\mathbf{r}_1, \mathbf{r}_2) \quad (13)$$

where $\Psi_{in}^{L,M=0}(\mathbf{r}_1, \mathbf{r}_2)$ is the initial state of the field-free system, i.e. the ground state of He in the present case.

2.2.2. *Calculation of the ionization, excitation and double-excitation.*
In the time-dependent configuration interaction approach (see section 2.2.1), the time-dependent solution is expanded on eigenfunction of He. Therefore, it is straightforward to extract the information about excitation and/or ionization from $\Psi_{v,l}(\mathbf{r}_1, \mathbf{r}_2, t = T/2)$ at the end of the pulse. Other quantities, e.g. the weight of the resonant part of an autoionizing state (R-AIS), can be easily obtained by projecting $\Psi_{v,l}(\mathbf{r}_1, \mathbf{r}_2, t = T/2)$ onto the solutions Φ_n^r of the equation (4). Another interesting quantity is the population left in doubly excited configurations, given by (expressed here in velocity gauge)

$$P_{de} = < \Psi_v(\mathbf{r}_1, \mathbf{r}_2, t = T/2)|Q|\Psi_v(\mathbf{r}_1, \mathbf{r}_2, t = T/2) > \quad (14)$$

It is important to keep in mind that the above quantity, referred to as "double-excitation", includes all configurations where both electrons are excited, that is double-excitation to R-AIS, double-ionization and ionization with excitation. The total contribution of configurations with at least one electron in $1s$ orbital is simply given by

$$P_{se} = < \Psi_v(\mathbf{r}_1, \mathbf{r}_2, t = T/2)|P|\Psi_v(\mathbf{r}_1, \mathbf{r}_2, t = T/2) > \quad (15)$$

The projection operators Q and P are defined in equations (5), (6) and (7).

2.2.3. *Evaluations of the populations above the double ionization threshold.*
Single and double ionization channels are present above the double ionization threshold (whose the energy is set to 0. a.u. here). Despite recent progress, an accurate evaluation of the populations in the different open channels is an extremely difficult task. This is due both to an infinite number of open channels and to the three-body nature of the double continuum problem. In order to obtain a qualitative description of dominant processes, as well as an evaluation of the channel populations, we calculate the following quantity (expressed here in velocity gauge)

$$P_{dit} = < \Psi_v'(\mathbf{r}_1, \mathbf{r}_2, t = T/2)|P'|\Psi_v'(\mathbf{r}_1, \mathbf{r}_2, t = T/2) > \quad (16)$$

where $\Psi_v'(\mathbf{r}_1, \mathbf{r}_2, t = T/2)$ refers to the part of expansion (12), at $t = T/2$, which contains only stationary states of He with positive energies $E_n > 0$

(see equation (1)). P' is the projection operator defined by

$$P'_{nl} = |\phi_{nl}(\mathbf{r_1}) >< \phi_{nl}(\mathbf{r_1})| \tag{17}$$

where $\phi_{nl}(\mathbf{r_1})$ is a (bound or continuum) hydrogenic orbital calculated for a nuclear charge $Z = 2$ in the B-spline basis set used for two-electron wavefunctions (see section 2.1.1). These orbitals are normalized to unity. P_{dit} is the probability to find the electron 1 in the hydrogenic orbital nl at $t = T/2$, after three-photon absorption. Note that P_{dit} gives, in the lowest order of the perturbation theory in $1/r_{12}$, an evaluation of the final (observable) products of the reaction. Due to the antisymmetrisation of the wavefunction the projection onto $|\phi_{nl}(\mathbf{r_2}) >< \phi_{nl}(\mathbf{r_2})|$ would lead to the same result for the electron 2.

3. RESULTS, DISCUSSION AND CONCLUSION

At the frequency $\omega = 1.1$ a.u., one-photon ionization is allowed and two-photon absorption leads to the excitation of autoionizing states located below the $He^+(2l)$ threshold (see Hasbani et al, 2000). Double ionization is allowed through three-photon absorption, it populates the continuum at $E \approx 0.4$ a.u. (the double ionization threshold is located at 0. a.u.). We use here 50 B-spline functions (N_b =50) of order k =7, they are placed on a linear knot sequence with R_{max} = 50 a.u.. With $l_i = 0 - 3$, this leads to about 5000 (i, j, l_1, l_2) combination terms (in equation 3) for $L = 0, 1, 3$, and about 6000 terms for $L = 2$. We have included in equation (3) all possible (l_1, l_2) pairs in the $^1S^e$ symmetry. For other symmetries we have considered all possible pairs $(l_1 = 0, l_2)$ and $(l_1 = 1, l_2)$ with $l_2 = 0, 1, 2, 3$. With this basis, the fundamental state of He is found at -2.90169 a.u.. The density of state in the $He^+(l = 0)$ continuum is shown in figure 1 as a function of the atomic energy of He^+. We note that this density varies monotonously from 0. to 4. a.u.. Beyond this latter value the continuum is certainly not well represented. The density of state $\rho(E)$ in the continuum is a crucial parameter in view of the convergence of the TDSE calculations. On the basis of the simple relation between the density of states $\rho(E)$ and the box length R_{max}

$$\rho(E) = \frac{R_{max}}{\pi\sqrt{2E}} \tag{18}$$

one can easily show that the condition $2/\rho(E) < \Delta\omega$ insures that the wavepacket does not reach the limit of the box during the laser-atom interaction time. In other words, the above condition is necessary to insure a correct build-up and propagation of the wavepacket during the laser-atom interaction time. In particular, it avoids artificial Rabi oscillations between

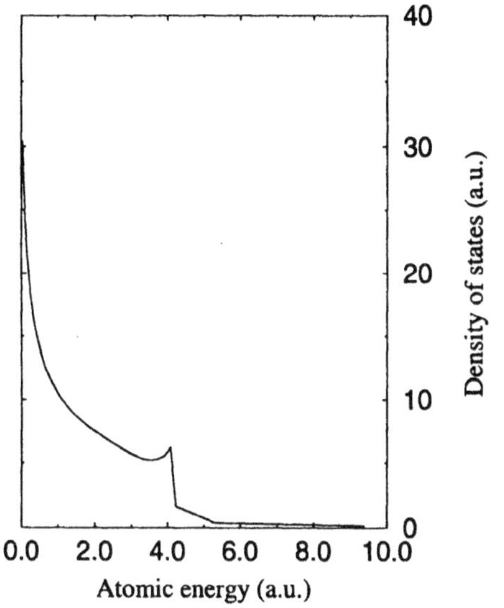

Figure 1. Density of continuum states of $He^+(l=0)$. The B-spline basis set is presented in section 3 (box of length 50 a.u., $N_b = 50$ B-splines and a linear knot sequence).

bound and continuum states. Note that the problem of setting an optimum value for $\rho(E)$ should appear in all $L2$ spectral methods, as well as in the grid approach where the wavefunction is determined on a lattice of finite dimension. As expected, for energies ranging from 0. to 4. a.u., the equation (18) is verified. It is also easy to check that, for energies ranging from 0. a.u. to 2.5 a.u., the above inequality ($2/\rho(E) < \Delta\omega$) holds for the pulse characteristics used here (a total pulse duration of 8 oscillations with $\omega = 1.1$ a.u.). As a matter of fact, the convergence must be numerically checked by varying the size of the box and the number of B-splines.

At $\omega = 1.1$ a.u., $I = 10^{14} W/cm^2$ and a total pulse duration of 8 oscillations, we have found that the channel $1skl$ plays a minor role in three-photon absorption. Therefore, three-photon absorption produces mainly ions He^+ in excited states and He^{++} ions. The calculation of the quantity P'_{nl} (see equation (17)) gives an insight into the relative importance of different open channels, as well as into their population. We show in figure 2 and figure 3 the probabilities P'_{nl} for $l = 1$ and $l = 2$ channels respectively (we have found that the $l = 0$ and $l = 3$ channels are much less populated). In abscissa we have reported the eigenenergies of the $He^+(nl)$ orbitals. In figure 2 we see that the population in orbital $2p$ (E = -0.5 a.u.) is important, we also note a peak in the continuum region at the energy $E \approx 0.2$ a.u.. There is also a peak in the population of the $l = 2$ channel

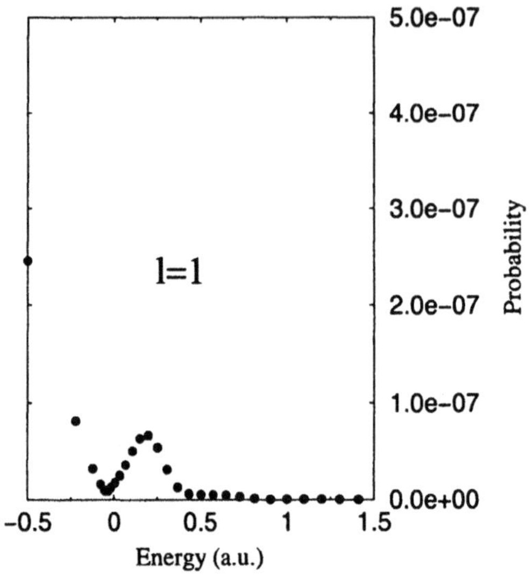

Figure 2. Probability to find the electron 1 in the orbital nl of He^+, with $l = 1$, after three-photon absorption (see formula 17). We have plotted in abscissa the nl eigenenergies.

(see figure 3) at the energy $E \approx 0.2$ a.u.. Therefore the channel leading to double electron ejection is dominated by two processes, (i) the absorption of one photon which produces 0.2 a.u. electrons in channel $l = 1$ and (ii) two-photon absorption from He^+ with the production of 0.2 a.u. electrons in channel $l = 2$. The other important three-photon absorption channel leads to the production of $He^+(2p)$ ions (see figure 2). Since the total energy shared by the electrons is about 0.4 a.u., we should observe in the channel $l = 2$ a corresponding peak of electrons at 0.9 a.u.. A second peak shows up in figure 3, but it is located at 0.6 a.u.. In fact, in the channel $2pkd$, the ejected electron should be calculated with the charge $Z = 1$ a.u., while $Z = 2$ a.u. is used in formula (17). We have verified numerically that, with continuum orbitals calculated with a correct asymptotic charge, the second peak in $l = 2$ channel is found at the expected position. Therefore our model must be refined by including relevant screening effects. A quantitative analysis of the probabilities given in figures 2 and 3 shows that the channels leading to double electron ejection (He^{++}) and single electron ejection (He^+) are equally populated.

We have been able to determine the dominant effects of multiphoton absorption above the second ionization threshold, at the end of the pulse. A precise determination of the final products of laser-atom interaction re-

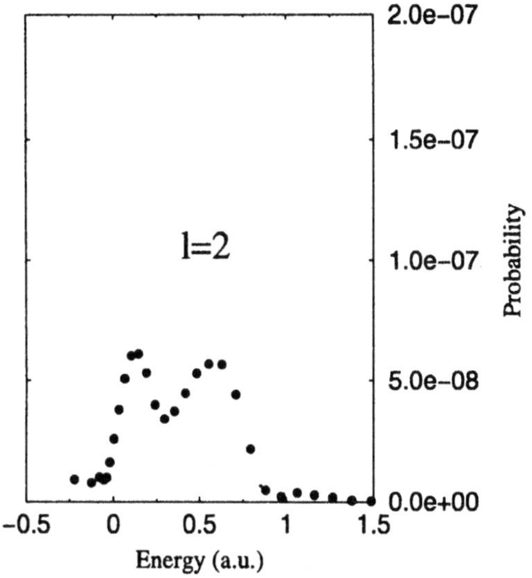

Figure 3. Probability to find the electron 1 in the orbital nl of He^+, with $l = 2$, after three-photon absorption (see formula 17). We have plotted in abscissa the nl eigenenergies.

quires more refined calculation. In particular the open (observable) channels must be described with appropriate asymptotic conditions (including channel couplings due to correlations). Added to our previous work (Hasbani et al, 2000), this contribution gives an illustration of the potential of a direct expansion on B-spline products when it is combined with the use of projection operators.

Acknowledgements

The authors thank Professor Gayet for a critical reading of the manuscript. Computer facilities from the CINES (Centre Informatique National de l'Enseignement Supérieur, Montpellier, France) are acknowledged.

References

Decleva, P., Lisini, A., and Venuti, M. (1995), *Int. J. Quantum Chem.* **56**, 27
de Boor, C. (1978), *A practical guide to splines* (Springer-Verlag, New-York)
Dundas, D., Taylor, K.T., Parker, J.S., and Smyth, E.S. (1999), *J. Phys. B: At. Mol. Opt. Phys.* **32**, L231
Fano, U. (1961), *Phys. Rev.* **124**, 1866
Feshbach, H. (1962), *Ann. Phys., NY* **19**, 287
Hasbani, R. (2000), Thèse de Doctorat, Université Bordeaux I, France, N°: 2236

Hasbani, R., Cormier, E., and Bachau, H. (2000), *J. Phys. B: At. Mol. Opt. Phys.* **33**, 2101

Horbatsch, M., Lüdde, H.J., and Dreizler, R.M. (1992), *J. Phys. B: At. Mol. Opt. Phys.* **25**, 3315

Kulander, K.C. (1987), *Phys. Rev.A* **36**, 2726

Kulander, K.C. (1988), *Phys. Rev.A* **38**, 778

Lagmago Kamta, G., Piraux, B., Hasbani, R., Cormier, E., and Bachau, H. (2000), *Proceedings of the* 8th *International Conference on Multiphoton Processes* (edited by L.F. DiMauro, R.R. Freeman and K.C. Kulander), American Institute of Physics, p. 219

Lambropoulos, P., Maragakis, P., and Zhang, J. (1998), *Phys. Rep.* **305**

Maragakis, P. and Lambropoulos, P. (1997), *Laser Phys.* **7**, 679

Moshammer, R., Feuerstein, B., Schmitt, W., Dorn, A., Schröter, C.D., Ullrich, J., Rottke, H., Trump, C., Wittman, M., Korn, G., Hoffmann, K., and Sandner, W. (2000), *Phys. Rev. Lett.* **84**, 447

O'Malley, T.F. and Geltman, S. (1965), *Phys. Rev.* **137**, 1344

Parker, J.S., Moore, L.R., Smyth, E.S., and Taylor, K.T. (2000), *J. Phys. B: At. Mol. Opt. Phys.* **33**, 1057

Smyth, E.S., Parker, J.S., and Taylor, K.T. (1998), *Comput. Phys. Commun.* **114**, 1

van der Hart, H. and Hansen, J.E. (1992), *J. Phys. B: At. Mol. Opt. Phys.* **25**, 41

Weber, Th., Weckenbrock, M., Staudte, A., Spielberger, L., Jagutzki, O., Mergel, V., Afaneh, F., Urbasch, G., Vollmer, M., Giessen, H., and Dörner, R. (2000), *Phys. Rev. Lett.* **84**, 443

Zhang, J. and Lambropoulos, P. (1995), *J. Phys. B: At. Mol. Opt. Phys.* **28**, L101

DYNAMICS OF A TWO-ELECTRON SYSTEM DRIVEN BY AN ULTRASHORT AND INTENSE LASER PULSE

B. PIRAUX
Laboratoire de Physique Atomique et Moléculaire
Université Catholique de Louvain
Chemin du Cyclotron 2, B-1348 Louvain-la-Neuve, Belgium

AND

G. LAGMAGO KAMTA
Department of Physics and Astronomy
The University of Nebraska
116 Brace Laboratory, Lincoln, NE 68588-0111, USA

1. Introduction

The behaviour of a two-active electron system exposed to a strong and ultrashort laser pulse remains a challenging theoretical problem which requires the non-perturbative solution of the corresponding time-dependent Schrödinger equation (tdse). *Ab-initio* calculations based on the numerical solution of the tdse are now becoming a necessary step towards a non-ambiguous interpretation of increasingly precise experiments that leave simple models far behind. In addition they allow a detailed study of the role of electron-electron correlation which is at the origin a genuine two-electron processes such as double excitation or direct double ionisation.

This contribution is a progress report on the development of a theoretical approach aimed at solving numerically the tdse for a two-electron atom exposed to an ultrashort (a few optical cycles) laser pulse. For the sake of illustration, we consider here the case of H^- and assume that the laser frequency is such that the absorption of one photon brings the system just above the double ionisation threshold. Results for the above-threshold detachment spectrum are discussed in detail. Our results clearly indicate evidence for both sequential and non-sequential double ionisation as well as single ionisation with excitation of the residual atom.

B. Piraux and K. Rzążewski (eds.), Super-Intense Laser-Atom Physics, 127–141.

This report is organized as follows. In sections 2 and 3, we give a brief account of our method. It is based on the expansion of the total wavefunction on eigenstates of the two-electron system in hand. These eigenstates are built as products of one-electron Coulomb Sturmian functions for the radial part and bipolar harmonics for the angular part. The time-propagation is performed by means of usual explicit method of Runge-Kutta type. The most difficult problems inherent to all *ab-initio* calculations based on the numerical resolution of the tdse are the control of the density of states in both the single or double continua as well as the extraction of a quantitative information from the total wave function. These problems are described in section 3 and 4. In the next section, we present results regarding the atomic structure of H^-. We then compare our results for the one-photon detachment of H^- exposed to an ultrashort laser pulse with those obtained with other approaches. Finally in the last section, we present and discuss our results of H^- in the regime of high frequencies and ultrashort pulse durations. Unless otherwise indicated, we use atomic units throughout this contribution.

2. Solution of the tdse in a discrete basis

The tdse for a two-electron system of nuclear charge Z in a laser field reads:

$$i\frac{\partial}{\partial t}\Psi(\mathbf{r}_1,\mathbf{r}_2,t) = [H + D(t)]\,\Psi(\mathbf{r}_1,\mathbf{r}_2,t), \tag{1}$$

where

$$H = \frac{\mathbf{p}_1^2}{2} + \frac{\mathbf{p}_2^2}{2} - \frac{Z}{r_1} - \frac{Z}{r_2} + \frac{1}{r_{12}}, \tag{2}$$

is the atomic Hamiltonian. r_1, r_2 give the distance between the position of the electrons and the (infinitely massive) nucleus while r_{12} is the electronic interdistance. $\mathbf{p}_1$ and $\mathbf{p}_2$ are the electron momenta. The Hamiltonian $D(t)$ describes the interaction of the system with the laser field and is given in the length and velocity gauges respectively by $D_L(t) = \mathbf{E}(t).(\mathbf{r}_1 + \mathbf{r}_2)$ and $D_V(t) = \mathbf{A}(t).(\mathbf{p}_1 + \mathbf{p}_2)$. $\mathbf{E}(t)$ denotes the electric field and $\mathbf{A}(t) = \hat{z}A_0 f(t)\sin\omega t$ the vector potential of the laser field, which is assumed to be linearly polarized along $\hat{z}$. $f(t)$ is the pulse envelope.

The solution of the tdse can be obtained as an expansion in terms of basis functions $\{|i\rangle\}$ as follows:

$$|\Psi(t)\rangle = \sum_i C_i(t)\,|i\rangle. \tag{3}$$

In this basis, the tdse is transformed into a matrix equation

$$i\frac{\partial}{\partial t}\mathbf{S}\boldsymbol{\Psi}(t) = \Big(\mathbf{H} + g(t)\mathbf{D}\Big)\boldsymbol{\Psi}(t), \tag{4}$$

where $\mathbf{\Psi}$ is a vector whose time-dependent components are the C_i's and which represents the wave function in the basis $\{|i\rangle\}$. $\mathbf{S}$, $\mathbf{H}$ and $\mathbf{D}$ are respectively the overlap matrix and the matrices associated to the atomic Hamiltonian and the dipole operator in the basis $\{|i\rangle\}$. $\mathbf{S}$ and $\mathbf{H}$ are symmetric, while $\mathbf{D}$ is symmetric in the length gauge and antisymmetric in the velocity gauge. $g(t)$ is a scalar function which equals $E(t)$ in the length gauge and $iA(t)$ in the velocity gauge. Lets $\mathbf{T}$ be the orthonormal matrix ($\mathbf{T}^t\,\mathbf{T}=\mathbf{T}\,\mathbf{T}^t=\mathbf{1}$) that diagonalises $\mathbf{S}$, i.e. $\mathbf{T}^t\,\mathbf{S}\,\mathbf{T}=\mathbf{s}$, where $\mathbf{s}$ denotes the diagonal matrix of eigenvalues of $\mathbf{S}$. Using basic matrix algebra, Eq. (4) can be rewritten (in the so-called "orthogonal" basis) as

$$i\frac{\partial}{\partial t}\tilde{\mathbf{\Psi}}(t) = \left(\tilde{\mathbf{H}} + g(t)\tilde{\mathbf{D}}\right)\tilde{\mathbf{\Psi}}(t), \tag{5}$$

where $\tilde{\mathbf{H}} = \mathbf{U}^t\,\mathbf{H}\,\mathbf{U}$, $\tilde{\mathbf{D}} = \mathbf{U}^t\,\mathbf{D}\,\mathbf{U}$, $\mathbf{U} = \mathbf{T}\,\mathbf{s}^{-1/2}$ and $\tilde{\mathbf{\Psi}} = \mathbf{U}^{-1}\,\mathbf{\Psi}$. If $\mathbf{P}$ denotes the orthogonal matrix that diagonalizes $\tilde{\mathbf{H}}$ (i.e. $\mathbf{P}^t\,\tilde{\mathbf{H}}\,\mathbf{P}=\mathbf{h}$, where $\mathbf{h}$ is the diagonal matrix of atomic eigenvalues), then Eq. (5) can be further transformed into the eigenstate representation or atomic basis as

$$i\frac{\partial}{\partial t}\mathbf{\Phi}(t) = \left(\mathbf{h} + g(t)\mathbf{W}\right)\mathbf{\Phi}(t), \tag{6}$$

where $\mathbf{W} = \mathbf{P}^t\,\tilde{\mathbf{D}}\,\mathbf{P}$ and $\mathbf{\Phi}(t) = \mathbf{P}^t\,\mathbf{\Psi}(t)$. $\mathbf{\Phi}(t)$ denotes the wave function in the atomic basis where it simply represents a linear superposition of atomic states resulting from the diagonalization. It is clear that Eqs. (4), (5) and (6) are equivalent and are first order differential equations which can be solved using Runge-Kutta methods. However, we solve these equations in the atomic basis. The reasons are the followings: (i) the resolution of the first order differential equations is a non-stiff problem [1] which can be solved using explicit Runge-Kutta methods; (ii) before the time-propagation, the accuracy of the eigenstates energies described by the basis can be easily assessed; (iii) during the time propagation, the evolution of the population of a given atomic state can be traced or even removed from the wave function. Moreover, we use the interaction picture via the transformation $\mathbf{\Phi}_I(t) = e^{i\mathbf{h}_r t}\mathbf{\Phi}(t)$, where $\mathbf{h}_r$ denotes the real part of $\mathbf{h}$. In this picture, the fast oscillations due to field free atomic energies are absorbed in the wave function. As a result, the time propagation is significantly speeded up. It is important to stress that explicit Runge-Kutta methods are very fast. In the present case, we use an embedded formula (see [2]) which allows an automatic control of the time step during the propagation. Finally, let us mention that for well known reasons [3], the time propagation is performed in the velocity gauge.

In practice the expansion (3) has to be truncated, the resulting wave function represents a restricted region of space, say a sphere of radius R_0.

As long as the system remains in this sphere over the time interaction with the laser field, the basis is adequate. But if the system breaks up during its evolution, as it is the case when ionisation occurs, a certain fraction of this system may leave the sphere. Nevertheless, provided that the basis functions $\{|i\rangle\}$ are real, the probability $\langle\Psi(t)|\Psi(t)\rangle$ of finding the system in the sphere is always unity. This means that if the system breaks up and one of its fragments reaches the surface of the sphere over the time interval of interest, this fragment must reflect from the surface and return to the interior of the sphere. Many techniques could be used to avoid these unphysical reflections. Here we consider the complex rotation technique which also allows to obtain doubly excited states energies and widths [11] in the field free case.

For a better efficiency, the choice of the trial functions $\{|i\rangle\}$ should adequately combine the following requirements: (i) The computation of **H** and **S** should be easy and fast in order to have a realistic computation time. (ii) The number of nonzero entries of **H** and **S** should be relatively small, and the structure of the resulting matrices should be adequate (e.g. a banded structure) so as to allow for an efficient and fast diagonalization. (iii) The eigenstates under interest should be accurately described with a relatively short basis expansion. In the following, we describe a Sturmian basis that meets these requirements.

3. The Sturmian basis approach

We search for the solution of the tdse as a configuration expansion

$$\Psi(\mathbf{r}_1,\mathbf{r}_2,t) = \sum_{L,M}\sum_{\lambda,\ell}\sum_{s}\sum_{\nu n} \alpha^{\lambda\ell}_{\nu n}\psi^{\lambda\ell LM}_{\kappa_s k_s \nu n}(t)\,\mathcal{A}\frac{S^{\kappa_s}_{\nu\lambda}(r_1)}{r_1}\frac{S^{k_s}_{n\ell}(r_2)}{r_2}\Lambda^{LM}_{\lambda\ell}(\hat{r}_1,\hat{r}_2); \tag{7}$$

$\psi^{\lambda\ell LM}_{\kappa_s k_s \nu n}(t)$ denotes the time-dependent coefficients and $\mathcal{A}$ a symmetrization operator. The radial part of the wave function is described by Sturmian functions given by

$$S^{\kappa}_{n\ell}(r) = N^{\kappa}_{n\ell} r^{\ell+1} e^{-\kappa r} L^{2\ell+1}_{n-\ell-1}(2\kappa r), \tag{8}$$

where κ is the nonlinear parameter and $L^{2\ell+1}_{n-\ell-1}(2\kappa r)$ the Laguerre polynomial. The index n of the Sturmian functions takes the values $\ell+1$, $\ell+2$, $\ell+3$, ... The normalization constant

$$N^{\kappa}_{n\ell} = \sqrt{\frac{\kappa}{n}}(2\kappa)^{\ell+1}\left(\frac{(n-\ell-1)!}{(n+\ell)!}\right)^{1/2} \tag{9}$$

is derived with the condition $\int_0^\infty S^{\kappa}_{n\ell}(r)^* S^{\kappa}_{n\ell}(r)dr = 1$. The bipolar spherical harmonics $\Lambda^{LM}_{\lambda\ell}(\hat{r}_1,\hat{r}_2)$ couple the individual angular momenta of the two electrons in the L-S scheme. The quantity $\alpha_{\nu n}^{\lambda\ell} = 1 + \left(\frac{1}{\sqrt{2}} - 1\right)\delta_{\lambda\ell}\delta_{\nu n}$ controls the redundancy which, from exchange, may occur in the basis. Note that in order to avoid clumsy subscripts, we respectively denote the non-linear parameter and the radial, orbital and magnetic quantum numbers by greek letters κ, ν, λ and μ for electron 1, and roman letters k, n, ℓ and m for electron 2. Eq. (7) for the total wave function contains several summations: on L and M the quantum numbers associated to the total angular momentum and its projection respectively, on λ and l, the angular quantum numbers of each individual electron, on ν and n the Sturmian indices and finally on s which defines a given set of non-linear parameters. The introduction of several sets of non-linear parameters takes into account the fact that in many different configurations and in particular the most asymmetric ones, both electrons occupy two rather distinct regions of space. If for instance one of the electrons is excited, the basis should contain a small non-linear parameter. Note that exchange effects are taken into account through the antisymmetrization of the wave function. In fact, it turns out that the use of several sets of non-linear parameters allows to span a larger region of space and permits a simultaneous description of many eigenstates. In addition, we also allow the Sturmian indices associated to each electron to vary within different ranges. Indeed, as the degree of excitation of one of the electrons increases, only one of the indices has to be increased. In practice, we introduce a combination $(\kappa_s,\nu_<,\nu_>;k_s,n_<,n_>)$ involving the Sturmian functions $S_{\nu,\lambda}^{\kappa_s}(r_1)$ with index $\nu=\lambda+\nu_<$, $\lambda+\nu_<+1$, ..., $\lambda+\nu_>$ associated to electron 1, and $S_{n,\ell}^{k_s}(r_2)$ with index $n=\ell+n_<$, $\ell+n_<+1$, ..., $\ell+n_>$ associated to electron 2. This way of optimizing the construction of the wave function does help keeping the basis size within reasonable limits. In addition, it allows an efficient control of the density of states in both the single and double continua. In fact, a good description of the dynamics of a two-electron system driven by a laser pulse requires that the distance between two consecutive eigenenergies is much smaller than the laser pulse bandwidth.

Finally, let us mention that within the present approach, the obtention of the wave function of the two-electron system in the momentum space is straightforward. This feature is particularly important in the perspective of a time-dependent study of this system in the presence of an intense laser field. Indeed, by taking the Fourier transform of Eq. (7), one obtains a similar expression for the wave function in the momentum space, but with the radial Sturmian function $S_{n,\ell}^{k}(r)$ replaced by its counterpart $\Sigma_{n\ell}^{k}(p)$ in the momentum space which is given by:

$$\Sigma_{n,\ell}^{k}(p) = \sqrt{\frac{2}{\pi}} \int_0^\infty r\, j_\ell(pr)\, S_{n\ell}^{k}(r)\, dr,$$

$$= \mathcal{N}_{n,\ell}^{k} \frac{p^{\ell}}{(p^2+k^2)^{\ell+2}} C_{n-\ell-1}^{\ell+1}\left(\frac{p^2-k^2}{p^2+k^2}\right). \tag{10}$$

$j_\ell(pr)$ denotes a spherical Bessel function and $C_{n-\ell-1}^{\ell+1}$ a Gegenbauer polynomial [4]. The normalization constant

$$\mathcal{N}_{n,\ell}^{k} = 2^{2\ell+2} k^{\ell+2} \ell! \sqrt{\frac{2kn}{\pi} \frac{(n-\ell-1)!}{(n+\ell)!}} \tag{11}$$

is derived from the normalization condition $\int_0^\infty \Sigma_{n,\ell}^{k}(p)^* \, \Sigma_{n,\ell}^{k}(p)\, p^2 dp = 1$.

Before addressing the difficult question of the calculation of the observables, it is important to mention two problems of the present approach. First, the inclusion of several sets of non-linear parameters in the expansion of the wave function brings near linear dependency in the basis set. This however, can be adequately solved (see [5]). In addition, the fact that beside the exponential factor, our trial functions are essentially (Laguerre) polynomials in r_1 and r_2 limits the accuracy of the ground state energy. This is due to the fact that the corresponding wave function has a cusp (a discontinuity in the derivative) which results from the divergence of the atomic Hamiltonian when $r_1 = r_2$ [6]. However, the presence of this cusp (which cannot be reproduced by polynomial functions of r_1 and r_2) is only important for extremely correlated states such as the ground state. Nevertheless, as we will see later, the accuracy we obtained is sufficient for a precise study of the dynamics of two-electron systems driven by laser pulses.

4. Calculation of the observables

The extraction of a quantitative information from the wave function is a difficult problem which, for some observables such as the double ionisation probability, is still unsolved. By contrast to grid methods, the present approach presents serious advantages. Because the time propagation of the wave function is performed in the atomic basis, the calculation of all bound and doubly excited state populations after the propagation is straigthforward. By selecting the energy components of the wave function between the first single ionisation threshold and the double ionisation threshold, it is easy to calculate a single ionisation energy spectrum. However, this spectrum is calculated as a function of the *total* energy of the system. It is therefore difficult in that case to distinguish between a single ionisation process with the residual ion left in its ground state and a similar process in which the residual ion is left excited (shake-up process). In order to remove this ambiguity, we project the corresponding energy components of the final wave function on two-electron wave functions built as antisymmetrized products of a one electron bound state wave function of the residual ion and a continuum state wave function associated to the coulomb potential of this ion. Although this projection is only valid in the asymptotic region where the electron-electron interaction is negligible, it provides reliable informations on the state of the residual ion after the ionisation. When the energy components of the wave function above the double ionisation threshold are selected, the additional problem of the entanglement of the single and double ionisation contributions arises. However, we can extract the single ionisation contribution by means of the same procedure as in the previous case. In this way, we can evaluate the total contribution of single ionisation. Indeed, the present method of projection is actually providing the single ionisation spectrum as a function the electron energy. If we integrate over this energy, we obtain the single ionisation probability. By substracting from 1, this probability and the probability to be in bound or resonant states, we obtain the probability of double ionisation. Single and double ionisation probability can also be obtained by means of the following procedure:

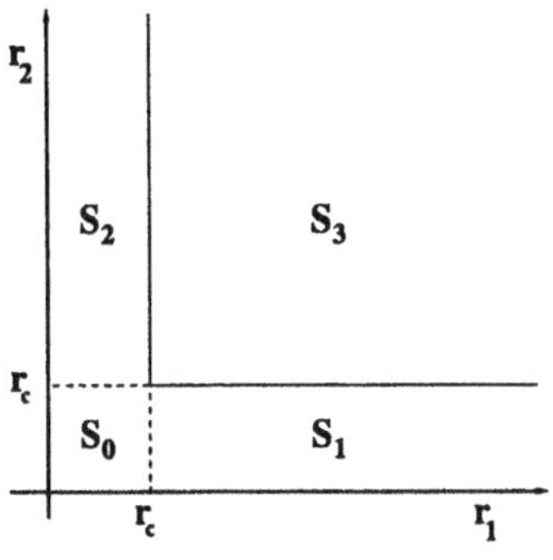

Figure 1. Grid in radial coordinates r_1 and r_2 of the two electrons.

We consider a grid in radial coordinates r_1 and r_2 of the two electrons (see Fig. 1). To obtain the double ionisation probability, we integrate the total probability found in the area S_3. However, to avoid any ambiguity, all energy components of the wave function which are below the double ionisation threshold are substracted from this wave function. Similarly, the single ionisation probability is obtained by integrating the total probability found in regions S_1 and S_2. As before, all the bound and resonance state contributions are substracted from the wave function. In Fig. 1, r_c represents a characteristic length of the system. In the case of H^-, $r_c = 3$ a.u. while for helium, $r_c = 8$ a.u.. Finally, let us mention that except for the bound and resonance state probability, the other observables are calculated without complex scaling.

5. Our method compared to others

Before considering the response of H^- to an intense high frequency laser field, It is important to evaluate what kind of accuracy may be achieved with the present method. We first compare our results for the atomic structure of H^- with those obtained with very accurate methods. Our result for E_g, the ground state energy, is $E_g = -0.52762$ a.u. differs by about 0.02% from the accurate value $E_g = -0.52775102$ a.u. obtained by Frankowski and Pekeris [7] who used a Hylleraas-type of wave function which explicitely includes the electron-electron correlation. As mentioned before, this difference (which is very small for our purpose) is due to the presence of a cusp in the wave function which cannot be reproduced within any configuration interaction approach like ours.

In the table, we consider the resonances of H^- which are around the $n = 3$ threshold. We compare our results for the position and the width with those of Lindroth [8] who used a very accurate method developed by Salomonson et al [9].

This work		Lindroth [8]	
Energy	width	Energy	width
-0.06271	0.00117	-0.06273	0.00119
-0.05857	0.0000089	-0.05857	0.0000088
-0.05612	0.0000026	-0.05612	0.0000022
-0.05591	0.000072	-0.05590	0.000071

TABLE 1. Energies and width (in a.u.) of the first resonances around the n=3 threshold of H^-. The reference data are from Lindroth [8].

We see in Table 1 that our results are in perfect agreement with those of Lindroth for both the position and the width of these autoionising states. It is important to stress that all states in our basis are generated by means of a *single* diagonalization.

We now turn to the dynamics of H^- in strong laser fields. Some results already exist. They concern the ionisation yield of H^- exposed to a cosine square shape pulse of frequency $\omega = 0.03$ a.u. and whose full width at half maximum (fwhm) is equal to 4 optical cycles. Our results (open circles) for the ionisation yield as a function of the peak intensity of the pulse are shown in Fig. 2 where they are compared to those obtained by other approaches.

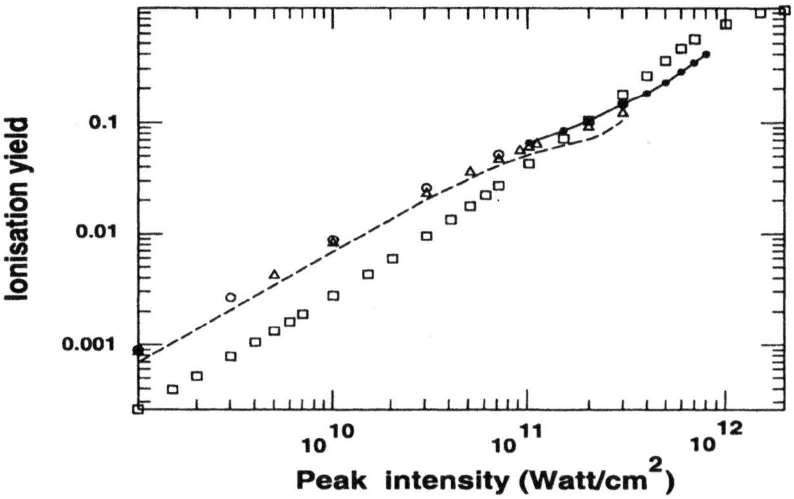

Figure 2. Ionisation yield as a function of the pulse peak intensity in Watt/cm^2 for H^- exposed to a cosine-square shape pulse whose frequency $\omega = 0.03$ a.u.. The pulse fwhm = 4 optical cycles. The open triangles are the results of Hasbani [10], the open circles are our results, the open squares are the results of Ref. [12] and the full circles are those of Ref. [11]. The dashed line is obtained by integrating over the pulse intensity profile, the R-matrix Floquet results of Ref. [13].

Except for the dashed line, all the other results are obtained by means of a spectral method in which the wave function built in a given basis is propagated in the atomic basis. The open triangles are the results obtained by Habani [10]. In this case, the eigenstates of H^- are built as products of B-splines. The full circles are the results of Scrinzi and Piraux [11] who expand the wave function of the system is in an explicitely correlated basis of Hylleraas type. The open squares are the results of Nikolopoulos and Lambropoulos [12]. In their approach, the eigenstates of H^- are built as products of hydrogenic functions. Finally the dashed line is obtained by integrating over the pulse intensity profile, the R-matrix Floquet results of Purvis *et al* [13]. The agreement between our results and the other ones is excellent except with those of Nikolopoulos and Lambropoulos. It is established [14] that this difference is due to a poor description of the ground state wave function of H^- in Nikolopoulos and Lambropoulos approach. In Fig. 2, we can distinguish between two regimes of intensities. For intensities lower than 10^{11} Watt/cm^2, the results are in agreement with the lowest order of perturbation theory (the slope of the curve is 1). Beyond this critical intensity, we observe a slight "knee" around 2×10^{11} Watt/cm^2. For that intensity and due to the ponderomotive shift of the single continuum threshold, the one-photon ionisation channel closes. For higher intensities, saturation sets in very quickly and non-perturbative effects become dominant. It is important to stress that in that case, all continuum states close to the threshold play an important role. We have checked that the density of states above the threshold is sufficient and that the results are stable with the choice of the complex scaling angle.

6. H^- in an intense high frequency laser field

Let us now consider the interaction of H^- with a high frequency field such that one photon is sufficient to bring the system above the double ionisation threshold. The results we show are obtained with a cosine-square pulse envelope of the form:

$$f(t) = \begin{cases} \cos^2(t/\tau), & -\pi\tau/2 \leq t \leq \pi\tau/2, \\ 0, & Otherwise. \end{cases} \quad (12)$$

The frequency $\omega = 0.6$ a.u. and the fwhm of the pulse is 10 cycles. The peak intensity is 3×10^{14} Watt/cm^2. The result for the Above Threshold De-

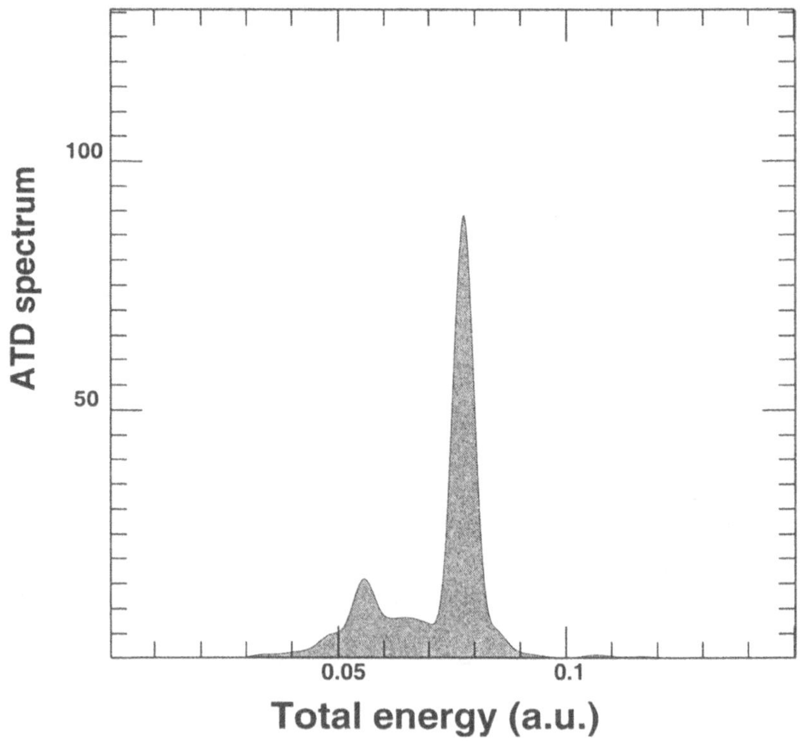

Figure 3. ATD spectrum as a function of the total energy in a.u. for H^- exposed to the cosine-square shape pulse of fwhm equal to 10 cycles. The frequency $\omega = 0.6$ a.u. and the peak intensity is equal to 3×10^{14} Watt/cm^2.

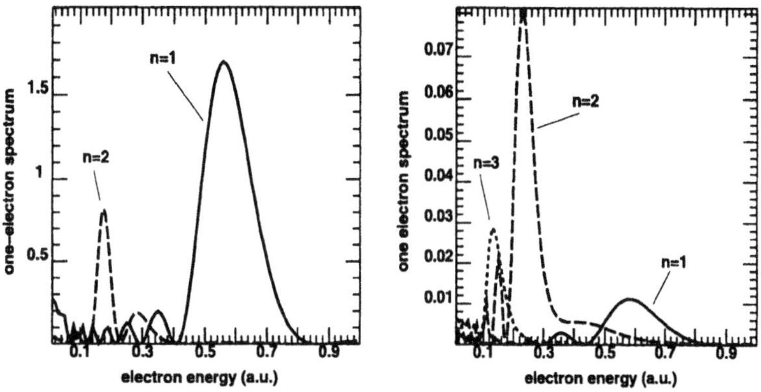

a. main peak   b. small peak

Figure 4. Projection of the energy components of the wave function of the system around the main peak (a) and around the small peak (b) of the ATD spectrum on antisymmetrized products of bound hydrogen states (labeled by their principal quantum number n) and plane waves. The modulus square of these projections is plotted as a function of the energy of one of the electrons in a.u.. along the transverse coordinate.

tachment (ATD) spectrum as a function of the total energy is shown in Fig. 3. This spectrum exhibits two well pronounced peaks: one around 0.08 a.u. which corresponds to the energy conservation peak ($0.08 \approx E_g + 0.6$) and one at 0.055 a.u.. The occurrence of two different peaks is only possible if they correspond to two different processes characterized by different shifts. In order to identify these processes, let us first select the energy components of the wave function between 0.07 a.u. and 0.09 a.u. (main peak). We then go back to the Sturmian basis and project on antisymmetrized products of hydrogen bound states and plane waves. The results are presented in Figs. 4. As far as the main peak is concerned (left hand plot of Fig. 4), we see that the one-electron spectrum exhibits a well pronounced peak at an electron energy of about 0.56 a.u.. This peak corresponds to a single ionisation process in which the residual atom stays in its ground state. In addition, we observe a second peak which is much smaller around an electron energy of 0.18 a.u.. This peak corresponds to a single ionisation process with the residual atom (hydrogen) in a n=2 state (shake-up process). Let us now consider the energy components of the wave function between 0.05 a.u. and 0.06 a.u. (smaller peak in the ATD spectrum of Fig. 3). After returning to the Sturmian basis and projecting on the same wave function as before, we obtain the results shown in Fig. 4 (right hand plot). Here, we see that single ionisation with the residual atom left in an excited state (n=2 and n=3) dominates. If we integrate the one-electron spectrum given in Fig. 4 (right hand plot) over the electron energy, we obtain the corresponding single ionisation probability which turns out to be much smaller than the area under

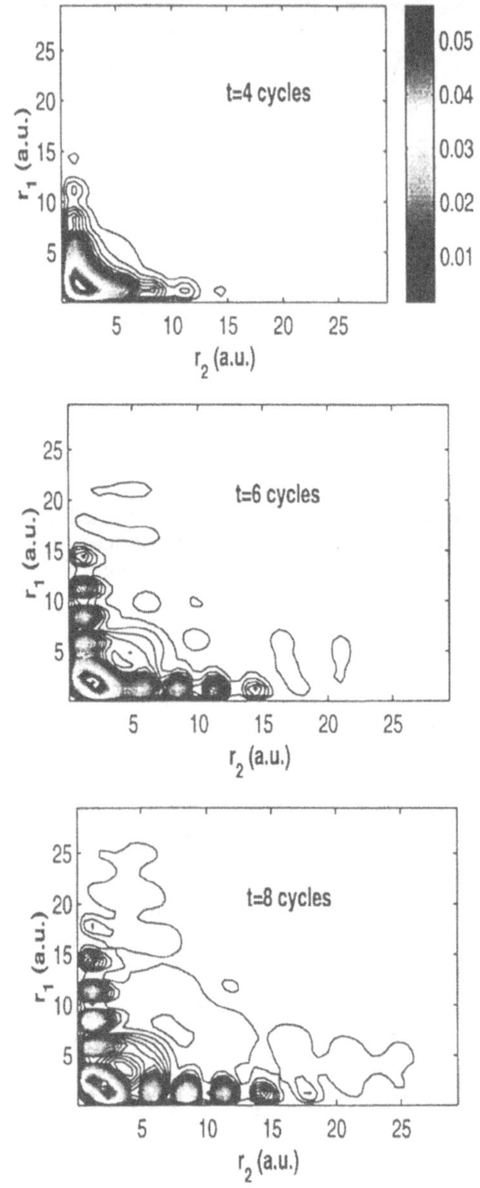

Figure 5. Radial plot of the angles-integrated two-electron probability density in the position space for H^- and the same laser pulse as in Fig. 3. r_1 and r_2 are in a.u. and 3 different times after the beginning of the propagation have been considered: 4, 6 and 8 cycles.

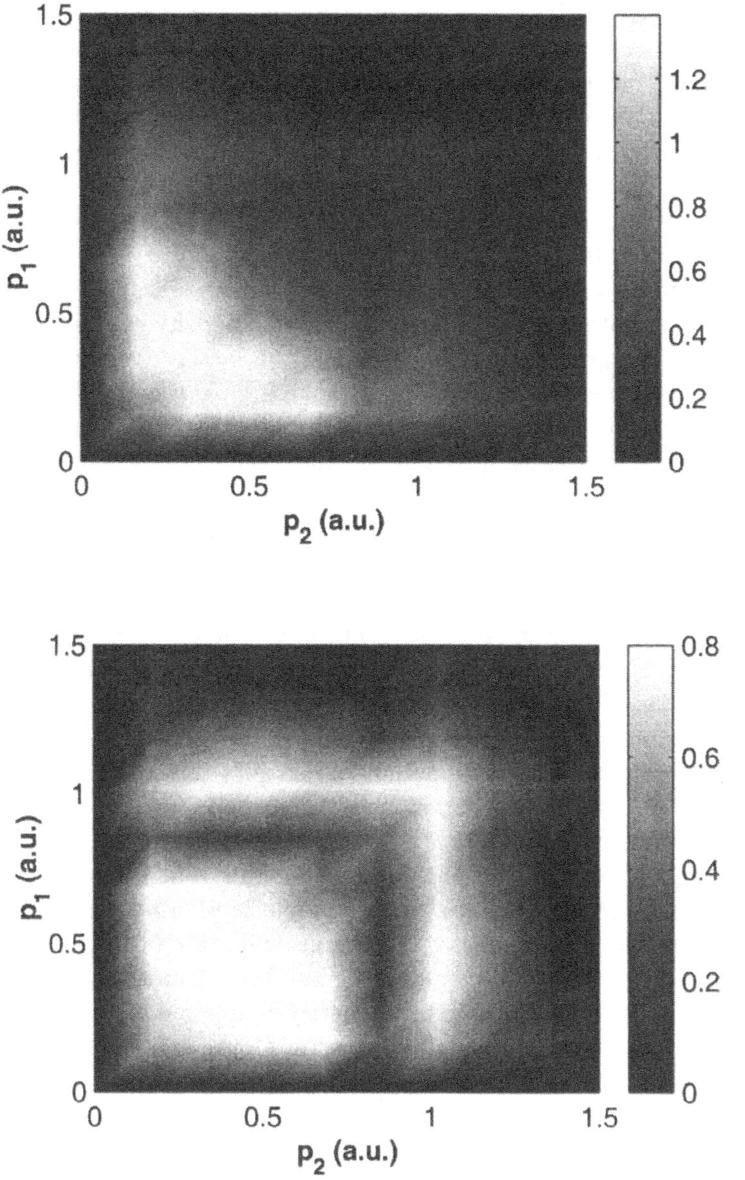

Figure 6. Radial plot of the angles-integrated two-electron probability density in the momentum space for H^- and the same laser pulse as in Fig. 3. p_1 and p_2 are in a.u. and 2 different times after the beginning of the propagation have been considered: 4, 6 and 8 cycles.

the small peak in the ATD spectrum (Fig. 3). This means that single ionisation with or without excitation is not sufficient to explain the amplitude of the small peak in the ATD spectrum. In order to elucidate this problem, let us now analyse the radial plot of the angles-integrated two-electron probability density in the position space for three different times during the propagation: 4, 6 and 8 optical cycles (see Fig. 5). 4 cycles after the beginning of the interaction (upper plot of Fig. 5), the radial probability is still dominated by the ground state probability density. However, we start to see maxima which develop along the r_1 and r_2 axes. This is single ionisation. In addition, it is also clear that some density of probability moves away from the ground state region along the bisector: we attribute this to direct (non-sequential) double ionisation. After 6 optical cycles (middle plot of Fig. 5), these trends are even more pronounced. Moreover, we now observe what we call sequential double ionisation: some density of probability moves parallel to the r_1 or r_2 axis. In the meantime, non-sequential ionisation keeps developping. It is interesting to note that along each axis, the position of the zeroes corresponds to those of the spherical Bessel functions associated to continuum states of momentum around 1 a.u. which is the momentum of the electron in the continuum when the system by absorbing one photon is singly ionised, the bound electron staying in the ground state of atomic two-electron probability density in the momentum space for two different times during the propagation: 4 and 6 optical cycles. 4 cycles after the beginning of the interaction (upper plot of Fig. 6), we see that the density of probability corresponds to the ground state probability density. After 6 optical cycles, two sharp lines parallel to the p_1 and p_2 axes appear around $p_1 = 1$ a.u. and $p_2 = 1$ a.u.. This corresponds to single ionisation: one of the electrons is emitted with a momentum around p_1 or p_2 equal to 1 a.u. while the other one stays in the ground state of hydrogen where his momentum distribution is very broad. This results indicate that despite the existence of sequential and non-sequential double ionisation, the ionisation process is dominated by single ionisation at the intensity, frequency and pulse duration considered here.

7. Conclusion

In this contribution we gave a brief account of a new spectral method of configuration interaction type which is aimed at solving numerically the Schrödinger equation associated to the interaction of a two active electron system with an intense ultrashort laser pulse. We then treated the problem of H^- exposed to an ultrashort laser pulse which brings this system above the double ionisation threshold. Despite the fact that single ionisation is the dominant process, shake-up and sequential as well as non-sequential double ionisation are present. We want to stress that this study is a first step towards a detailed study of the interplay between these various processes. More quantitative evaluation of their respective probability will be necessary in the future.

Acknowledgments

We would like to thank A. Scrinzi and H. Bachau for fruitfull discussions. G.L.K acknowledges a support from the Université Catholique de Louvain and from Grant No. DE-FG-03-96ER14646 from the U.S. Department of Energy, Office of Basic Energy Sciences, Division of Chemical Sciences.

References

1. Huens, E., Piraux, B., Bugacov, A. and Gajda, M. (1997) *Phys. Rev. A* **55**, 2132.
2. Hairer, E., Norsett, S.P. and Wanner, G. (1987) *Solving ordinary differential equations I: non-stiff problems*, Springer Verlag, Berlin.
3. Cormier, E. and Lambropoulos, P. (1996) *J. Phys. B* **29**, 1667.
4. Abramowitz, M. and Stegun, I.A. (1972) *Handbook of Mathematical Functions*, Dover Publication, New York.
5. Lagmago Kamta, G. (1999) *Semiclassical approaches to dipole radial integrals for nonhydrogenic ions and two-active electron systems in an intense laser field*, PhD Thesis, Université Nationale du Bénin.
6. Goldman, S.P. (1994) *Phys. Rev. Lett.* **73**, 2547; Carrol, D.P., Silverstone, H.J. and Metzger, R.M. (1979) *J. Chem. Phys.* **71**, 4142; Kutzelnigg, W. and Morgan, J.D. (1992) *J. Chem. Phys.* **96** 4484.
7. Frankowski, K. and Pekeris, C.L. (1966) *Phys. Rev.* **146**, 46.
8. Lindroth, E. (1995) *Phys. Rev. A* **52**, 2737.
9. Salomonson, S. and Öster, P. (1989) *Phys. Rev. A* **40**, 5559.
10. Hasbani, R. (2000) *Processus non-linéaires associés à l'interaction entre un atome et un champ laser intense: étude théorique de l'ionisation multiple*, PhD Thesis, University of Bordeaux I.
11. Scrinzi A. and Piraux B. (1997) *Phys. Rev. A* **56**, R13; Scrinzi A. and Piraux B. (1998) *Phys. Rev. A* **58**, 1310.
12. Nikolopoulos, L.A.A. and Lambropoulos, P. (1999) *Phys. Rev. Lett.* **82**, 3771.
13. Purvis, J., Dörr, M., Terao-Dunseath, M., Joachain, C.J. Burke, P.G. and Noble, C.J. (1997) *Phys. Rev. Lett.* **71**, 3943.
14. Lagmago-Kamta, G., Grosges, Th., Piraux, B., Hasbani, R., Cormier, E. and Bachau, submitted for publication to *J. Phys. B.*.

ANGULAR DISTRIBUTIONS FOR DOUBLE IONIZATION BY AN ULTRASHORT, INTENSE LASER PULSE: THE CASE OF LI$^-$

G. LAGMAGO KAMTA AND ANTHONY F. STARACE

Department of Physics and Astronomy
The University of Nebraska
116 Brace Laboratory, Lincoln, NE 68588-0111, USA

1. Introduction

Angular distributions for double ionization of atomic systems by photon impact have attracted much interest from both theorists and experimentalists [1]. Such studies have primarily been limited to double ionization by a single photon in which case double ionization occurs only via correlation between the two electrons. In this case, most theoretical approaches evaluate triply differential cross section (TDCS) via the transition matrix coupling initial and final states. *Ab initio* approaches to this problem have also appeared [2]. Obtaining angular distributions for an ultrashort pulse is a challenging problem because the time-dependent Schrödinger equation (TDSE) has to be solved in its full dimensionality, and because the finiteness of the pulse introduces an uncertainty into the sharing of the excess energy. Many approaches for solving the TDSE for two-electron systems in interaction with an ultrashort pulse have been developed [3, 4], but we are not aware of any studies of angular distributions in this case. For double ionization by intense fields (in the multiphoton regime), very little is known concerning angular distributions of electrons, despite recent electron coincidence [5] and COLTRIMS [6] measurements for the rare gases. In this work, we analyze angular distributions in double ionization by an ultrashort laser pulse. Following a direct numerical solution of the TDSE for Li$^-$, treated as a two-active electron system, we present angular distributions from the lower to the higher intensity regime.

B. Piraux and K. Rzążewski (eds.), Super-Intense Laser-Atom Physics, 143–152.

2. Doubly differential double ionization probability

Explicitly or implicitly, the time propagation of the TDSE for an ultrashort laser pulse is always performed over finite distances. In this case, there is no exact definition for the double ionization probability, due to the difficulty of desentangling single and double ionization contributions from the time propagated wave function. However, to obtain the double ionization probability P, one usually integrates the total probability found in the area $\mathbf{S_3}$ in Fig. 1, which means evaluating the integral

$$P = \int d\Omega_1 \int d\Omega_2 \int_{r_1 > r_c} dr_1 \int_{r_2 > r_c} dr_2 \, |\Psi(\mathbf{r}_1, \mathbf{r}_2, T)|^2, \tag{1}$$

where $\mathbf{r}_j \equiv (r_j, \theta_j, \phi_j)$ and $d\Omega_j \equiv \sin\theta_j d\theta_j d\phi_j$ $(j = 1, 2)$ denote respectively the coordinates and the differential solid angles for the two electrons. $\Psi(\mathbf{r}_1, \mathbf{r}_2, T)$ represents the antisymmetrized wave function at time T, the end of the laser excitation. The cutoff radius r_c should be chosen such that the region $\mathbf{S_0}$ contains most of the probability of the initial state. Of course, this definition of the double ionization probability is not exact because, for example, residual bound and doubly excited states may have contributions in the region $\mathbf{S_3}$. These contributions may be small but nevertheless non-negligible compared to the double ionization probability, which is usually small as well. We also use Eq. (1) as our definition for the double ionization probability, with the difference that our time-propagation technique allows us to exclude the contribution of bound and doubly excited states from the final wave function. We shall return to this point later.

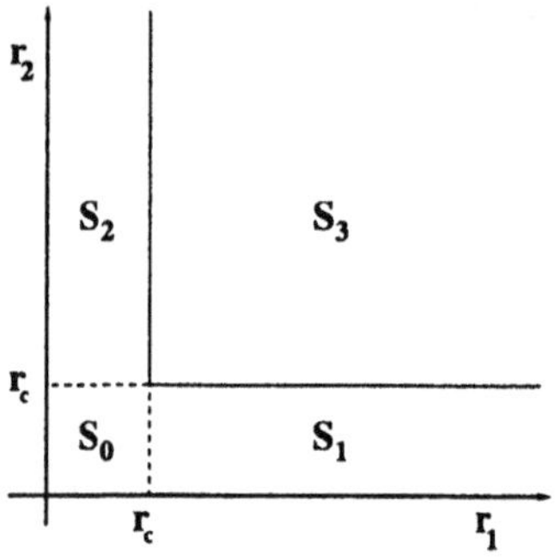

Figure 1. Spatial grid in the radial coordinates r_1 and r_2 of the two electrons.

To obtain electron angular distributions for double ionization, we skip the integration over the solid angles in Eq. (1), so that the quantity

$$\frac{d^2P}{d\Omega_1 d\Omega_2} = \int_{r_1 > r_c} dr_1 \int_{r_2 > r_c} dr_2 \, |\Psi(\mathbf{r}_1, \mathbf{r}_2, T)|^2 \tag{2}$$

may be interpreted as a *doubly differential double ionization probability* (DDDIP) for electron 1 to emerge in the solid angle $d\Omega_1$ and electron 2 in the solid angle $d\Omega_2$. It is a function of the four spherical angles θ_1, ϕ_1, θ_2 and ϕ_2. The resulting DDDIP is not differential in energy and therefore accounts for all possible energy transfers to the electrons from the laser field pulse as well as for all possible energy sharing configurations between the two ejected electrons. In this work, atomic units are used and angles are expressed in radians, unless otherwise stated.

3. Solution of the TDSE

We treat the Li^- ion as a two-active electron system, where each electron moves in the following core potential describing the Li^+ core [7, 8]:

$$V(r) = -\frac{1}{r}\left[Z_c + (Z_n - Z_c)e^{-a_1 r} + a_2\, r\, e^{-a_3 r}\right] - \frac{\alpha_c}{2r^4}\left(1 - e^{-(r/r_c)^3}\right). \quad (3)$$

Z_n and Z_c respectively denote the nuclear and the Li^+ core charges. $\alpha_c = 0.1894$ is the polarizability of the Li^+ core [9], and the empirical parameters a_1, a_2 and a_3 are fitted so that $V(r)$ reproduces the experimentally measured energy levels of the Li atom. The atomic Hamiltonian of Li^- is

$$H_0 = h(r_1) + h(r_2) + \frac{1}{r_{12}}, \quad h(r_j) = -\frac{\nabla_j^2}{2} + V(r_j), \quad (4)$$

where $h(r_j)$ describes the interaction of each electron with the Li^+ core.

The TDSE equation describing the interaction of the atomic system with a linearly polarized laser field is

$$i\frac{\partial}{\partial t}\Psi(\mathbf{r}_1, \mathbf{r}_2, t) = \left[H_0 + D(t)\right]\Psi(\mathbf{r}_1, \mathbf{r}_2, t), \quad (5)$$

where $D(t)$ represents the interaction of the system with the laser field. In the dipole approximation, $D(t)$ is given in the length and velocity forms by either $D_L(t) = \mathbf{E}(t).(\mathbf{r}_1 + \mathbf{r}_2)$ or $D_V(t) = i\mathbf{A}(t).(\nabla_1 + \nabla_2)$, where $\mathbf{E}(t) = -\frac{\partial}{\partial t}\mathbf{A}(t)$ is the electric field and $\mathbf{A}(t) = \hat{\mathbf{z}}A_0 f(t)\sin\omega t$ the vector potential. ω is the laser frequency and $\hat{\mathbf{z}}$ the unit vector along the polarization axis of the field. The pulse envelope used throughout this work is given by

$$f(t) = \begin{cases} \cos^2(\pi t/2T), & -T \le t \le T; \\ 0, & |T| < t. \end{cases} \quad (6)$$

We integrate the TDSE in a box by expanding the wave function as

$$\Psi(\mathbf{r}_1, \mathbf{r}_2, t) = \sum_{L,M}\sum_{\ell_1,\ell_2}\sum_{n_1,n_2} \psi_{n_1 n_2}^{\ell_1 \ell_2 L M}(t)\mathcal{A}\frac{R_{n_1,\ell_1}(r_1)}{r_1}\frac{R_{n_2,\ell_2}(r_2)}{r_2}\mathcal{Y}_{l_1,l_2}^{L,M}(\hat{\mathbf{r}}_1, \hat{\mathbf{r}}_2). \quad (7)$$

The expansion coefficients are denoted by $\psi_{n_1 n_2}^{\ell_1 \ell_2 LM}$ and the bipolar spherical harmonics $\mathcal{Y}_{l_1,l_2}^{L,M}$ couple the individual angular momenta of the two electrons in the L-S coupling scheme. The radial functions $R_{n,\ell}$ are obtained by numerically solving the one-electron radial Schrödinger equation

$$\left(-\frac{1}{2}\frac{d^2}{dr^2}+\frac{\ell(\ell+1)}{2r^2}+V(r)\right)R(r)=E\,R(r) \tag{8}$$

in a radial box of size $r=r_0$ with boundary conditions $R_{n,\ell}(r_0)=0$. Thus the time dependent wave function is subjected to similar boundary conditions: $\Psi(r_1, r_2 = r_0, t) = \Psi(r_1 = r_0, r_2, t) = 0$. In order to minimize flux reflections that may occur at the boundaries of the box, its size as well as the laser intensity and time duration are adjusted such the wave function remains negligeable at the edge of the box. The linearly polarized laser field considered, and the fact that the ground state of Li^- is ^1S allows one to set $M=0$ in Eq. (7). The TDSE is projected onto the atomic basis following the procedure of Ref. [4], and an embedded Runge-Kutta method of order 5 is used to solve the corresponding first order differential equation.

4. Results

The results presented here are obtained with 8 total angular momenta ($L = 0,1,2,3,4,5,6,7$) and a box of 250 a.u. size. From 2200 to 3700 configurations are used for each total angular momentum, leading to a system of about 21,000 ordinary differential equations to be solved. Typical run times are about 24 hours on a 667MHz Alpha EV6 workstation. We have varied the size of the box from 180 a.u to 300 a.u. as well as the number of angular momenta to check the stability of our results. The binding energy obtained for Li^- is $E_g = -0.02251$, which is in good agreement with the reference value -0.02269 [10]. This means that one needs 6 eV to eject the two active electrons in Li^-, and only 0.6 eV to eject one electron.

For a laser of frequency $\omega=0.023$ a.u., which is just above the detachment threshold, Fig. 2 shows the intensity dependence (up to saturation) of the detachment yield given by $1-P_0$, where P_0 is the projection of the full wave function at the end of pulse onto the ground state of Li^-. The length and the velocity gauge results are in very good agreement. At lower intensities, where the one photon detachment is predominant, one observes a linear behaviour with a slope close to 1, in agreement with the lowest-order perturbation theory. As intensity increases, this linear behavior is altered by Stark and ponderomotive shifts that close the one-photon detachment channel, leaving the two-photon one open. This effect is also present in laser ionization of the H^- ion [11, 12].

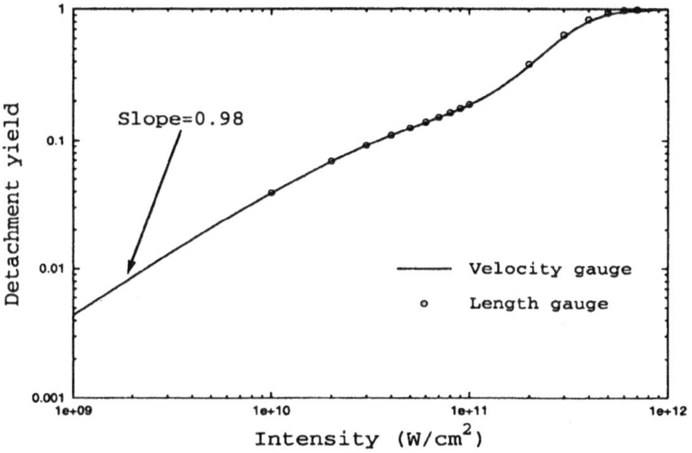

Figure 2. Photodetachment of Li^- w.r.t the peak laser intensity. A laser pulse of frequency 0.023 a.u. and fwhm of 4 cycles is used.

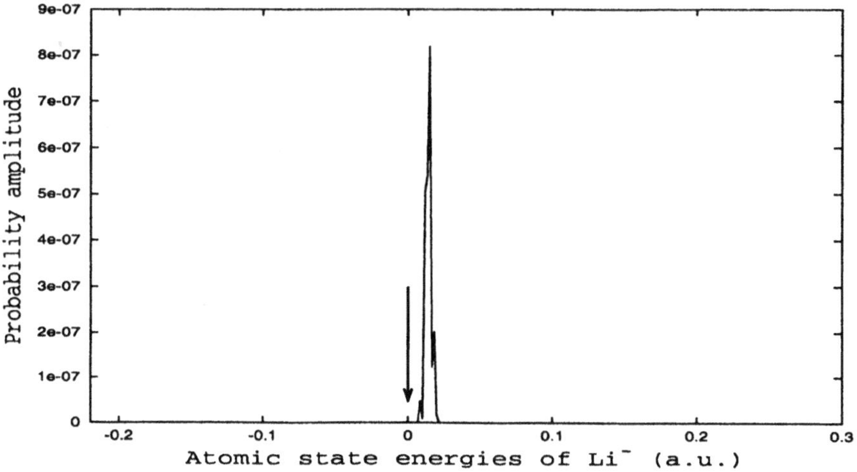

Figure 3. Probability amplitude of the wave function at the end of the laser pulse w.r.t. atomic energies of Li^-. The laser pulse has peak intensity $I = 10^9$ W/cm^2, frequency $\omega = 0.235$ a.u., and fwhm of 24 fsec. The arrow indicates the double ionization threshold.

As mentioned in Ref. [4], the wave function $\Psi(T)$ in the basis expansion (7) (which is employed in Eq. (2)) can be deduced from the wave function $\Phi(T)$ in the atomic basis by matrix-vector operations. When performing these operations, we set to zero all components of $\Phi(T)$ corresponding to atomic states below the double ionization threshold, thereby eliminating any contribution to the DDDIP from the ground, doubly excited, and singly ionized states below the double ionization threshold. The radial integration above the cutoff radius r_c in Eq. (2) is then used to approximately separate the doubly-ionized continua from the remaining singly-ionized continua. A cutoff radius $r_c = 25$ is used to evaluate the DDDIP since the ground state of Li^- extends to approximately 20 a.u.. Varying r_c between 20 and 25 a.u., we found a 9% maximum difference in the magnitude of DDDIP, but the angular distributions remained unchanged.

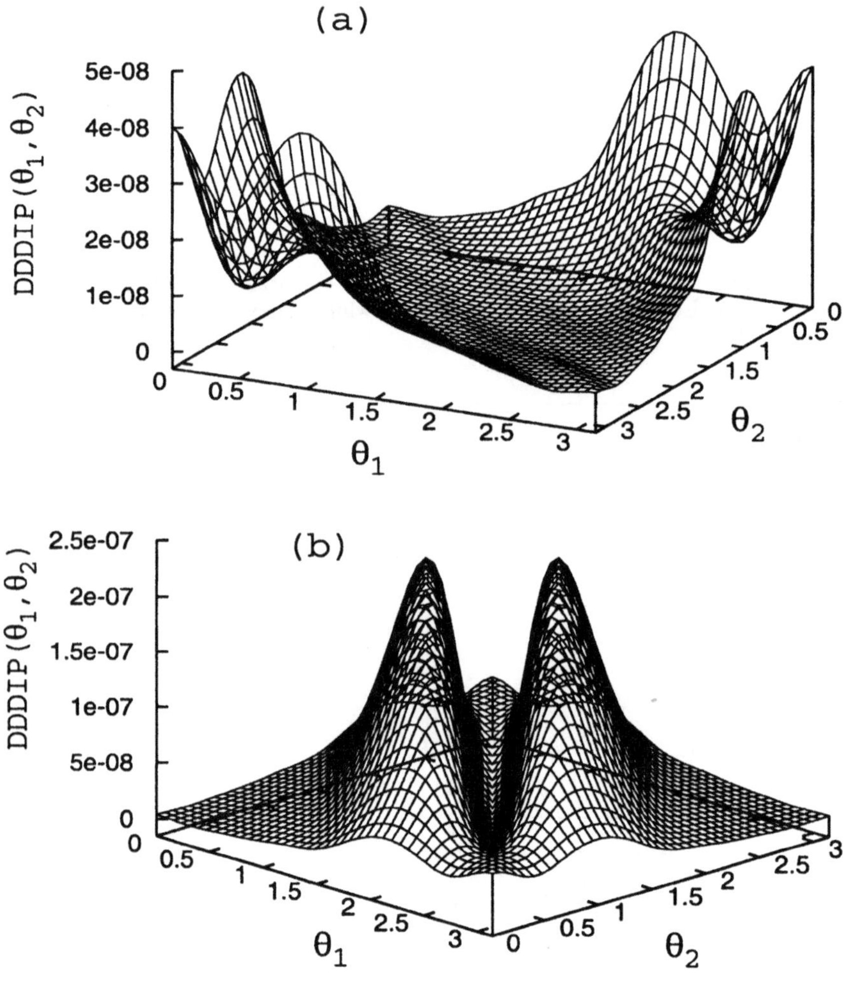

Figure 4. Single photon doubly differential double ionization probability (DDDIP) for Li^- for : **(a)** $\phi_1 = 0$ and $\phi_2 = 0$; **(b)** $\phi_1 = 0$ and $\phi_2 = \pi$. Laser parameters are as in Fig. 3.

We first consider angular distributions in double ionization by a weak field. The purpose here is to verify that our approach to angular distributions agrees with general features already well established for double ionization by a single photon. The laser pulse parameters are: $I = 10^9$ W/cm^2,

$\omega = 0.235$ a.u., and 38 cycles within the full width at half maximum (fwhm), corresponding to about 24 fsec. The low intensity used makes the absorption of a single photon predominant, but most of the population remains in the ground state. The frequency chosen is near the double ionization threshold, giving about 0.4 eV of excess energy above threshold. A plot in Fig. 3 of the probability amplitudes of the wave function w.r.t. the atomic energies in the basis shows a single, well resolved peak above the double thresold, indicating the absorption of a single photon. Three-dimensional plots of the DDDIP w.r.t. the polar coordinates θ_1 and θ_2 are given in Fig. 4(a) for an emission of the two electrons in the same half plane (i.e., at azimuthal angles $\phi_1 = \phi_2 = 0$) and in Fig. 4(b) for an emission back-to-back ($\phi_1 = 0$, $\phi_2 = \pi$). An analysis of these plots indicates that due to the Coulomb repulsion between the two electrons: (i) their emission in opposite directions is higher in magnitude than their emission in the same direction; (ii) The DDDIP is negligible in the vicinity of configurations where the two electrons are emitted at zero relative angles (i.e., $\theta_1 = \theta_2 = 0$ and $\theta_1 = \theta_2 = \pi$ in both figures, as well as in the vicinity of $\theta_1 \approx \theta_2$ in Fig. 4(a)). It follows that electrons are predominantly emitted at large relative angles. Fig. 5 shows DDDIP patterns (extracted from Fig. 4) plotted in polar coordinates, for various emission angles θ_1 of electron 1. This figure gives a better illustration of correlation effects preventing the two electrons from emerging in the same half plane. Note the node in the DDDIP pattern at angles $\theta_1 = \theta_2 = \pi/2$ (clearly visible in Figs. 4(b) and 5(c), but barely visible in Fig. 4(a)), indicating that there is no emission of both electrons simultaneously in the direction perpendicular to the polarization axis. This agrees with a selection rule valid for the TDCS for any sharing of the excess energy [13].

Fig. 5(a)) contains a signature of the fact that our DDDIP accounts for all possible energy sharing configurations between electrons. Indeed, a selection rule [13] for the TDCS for equal energy sharing states that there should be a node at the angles $(\theta_1, \theta_2) = (0, \pi)$ and $(\theta_1, \theta_2) = (\pi, 0)$; in other words, the lobe along the polarization axis in Fig. 5(a)) should be absent. Here this lobe arises from contributions of unequal energy sharing configurations. Except for this particular situation, the angular distributions in Figs 5(b), (c) and (d) consist essentially of two lobes whose relative size is a function of the angle between $\mathbf{k}_1$ and $\mathbf{z}$. Similar features have been obtained for He by perturbative calculations [13] and by a "Wannier" analysis [14].

We now move to angular distributions in the multiphoton regime by considering a laser pulse with characteristics: $I = 3 \times 10^{11}$ W/cm^2, $\omega = 0.037$ a.u., and 2 cycles at fwhm, corresponding to about 9 fsec. At this frequency at least six photons are necessary to reach the double ionization threshold. The DDDIP are presented in Fig. 6(a) for electron emission with azimuthal angles ($\phi_1 = 0, \phi_2 = 0$) and in Fig. 6(b) for an emission with azimuthal angles

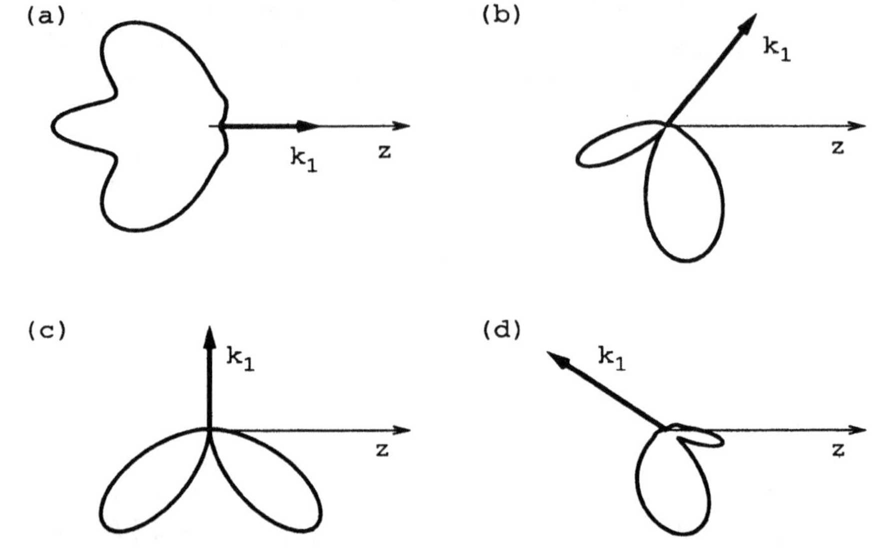

Figure 5. Single photon DDDIP patterns in polar coordinates for four emission angles θ_1 (with directions given by the unit vector $\mathbf{k}_1$) of electron 1: (a) $\theta_1 = 0$; (b) $\theta_1 = \pi/3$; (c) $\theta_1 = \pi/2$; (d) $\theta_1 = 3\pi/4$. Laser parameters are as in Fig. 3.

($\phi_1 = 0, \phi_2 = \pi$). One observes very different angular distributions from those for single photon double ionization, with the possibility for electrons to be emitted in all directions. The DDDIP's distribution has four prominent peaks, corresponding to the four possible configurations for ejection of the two electrons along the polarization axis: (i) both along positive $\mathbf{z}$; (ii) both along negative $\mathbf{z}$; (iii) electron 1 along positive $\mathbf{z}$ and electron 2 along negative $\mathbf{z}$; (iv) the configuration (iii) with electrons 1 and 2 exchanged. (Note that (iii) and (iv) are identical in magnitude because of antisymmetry requirements.) The fact that the configurations (i) and (ii), which were not observed at low intensity because of the Coulomb repulsion, are now prominent illustrates the decreasing influence of the Coulomb interaction vis-a-vis the laser-atom interaction at higher laser intensity. It is also a signature of nonsequential double ionization. However, peaks corresponding to configurations in which electrons are emitted in opposite directions along the polarization axis are more intense.

Contrary to the one-photon case, the differences in shape and magnitude between the DDDIP for double ejection with azimuthal angles $(\phi_1, \phi_2) = (0, 0)$ and $(\phi_1, \phi_2) = (0, \pi)$ is now less significant. Major differences between the two configurations appear away from the polarization axis, with a particularly enhanced difference at $\theta_1 = \theta_2 = \pi/2$. Indeed, due to the contributions of higher order total angular momentum channels in the process, the node observed in the one photon case at $\theta_1 = \theta_2 = \pi/2$ disappears. As

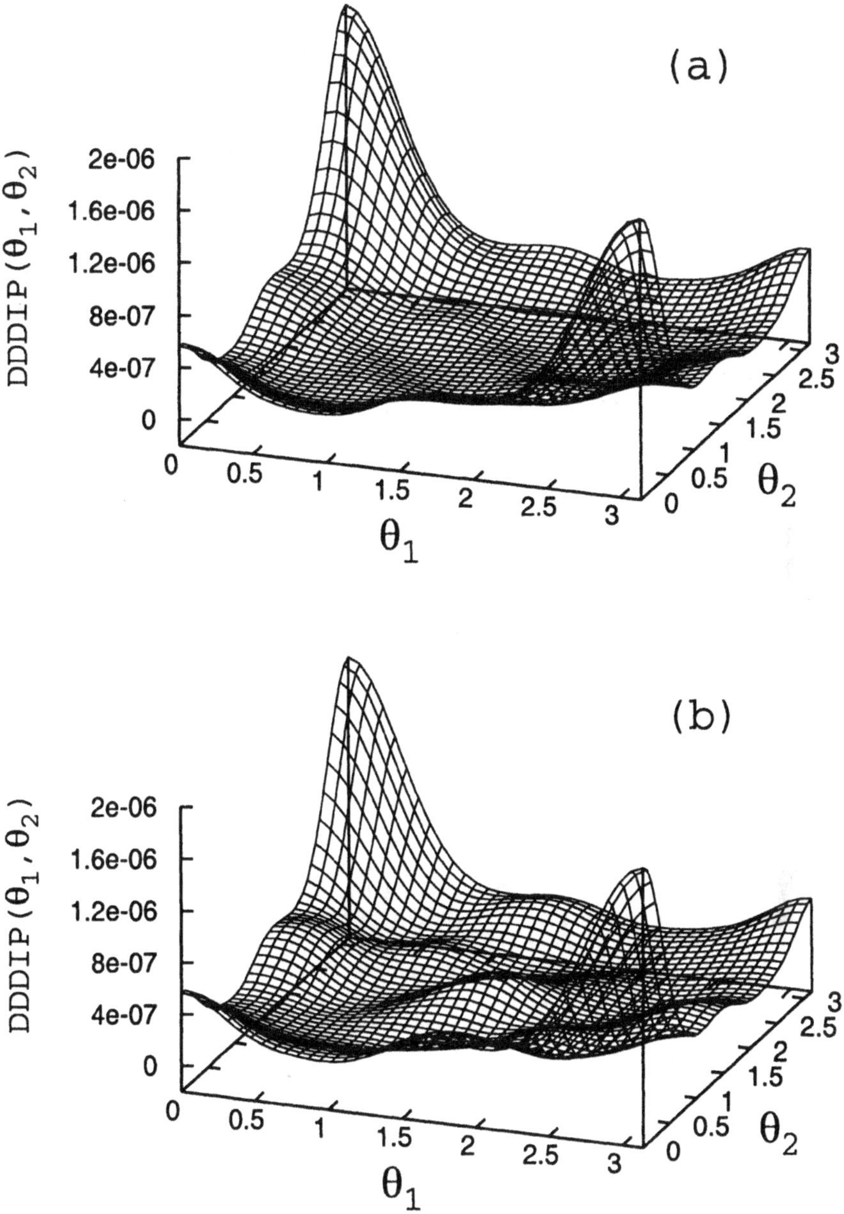

Figure 6. Multiphoton DDDIP for Li^- for: (a) $\phi_1 = 0$ and $\phi_2 = 0$; (b) $\phi_1 = 0$ and $\phi_2 = \pi$. An intense laser pulse of peak intensity 3×10^{11} W/cm^2, frequency $\omega = 0.037$ a.u., and fwhm of 9 fsec is used.

one can see from Fig. 6(b), there is a strong local maximum at these angles for the back-to-back double ejection case. In Fig. 6(a), one can barely observe a much smaller maximum for emission of both electrons in the same direction at these angles.

5. Conclusion

We have solved the three dimensional time-dependent Schrödinger equation for the Li^- ion interacting with an ultrashort, intense, linearly polarized laser pulse, using a two-active electron approximation that accounts for polarization effects of the Li^+ core. The technique used for obtaining angular distributions for double ionization by ultrashort laser pulses applies for both weak and intense pulses, and could be extended to treat any two-electron system. In the weak field regime, we recover essential features of angular distributions for the one-photon double ionization, namely selection rules and strong correlation effects between electrons. In the multiphoton regime, we find that electrons may possibly be ejected in all directions, but with peaks along directions parallel and perpendicular to the polarization axis of the field. For intensities used in the multiphoton regime, configurations where electrons are ejected in opposite directions are more likely that those for ejection in the same direction.

Acknowledgments

We thank Chien-Nan Liu and Bernard Piraux for fruitful discussions. This work was supported in part by the Dept. of Energy, Office of Basic Energy Sciences, Div. of Chemical Sciences, under Grant No. DE-FG03-96ER14646. G.L.K. acknowledges also support from the Univ. Catholique de Louvain.

References

1. J. S. Briggs and V. Schmidt, J. Phys. B **33**, R1 (2000); G. C. King and L. Avaldi, J. Phys. B **33**, R215 (2000). Phys. Rev. A **60**, 3667 (1999);
2. L. Malegat *et al.*, Phys. Rev. A **60**, 3667 (1999); A. S. Kheifets and I. Bray, J. Phys. B **31**, L447 (1998); M. Pont and R. Shakeshaft, Phys. Rev. A **51** R2676 (1995).
3. X. Tang, H. Rudolph, and P. Lambropoulos, Phys. Rev. A **44**, R6994 (1991); C. A. Nicolaides, S Dionissopoulou, and Th. Mercouris, J. Phys. B **29**, 231 (1996); J. Parker *et al.*, J. Phys. B **29**, L33 (1996); M. S. Pindzola and F. Robicheaux, Phys. Rev. A **57**, 318 (1998); A. Scrinzi and B. Piraux, Phys. Rev. A **58**, 1310 (1998); R. Hasbani, E. Cormier, and H. Bachau, J. Phys. B **33**, 2101 (2000).
4. G. Lagmago Kamta and B. Piraux, Article in this volume.
5. B. Witzel *et al.*, Phys. Rev. Lett. **85**, 2268 (2000).
6. Th. Weber *et al.*, Phys. Rev. Lett. **84**, 443 (2000); R. Moshammer *et al.*, Phys. Rev. Lett. **84**, 447 (2000).
7. F. Robicheaux and C. H. Greene, Phys. Rev. A **46**, 3821 (1992).
8. C. Pan, A. F. Starace and C. H. Greene, *Phys. Rev. A* **53**, 840 (1996).
9. W.R. Johnson *et al.*, At. Data Nucl. Data Tables **28**, 333 (1983).
10. K.T. Chung and P. Fullbright, Phys. Scr. **45**, 445 (1992).
11. A. Scrinzi and B. Piraux, Phys. Rev. A **58**, 1310 (1998).
12. G. Lagmago Kamta *et al*, in *Multiphoton Processes*, ed. by L. F. DiMauro, R, R, Freeman and K. C. Kulander (A.I.P., Melville, N.Y., 2000), p 219.
13. F. Maulbetsch and J. S. Briggs, J. Phys. B **28**, 551 (1995); *ibid* **26**, L647 (1993).
14. A. Huetz *et al.*, J. Phys. B **24**, 1917 (1991).

TWO- AND- THREE ELECTRON ATOMS IN STRONG LASER FIELDS

P. LAMBROPOULOS[1,2], *L. A.A. NIKOLOPOULOS*[2], *L.B. MADSEN*[3] *and M.G. MAKRIS*[1]
[1] *University of Crete, Physics Department, Heraklion - Crete, Greece*
[2] *Institute of Electronic Structure and Laser - FORTH, P.O. BOX 1527, Heraklion-Crete 71 110, Greece*
[3] *Institute of Physics and Astronomy, Univesity of Aarhus, DK-8000 Aarhus C, Denmark*

1. Introduction

Although single-photon double ionization of Helium is well understood theoretically as well as experimentally [18, 14], much remains to be done in the multiphoton regime, especially when it comes to ab initio approaches. Even two-photon double ionization within perturbation theory represents practically unchartered territory, with formidable difficulties. We review here recent progress formulated in terms of discretized bases constructed as linear combinations of B-splines. After a brief outline of the main aspects of the method, we present a number of illustrative results including an attempt at solving the time-dependent Schrodinger equation. Our work on He is complementary to that of Taylor and collaborators [19, 6]. Finally, we discuss the application of the same techniques to the study of the strong coupling between triply excited hollow states in Lithium.

2. Method

2.1. MULTICHANNEL TWO-ELECTRON STATES

In low-energy scattering theory, the final state wavefunction of an $N+1$ electron atom, with angular quantum numbers $\Lambda = (L, S, M_L, M_S)$, is

B. Piraux and K. Rzążewski (eds.), Super-Intense Laser-Atom Physics, 153–168.

expanded as (close-coupling expansion) [24]:

$$\Psi^{\Lambda} = A_{\Lambda} \sum_i \!\!\!\!\!\!\!\int \phi_i(\mathbf{r}_1, ..\mathbf{r}_N, \hat{r}_{N+1}) F_i(r_{N+1}) + \sum_i c_i \chi_i(\mathbf{r}_1, ..\mathbf{r}_{N+1}), \qquad (1)$$

where $\mathbf{r}_i$ includes both space and spin coordinates of the i-th electron, while the operator A_{Λ} denotes the necessary antisymmetrization, as well as the relevant angular momentum coupling between the individual angular momentum of each electron resulting in the set of quantum numbers denoted by Λ. The functions ϕ_i, known as target states, are eigenstates of the residual system after one electron has escaped. The index i, in principle, runs over both the bound and continuum eigenstates of the total spectrum of the residual system. As long as these eigenstates are computed in terms of basis functions in a finite radial space, the resulting continuous spectrum is discrete and therefore the index i reduces to a finite set of integers which represent both the negative (bound) and positive (continuous) energy target eigenstates. The χ_i functions are known as correlation functions, which allow for a better computation of the short-range electron-electron interactions. Note that these functions are solutions of the total system, core plus outgoing electron.

Specifically, for the case of two-electron atoms, we derive the matrix equations resulting from the two-electron Hamiltonian using basis states of the type:

$$\Psi^{\Lambda}(\mathbf{r}_1, \mathbf{r}_2) = \sum_{nll',n'}^{N_c,N_s} c_{nll'n'} \Phi^{\Lambda}_{nln'l'}(\mathbf{r}_1, \mathbf{r}_2) + \sum_i^{N_b} c_i \chi_i^{\Lambda}(\mathbf{r}_1, \mathbf{r}_2) \qquad (2)$$

where the basis channel wavefunctions $\Phi^{\Lambda}_{nln'l'}$ are LS-coupled antisymmetrized products of the one-electron target radial functions $P_{nl}(r), n = 1, 2, ..N_c$, constructed in a spherical box of radius R with vanishing asymptotic conditions at the boundaries and the B-spline functions $B_{n'}(r), n' = 1, 2, ...N_s$ of order k, defined in the region $[0, R]$ [5]:

$$\Phi^{\Lambda}_{nln'l'}(\mathbf{r}_1, \mathbf{r}_2) = A_{12} \frac{P_{nl}(r_1)}{r_1} \frac{B_{n'}(r_2)}{r_2} Y_{LM_L}(\hat{r}_1, \hat{r}_2; l, l') Y_{SM_S}(s, s'), \qquad (3)$$

where A_{12} is the two-electron antisymmetrization operator. Classification of the states through the index n, denotes the discretization of the spectrum of the eigenstates of the core ϕ_i and should be understood as characterizing both, negative energy states (bound) when $\epsilon_n \to \epsilon < 0$ and positive states (continuum) when $\epsilon_n \to \epsilon > 0$. The two-variable angular function Y_{LM_L} is just the projection of the total angular momentum state $|LM_L\rangle$ onto the independent electron states $|lm, l'm'\rangle$ [17].

The field-free Hamiltonian H for a two-valence-electron atom is of the form:

$$H_0 = \sum_{j=1}^{2}\left[-\frac{1}{2}\nabla_j^2 + V_0(\mathbf{r}_j)\right] + \frac{1}{|\mathbf{r}_1 - \mathbf{r}_2|}, \tag{4}$$

where $V_0(\mathbf{r}_j)$ represents an effective potential for the jth electron and the core (nuclear charge and inner closed shell electrons). Expanding now the Hamiltonian on the basis (3), the variational method transforms the coupled-channel Schrödinger equations of the system to a linear algebraic equation problem [1, 13], with coefficients depending on the energy of the one-electron core states and interchannel coupling potentials [17]. For an energy of the whole system E, such that N_o channels are open ($\epsilon_n \leq E, n = 1, 2, .., N_0$), for each channel i, N_o linearly independent solution vectors result. Physically, this reflects the N_o different asymptotic boundary conditions, corresponding to each channel. The discretized channel vectors, we obtain in this procedure, are not normalized, satisfying thus arbitrary boundary conditions. Recalling the correct boundary conditions, which continuum channels resulting from a photoionization process should satisfy, allows us to normalize the discrete states of positive energy in order to properly represent the physical states of the system. Then dipole matrix elements can be calculated properly.

2.2. NORMALIZATION OF DIPOLE MATRIX ELEMENTS

In this section, we present the method we use for computing the two-electron dipole matrix element between eigenstates of the system. Assume that the system, in its initial state $|\Psi_{in}\rangle$, undergoes a transition through photoabsorption to the final state $|\Psi_f\rangle$. The strength of this transition is given by the dipole matrix element $D_{if} = \langle\Psi_{in}|D|\Psi_f\rangle$, with D the dipole transition operator, either in the velocity $-(1/c)\hat{e}(\mathbf{p_1} + \mathbf{p_2})$ or in the length gauge $-\hat{e}(\mathbf{r}_1 + \mathbf{r}_2)$, with $\mathbf{p}_1, \mathbf{p}_2$ being the momentum operators corresponding to $\mathbf{r}_1$ and $\mathbf{r}_2$, $\hat{e}$ the polarization of the external electromagnetic field and c the velocity of light. Assuming the final energy of the system to be E, with N_f channels open, the final state formally may be rewritten as, $\Psi_f = \sum_{ij} c_{ij}(E) F_{ij}(r)$ where the $F_{ij}(r)$ functions represent the radial wavefunction of the j-th solution (out of N_o) of the i-th channel.

The $F_{ij}(r)$, will be normalized by simply invoking the asymptotic conditions which should be satisfied by an outgoing channel wave. At the asymptotic region any such solution $F_{ij}(r)$ should be written as a linear combination of the asymptotic solutions of the SE, namely the energy normalized regular and irregular Coulomb functions, $F_{ij}(r) = A_{ij}P_i(r) + B_{ij}Q_i(r)$ where, $P(r) = \sqrt{2/\pi\zeta(r)} \cdot \sin\phi(r), Q(r) = \sqrt{2/\pi\zeta(r)} \cdot \cos\phi(r)$ with ζ, ϕ, r-dependent functions [2]. Matching the solutions F_{ij}, when $r \to R$, the

$N_o \times N_o$ matrices $\mathbf{A}, \mathbf{B}$, with elements A_{ij}, B_{ij} are obtained. On the other hand, the scattering asymptotic conditions of the final state demand outgoing wave in the i-th channel and incoming wave in all other channels [21]:

$$F_{ij}^{-} \overset{r\to\infty}{\longrightarrow} \frac{1}{r}\sqrt{\frac{1}{2\pi k_j}}\left(\delta_{ij}e^{i\theta_j(r)} - S_{ij}^{\dagger}e^{-i\theta_j(r)}\right), \tag{5}$$

where $S_{ij}^{\dagger}$ are the adjoint matrix elements of the S-matrix between the channels i, j and and θ_j is the total phase of the one-electron wavefunction in a central field:

$$\theta_j(r) = k_j r - \frac{1}{2}l_j\pi - \frac{z_{eff}}{k_j}\ln(2k_j r) + \arg\left[\Gamma(l_j + 1 - i\frac{z_{eff}}{k_j})\right], \tag{6}$$

The normalized solution vector $\mathbf{F}^{-}$ is therefore calculated in terms of the unnormalized solutions $\mathbf{F}$ and the obtained matrices $\mathbf{A}, \mathbf{B}$ as $\mathbf{F}^{-} = \mathbf{F}\cdot\mathbf{N}$, where the transformation matrix $\mathbf{N}$ in terms of $\mathbf{A}, \mathbf{B}$ is given by $\mathbf{N} = (\mathbf{A} + i\mathbf{B})^{-1}$. The relevant transformation for the dipole transition reads:

$$D^{-}(j) = \langle\Psi_{in}|D|\Psi_f^{-}(j)\rangle = \sum_i D(i)N_{ij}, \tag{7}$$

with $D(i) \equiv \langle\Psi_{in}|D|\Psi_f(i)\rangle$ the dipole matrix element between the initial state and the unnormalized i-th solution. The present discussion refers to transition matrix elements where the initial state belongs to the bound spectrum while the final to the continuus one of the atom. In a similar way we normalize free-free dipole matrix elements, the case where both states (initial and final) belongs to the continuus spectrum [17].

3. Lowest-order perturbation theory

The N-photon generalized cross section is written as [9, 17]:

$$\hat{\sigma}_N(E) = 2\pi(2\,\pi\,\alpha)^N\omega^N \sum_{\mu}^{\epsilon_n} |D^{(N)}(\mu)|^2, \tag{8}$$

We have assumed the system in energy ($E = E_g + N\hbar\omega$), with the core state characterized by the quantum numbers ϵ, l and the outgoing electron in state $|\mathbf{k}', m_s'\rangle$ with kinetic energy $\epsilon' = k'^2/2$. In addition, within the discretization method we assume that $\epsilon \to \epsilon_n$ even for the continuum states of the core in order to include the case of the doubly ionized states. The upper energy limit of ϵ_n determines wether we consider single ($\epsilon_n < 0$) or double ejection ($0 < \epsilon_n < E$). Finally, we have compressed the sum over Λ, ϵ, l, l' to the channel index $|\mu\rangle$.

Here $D^{(N)}(\mu)$ is the transition amplitude of an N-photon process from the initial state $|g\rangle$ the final channel state $|\Lambda;\epsilon ll'\rangle$, through the intermediate channel states $|\nu_i\rangle\, i = 1, 2, ..., N-1$,

$$D^{(N)}(\mu) \equiv \sum\!\!\!\!\!\!\int_{\nu_{N-1}} \cdots \sum\!\!\!\!\!\!\int_{\nu_1} \frac{\langle\mu|D|\nu_{N-1}\rangle \cdots \langle\nu_1|D|g\rangle}{\Delta\omega(\nu)}, \tag{9}$$

From the above expression, it is obvious that the quantities to be calculated are the dipole matrix elements $\langle\nu|D|\nu'\rangle$, where ν, ν' could represent bound or continuum two-electron states. Note here that in the general case various ionization stages of the atom are accessible. The final state may correspond to single ionization of the atom, leaving the core in the ground or in an excited bound state, or to double ionization where both electrons leave the atom.

Within the discretized basis approach, there is a systematic way to distinguish between the different types of ionization. The different channels $|ELSM_LM_S;\epsilon ll'\rangle$ contributing to the final state are characterized by the quantum number of the core target state (ϵ, l) and the quantum numbers (ϵ', l'), subject to the relation $\epsilon + \epsilon' = E$, and with l and l' restricted by angular and parity addition rules so as to result to a state with L, S angular and spin momenta. Since we construct the target (one-electron) states $|\epsilon l\rangle$ as solutions of the He^+ $(z = 2)$ Hamiltonian, the resulting discrete eigenergies ϵ are negative and positive. The negative ones correspond to the bound spectrum while the positive ones to the continuous spectrum. We separate single from double ejection final states as follows: Final channel states with one-electron core states ϵ negative $(\epsilon < 0)$ contribute to single ejection, while all of the rest $(\epsilon > 0)$ contribute to double ejection. With the above in mind, the procedure for calculating the above formula is the following: We first construct the bound states of the system using fixed boundary conditions, namely the two electron wavefunctions assumed confined in a spherical box of radius $R = 40$ $a.u.$ with nodes at the two edges $\psi(0,0) = \psi(0,R) = \psi(R,R) = 0$. For the ground state energy, we obtain $E_g = -78.467$ eV to be compared with the exact value $E = 79.002$ eV calculated by Pekeris. Next, the multichannel continuum states are calculated with no assumed boundary conditions. For this we allow only channels with $l, l' < 4$ and energies corresponding to the bound and continuum spectrum of He+. Representing the final states in this way, one and two-photon multichannel single ionization cross sections can be calculated, as well as double ionization cross sections.

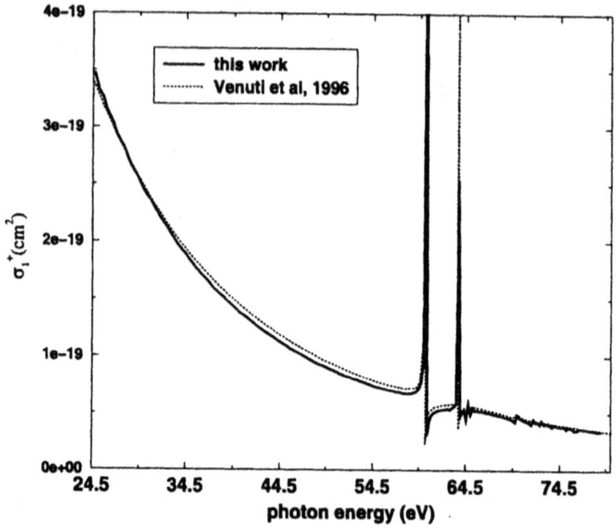

Figure 1. Comparison between present results and theoretical calculations by Venuti et al below the N=2 threshold for He single-photon ionization.

3.1. ONE-PHOTON IONIZATION OF HELIUM

In order to test the accuracy of the multichannel wavefunctions obtained with the basis we have chosen, we first compare the one-photon ionization cross section from the Helium ground state with the benchmark calculations by Venuti et al [22] (see Fig. 1). The energy range has been chosen such that only the $|2s2p\ ^1P\rangle, |2s3p\ ^1P\rangle$ resonant states are resolved. The agreement is in general quite good for the whole photon energy range. In principle, this agreement can be improved, if we enlarge appropriately the basis parameters we have chosen.

3.2. TWO-PHOTON IONIZATION OF HELIUM

From equation (8) it follows that, for the calculation of a two-photon ionization (single or double) cross section from an initial state $|g\rangle$ to a final state of the atom $|\Psi_f\rangle$ of energy E_f, summation over the final channels of the two-photon transition amplitude should be performed in addition to that over intermediate states. Those final state channels $|\mu\rangle$, which contribute to the total wavefunction $|\Psi_f\rangle$ of the system, are needed as expressed in equation (10) below,

$$D^{(2)}(\mu) \equiv \sum_{\nu} \frac{\langle\mu|D|\nu\rangle\langle\nu|D|g\rangle}{\omega_g + \omega - \omega_\nu}, \tag{10}$$

We consider now as initial state, the ground state of Helium $|g\rangle = |1s^2\,{}^1S\rangle$ and photoionization with linearly polarized light. Adopting the dipole approximation, the selection rules for transitions leave the atom in final states where two total angular momenta contribute, namely $L = 0$ and $L = 2$. The whole manifold of states with total angular momenta $L = 1$ serve as intermediate states. Therefore in equation (10) channels $|\mu\rangle$ are written more specifically as $|EL = 0, S = 0, M_L = 0, M_S = 0; \epsilon ll'\rangle$ and $|EL = 2, S = 0, M_L = 0, M_S = 0; \epsilon ll'\rangle$ while the intermediate states $|\nu\rangle$ are two-electron states with $L = 1$. The total number of channels included in the calculation for the S, D continua was 39, with the number of them which corresponds to single electron ejection being 9 for the S and 13 for the D continuum.

A further complication arises, however, due to the large photon energy needed to reach the double ionization threshold. The double ionization threshold for Helium is about $\sim$ 79 eV above the ground state which means that photons of energies of at least $\sim 40eV$ must be considered. On the other hand, the first ionization threshold for He is about $\sim 24.4eV$. Therefore, in performing the integral for the calculation of the two-photon absorption cross section above the double ejection threshold, a pole arises somewhere in the continuum, depending on the exact value of the photon energy. Methods for handling this problem exist [4, 16].

In figure (2) we present two-photon ionization cross sections of Helium for photon energy between $\omega \sim (42.5-46.5)$ eV. This is the energy range for which we have converged results for the moment. For photons of energies $\omega \sim (39-41)$ eV which raise the atom close to the double ejection threshold, reliable calculations demand a sufficient number of He^+ continuum core states with energy less than 1 eV. For $R = 40$ a.u. with the present basis this is not the case. From this figure, it is evident that the He^{++} cross section is about 10% of the He^+ cross section of a two-photon absorption. In an experiment, however, He^+ ions (in the ion ground state) are also produced by single-photon ionization. It is expected that this ionization path, for low peak intensities, will give the dominant contribution to the He^+ signal. As an example, consider photons of energy $\omega = 45$ eV. Then the single photon He^+ ionization cross section is $\sigma_1(He^+) \sim 1.1 \times 10^{-19}$ cm^2 while for the two-photon He^+ ionization cross section we obtain $\sigma_2(He^+) \sim 1.2 \times 10^{-53}$ cm^{4}s.

Earlier works of a semiquantitative nature on two-photon double ejection of Helium are those by Crance and Aymar [3], obtained some years ago, and more recently those of Kornberg and Lambropoulos [8], with which we find reasonable agreement, considering the difference in the approaches.

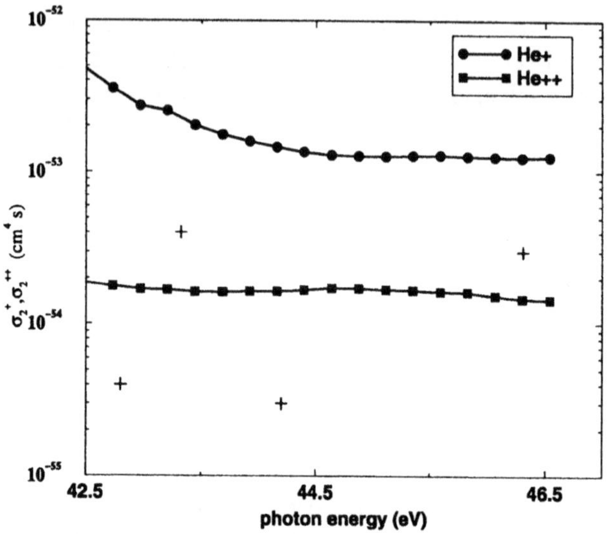

Figure 2. Two-photon single and double ionization cross sections for Helium.

4. Time - dependent Schröndiger equation (TDSE).

The time-dependent wavefunction of the two-electron atom (or ion) is expanded in terms of the field-free multichannel states

$$\Psi(t) = \sum_i (\sum_n C_n^i(t)\psi_{E_n}^i + \int C_E^i(t)\psi_E^i dE), \tag{11}$$

where i indicates a channel associated with a certain ionic eigenstate Φ_i named i. This ionic eigenstate can be either a bound or a continuous eigenstate of the ion.

The states $\psi_{E_n}^i$ are solutions of the field-free three-body Hamiltonian, namely H_0, $i\frac{\partial}{\partial t}\Psi(t) = [H_0 + D(t)]\,\Psi(t)$, where $D(t)$ is the field - atom interaction operator. The time-independent three-body Schrödinger equation cannot be solved exactly, hence approximate solutions will be used in the calculations. The procedure for obtaining these solutions is described in detail elsewhere [13, 17]. The states $\psi_{E_n}^i$ represent the bound states of the total system with energies E_n, while the states ψ_E^i correspond to the continuum states of the total system. In the case where the ionic state i is a bound state of the ion, the state ψ_E^i has an electron in the continuum with kinetic energy $k_i^2 = E - E_i$, where E_i is the ionization threshold of channel i. Similarly, when the state Φ_i belongs to the continuum spectrum of the ion, the state ψ_E^i represents a state where both electrons are in the continuum. In this case the two electrons share the total energy according to

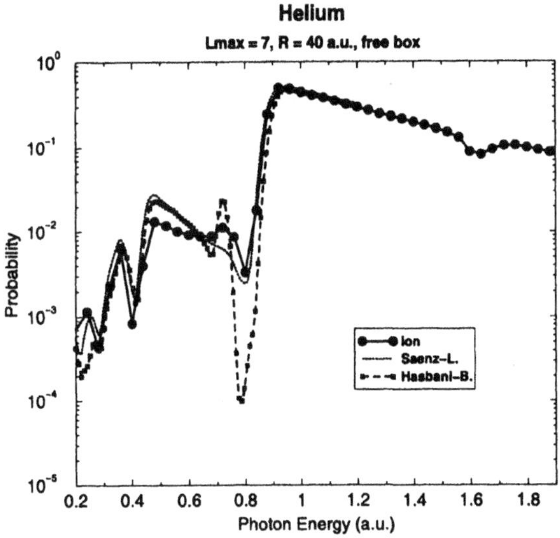

Figure 3. Ionization yield as function of photon frequency for constant peak intensity.

$\epsilon_i + \epsilon_i' = E$, where ϵ, ϵ' are the kinetic energies of the electrons. The positive energy solutions of the Hamiltonian H_0, however, provided that more than one channels are open, are not unique since they can satisfy a number of alternative boundary conditions, due to the various final ionic states which are accesible. In order to determine them uniquely, we must pose the proper conditions conforming to the physical process under consideration. In the case of photoionization, the proper asymptotic behaviour of the final state, in the absence of the laser field or in the perturbative regime, (known also as incoming-wave boundary conditions) is given by [21],

$$\psi_E^{i-} \overset{r\to\infty}{\longrightarrow} \sum_j \phi_j \sqrt{\frac{1}{2\pi k_j}} \frac{1}{r} (\mathrm{e}^{\mathrm{i}\theta_j} \delta_{ij} - S_{ij}^{\dagger} \mathrm{e}^{-\mathrm{i}\theta_j}), \tag{12}$$

where $S_{ij}^{\dagger}$ are the adjoint matrix elements of the S-matrix between the channels i, j determining their interaction and ϕ_i representing the corresponding ionic core and the angular factors and θ_{k_i} is the total phase of the one-electron wavefunction in a central field. The time-dependent Schrödinger equation when using the multichannel expansion (11), is rewritten in terms of the unknown coefficients $C_E^i(t)$:

$$i\frac{d}{dt}C_E^i(t) = C_{E_n}^i(t)E_n + \sum_{i', E_n'} \langle \psi_{E'}^{i'} | D(t) | \psi_{E'}^{i'} \rangle C_{E'}^{i'}(t) \tag{13}$$

Integration of the above system of differential equations, subject to the proper initial conditions provides the coefficients at the end of the pulse,

$t \to \infty$. Using as multichannel continuum states in (11) the states ψ_E^{i-},

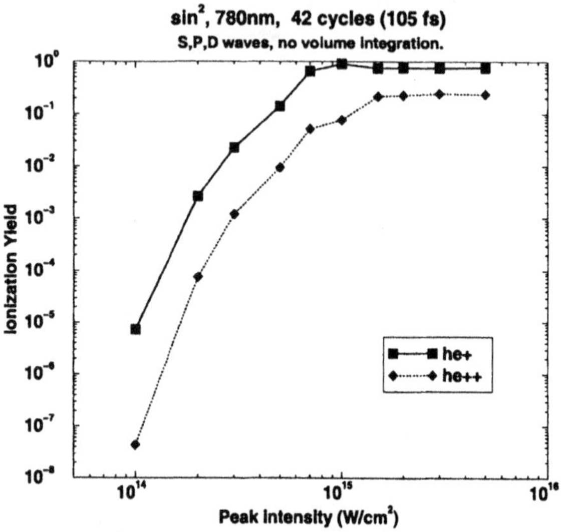

Figure 4. Ionization yield as a function of the peak intensity for wavelength of 780nm with an 105 fs pulse duration.

we ensure that in the long time limit $t \to \infty$ (after the passage of the laser pulse) each term of $\psi(t)$ will contain only one outgoing wave (first part of (12)) associated with the ionic state i. All incoming waves (second part of (12)) will give zero contribution in this limit. In other words, each state ψ_E^{i-} is a positive-energy eigenstate of the total Hamiltonian with energy $k_i^2/2$. The physical meaning of the coefficients C_E^i becomes obvious and rigorous: For finite laser pulses $E_\omega(t \to \infty) \to 0$, the quantity $|C_E^i(t \to \infty)|^2$ gives the probability of finding the ion in the state ϕ_i and electron ejected with kinetic energy $k^2/2 = E - E_i$. Needles to say that if ϕ_i is a continuum ionic state, we have the case of double ejection. The coefficients C_n^i in (11) have similar interpretation. Their square amplitude for $t \to \infty$ gives the probability of excitation of the system to the various bound states $\psi_{E_n}^i$ by the laser field. The time-dependent laser field is modelled as: $I(t) = I_0 f(t) \sin\omega t$, where, $f(t)$ denotes the evelope of the pulse (normalized to unity), and can be chosen such that to represent a sine squared ($\sim \sin^2$), quassian ($\exp(-t * t)$) or squared pulse (~ 1). In the present paper the envelope of the pulse is a sine-squared $f(t) = sin^2(\pi t/\tau)$, where $\tau = n_c(2\pi/\omega)$ is the total duration of the pulse and n_c the number of cycles within the duration of the pulse.

As a first example we have caclulated the single ionization yield of Helium as a function of the photon frequency ranging from about 5.5 eV up to 50 eV, at peak intensity $I_0 = 2.97 \times 10^{14} W/cm^2$ (see Fig. 3). In this case the total pulse duration in each of the frequencies has been kept constant equal to $\sim 3.8fs$. The agreement is quite good for the whole range of frequencies calculated by Lambropoulos et al [10] and Hasbani et al [7]. The large dip at frequency of about 21.5 eV given by Hasbani et al is due to technical details of their approach. Note that similar calculations have been done by Scrinzi and Piraux [20], not presented here. Their calculations also compare very well with ours. Our calculations, except testing our method with others results, had as main target the range of frequencies where excitation channels are open as well as double ionization of Helium is present. In the figure, a small dip around frequency of 41 eV is due to the fact that the double ionization channels open, and for the chosen intensity double ionization is increased significantly.

We have also performed calculations of single and double ionization yield as a function of intensity for photon wavelength of $780nm$ and pulse duration of $108fs$. For those calculations we kept only total S,P,D angular momenta and up to $l, l' < 4$ for individual one-electron angular momenta. This is a rather severe approximation since for this wavelength ($\sim 1.59eV$, according the lowest-order perturbation theory(LOPT), about 15 photons, at least, are needed for the He^+ production, while for He^{++} production about 50 photons are needed. This means that the minimum number of total angular momenta demanded even within LOPT is L = 50. Note that the related experiment reported by Walker et al [23] uses the same wavelength, with a pulse duration of 160 fs. Given the drastic approximation of limiting the number of angular momenta to only three, we can not expect quantitative agreement with experiment. On the other hand, one-dimensional models contain no angular momenta. The purpose of the exercise therefore was to see how far such a severely limited three-dimentional model with correlation can go in producing some of the essential features, especially since we have been able to use a fairly long pulse duration. As seen in the figure, the difference with the experiment is quantitative and not qualitative. A shallow knee structure does appear at the right intensity and the ratio of the single to double ionization signal differs by less than one order of magnitude from the experimental. Further and longer calculations will clearly be necessary for a quantitative assessment of the method in the long-wavelength limit.

5. Angular distribution of photoelectrons in two-photon double ejection.

We denote the state of the system with both electrons ejected as, $|\mathbf{k}, m_s; \mathbf{k}', m_s'\rangle$. The LS-representation of the double-continuum state is written as [17]:

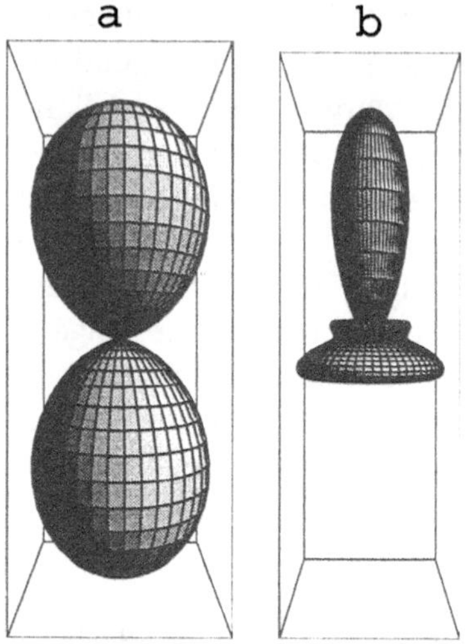

Figure 5. Relative angular distribution of the ejected electrons in a two-photon double ionization process for kinetic energy of the two electrons 3.54ev and 7.66eV, respectively

$$|\mathbf{k}, m_s; \mathbf{k}', m_s'\rangle = \sum_{\Lambda lml'm'} i^{l+l'} e^{-i\delta_l - i\delta_{l'}} Y^{\star}_{lm}(\hat{k}) Y^{\star}_{l'm'}(\hat{k}') C_{LM_L} C_{SM_S} |\Lambda; \epsilon ll'\rangle, \tag{14}$$

where the energy E of the state is the sum of the kinetic energies of the ejected electron, $E = \epsilon + \epsilon' = k^2/2 + k'^2/2$. The channel functions $|\Lambda; \epsilon ll'\rangle$ should be understood as, $|\mu\rangle = |E\Lambda; \epsilon ll'\rangle \equiv |ELSM_LM_S; \epsilon ll'\rangle$. The sum over Λ denotes sum over the angular quantum numbers LM_LSM_S. The Glebsch-Gordan coefficients C_{LM}, C_{SM_S} project the total orbital and spin angular momenta $|LM\rangle, |SM_S\rangle$ onto the independent electron states $|lmm_s; l'm'm_s'\rangle$. The N-photon cross section is written now as,

$$M^{(N)}(\hat{k}; \hat{k}') = \sum_{\Lambda lml'm'} i^{l+l'} e^{-i\delta_l - i\delta_l'} C_{LM_L} C_{SM_S} Y^{\star}_{lm}(\hat{k})\, Y^{\star}_{l'm'}(\hat{k}') D^{(N)}(\mu) \tag{15}$$

, with $D^{(N)}(\mu)$ given by equation (9

The triply differential cross section is defined by the following formula:

$$\frac{d\hat{\sigma}_N^{++}(E)}{d\epsilon d\Omega d\Omega_{k'}} = 2\pi(2\pi\alpha)^N \omega^N \left| M^{(N)}(\hat{k}, \hat{k}')\right|^2, \tag{16}$$

In the case of two-photon absoprtion ($N = 2$) from the ground state of Helium $1s^2(^1S)$, only the S, D (total orbital angular momentum) final waves are accesible. Their projection is zero since we start from the ground state of Helium and we assume linearly polarized light. The total spin is also zero. The one-electron angular momenta l, l' included in the calculation are s,p,d,f waves. In the figure (5) we have plotted angular distributions of the ejected electrons for photon energy of 45 eV, giving thus a total energy of 11.2 eV to be shared between the electrons. In this figure the energy of electron '1' is 3.54 eV while for the other is $(11.2 - 3.54)eV = 7.66$eV. The set up of the figure is the following: We have fixed the angles $(\theta_1 = 0, \phi_1 = 0)$ for the electron '1', as well as the azimuthal angle for the electron '2', namely $\phi_2 = 0$, and then we plot the probability for the first electron to be ejected in angle θ_2. Since $\theta_1 = 0$ this is the relative angle between the electrons. In the figure, we have plotted two different cases. The plot (a) is calculated when we have kept in the final state only the partial one-electron *pp* waves, or in other words $l = l' = 1$. This situation correspond to assuming that the two electrons ionize independently of each other with higher probability to leave in the same direction $\delta\theta = \theta_2 = 0$, or the opposite one $\delta\theta = \theta_2 = \pi$. It is also a symmetric distribution around the polar axis $\theta_2 = 0$. The relative angular distribution, however, changes when other angular momenta are allowed as it should happen when electrons are able to interact each other during ionization (see plot (b)). In this case a clear non-symmetric shape appears and electrons have higher probability for leaving the core with the same direction with $\delta\theta = \theta_2 = 0$ instead of to leaving in opposite directions. This is more apparent when the two electrons have comparable energies. This influence, of the relative energy sharing between the two electrons, to the relative angular distributions will not be shown here, but in a forthcoming work [15].

6. Triply excitation of Lithium

The approach and technique outlined in the previous sections have been used very recently in the study of laser-induced coupling between triplet excited hollow states of Lithium. The term refers to multiple electron excitations, the lowest of which is the $2s^22p(^2P^o)$ whereby both electrons of the innermost shell $(1s^2)$ have been promoted to 2s and 2p. Higher enery states are $2s2p^2$ and $3s^2p$ and so on. These are highly correlated configurations which pose considerable challenge in their theoretical description. As far as structure is concerned, much effort has been devoted to the calculation of their energies and widths. Experimentally, odd parity, J = 1 states, which can be reached via single photon-absorption, have been studied using synchrotron radiation. In perharps the only exception to this so far, an even

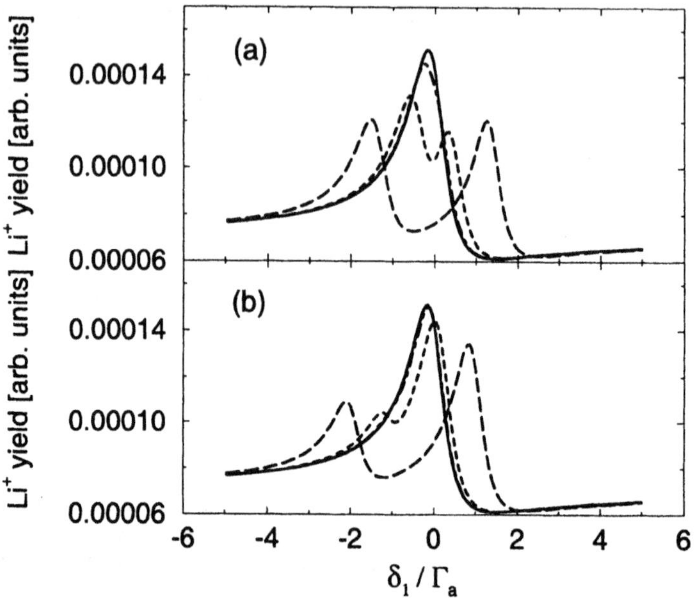

Figure 6. Li^+ photoion yield as a function of the detuning associated with the transition between the ground state and the $2s^2 2p(^2P^o)$ triply excited state. The laser coupling the $2s^2 2p(^2P^o)$ $2s2p^2(^2D^e)$ is (a) on resonance with the transition and (b) detuned off resonance by the width of $2s^2 2p(^2P^o)$. The full, dot-dashed, dashed, and long-dashed curves correspond to $I_2 = 1.4 \times 10^7, 1.4 \times 10^{10}, 1.4 \times 10^{11}$ and $1.4 \times 10^{12} W/cm^2$, respectively. The intensity of the 10ps synchrotron pulse is $I_1 = 4 \times 10^8 W/cm^2$. We clearly see that the laser coupling induces an ac splitting of the line, the peak separation of which equals twice the rabi frequency.

parity state has been reached with synchrotron radiation from the bound singly excited state $1s^2 2p$, which was excited by a laser. With improving synchrotron sources, as well as the upcoming short-wavelentgh light sources based on accelerator driven free electron lasers, the relevant studies are continuing and expected to expand in new directions, such as for example, the direct electromagentic coupling between such hollow states. It is some new results [11, 12] along those lines that we present in this section.

Assume that the hollow state $2s^2 2p(^2P^o)$ has been excited by synchrotron radiation of photon energy $\hbar\omega_1 = 142.3$ eV, in the presence of a laser of frequency ω_2 and pulse duration longer than that of synchrotron. Both radiation sources are assumed linearly polarized along the same direction. The frequency ω_2 is to be tunable around the transition $2s^2 2p(^2P^o) \rightarrow 2s2p^2(^2D^e)$ which corresponds to about 2.5 eV. Note that the laser would not excite the bound states $1s^2 2p(^2P^o)$ and $1s^2 3p(^2P^o)$ of the atom as they lie about 1.85 eV and 3.8 eV, respectively, from the ground state, being thus far from resonance. The transition induced by the relatively weak radiation at ω_1 can be described by a single rate, but the transition to the higher hollow state must be treated non-perturbatively, as the objecture is to explore the effects of this coupling to the autoionization line-shape of the lower resonance, thereby exctracting information about the strength of the coupling between the two hollow states. The coupling represents a parameter very sensitive to the accuracy of the wavefunctions of the highly correlated states, which are coupled to several continua. One representative result is shown in figure (Fig. 6) depicting the change (splitting) of the line shape for various values of the intensity of the coupling laser, as the weak ω_1 is tuned around the resonance. From the value of the splitting between the two peaks, one can essentially read off the effective Rabi frequency of the transition between the hollow states thereby extracting the coupling strength; i.e the dipole matrix element. The uselfuness of this information relies on having a reliable calculation to compare with, testing thus the validity of the theoretical description from which further understanding of the underlying physics can be pursued.

Space does not allow elaboration of the many aspects of the theory and details of the calculation. We will have to refer the reader to the relevant papers limiting ourshelves here to a few pertinent comments. The structure part of the calculation was implemented in terms of B-splines in a fixed box. The widths of the resonances were calculated through complex scaling, as it was found more expedient. Although, as mentioned above, a number of accurate approaches to the structures of these states exist, we needed to implement one that lends itself readily to the time-dependent formulation necessary for the processes under consideration. It was, nevertheless improtant to ensure that the accuracy of the structure was at the level of the best available. This problem has anumber of further dimensions not discussed here. For example, hollow states at energies above about 147 eV can decay via single or double electron ejection. The treatment and quantitative understanding of such branching ratios, and their manipulation through judiciously chosen couplings by a laser provide further and deeper insight into the physics of these four-body systems.

References

1. Tomas Brage, Charlotte Froese Fischer, and Gregory Miecznick. Non-variatonal, spline-Galerkin calculations of resonance positions widths and photodetachment cross sections for H^- and He. *J. Phys. B*, 25:5289, 1992.
2. A. Burgess. The deternination of phases and amplitudes of wave functions. *Proc. Phys. Soc.*, 81:442, 1963.
3. M. Crance and M. Aymar. Two-photon detachment of H^-. A model calculation. *J. Phys. B*, 18:3529, 1985.
4. E. Cormier and P. Lambropoulos. Extrapolation method for the evaluation of above threshold ionization cross sections for the one- and two-electron systems. *J. Phys. B*, 28:5043–5055, 1995.

5. Carl de Boor. *A Practical Guide to Splines.* Springer - Verlag, New York, 1978.
6. Daniel Dundas, K. T. Taylor, Jonathan S. Parker, and Edward S. Smyth. Double-ionization dynamics of laser-driven helium. *J. Phys. B*, 32:L231, 1999.
7. R. Hasbani, E. Cormier, and H. Bachau. Resonant and non-resonant ionization of helium by xuv ultrashort and intense laser pulses. *J. Phys. B*, 33:2101, 2000.
8. M. A. Kornberg and P. Lambropoulos. Photoelectron energy spectrum in 'direct' two-photon double ionization of helium. *J. Phys. B*, 32:L603–613, 1999.
9. P. Lambropoulos. Topics on multiphoton processes. *Adv. At. Mol. Phys.*, 12:67, 1976.
10. P. Lambropoulos, M. A. Kornberg, L. A.A. Nikolopoulos, and A. Saenz. Ab initio approaches to multiphoton processes in two electron atoms. In Louis F. DiMauro, Richard R. Freeeman, and Kenneth C. Kulander, editors, *Multiphoton Processes*, volume 525 of *AIP conference proceidings*, page 231, Monterey, California, 3-8 October 1999. American Institute of Physics.
11. P. Schlagheck L.B. Madsen and P. Lambropoulos. Laser-induced transitions between triply excited states. *Phys. Rev. Lett.*, 85:42–45, 2000.
12. P. Schlagheck L.B. Madsen and P. Lambropoulos. Laser-induced transitions between triply excited states. *Phys. Rev. A*, in press, 2000.
13. P. Lambropoulos, P. Maragakis, and Jian Zhang. Two-electron atoms in strong fields. *Physics Report*, 305:203, 1998.
14. Kurt W. Meyer, Chris H. Greene, and Brett D. Esry. Two-electron photoejection of he and h^-. *Phys. Rev. Lett.*, 78:4902, 1997..
15. M. G. Makris, L. A. A. Nikolopoulos, and P. Lambropoulos. Electron angular distributions in two-photon double ionization of helium. *Europhys. Lett.*, submitted, 2000.
16. L. A. A. Nikolopoulos and P. Lambropoulos. Above-threshold Ionization of negative hydrogen. *Phys. Rev. A*, 56:3106–3115, 1997.
17. L. A. A. Nikolopoulos and P. Lambropoulos. Multichannel theory of two-photon single and double ionization of helium. *J. Phys. B*, submited, 2000.
18. M Pont and R Shakeshaft. Absolute cross sections for double photoionization of helium at energies from 0 to 80 ev above threshold. *J. Phys. B*, 28:L571–L577, 1995.
19. Jonathan Parker, K. T. Taylor, Charles W. Clark, and Sayoko Blodgett-Ford. Intense-field multiphoton ionization of a two-electron atom. *J. Phys. B*, 29:L33, 1996.
20. Armin Scrinzi and Bernard Piraux. Ionization and double excitation of helium by short intense laser pulses. *Phys. Rev. A*, 56:R13, 1997.
21. Antony F. Starace. *Corpuscles and Radiation in Matter I*, volume 31 of *Handbuch der Physik*, chapter Theory of Atomic Photoionization, pages 1–121. Springer-Verlag, Berlin, 1982.
22. M Venuti, P Decleva, and A Lisini. Accurate multichannel continuum states by a general configuration interaction expansion in a b-spline basis: application to he photoionization. *J. Phys. B*, 29:5315, 1996.
23. B. Walker, B. Sheehy, L. F. DiMauro, P. Agostini, K. J. Schafer, and K. C. Kulander. Precision measurement of strong field double ionization of helium. *Phys. Rev. Lett.*, 73:1227, 1994.
24. Oleg Zatsarinny and Charlotte Froese Fischer. The use of basis splines and non-orthogonal orbitals in r-matrix calculations: application to li photoionization. *J. Phys. B*, 33:313, 2000.

PART II

IONIZATION AND DISSOCIATION DYNAMICS OF MOLECULES

TUNNELLING IONISATION AND THE FRANCK–CONDON PRINCIPLE

J.H. POSTHUMUS
Sektion Physik, Universität München, Am Coulombwall 1, D-85748 Garching, Germany

AND

L.J. FRASINSKI AND K. CODLING
Physics Department, University of Reading, Reading RG6 6AF, United Kingdom

According to the Franck–Condon principle, transition probabilities between two molecular states are determined by the electronic transition probability, weighted by the overlap of the two vibrational states involved. In this paper we discuss a situation where this approximation is invalid, namely the tunnelling ionisation of H_2. The fragmentation of H_2 in intense laser fields of visible and infrared wavelengths has been studied experimentally for more than a decade [2]. It is usually assumed that the complete fragmentation occurs in three, more or less independent, steps. First there is the ionisation of H_2. Due to the small photon energy, the high ionisation potential and the absence of appropriate resonances, the ionisation takes place in the tunnelling or even the over-the-barrier regime [7]. This type of ionisation occurs at the peaks and troughs of the coherent electromagnetic wave, i.e. in short periods of less than 1 fs during which the vibrational motion is essentially frozen. Quite naturally this leads to the idea of vertical transitions centered on the equilibrium distance of H_2, see e.g. Walsh *et al.* (1998). The population distribution amongst the vibrational levels of H_2^+ is then given by the standard Franck–Condon factors [3], which are shown in figure 1. However, it would be more accurate to take the large deformation of the H_2^+ electronic ground state in the strong laser field into account. This would be expected to shift the population towards higher vibrational levels [14], but we have strong reasons to believe that the population distribution is actually shifted towards lower levels.

After ionisation, the second step is the photodissociation of H_2^+. High laser powers allow non-resonant photon absorption and a process called

B. Piraux and K. Rzążewski (eds.), Super-Intense Laser-Atom Physics, 171–178.

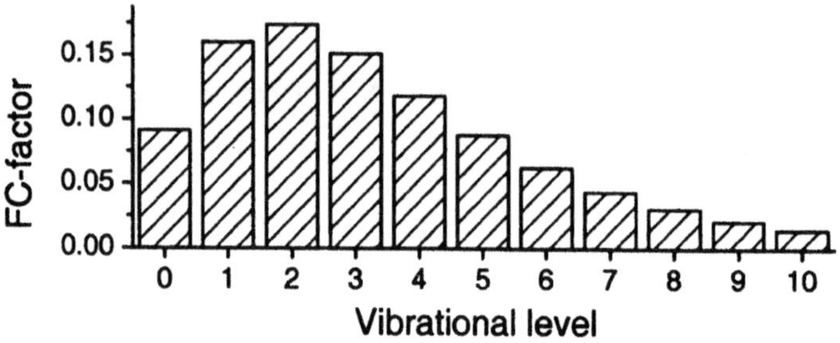

Figure 1. The standard Franck–Condon overlap between the $v = 0$ level of H_2 and the v' levels of H_2^+.

'bond-softening' (for a review of this and related processes, see e.g. Giusti-Suzor *et al.*, 1995). Naive energy conservation considerations suggest that the higher vibrational levels, $v' \geq 5$, may dissociate via one-photon absorption at $\lambda = 800$ nm, yielding low-energy protons. The lower levels of H_2^+ require at least two photons for their dissociation. The kinetic energy release of the two-photon channel, occasionally (though imprecisely) called 'above threshold' dissociation', is therefore nearly one photon energy higher than the other. Let us call these dissociation channels 1ω and 2ω.

The final step in the fragmentation of H_2 is the ionisation of the dissociating H_2^+ molecular ion, followed by Coulomb explosion of the remaining protons. The field ionisation probability of H_2^+ depends strongly on the internuclear distance. The ionisation rate is a maximum around the distance where the electron is on the brink of localising at either of the two ion cores. The effect is usually called (charge resonance) enhanced ionisation and appears to be a general phenomenon in molecules. The internuclear distance where the ionisation rate maximises is usually called 'critical distance'. More details can be found in e.g. Thompson *et al.* (1997) and Walsh *et al.* (1998).

This, then, was the situation until a number of interesting measurements called for a re-examination of the facts. Firstly, Rottke *et al.* (1999) used a supersonic H_2 beam, thus removing the Doppler-broadening from the ion kinetic energy spectra. Using ultrashort pulses from a Ti : sapphire laser, they found that the dissociation channels had very narrow ion kinetic energy distributions. The 1ω peak, for example, could accommodate only one vibrational level, namely $v' = 5$.

Using similar laser pulses at lengths varying from 45 to 540 fs, the Reading group discovered that the kinetic energy release of the 1ω dissociation channel shifted from about 0.4 eV for the shortest pulses, to about 0.0 eV for the longest pulses [4]. The interpretation of this phenomenon involves potential curves that are greatly deformed by the laser field. One finds, for example, that the dissociative potential curve develops a shallow minimum in a strong laser field that may trap a fraction of the vibrational population. On the falling edge of the laser pulse this potential curve flexes upwards to create a repulsive shape. Thus the dissociating wavepacket receives additional energy – the amount depending on the shortness of the falling edge – on top of the energy that it gained from the absorption of one photon. It may even happen that a molecule, which was initially in the $v' = 4$ vibrational state, dissociates to the one-photon limit, even though the nominal photon energy is insufficient. Only this 'dynamic Raman effect' allows the wavepacket to reach the dissociation continuum with an energy apparently corresponding to the $v' = 5$ level. It is therefore appropriate to call the entire process 'below-threshold dissociation'. The extent of the shift led us to conclude that it was primarily this process that we were observing[1]. However, this can only be correct if the population of the $v' \geq 5$ levels of H_2^+ is small relative to the population of the $v' = 4$ level. For the standard Franck–Condon distribution, see figure 1, this is clearly not the case.

A very interesting recent development has been the introduction of molecular ion beams to this field of research [18]. Two groups have used H_2^+ ion beams to study the dissociation and ionisation of this molecular ion under the influence of ultrashort, intense Ti : sapphire laser pulses [17, 13]. In both cases, the ions were extracted from an ion source for which it is quite well established that the vibrational population possesses the standard Franck–Condon distribution. The fragment kinetic energy spectra confirmed this expectation. Williams *et al.* (2000) showed that the experimental population of vibration levels followed the Franck–Condon distribution for $v' \geq 9$. (Vibrational levels below $v' = 9$ are progressively attenuated, presumably due to decreasing tunnelling rate through the lower branch of the 1-photon anticrossing.) Saending *et al.* (2000) were able to partially resolve the vibrational levels, showing interesting details of the dissociation dynamics. It was possible to identify processes starting from vibrational levels as high as $v' = 11$. These results are in contrast to the gas target experiments where virtually no population seems to be present in such high vibrational levels.

For the solution to this disparity in population distribution we have had another look at the Franck–Condon principle[2]. Using the Born-Oppenheimer approximation, we write the rate for the transition from the $v = 0$

[1] In a more recent experiment with the third harmonic of the Ti : sapphire laser, we have observed zero-photon dissociation, i.e. dissociation due to dynamic deformation only and without the absorption of a nett number of photons [8].

[2] At this conference we found that A. Saenz had developed quite similar ideas several months earlier [10].

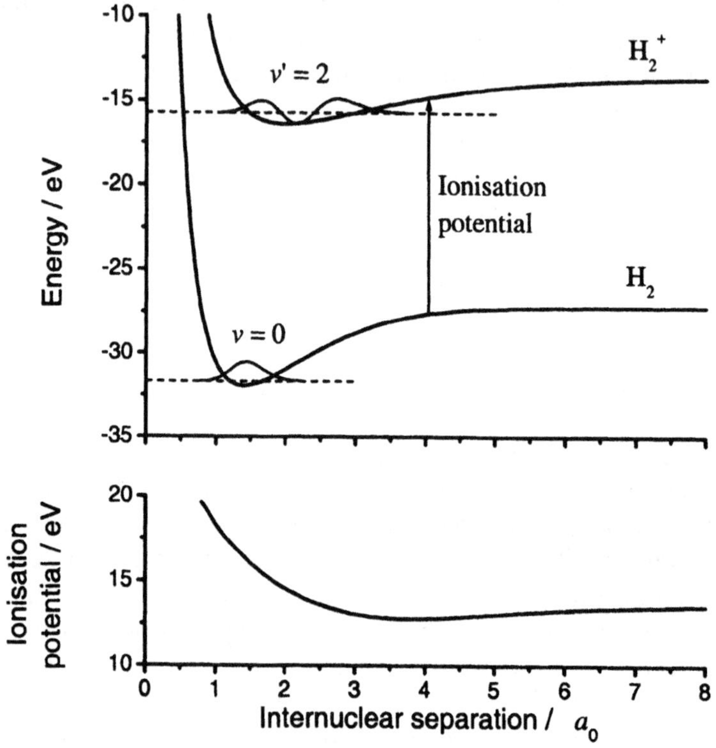

Figure 2. The unperturbed, ground-state potential curves of H_2 and H_2^+ (top) and the ionisation potential of H_2 as a function of R (bottom). Also shown are the amplitudes of the $v = 0$ and the $v' = 2$ levels.

level of H_2 to the v' level of H_2^+ as

$$\Gamma_{v'} = | < \phi_f \, \chi_{v'} | A \, | \phi_i \, \chi_{v=0} > |^2, \tag{1}$$

where ϕ_f and ϕ_i are the final and initial electronic states, $\chi_{v'}$ and $\chi_{v=0}$ are the ion and neutral vibrational states and A is the interaction operator responsible for the ionisation. When the interaction is only weakly dependent on the internuclear separation, R, the Franck–Condon approximation can be applied and (1) reduces to

$$\Gamma_{v'} = \Gamma_i \, | < \chi_{v'} | \chi_{v=0} > |^2, \tag{2}$$

which states that the rate for producing H_2^+ in a particular vibrational level is simply the ionisation rate, Γ_i, weighted by the Franck–Condon factor. However, in the case of tunnelling ionisation, the assumption of a weak

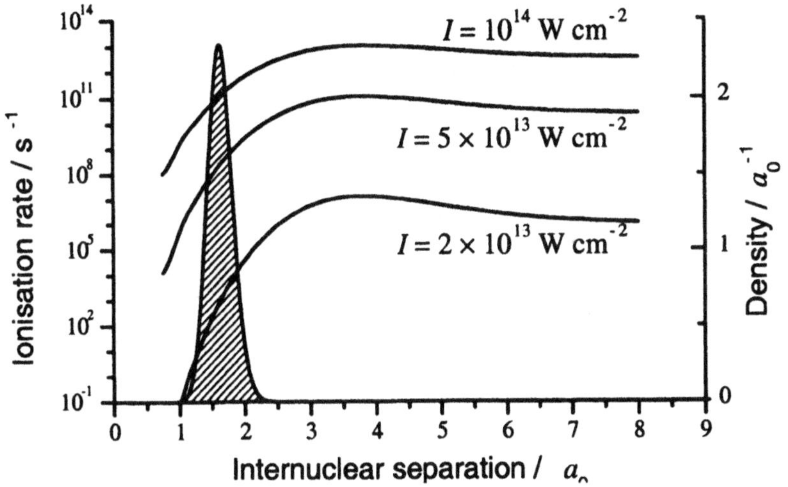

Figure 3. Tunnelling ionisation rates for H_2 as a function of R for three different laser intensities. On the same horizontal scale: the density of the $v = 0$ level of H_2.

dependence on R is totally incorrect. One is therefore left with the more general expression

$$\Gamma_{v'} = | < \chi_{v'}|A_i(R)|\chi_{v=0} > |^2, \tag{3}$$

where $A_i(R)$ is the R-dependent tunnelling ionisation amplitude. For simplicity we employ the ADK tunnelling amplitude for s-electrons [1] and the R-dependent ionisation potential of H_2 to determine $A_i(R)$. Furthermore, we tentatively assume that the ionisation potential is given by the energy difference between the unperturbed electronic ground states of the H_2 and H_2^+ molecules. The pertinent potential curves as well as the ionisation potentials are illustrated in figure 2. We note that in the small-R region the ionisation potential decreases substantially with R. The tunnelling rate, $\Gamma_i(R)$, which is very sensitive to the ionisation potential, is therefore a steep function of R in this region, as shown in figure 3 for three different laser intensities (for ionisation of H_2, see also Saenz, 2000 B and Kawata *et al.*, 2000). On the same horizontal scale is shown the density of the $v = 0$ level of H_2. Clearly, the tunnelling ionisation rate varies enormously across the extent of the H_2 ground state. In classical terms, the molecule is most likely to ionise at the outer turning point of its vibration, i.e. somewhere in the vicinity of $R = 1.8\, a_0$. Since this value is quite close to the equilibrium distance of the ion, one may expect an increased population of the low vibrational levels of H_2^+.

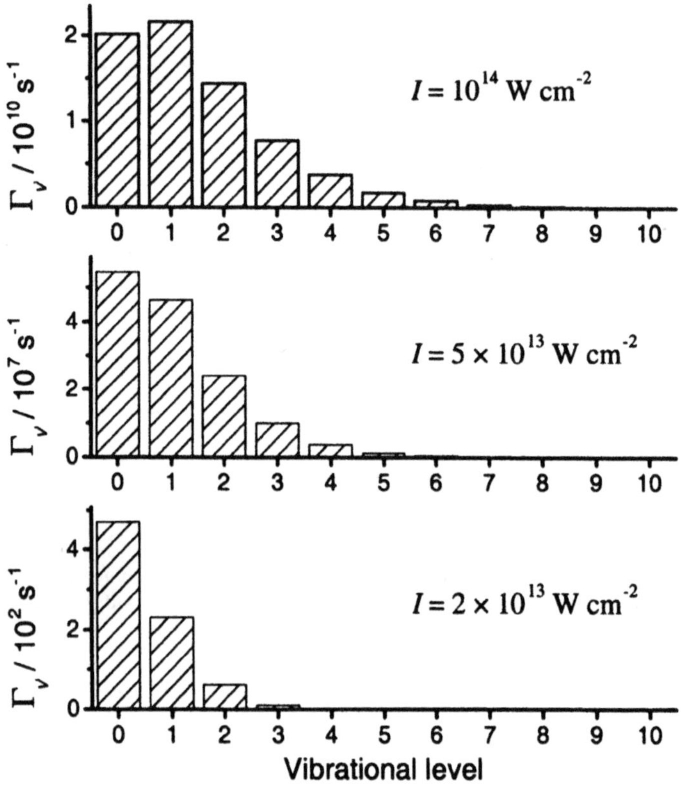

Figure 4. The tunnelling ionisiation rates to the vibrational levels of H_2^+, calculated according to equation (3), for three different laser intensities.

In figure 4 we show a few distributions of $\Gamma_{v'}$, which have been calculated according to (3). Compared to the standard Franck–Condon distribution, we note a considerable shift towards lower vibrational levels. We note further that the lower the intensity, the more likely it is that the low vibrational states are populated, which is understandable in view of the steepness of the ionisation rates, see figure 3. Since the saturation intensity of the ionisation depends the pulse length, this implies that tunnelling ionisation with relatively long pulsed lasers, such as the Nd : YAG and the CO_2 lasers, will favour the low levels even more than the ultrashort pulsed Ti : sapphire laser.

With respect to the experiment on below-threshold dissociation [4], we stated earlier that the population of the $v' \geq 5$ levels of H_2^+ has to be small relative to the population of the $v' = 4$ level. Although the situation is now far better than in the standard Franck–Condon case, the distribution at $I = 10^{14}$ W cm^{-2} still shows non-negligible contributions from $v' \geq 5$.

However, we have yet to take into account the deformation of the potential curves in the strong laser field. Calculations of the ground-state potential curve of H_2 in a static electric field [11], suggest that the $v = 0$ level might shift by 0.1 or 0.2 a_0 towards higher R in a laser field of $I = 10^{14}$ W cm^{-2}. This shift is probably enough to suppress the $v' \geq 5$ levels in H_2^+. Thus the small population of the higher vibrational levels is a combination of R-dependent tunnelling ionisation amplitudes and bond-stretching of H_2 in the intense laser field. Calculations that take both effects into account are in progress [12].

Finally, we reconsider the time-independent formalism represented by equation (1). It implies that the interaction is long compared to all other timescales. However, experimentally it has become quite realistic recently to have laser pulses which are shorter than the vibrational period of H_2. Under this condition the vibrational motion can be considered frozen. The preferential ionisation at the outer turning point leaves a reduced wavepacket behind, close to the inner turning point. This vibrational wavepacket is not an eigenstate, but begins to oscillate in the potential well. In other words, the surviving H_2 molecules will be vibrationally and coherently excited.

To summarise, we have briefly reviewed some experimental results which indicate that the tunnelling ionisation of H_2 does not produce the standard Franck– Condon distribution amongst the vibrational levels of H_2^+. Instead, the data suggest that the population of the levels $v' \geq 5$ is negligible. By taking into account the profound variation of the tunnelling amplitude with internuclear separation, we are able to reproduce a distribution close to the expected one. Improved agreement is anticipated when the bond-stretching of H_2 in the strong laser field is included.

Acknowledgments

The authors would like to express their gratitude to H. Walther for his support of this work. The first author also thanks A. Saenz and A. Apalategui for friendly and stimulating discussions on this and other subjects.

References

1. Ammosov, M.V., Delone, N.B. and Kraǐnov, V.P. (1986) *Sov. Phys.-JETP* **64** 1191
2. Codling, K., Frasinski, L.J. and Hatherly, P.A. (1988) *J. Phys. B* **21** L433
3. Dunn, G.H. (1966) *J. Chem. Phys.* bf 44 2592
4. Frasinski, L.J., Posthumus, J.H., Plumridge, J., Codling, K., Taday, P.F. and Langley, A.J. *Phys. Rev. Lett.* **83** 3625 (1999).
5. Giusti-Suzor, A., Mies, F.H., DiMauro, L.F., Charron, E. and Yang, B. (1995) *J. Phys. B* **28** 309 (topical review)
6. Kawata, I., Kono, H., Fujimura, Y. and Bandrauk, A.D. (2000) *Phys. Rev.* A **62** 031401(R)
7. Keldysh, L.V. (1965) *Sov. Phys.-JETP* **20** 1307

8. Posthumus, J. H., Plumridge, J., Frasinski, L. J., Codling, K., Divall, E., Langley, A.J. and Taday, P.F. (2000) *J. Phys. B* **33** L563
9. Rottke, H., Trump, C. and Sandner, W. (1999). *Las. Phys.* **9** 171
10. Saenz, A. (2000) *J. Phys. B* **33** 4365
11. Saenz, A. (2000) *Phys. Rev.* A **61** 051402(R)
12. Saenz, A. and Posthumus, J.H. (2001) to be published
13. Sändig, K., Figger, H. and Hänsch, T.W. (2000) *Phys. Rev. Lett.* in print
14. Sukharev, M.E. and Krainov, V.P. (1996) *Sov. Phys. JETP* **83** 457
15. Thompson, M.R., Thomas, M.K., Taday, P.F., Posthumus, J.H., Langley, A.J., Frasinski, L.J. and Codling, K. (1997) *J. Phys. B* **30** 5755
16. Walsh, T.D.G., Ilkov, F.A., Chin, S.L., Châteauneuf, F., Nguyen-Dang, T.T., Chelkowski, S., Bandrauk, A.D. and Atabek, O. (1998) *Phys. Rev.* A **58** 3922
17. Williams, I.D., McKenna, P., Srigengan, B., Johnston, I.M.G., Bryan, W.A., Sanderson, J.H., El-Zein, A., Goodworth, T.R.J., Newell, W.R., Taday, P.F. and Langley, A.J. (2000) *J. Phys. B* **33** 2743
18. Wunderlich, C., Kobler, E., Figger, H. and Hänsch, T.W. (1997) *Phys. Rev. Lett.* **78** 2333

DISSOCIATIVE IONIZATION OF FEW-ELECTRON MOLECULES IN INTENSE LASER FIELDS

DANIEL DUNDAS, J.F. MCCANN AND K.T. TAYLOR
(e-mail: d.dundas@am.qub.ac.uk)
Department of Applied Mathematics and Theoretical Physics,
Queen's University of Belfast, Belfast BT7 1NN. Northern Ireland

1. Introduction

The energy and angular momentum carried by laser radiation readily disturbs the electronic states of molecules. Electronic excitations produced in this interaction are transferred to the nuclear motion and result in rotational heating and large-amplitude molecular vibration of the system often leading to dissociation. With high frequency or high intensity radiation electron ejection becomes a dominant process. Photoionization will occur rapidly over timescales much shorter than vibrational relaxation times, and the chemical stability of the molecule is destroyed in a *Coulomb explosion.*

Although the molecule in unstable under such conditions, the interaction with the oscillating radiation field gives rise to laser-molecule dressed states which have characteristics of both the laser and the molecular bound and continuum states. The decay of these states, observed by particle energy spectroscopy, is a key to understanding their nature and behaviour. However, even in simple diatomic molecules the complexity of the interactions leads to rich variety in ionic fragment charges and energies [1, 2] which are dependent on laser pulse intensity and duration. Inevitably the analysis of the data produced in experimental studies is obscured by the light intensity and particle density variations over the focal volume of the interaction region. And although a great deal of data has been acquired for large molecules [3], an understanding of the detailed underlying mechanisms in simple diatomic systems has remained elusive.

The problem is inherently difficult, for experimentalists and theorists, owing to the proliferation of many-body continuum channels. So far, laboratory studies have concentrated on simple diatomic systems through the observation of fragment ions using techniques such as covariance mapping [1]. While theory has been less well developed, nonetheless one can make some progress with the inter-

B. Piraux and K. Rzążewski (eds.), Super-Intense Laser-Atom Physics, 179–188.

pretation of such data with simplified models which treat the ionization process as a sequence of electron release processes, neglecting the electron-electron interactions. The treatment of multielectron systems is a formidable task and only recently have the details of the dynamics begun to emerge from full-dimensionality simulations for atoms [4]. Examples of the laser-driven multielectron dynamics of the helium atom are presented elsewhere in this volume [5]. The dynamics of multiphoton dissociation of simple molecules has been extensively studied theoretically and is reviewed in other papers [2, 6]. In this paper we study dissociative ionization by ultrashort ultra-intense laser pulses, with a view to the important mechanisms and to produce quantitatively accurate data that can be used to guide experimental studies. We consider the simplest molecule, the hydrogen molecular ion, and approach the problem by numerical solution of the time-dependent Schrödinger equation (TDSE).

2. Model

We solve the TDSE for the interaction between an ultrashort ($<$ 100fs) intense ($>$ 10^{14} Wcm^{-2}) laser pulse with the hydrogen molecular ion. The electronic motion is considered in its full dimensionality and the nuclear vibrational motion is described quantally. In this paper, we consider *parallel* electronic transitions from the ground state of the ion, that is, we consider the laser linearly polarized along the internuclear axis. This allows a simpler electronic description since only Σ symmetries are coupled. Thus the electronic wavefunction has cylindrical symmetry and the evolution occurs in only two-dimensions.

2.1. GRID METHOD FOR z AND R

Specifically we construct a 3D-grid to span the two electronic and one nuclear degrees of freedom and discretize the wavefunction on this grid. Following Fujimura and co-workers [7], we solve the electronic problem in scaled cylindrical coordinates. Thus, adopting atomic units, the electron position vector is given by

$$\boldsymbol{r} = \rho^{\ell} \cos\varphi \boldsymbol{i} + \rho^{\ell} \sin\varphi \boldsymbol{j} + z\boldsymbol{k}, \tag{1}$$

where ℓ is the scaling parameter. Since, for Σ-symmetry, the azimuthal (φ) dependence does not arise and the Hamiltonian has the form

$$H = -\frac{1}{m_p}\frac{\partial^2}{\partial R^2} - \frac{1}{2\mu}\left\{\frac{\partial^2}{\partial z^2} + \frac{1}{\ell^2\rho^2}\frac{\partial}{\partial\rho}\rho\frac{\partial}{\partial\rho}\right\} + V(\rho, z, R) + U(z,t), \tag{2}$$

where $\mu = 2m_p m_e/(2m_p + m_e)$ is the reduced mass, m_e and m_p are respectively the electron and nuclear masses, R the inter-nuclear distance, $V(\rho, z, R)$ represents the coulomb interactions and $U(z,t)$ represents the interaction between the electron and the electric field, $E(t)$, in the length gauge. Writing the wavefunction as

$$\Psi(\rho, z, R, t) = (\ell\rho^2)^{-1/2}\psi(\rho, z, R, t) \tag{3}$$

means the TDSE has the form

$$\begin{aligned} i\frac{\partial}{\partial t}\psi(\rho, z, R, t) &= \left[-\frac{1}{m_p}\frac{\partial^2}{\partial R^2} - \frac{1}{2\mu}\left\{\frac{\partial^2}{\partial z^2} + D_\rho\right\}\right. \\ &\quad \left. + V(\rho, z, R) + U(z, t)\right]\psi(\rho, z, R, t), \end{aligned} \tag{4}$$

where

$$D_\rho = \frac{1}{\ell^2\rho^{2\ell}}\left[\rho^2\frac{\partial^2}{\partial\rho^2} - 2(\ell - 1)\rho\frac{\partial}{\partial\rho} + \left(\ell - \frac{1}{2}\right)^2\right]. \tag{5}$$

In order to solve Equation 4 we employ a five-point finite-difference grid method for the R and z coordinates, the spacings of the grid being $\delta R = 0.2$ a.u. and $\delta z = 0.2$ a.u.

2.2. FINITE-DIFFERENCE GRID METHOD FOR ρ

The simplest grid method for treating the ρ coordinate is a finite difference grid. In this case we use scaled cylindrical coordinates with $\ell = 3/2$ such that $\psi(\rho, z, R, t) \sim \rho$ at $\rho = 0$. The use of a 5-point rule for differentiation requires the wavefunction to be assigned a value at the point $\rho = -\delta\rho$. A certain amount of freedom is available in the choice of this value and this enables us to substantially improve the degree to which the grid represents the lowest energy states which play a vital role in the problem [8].

When using the scaling parameter $\ell = 3/2$, the finite difference matrices for the ρ derivatives given by Equation 5 are non-hermitian meaning that the time-evolution operator will be non-unitary. The effect of this non-unitarity can be reduced by using small grid spacings in ρ. In the calculations presented in this paper a grid in ρ having extent $0 \leq \rho \leq 50$ a.u. with a spacing of $\delta\rho = 0.15$ a.u. is used corresponding to 90 grid points. Time propagation is performed using a 4th-order Taylor series propagator [8].

2.3. LAGRANGE MESH METHOD FOR ρ

The Lagrange mesh method is a special case of the Discrete Variable Representation [10]. It is a basis set method which leads to a set of grid equations and has been extensively applied to both time-independent problems [11, 12] and time-dependent problems [13, 14]. For a fuller description of the method, the reader is referred to the paper of Baye and Heenan [15]. Briefly, an orthonormal basis is chosen which possesses an underlying Gauss quadrature rule with pivots x_i and weights λ_i. This basis is then used to construct another basis, the Lagrange basis, where the Lagrange basis functions, $f_i(x)$, satisfy

$$f_i(x_j) = \lambda_i^{-1/2}\delta_{ij}, \tag{6}$$

$$\langle f_i | f_j \rangle = \delta_{ij}. \tag{7}$$

As an example we show how this can be applied to a one-dimensional problem to obtain the mesh equations. Starting from the TDSE

$$[T + V(x)]\,\Psi(x,t) = i\frac{\partial}{\partial t}\Psi(x,t), \tag{8}$$

where T is the kinetic energy operator, $V(x)$ a potential and $\Psi(x,t)$ the wavefunction, we expand the wavefunction in the Lagrange basis

$$\Psi(x,t) = \sum_{i=1}^{N} \psi(x_i,t) f_i(x), \tag{9}$$

where N is the number of basis functions used. Taking the inner product of Equation 8 with an arbitrary basis member $f_j(x)$, using Equations 6 and 7 and introducing the linear scaling $x_i \to hx_i$ leads to

$$\sum_{j=i}^{N} \left[\frac{1}{h^2}T_{ij} + V(hx_i,t)\delta_{ij} - i\frac{\partial}{\partial t}\delta_{ij}\right] \lambda_i^{1/2}\psi(hx_j,t) = 0, \tag{10}$$

where $T_{ij} = \langle f_i | T | f_j \rangle$.

We now apply the method to the treatment of the ρ coordinate. Using an orthonormal set of generalised Laguerre polynomials, $L_n^{\alpha}(\rho)$ to form the Lagrange basis [15] will allow us to treat the kinetic energy operator

$$D_\rho = -\frac{\partial^2}{\partial\rho^2} + \frac{\alpha(\alpha-2)}{4\rho^2}, \tag{11}$$

which corresponds to Equation 5 with $\ell = 1$ when $\alpha = 1$. The matrix elements of D_ρ in the Lagrange basis will be hermitian meaning that the evolution operator will be unitary. Therefore, we use a sixth order Arnoldi propagator [16] for time propagation. In the calculations presented in this paper $N = 30$ basis functions are using with $h = 0.4$. This results in a mesh having extent $0 \le \rho \le 50$ a.u.

2.4. CALCULATION OF THE GRID ENERGY SPECTRUM

We can obtain the field-free ground electronic and vibrational states of the grid by propagation of a trial wavefunction in imaginary time. In this case the TDSE becomes a diffusion equation and time-propagation causes suppression of the excited state components. This method cannot therefore tell us how accurately the excited states of the molecular ion are represented on the grid.

In order to calculate the excited state energies we use a time auto-correlation technique. This consists of calculating the time auto-correlation function

$$C(t) = \langle \psi(t) | \psi(0) \rangle \,, \tag{12}$$

by propagating a trial wavefunction, $\psi(0)$, in real time in the absence of the laser field. The Fourier transform of $C(t)$ gives us the energies supported by the grid. The trial wavefunction can be any function which includes even and odd parities. We use the discretized form of a trial function

$$\psi(0) = \psi_g \left(1 + z + z^2\right) \left(1 + \rho + \rho^2\right) e^{-z^2}, \tag{13}$$

where ψ_g is the ground state on the grid obtained from time-propagation of the diffusion equation.

Figure 1 presents the electronic energy spectrum of the grid for a fixed nuclei calculation at $R = 2$ a.u. using both the finite difference treatment of ρ (Figure 1(a)) and the Lagrange mesh treatment (Figure 1(b)). A range of states are present up to the ionization limit at 0.5 a.u. Table I presents a comparison with exact results given by Sharp [9]. The results are in quite good agreement although there are slight inaccuracies in certain excited state energies which is to be expected due to the discretizion and finite extent of the grids in each case. We see that on the whole the energies using the Lagrange mesh approach are more accurate than those from the finite difference method for the parameters chosen.

3. Results

3.1. FLOQUET COMPARISON

The hydrogen molecular ion has for some years been the benchmark for simulations of molecular dynamics in intense laser fields and in particular molecular multiphoton ionization. High-frequency results ($\omega = 0.2a.u.$) have been obtained for laser intensities in the range $1.0 - 6.0 \times 10^{14}$ Wcm^{-2}. In Figure 2 we compare our time-dependent results for H_2^+ at $R = 2a_0$ using the Lagrange mesh and the finite-difference methods with results from the Floquet method of Madsen and Plummer [17]. In our time-dependent calculations, a pulse which ramps up over 6 optical cycles, remains constant in intensity for a further 8 cycles and then ramps off over 6 further cycles is used. Results from both approaches show good agreement with Floquet theory except at the highest intensity where the Floquet results show a decrease in ionization rate. This drop in rate is attributed to channel closing—that is the Stark shifting of the ground state with respect to the continuum being sufficient at this intensity for seven photons to be required

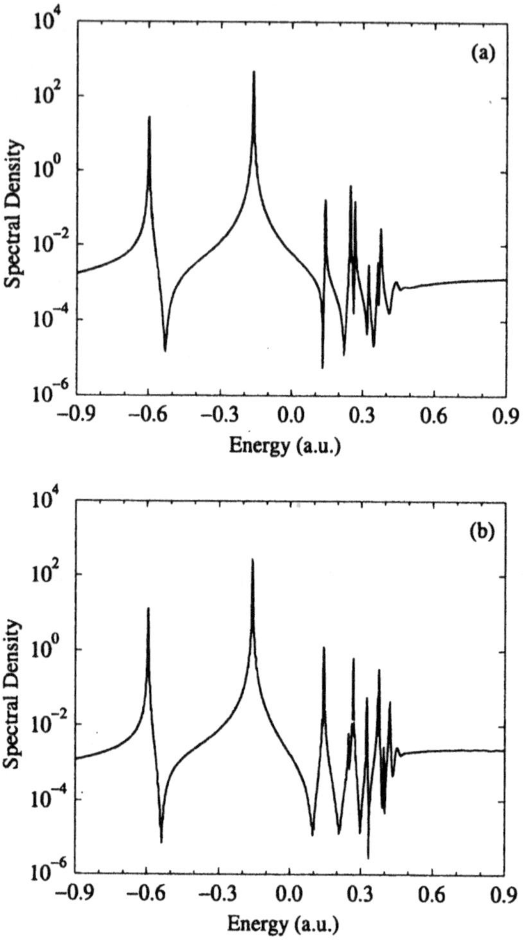

Figure 1. Electronic energy spectrum of fixed nuclei H_2^+ at $R = 2$ a.u. using (a) the finite difference model and (b) the Lagrange mesh model. In both cases a finite difference grid in z is used with extent $-50 \leq z \leq 50$ a.u.

for ionization instead of six. The general trend as a function of intensity is very well reproduced and the absolute ionization rates are in good agreement. In this regime of frequency and intensity (Keldysh parameter $\gamma \approx 2$) the process is dominated by resonant enhancement, and the small discrepancy we can attribute to the slight inaccuracy in the excited state spectrum. The present results seem to be the most accurate wavepacket simulations yet obtained when benchmarked against converged Floquet calculations.

TABLE I. Energies of electronic states supported by the grid using the finite difference and Lagrange mesh calculations of fixed nuclei H_2^+ at $R = 2$ a.u. compared with energies given by Sharp [9].

Possible States		Peaks in Graph	
Sharp Data		Finite Difference	Lagrange Mesh
State	Energy (a.u.)	Energy (a.u.)	
$\Sigma_g 1s$	-0.602	-0.599	-0.599
$\Sigma_u 1s$	-0.167	-0.163	-0.161
$\Sigma_g 2s$	0.139	0.141	0.141
$\Sigma_u 2s$	0.245	0.247	0.247
$\Sigma_g 2p$	0.264	0.266	0.264
$\Sigma_g 3s$	0.315	0.326	0.323
$\Sigma_g 3p$	0.366	0.364	
$\Sigma_u 2p$	0.373	0.376	0.372
$\Sigma_g 4s$	0.394		0.394
$\Sigma_u 4s$	0.414		↑
$\Sigma_u 3p$	0.416		0.418
$\Sigma_g 3d$	0.417		
$\Sigma_g 4p$	0.417		↓
$\Sigma_u 3d$	0.441	0.440	↑
$\Sigma_u 4p$	0.444		
$\Sigma_g 4d$	0.444		
$\Sigma_u 4d$	0.459		0.449
$\Sigma_g 4f$	0.459		
$\Sigma_u 4f$	0.468		↓

3.2. VIBRATING NUCLEI RESULTS

In this subsection we present results for vibrating nuclei H_2^+ using a finite difference treatment of the ρ coordinate. Figure 3 presents visualizations of the probability density in the $R - z$ plane for various instants in time during the interaction with the pulse. These plots are obtained by integrating the probability density over the ρ coordinate

$$P(z, R, t) = \int d\rho\, \psi^{\star}(\rho, z, R, t)\psi(\rho, z, R, t), \qquad (14)$$

and are valuable because they allow us to study the interplay between the dis-

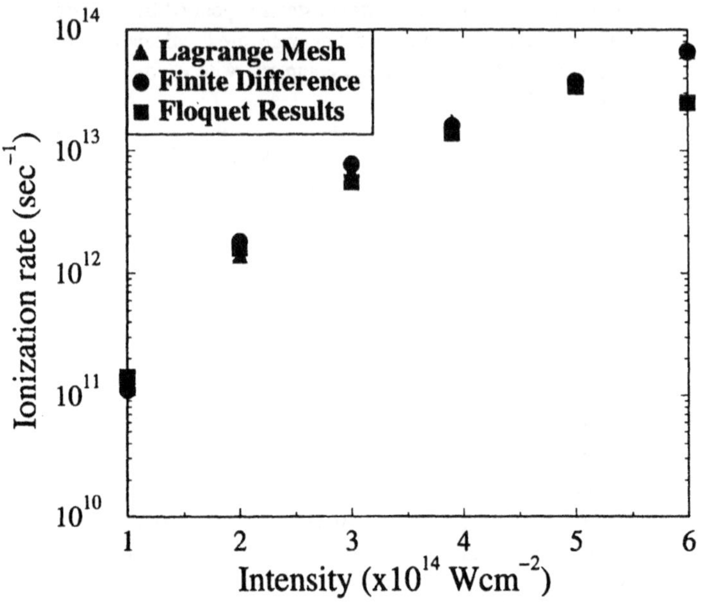

Figure 2. Comparison of fixed nuclei H_2^+ ionization rates at R=2a$_0$ for a wavelength of λ = 228nm using finite-difference grid and Lagrange mesh techniques with Floquet calculations [17].

sociation dynamics and the electron dynamics with emphasis on the laser field polarization axis. In these calculations z has extent $-50 \leq z \leq 50$ a.u. while R has extent $0 \leq R \leq 20$ a.u. We only plot that part $-10 \leq z \leq 10$ a.u. and $0 \leq R \leq 10$ a.u. since this is the most interesting region. In Figure 3(a), near the end of the ramp-on, we see that the electron has clearly responded to the field whereas the nuclear dynamics show no more than a slight distortion. However, later in the pulse as shown by Figures 3(b)–(d) we see nuclear motion occurring with a breakup of the nuclear wavepacket (Figure 3(b)) followed by the formation of small *horns* along $z = \pm R$, a signature of dissociative ionization.

4. Conclusions and Outlook

We have described a grid method in cylindrical coordinates for solving the TDSE for a one electron molecular ion including nuclear vibration effects. Two approaches for treating the ρ coordinate were discussed which gave results of comparable accuracy. Not only did the Lagrange mesh approach yield a unitary time evolution operator, but it also had the desirable property of requiring many fewer grid points to cover the same region of space whilst giving the same accuracy as the finite difference method (30 points compared to 90 points).

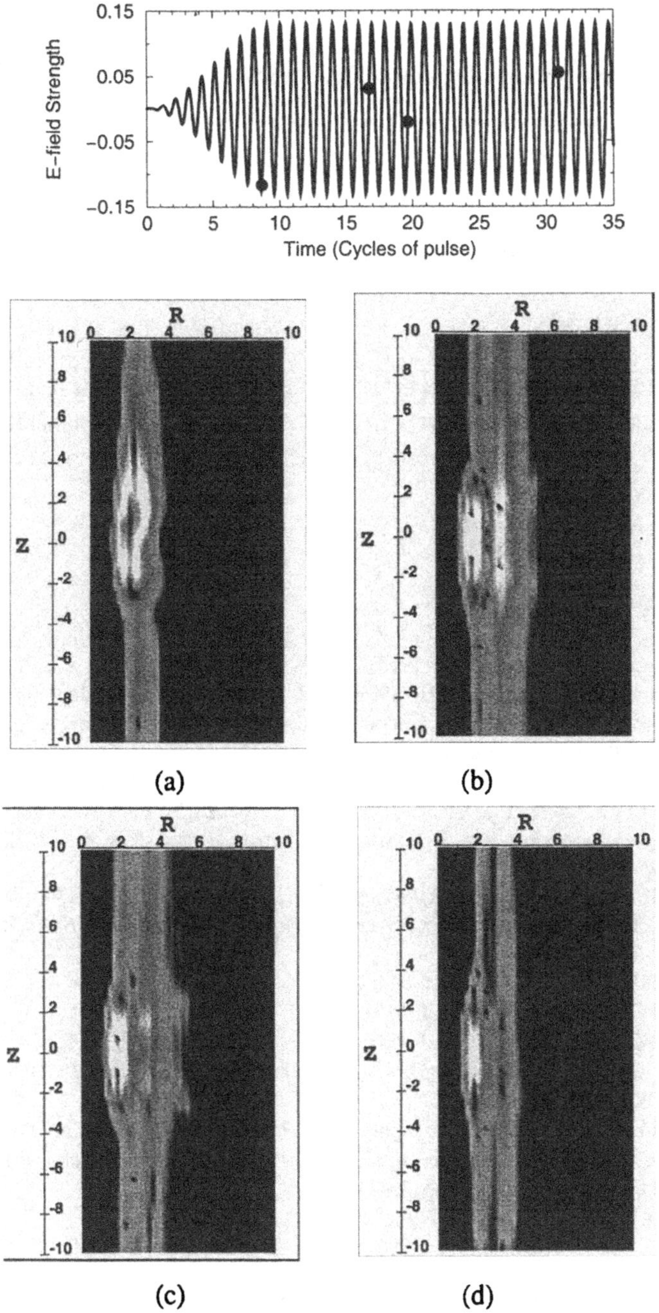

Figure 3. $R - z$ plots of the electron probability density of vibrating nuclei H_2^+ for a laser pulse having a wavelength of 248nm and a peak intensity of 6.0×10^{14} Wcm^{-2}. The picture at the top shows a profile of the laser pulse with the circles corresponding (from left to right) to the location of frames (a)-(d).

We are presently extending the Lagrange mesh method to treat molecular hydrogen including nuclear vibration. This involves the solution of a six-dimensional TDSE; where the various degrees of freedom will be treated as follows

- two z electronic coordinates using finite difference grids,
- two ρ electronic coordinates using Lagrange-Laguerre meshes,
- one azimuthal electronic coordinate using a discrete variable representation,
- one nuclear vibrational coordinate using a Lagrange-Laguerre mesh.

Acknowledgements

The work reported in this paper is supported in part by the UK Engineering and Physical Sciences Research Council by provision of financial support for DD as well as computational resources used at Computer Services for Academic Research, Manchester.

References

1. Codling, K. and Frasinski, L.J. (1993), *J. Phys. B: At. Mol. Opt. Phys.*, **26**, 783.
2. McCann J.F. and Posthumus J.H. (1999), *Phil. Trans. Roy. Soc.* (London), **357**, 1309.
3. Sanderson, J.H., Thomas, R.V., Bryan, W.A., Newell, W.R., Langley, A.J. and Taday, P.F. (1998), *J. Phys. B: At. Mol. Opt. Phys.*, **31**, L599.
4. Taylor, K.T. and Dundas D. (1999), *Phil. Trans. Roy. Soc.* (London), **357**, 1331.
5. Moore L. R., Parker J.S., Dundas D., Meharg K.J. and Taylor K. T. This volume.
6. Giusti-Suzor, A., Mies, F.H., DiMauro, L.F., Charron, E., and Yang, B. (1995), *J. Phys. B: At. Mol. Opt. Phys.*, **28**, 309.
7. Kono H., Kita A., Ohtsuki Y. and Fujimura Y. (1997), *J. Comp. Phys.* **130**, 148.
8. Dundas D., McCann J.F., Parker J.S. and Taylor K.T. (2000), *J. Phys. B: At. Mol. Opt. Phys.*, **33**, 3261.
9. Sharp T.E. (1971), *Atomic Data* **2**:119.
10. Light J.C., Hamilton I.P. and Lill J.V. (1985), *J. Chem. Phys.*, **82**, 1400.
11. Hesse M. and Baye D. (1999), *J. Phys. B: At. Mol. Opt. Phys.*, **32**, 5605.
12. Baye D., Hesse M., Sparenberg J.M. and Vincke M. (1998), *J. Phys. B: At. Mol. Opt. Phys.*, **31**, 3439.
13. Muckerman J.T., Weaver R.V., Kennedy T.A.B. and Uzer T. (1993), in C.Cerjan (ed.), *Numerical Grid Methods and Their Applications to Schrödinger's Equation*, Kluwer Academic Publishers, The Netherlands, pp. 89–119.
14. Melezhik V.S. and Baye D. (1999), *Phys. Rev. C*, **59**, 3232.
15. Baye D. and P-H Heenen P-H. (1986), *J. Phys. A: Math. Gen.*, **19**, 2041.
16. Smyth E.S., Parker J.S. and Taylor K.T. (1998), *Comp. Phys. Commun.* **114**, 1.
17. Madsen L.B. and Plummer M. (1998), *J. Phys. B: At. Mol. Opt. Phys.*, **31**, 87.

ONE-PHOTON BREAKUP OF H_2^+ IN A STRONG DC FIELD

ZUFAR MULYUKOV[1] AND ROBIN SHAKESHAFT
Physics Department,
University of Southern California,
Los Angeles, California 90089-0484

1. Introduction

Recent studies of the dissociation and ionization of the one-electron molecular ions H_2^+ and HD^+ by strong ac fields have revealed a wide variety of interesting effects that distinguish the breakup of molecular ions from the breakup of atoms (see e.g. [1]-[7]). For an informative review see Giusti-Suzor *et al.*[8] Here we report on calculations of the frequency profile for breakup of H_2^+ by one photon in the presence of a strong dc electric field whose direction coincides with the (linear) polarization axis of the photon. The frequency range is chosen so that the photon can excite the molecular ion from its ground $1s\sigma$ electronic state to the $2p\sigma$ electronic state. After electronic excitation, H_2^+ begins to dissociate along the repulsive $2p\sigma$ potential curve, but the dc field results in a barrier which impedes dissociation; nevertheless, the molecular ion breaks up by resonant dc-field ionization.

The most severe approximation made in our calculations is the restriction of the motion of the nuclei to one dimension, along the electric field axis. There is evidence that a strong laser field quickly aligns a molecular ion along the polarization axis.[8] Hence it is reasonable to assume that in the presence of the dc field, but prior to excitation, the H_2^+ ion is aligned with its dipole moment antiparallel to the dc field. However, if the orientation of the molecular ion were not to change after excitation to the $2p\sigma$ electronic state its dipole moment would be parallel to the field, an unstable configuration. Therefore the question arises as to whether the orientation of the H_2^+ ion can remain fixed during the time it takes for dissociative ionization. A

[1]Present Address: Department of Computer Science and Engineering, University of California, La Jolla, CA 92093-0114

B. Piraux and K. Rzążewski (eds.), Super-Intense Laser-Atom Physics, 189–201.

definitive answer would require a formidable calculation, but we can make a rough assessment as follows: At a moderately large internuclear separation R the dipole moment is $eR/2$ and the frequency of rotation of the dipole in a dc field of strength $F_{\rm dc}$ is $\sqrt{2MR/F_{\rm dc}e}$ where M is the proton mass. The period of rotation must be larger than the time it takes for the H_2^+ ion to separate to the point, $R_{\rm res}$, where resonant ionization leads to rapid breakup. Roughly speaking the latter time is $R_{\rm res}/v$, where $Mv^2/2 = \hbar\Omega$ and Ω is the photon frequency. Thus we require $(\hbar\Omega/eF_{\rm dc}R_{\rm res}) > 1$. In section III we show that $eF_{\rm dc}R_{\rm res} \approx (3/8)$ a.u., and since $\hbar\Omega$ is about 0.7 a.u. our condition is only barely satisfied, if at all. Therefore our calculation may only give a qualitative picture. Our remaining approximations are more reliable: We place no restriction on the degrees of freedom of the electron motion. Nor do we place any limit on the number of electronic levels, aside from the limit implicitly imposed by the size of the basis on which the electronic wavefunction is expanded. We include vibronic coupling, i.e. the explicit coupling of the nuclear and electronic motions, through leading order in $(m_e/\mu_n)^{1/4}$, where m_e is the electron mass and μ_n is the reduced mass of the nuclei; this results in an error of order (m_e/μ_n) in the molecular wavefunction.

2. Calculation

Units in which $\hbar = 1$ and the electron charge is -1 are used throughout. The nuclei are labelled by a and b and the electron by e. The masses of a and b are M_a and M_b, their charges are Z_a and Z_b, and the electron mass is m_e. Let $\mathbf{R}$ locate the position of a relative to b, and let $\mathbf{r}_c$ locate the position of e relative to the midpoint of the nuclei. We define the following masses and mass ratios:

$$\mu_e = m_e(M_a + M_b)/(m_e + M_a + M_b), \quad (1)$$
$$\mu_n = M_a M_b/(M_a + M_b), \quad (2)$$
$$\alpha = \mu_n/M_a, \quad (3)$$
$$\beta = \mu_n/M_b. \quad (4)$$

The electronic Hamiltonian $H_e(R)$ can be simplified by using prolate spheroidal coordinates (λ, μ, ϕ), with origin at $\mathbf{r}_c = 0$. Here, ϕ is the azimuthal angle (with polar axis along the internuclear axis) and

$$\lambda = (r_a + r_b)/R, \quad (5)$$
$$\mu = (r_a - r_b)/R, \quad (6)$$

where r_a and r_b are the distances of the electron from a and b, respectively, and where $1 \leq \lambda \leq \infty$ and $-1 \leq \mu \leq 1$. We have

$$H_e(R) = \frac{-2}{m_e R^2(\lambda^2 - \mu^2)} \left[\frac{\partial}{\partial \lambda} \left((\lambda^2 - 1) \frac{\partial}{\partial \lambda} \right) + \frac{\partial}{\partial \mu} \left((1 - \mu^2) \frac{\partial}{\partial \mu} \right) \right.$$
$$\left. + \left(\frac{1}{(\lambda^2 - 1)} + \frac{1}{(1 - \mu^2)} \right) \frac{\partial^2}{\partial \phi^2} + 2 m_e R \lambda \right]. \quad (7)$$

For the purpose of illustration only, a specific gauge, the length gauge, is now adopted; we have, neglecting a term in $(1/\mu_n)\nabla^2_{\mathbf{r}_c}$ (which is of order m_e/μ_n)

$$H(t) = -\frac{1}{2\mu_n}\nabla^2_{\mathbf{R}} - \frac{(\beta - \alpha)}{2\mu_n}\nabla_{\mathbf{R}} \cdot \nabla_{\mathbf{r}_c} + \frac{1}{R} + H_e(R) + \frac{(\beta - \alpha)}{2}\mathbf{F}(t) \cdot \mathbf{R} + \mathbf{F}(t) \cdot \mathbf{r}_c. \quad (8)$$

We now make the approximation that the internuclear line is oriented along the electric field axis, $\hat{\mathbf{e}}$; this gives

$$\nabla_{\mathbf{R}} \approx \hat{\mathbf{e}} \frac{\partial}{\partial R}, \quad (9)$$
$$\mathbf{F}(t) \cdot \mathbf{R} = F(t) R, \quad (10)$$
$$\mathbf{F}(t) \cdot \mathbf{r}_c = F(t)(R\lambda/2)\mu. \quad (11)$$

In Eq. (9) differentiation with respect to R is to be carried out holding $\mathbf{r}_c$ constant. However, to use prolate spheroidal coordinates, differentiation with respect to R must be carried out holding λ, μ, and ϕ constant; hence Eq. (9) must be replaced by:

$$\nabla_{\mathbf{R}} \approx \hat{\mathbf{e}} \left(\frac{\partial}{\partial R} - \frac{1}{R} \frac{\lambda(\lambda^2 - 1)}{(\lambda^2 - \mu^2)} \frac{\partial}{\partial \lambda} - \frac{1}{R} \frac{\mu(1 - \mu^2)}{(\lambda^2 - \mu^2)} \frac{\partial}{\partial \mu} \right). \quad (12)$$

In reexpressing $(1/\mu_n)\nabla^2_{\mathbf{R}}$ and $(1/\mu_n)\nabla_{\mathbf{R}} \cdot \nabla_{\mathbf{r}_c}$ in prolate spheroidal coordinates, cross terms involving $\partial/\partial R$ and either $\partial/\partial\lambda$ or $\partial/\partial\mu$ are retained — such cross terms are of the order of the ratio of the momentum of the electron and the relative momentum of the nuclei, i.e. of order $(m_e/\mu_n)^{1/4}$ or higher — but cross terms involving $\partial/\partial\lambda$ and $\partial/\partial\mu$ — which are of order (m_e/μ_n) — are neglected. Thereby we arrive at

$$\nabla^2_{\mathbf{R}} \approx \frac{\partial^2}{\partial R^2} - \frac{2}{R(\lambda^2 - \mu^2)} \left(\lambda(\lambda^2 - 1) \frac{\partial}{\partial \lambda} + \mu(1 - \mu^2) \frac{\partial}{\partial \mu} \right) \frac{\partial}{\partial R}, \quad (13)$$

$$\nabla_{\mathbf{R}} \cdot \nabla_{\mathbf{r}_c} \approx \frac{2}{R(\lambda^2 - \mu^2)} \left((\lambda^2 - 1)\mu \frac{\partial}{\partial \lambda} + (1 - \mu^2)\lambda \frac{\partial}{\partial \mu} \right) \frac{\partial}{\partial R}. \quad (14)$$

When $|\mathbf{r}_c| \gg R$, we have $|\mathbf{r}_c| \approx R\lambda$, but since the asymptotic distance of the electron from the midpoint of the nuclei is of physical importance (it is used to specify the ionization channel boundary condition) it should be expressed in terms of a single independent variable; a suitable variable is

$$\xi \equiv R(\lambda - 1), \tag{15}$$

which ranges from 0 to ∞. Therefore, one final change of variables is necessary, i.e. replacement of λ by ξ. Since differentiation with respect to R must now be carried out holding ξ, μ, and ϕ constant, we must make the replacements:

$$\frac{\partial}{\partial\lambda} = R\frac{\partial}{\partial\xi}, \tag{16}$$

$$\frac{\partial}{\partial R} \to \frac{\partial}{\partial R} + \frac{\xi}{R}\frac{\partial}{\partial\xi}, \tag{17}$$

$$\frac{\partial^2}{\partial R^2} \to \frac{\partial^2}{\partial R^2} + 2\frac{\xi}{R}\frac{\partial}{\partial\xi}\frac{\partial}{\partial R}, \tag{18}$$

where in Eq. (18) we have omitted a term in $\partial^2/\partial\xi^2$, as this is consistent with neglecting terms of order (m_e/μ_n); for the same reason we can drop the second term on the right side of Eq. (17). Combining Eqs. (8), (7), and (10)-(18), yields an expression for $H(t)$ in terms of the *entirely independent variables* ξ, μ, ϕ, and R. Since the electron motion is cylindrically symmetric about $\hat{\mathbf{e}}$, the ϕ-dependence can be separated out by substitution of $\exp(im\phi)$, where m is the magnetic quantum number of the electron. The final expression for $H(t)$ in terms of ξ, μ, and R is

$$\begin{aligned} H(t) = &-\frac{1}{2\mu_n}\left[\frac{\partial^2}{\partial R^2} + 2\frac{\xi}{R}\frac{\partial}{\partial\xi}\frac{\partial}{\partial R} - \right. \\ &\left. \frac{2}{R[(\xi+R)^2-(R\mu)^2]}\left(\xi(\xi+R)(\xi+2R)\frac{\partial}{\partial\xi} + R^2\mu(1-\mu^2)\frac{\partial}{\partial\mu}\right)\frac{\partial}{\partial R}\right] - \\ &\frac{1}{\mu_n}\left(\frac{\beta-\alpha}{[(\xi+R)^2-(R\mu)^2]}\right)\left(\xi(\xi+2R)\mu\frac{\partial}{\partial\xi} + (1-\mu^2)(\xi+R)\frac{\partial}{\partial\mu}\right)\frac{\partial}{\partial R} + \\ &H_e(R) + \frac{1}{2}[(\beta-\alpha)R + (R+\xi)\mu]F(t), \end{aligned} \tag{19}$$

where $H_e(R)$ differs from the right side of Eq. (7) only by the change of the variable λ to $\lambda = (\xi + R)/R$, accomplished using Eq. (16), and by the replacement of $\partial^2/\partial\phi^2$ by $-m^2$. Hereafter, we consider only excitation between σ states, and we put $m = 0$ and $\alpha = \beta = 1/2$.

We represent $H(t)$ on a basis composed of functions of the form

$$u_p(\xi)v_q(\mu)w_s(R),$$

where, with $L_p(x)$ and $P_q(\mu)$ Laguerre and Legendre polynomials, respectively, and with p, q, and s nonnegative integers,

$$u_p(\xi) = e^{i\kappa\xi}L_p(-2i\kappa\xi), \tag{20}$$

$$v_q(\mu) = P_q(\mu), \tag{21}$$

$$w_s(R) = Re^{iKR}L_s(-2iKR). \tag{22}$$

Here κ and K lie in the upper right quadrant of the complex plane so that the functions $u_p(\xi)$ and $w_s(R)$ can represent both closed (bound) and open (outgoing-wave) channels. Note the prefactor of R in $w_s(R)$, which conforms with the boundary condition that, once the nuclear motion is restricted to one dimension, the wavefunction of the molecular system vanishes at $R = 0$. The volume element, used in constructing the integral expressions for the matrix elements of $H(t)$, is

$$dV = (\pi/4)[(\xi + R)^2 - (R\mu)^2]d\xi d\mu dR. \tag{23}$$

The singular factors $1/R$ and $1/[(\xi+R)^2-(R\mu)^2]$, which premultiply derivatives in $H(t)$, disappear in the integrands of the matrix elements since $H(t)$ is premultiplied by $w_s(R)dV$.

3. Results and Discussion

The dc quasienergy of the electron, denoted by $E^{(e)}_{\rm dc}(R)$, is a complex eigenvalue of $H^{(e)}_{\rm dc}(R)$; its imaginary part is $-\Gamma^{(e)}_{\rm dc}(R)/2$ where $\Gamma^{(e)}_{\rm dc}(R)$ is the rate for ionization when the nuclei are clamped in place at the internuclear separation R. Values of $E^{(e)}_{\rm dc}(R)$ have been calculated previously [9]-[11] for H_2^+ for various values of R and $F_{\rm dc}$.

The electronic potential, along the direction of $\mathbf{F}_{\rm dc}$, consists of two wells, each one centered about a different nucleus, with one well higher than the other.[12] The electron is localized (for large R) in one of these wells, depending on its state. If the electron is in the $2p\sigma$ state it is localized in the upper well, shifted upwards by $F_{\rm dc}R/2$, while if the electron is in the $1s\sigma$ state it is localized in the lower well, shifted downwards by $F_{\rm dc}R/2$. Although the $2p\sigma$ energy level shifts upwards as $F_{\rm dc}$ increases, other more highly-excited-state levels shift downwards. Resonances in the rate for ionization from the $2p\sigma$ state occur at those values of $F_{\rm dc}$ where a highly-excited-state level crosses the $2p\sigma$ level. The importance of such resonances in enhancing molecular ionization was discussed earlier by Seideman *et al*, [13] who carried out calculations for a model one-dimensional molecular ion. The crossing points vary with both $F_{\rm dc}$ and R; if $F_{\rm dc}$ is held fixed (but nonzero) crossing points always occur for some values of R. In Fig. 1 we show Re $E^{(e)}_{\rm dc}(R)$; the solid squares and triangles, respectively, mark the

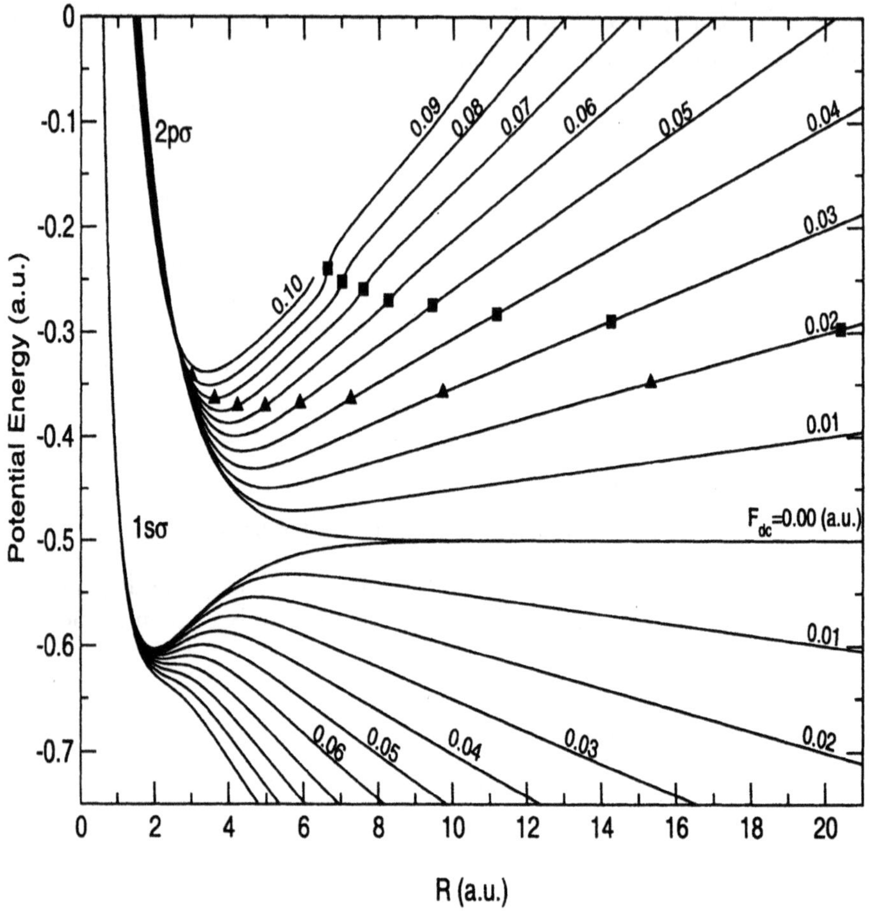

Figure 1. The 1sσ and 2pσ potentials for nuclear motion of H_2^+ when dressed by a dc field of various strengths. We have indicated the internuclear separation at which the rate for ionization from the (dressed) 2pσ state exhibits its most prominent (solid square) and and next most prominent (solid triangle) resonance.

internuclear separations at which the rate for ionization from the 2pσ state exhibits its most prominent, and next most prominent, resonances.

We can make a crude estimate of the internuclear separations at which the two most prominent resonances occur as follows: For large R, and in the absence of the dc field, the energy of the electron in the 2pσ or 1sσ state is approximately $-0.5 - 9/(4R^4)$ in a.u., where the correction of order $1/R^4$ is due to the polarization of the ground-state hydrogen atom, whose polarizability is $9/2$ a.u., by the distant isolated proton. For simplicity we neglect corrections of order $1/R^4$, and, when the dc field is included, we also neglect corrections of order $F_{\rm dc}^2$ and $F_{\rm dc}/R^2$. Thus the 2pσ electronic energy, in the presence of the dc field, is

$$E^{(e)}_{\rm 2p,dc}(R) \approx -0.5 + F_{\rm dc}R/2 \tag{24}$$

in a.u. The levels which shift downwards and cross the 2pσ level correspond, at asymptotically large R, to channels in which the isolated hydrogen atom is in an excited state. Now, in the presence of a dc field an excited hydrogen atom has a permanent dipole moment, due to the mixing of degenerate levels with opposite parities, and it therefore undergoes a linear Stark shift. The levels that are closest to the 2pσ level correspond to asymptotic channels in which the isolated hydrogen atom is in a mixture of 2s and 2p states, with a Stark shift of $\pm 3F$ in a.u. The energies of these more highly excited levels are therefore

$$E^{(e)}_{\rm ex,dc}(R) \approx -0.125 - F_{\rm dc}R/2 \pm 3F_{\rm dc} \tag{25}$$

in a.u. The resonances occur where $E^{(e)}_{\rm ex,dc}(R) = E^{(e)}_{\rm 2p,dc}(R)$, i.e. where, in a.u.,

$$R \approx (3/8F_{\rm dc}) \pm 3. \tag{26}$$

According to this rather crude estimate, the two most prominent resonances occur at internuclear separations that (i) decrease as $F_{\rm dc}$ increases, and (ii) differ by 6 a.u. We can assess the validity of these properties by perusal of Fig. 1. Both properties are confirmed, although not surprisingly the second property is more accurate at smaller field strengths, i.e. at larger R. The most prominent of the two resonances is the one occuring at the larger internuclear separation, i.e. close to $(3/8F_{\rm dc})+3$ a.u., since the outer hump of the lower well of the electronic potential, over which the electron must pass as it escapes, becomes narrower as R increases, and the likelihood of the electron backscattering from this hump diminishes.

To obtain the dc quasienergy, $E_{\rm dc}$, of the molecular ion we solved the eigenvalue problem

$$H_{\rm dc}|\Phi_{\rm dc}\rangle = E_{\rm dc}|\Phi_{\rm dc}\rangle, \tag{27}$$

where $H_{\rm dc}$ is the Hamiltonian of the complete molecular ion dressed by the dc field; we took $H_{\rm dc}$ to be the right side of Eq. (19) with $F(t)$ replaced by $F_{\rm dc}$. The probability density for the nuclei to be at the separation R is

$$(\pi/4)\int_0^\infty d\xi \int_{-1}^1 d\mu\ [(\xi+R)^2 - (R\mu)^2]|\Phi_{\rm dc}(\xi,\mu,R)|^2.$$

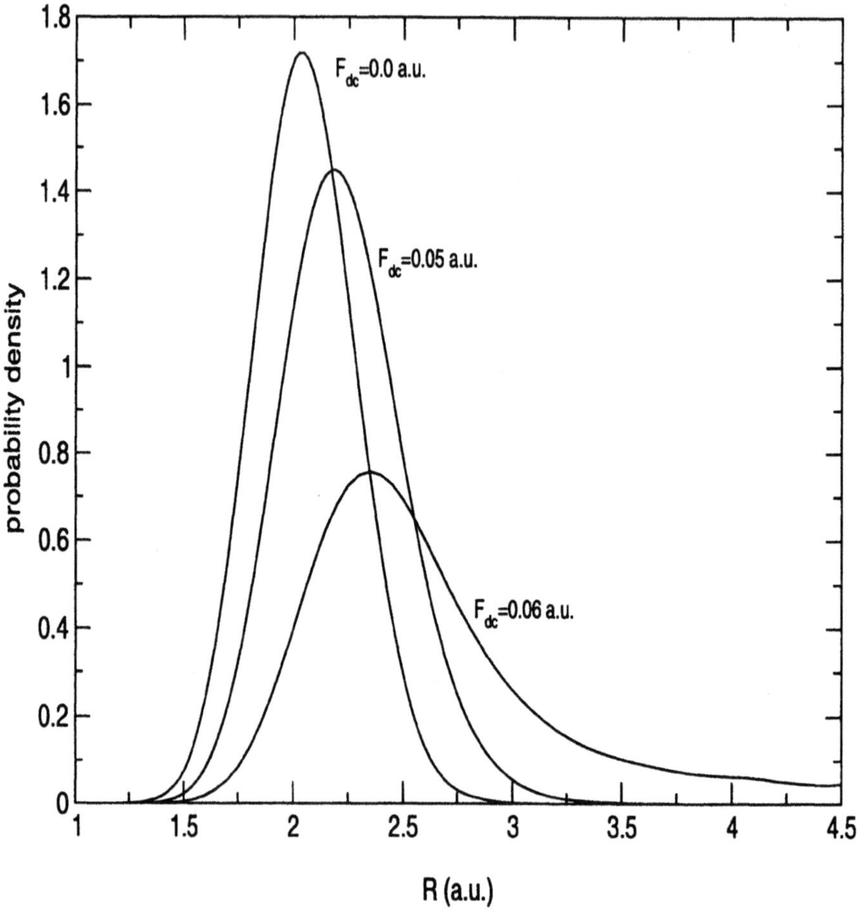

Figure 2. The normalized probability density for the nuclei to be separated a distance R in the initial $1s\sigma$ state, when dressed by a dc field of various strengths.

In Fig. 2 we show the probability density for the nuclei to be at a given separation in the initial $1s\sigma$ state, for various dc field strengths. The "equilibrium" separation, R_0 say, of the nuclei is the point at which this probability density has its maximum. Note that as the dc field strength increases, R_0 increases and the probability distribution of the nuclei broadens, corresponding to the change in shape of the $1s\sigma$ potential well. The rate at which the molecular ion breaks up due to a dc field is $\Gamma_{1s,dc} = -2\mathrm{Im}\, E_{1s,dc}$ where the label 1s has been added as a reminder that the initial state is the ground state.

When a moderately weak monochromatic ac field $\mathbf{F}_{ac}(t) \equiv F_{ac}\cos(\Omega t)\hat{\mathbf{e}}$

is also included, the total rate for breakup by one photon in the presence of the dc field is, treating the ac field as a perturbation,

$$\Gamma_{1s,tot} = \Gamma_{1s,dc} + \Gamma_{1s,ac}, \tag{28}$$

where $\Gamma_{1s,ac}$ is the contribution due to the ac field:

$$\Gamma_{1s,ac} = -2 \text{ Im } \langle \Phi_{1s,dc} | V_- G_{dc}(E_{1s,dc} + \Omega) V_+ | \Phi_{1s,dc} \rangle, \tag{29}$$

where $G_{dc}(E) = 1/(E - H_{dc})$ and where, adopting the length gauge,

$$V_\pm = F_{ac} \hat{\mathbf{e}} \cdot \mathbf{r}/2. \tag{30}$$

In Fig. 3 we show the frequency profile of $\Gamma_{1s,tot}$ for various dc field strengths. The small hump on the high-frequency side of this profile is due to electronic transitions to σ states that are more highly excited than the $2p\sigma$ state.

As F_{dc} increases, the frequency profile shifts to lower frequencies, at least until F_{dc} reaches the strength, about 0.06 a.u., at which over-the-barrier dissociation from the $1s\sigma$ well becomes possible. The shift in the profile as F_{dc} varies below the threshold for over-the-barrier dissociation can be understood from the condition for the $1s\sigma \rightarrow 2p\sigma$ electronic transition to be on resonance. This condtion — see Fig. 4 — is that $\Omega = \Omega_0$ where Ω_0 is the energy splitting of the dressed $1s\sigma$ and $2p\sigma$ electronic energies at R_0, the "equilibrium" (i.e. most probable) separation of the nuclei. While Ω_0 overestimates the frequency at which the profile has its peak, [14] the trend in Ω_0 as F_{dc} increases is consistent with the shift in the profile (below the threshold for over-the-barrier dissociation).

The width of the frequency profile remains almost constant, decreasing slightly, as F_{dc} increases, until F_{dc} reaches about 0.05 a.u. This is clear from Fig. 11. The contribution to the width of the profile from the dc width of the $1s\sigma$ energy level is negligible over the range of field strengths considered. Rather, the width of the profile is related, primarily, to the width of the probability distribution of the nuclei. The variation in Ω_0 that is allowed by the resonance condition is shown as $\Delta\Omega_0$ in Fig. 4. Just as Ω_0 overestimates the frequency at which the profile has its peak, so does $\Delta\Omega_0$ overestimate the width of the profile, [14] but the trend is again consistent. However, the absence of broadening of the frequency profile for $F_{dc} < 0.05$ a.u. is perhaps surprising since it is evident from Fig. 2 that the probability distribution of the nuclei does broaden significantly as F_{dc} increases. It turns out, fortuitously, that the shift in R_0 is such that $\Delta\Omega_0$ hardly changes. The broadening of the frequency profile for $F_{dc} > 0.05$ a.u. is due primarily to the merging of the main part of the profile with the hump on the high-frequency side, reflecting the increasing importance of

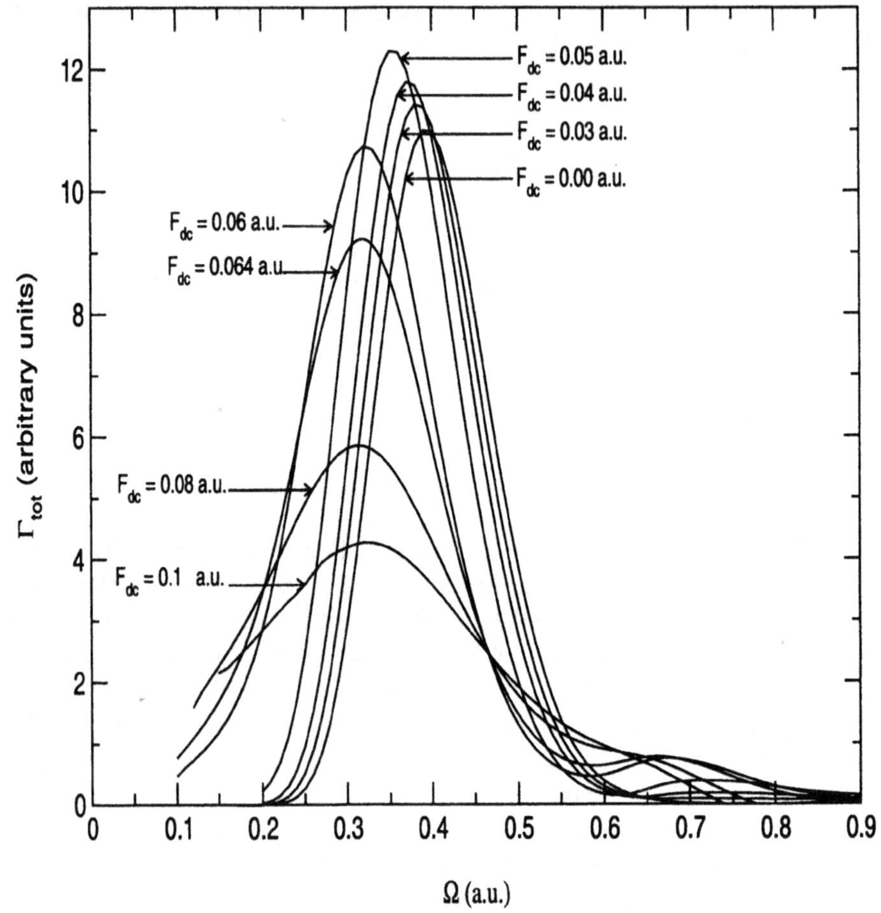

Figure 3. The frequency profile of the total breakup rate for H_2^+ for various dc field strengths.

electronic transitions to σ states that are more highly excited than the 2pσ state.

The height of the profile at first increases slightly as F_{dc} increases, and then decreases rapidly. We believe that the initial slight increase in the profile's height is due to the shift to lower frequencies, which results in a larger oscillator strength for the electronic 1sσ-2pσ excitation. To explain the sharp falloff in the profile's height when F_{dc} increases beyond about 0.05 a.u. we note that the relative height of the initial probability distribution of the nuclei in the 1sσ potential well decreases immediately as F_{dc} increases (due to the broadening of the distribution) and the effect is to reduce the

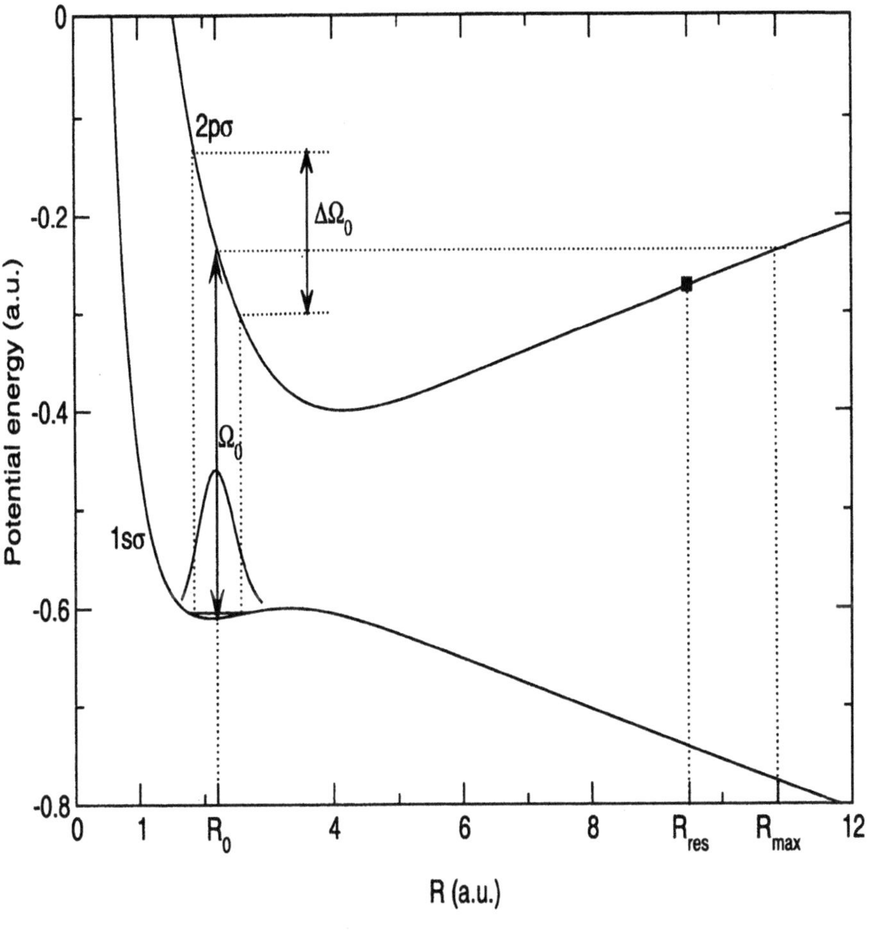

Figure 4. Schematic diagram showing resonant electronic excitation from the 1sσ state to the 2pσ state by a photon of frequency Ω_0 in the presence of a dc field. Here R_0 is the "equilibrium" separation of the nuclei in the initial (dressed) 1sσ state, $R_{\rm max}$ is the maximum separation to which the nuclei can dissociate after photoexcitation, and $R_{\rm res}$ is the internuclear separation at which the the rate for ionization from the 2pσ state exhibits its most prominent resonance. The width of the frequency profile, when determined from the condition for resonant excitation, is $\Delta\Omega_0$.

oscillator strength for photoexcitation. At first the decrease in the height of this probability distribution is gentle, and is more than compensated for by the shift to lower frequencies, but as $F_{\rm dc}$ approaches the threshold for over-the-barrier dissociation the dispersion of the nuclei becomes severe, and dominates at $F_{\rm dc} \approx 0.05$ a.u.

We have ignored the rotation of the H_2^+ ion, and, to the extent that this

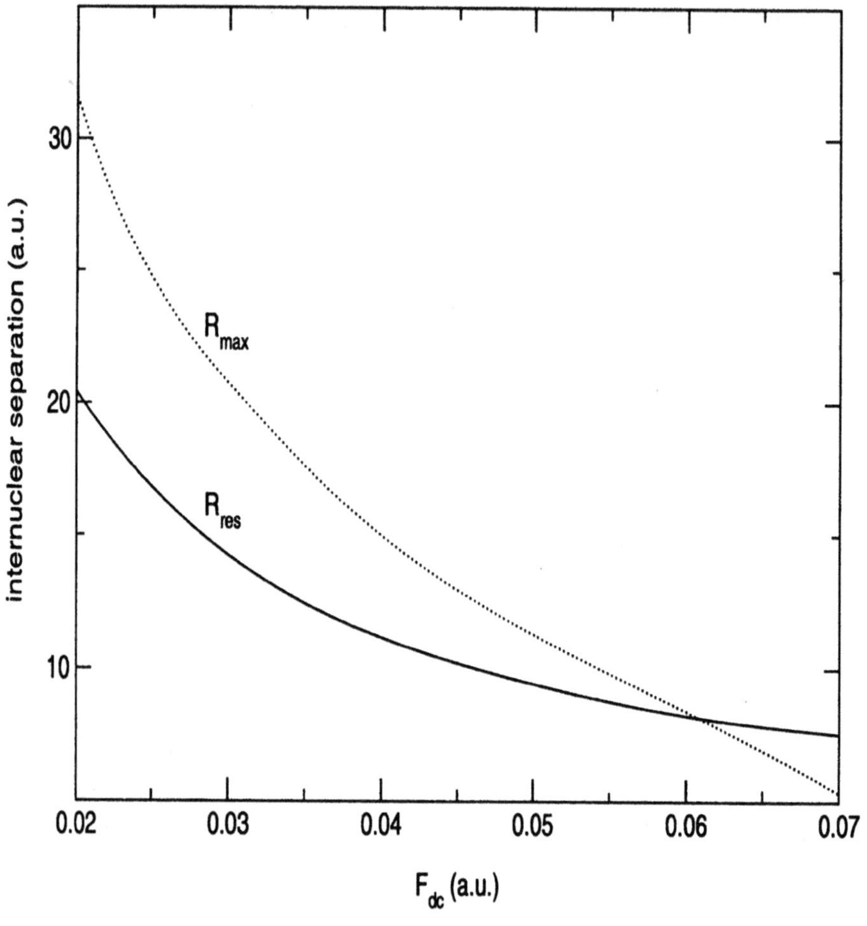

Figure 5. The internuclear separation, R_{res}, at which the the rate for ionization from the $2p\sigma$ state exhibits its most prominent resonance. The dashed line is the maximum separation, R_{max}, to which the nuclei can dissociate after photoexcitation.

is legitimate, dissociation is impeded by the field-induced barrier, which opposes the movement of the nuclei beyond the turning point, R_{max} say. In order for the nuclei to separate beyond R_{max} it is necessary that the most prominent resonance occur at an internuclear separation, R_{res} say, which is less than R_{max}. The closer R_{res} is to R_{max}, while remaining less than R_{max}, the longer the time the resonance exerts its influence since the nuclei slow down as they approach the field-induced barrier. In Fig. 5 we show R_{max} and R_{res} *vs* F_{dc}. We see that R_{max} approaches, but remains above, R_{res} until F_{dc} reaches about 0.06 a.u., at which point over-the-barrier dissociation from the initial state begins to play an important role.

Acknowledgments

This work was supported by the NSF under Grant No. PHY-0070936.

References

1. P. H. Bucksbaum, A. Zavriyev, H. G. Muller, and D. W. Schumacher, Phys. Rev. Lett. **64**, 1883 (1990).
2. G. Yao and S. -I. Chu, Chem. Phys. Lett. **197**, 413 (1992).
3. A. Giusti-Suzor and F. Mies, Phys. Rev. Lett. **68**, 3869 (1992).
4. E. E. Aubanel, J. M. Gauthier, and A. D. Bandrauk, Phys. Rev. A **48**, 2145 (1993).
5. O. Atabek, M. Chryos, and R. Lefebvre, Phys. Rev. A **49**, R8 (1994).
6. B. Sheehy, B. Walker, and L. F. DiMauro, Phys. Rev. Lett. **74**, 4799 (1995).
7. E. Charron, A. Giusti-Suzor, and F. Mies, Phys. Rev. Lett. **75**, 2815 (1995).
8. A. Giusti-Suzor, F. H. Mies, L. F. DiMauro, E. Charron, and B. Yang, J. Phys. B **28**, 309 (1995).
9. M. Plummer and J. F. McCann, J. Phys. B bf 29, 4625 (1996).
10. T. Zuo and A. D. Bandrauk, Phys. Rev. A **52**, R2511 (1995).
11. Z. Mulyukov, M. Pont, and R. Shakeshaft, Phys. Rev. A **54**, 4299 (1996).
12. K. Codling, L. J. Frasinski, and P. A. Hatherly, J. Phys. B **22**, L321 (1989).
13. T. Seideman, M. Yu. Ivanov, and P. B. Corkum, Phys. Rev. Lett. **75**, 2819 (1995).
14. The oscillator strength of the electronic transition is larger on the lower-frequency side of the profile, so this side should be more heavily weighted.

PART III

INTERACTION OF CLUSTERS WITH VERY INTENSE FEMTOSECOND LASER PULSES

NUCLEAR FUSION IN GASES OF DEUTERIUM CLUSTERS AND HOT ELECTRON GENERATION IN DROPLET SPRAYS UNDER IRRADIATION WITH AN INTENSE FEMTOSECOND LASER

T. Ditmire, J. Zweiback, T. E. Cowan, G. Hays, K. B. Wharton, J. K Crane, S. C. Wilks
Laser Program
Lawrence Livermore National Laboratory
Livermore, CA 94550 USA

R. A. Smith
Imperial College of Science Technology and Medicine,
Blackett Laboratory, London, SW7 UK

T. D. Donnelly, M. Rust, I. Weiner
Department of Physics
Harvey Mudd College
Claremont, CA USA

M. Allen
Department of Nuclear Engineering
University of California
Berkeley, CA USA

1. Nuclear Fusion from Exploding Deuterium Clusters

1.1 INTRODUCTION

A number of experiments have been conducted in recent years examining the interactions of intense femtosecond laser pulses with large van der Waals bonded clusters. Most experiments indicate that these interactions can be very energetic, with a variety of experimental manifestations. Studies from a few years ago revealed that very bright x-ray emission in the 100 to 5000 eV range resulted from the plasmas produced from the interaction of pulses focused to intensity between 10^{16} and 10^{18} W/cm^2 into gas jets that contained rare gas atomic clusters of a few hundred to a few thousand atoms each [1, 2]. This bright x-ray production resulted from the efficient coupling of intense laser light into the cluster gas, strong coupling that was found to be absent when irradiating monatomic gas of similar average density at these intensities. This efficient coupling

B. Piraux and K. Rzążewski (eds.), Super-Intense Laser-Atom Physics, 205–218.

was confirmed by subsequent measurements of laser energy absorption in clustering gases [3, 4]. In fact, these experiments illustrated that nearly 100% of an intense laser pulse could be absorbed by a modest average density gas jet (~10^{19} cm^{-3}) within a few mm of propagation length when clusters were present in the gas. This greatly enhanced absorption arises from the fact that the density within the clusters is high, nearly that of a solid, and can therefore exhibit the large absorption efficiency usually associated with solid targets [5].

The laser absorption measurements indicated that many keV of energy per atom were being deposited in the clustering gas. The mechanisms and dynamics of this energy deposition by femtosecond pulses in individual clusters has also been examined by a number of groups. These studies have included measurement of ionization charge states produced in the cluster [6-8], pump-probe and pulsewidth experiments which have shown that the clusters disassemble on a picosecond time scale [9], and photo electron energy measurements which indicated that multi-keV electrons were ejected from large (>1000 atom) Xe clusters [10]. Perhaps most remarkable, has been the discovery that these large clusters, when irradiated at intensity above 10^{15} W/cm^2, also eject ions with substantial kinetic energy. In fact, ions with energy as high as 1 MeV have been seen from exploding Xe clusters [8].

This large release of kinetic energy in fast ions can be harnessed to drive nuclear fusion between deuterium ions if deuterium clusters are irradiated in a gas of sufficient average density to permit collision between ions ejected from different clusters in the gas. Indeed, we recently observed these fusion reactions in gases containing deuterium clusters of roughly 5 nm diameter [11]. In this section we report on a detailed series of measurements of fusion yield from these deuterium cluster explosions and show that, while total fusion yield is modest, the fusion efficiency can be substantial, over 10^5 fusion neutrons per joule of laser energy.

In our experiment a high intensity, ultrafast laser pulse is focused into a gas jet of deuterium clusters, rapidly heating the inertially confined clusters before they can expand. These clusters subsequently explode, ejecting energetic ions. This process creates a plasma filament with a diameter roughly that of the laser focus (~100 μm) and a length comparable to the extent of the gas jet plume (~ 2 mm). The fast deuterium ions ejected from the exploding clusters can then collide with ions ejected from other clusters in the plasma. If the ion energy is high enough (greater than a few tens of keV), D+D nuclear fusion events can occur with high probability. The well known signature of this process arises from one branch of the fusion reaction, $D + D \rightarrow He^3 + n$, in which a neutron is released with 2.45 MeV of energy.

The previous experiments on cluster explosions [8] indicated that large deuterium clusters ionized at high intensity should produce the high ion energies required for fusion. Though large, high Z clusters typically exhibit a "plasma-like" hydrodynamic explosion due to the fact that space charge forces confine many of the photo ionized electrons to the body of the cluster [5], low-Z, medium size (~1000 atoms per cluster) D_2 clusters can be stripped of all of their electrons during the laser pulse, before the cluster explodes. In this case, essentially a pulse Coulomb explosion of the deuterium cluster result. With deuterium cluster sizes of greater then a few tens of angstroms, fully stripping the cluster with the laser field will result in the explosion of ions with energies greater than a few keV.

1.2 EXPERIMENTAL APPARATUS

Our experiment uses a 10 Hz, Ti:sapphire, chirped pulse amplification laser delivering 120 mJ of laser energy per pulse with 35 fs pulse width and wavelength of 820 nm. This laser was focused into the exit of a deuterium gas jet with an f/12 lens. Because the van der Waals forces between deuterium molecules are weak, the gas jet was cryogenically cooled by flowing cooled nitrogen through a jacket surrounding the gas jet body to -170° C [12], producing large clusters in the D_2 gas. The laser spot size within the gas jet varied from roughly 20 to 130 μm depending on the focal spot position, with respect to the gas jet nozzle. Consequently the peak laser intensity was between 5 x 10^{17} W/cm^2 and 2 x 10^{16} W/cm^2 for all of the experiments described here.

To estimate the deuterium cluster size in our gas jet, we conducted Rayleigh scattering measurements in the jet by passing low energy (< 0.1 mJ) 532 nm pulses from a frequency doubled, Q-switched Nd:YAG laser (pulse width 10 ns) through the plume of the gas jet. The Rayleigh scattered light from cluster formation was imaged 90 degrees from the laser propagation axis. From the observed onset of Rayleigh scattering, we can determine the onset of large cluster formation in the jet. In addition, using independent measurements of the average gas density (see below), estimates for the throughput and detection sensitivity, and the Rayleigh scatter cross section, we are able to estimate the average cluster size in the jet. Measured scattering as a function of gas jet pressure is illustrated in Figure 1 when the gas jet backing temperature is 100K. From these data it can be seen that the scattering signal is a strong function of the jet pressure and that no significant cluster formation is observed until the jet pressure increases above 20 atm. A strong variation with gas jet backing pressure is also observed.

1.3 FUSION EXPERIMENTAL RESULTS

To determine the efficiency with which the 120 mJ, femtosecond laser pulse was coupled into the deuterium clusters we measured the absorption efficiency of the laser pulse within the deuterium cluster gas jet. To do this, we measured the transmitted laser energy collected within an f/3 cone (to collect all potentially refracted light from the f/12 cone) with a large aperture lens and calorimeter. We monitored backscattered light with a photodiode (very little was observed in these experiments). We also estimated scattered light using a CCD camera and collection lens monitoring light scattered 90° from the laser propagation direction perpendicular to the laser polarization direction (the direction of maximum Rayleigh scatter). Extrapolating into 4π we determined that the scattered light was negligible (< 1 mJ),

The absorption as a function of gas jet reservoir backing pressure is illustrated in Figure 2 when the gas has been initially cooled to -170° C (the case in which we observe large D_2 cluster formation). As can be seen, when the deuterium gas jet pressure rises to

>30 atm, nearly 90% of the laser energy is deposited in the plasma. This is in sharp contrast to the energy deposition efficiency of the laser in pure D_2 molecular gas (the case when the gas jet reservoir is at 20° C), a case also plotted in Figure 2. In this case very little (< 5%) of the laser energy is deposited in the plasma, illustrating the importance of the laser/cluster interaction in heating the D_2 plasma.

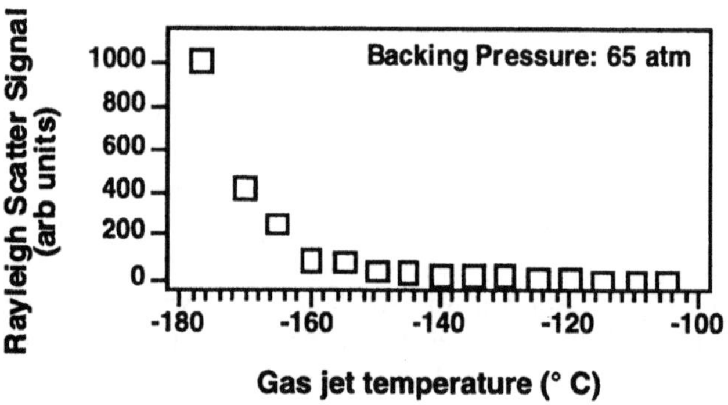

Figure 1: Measured Rayleigh scatter signal in the deuterium gas jet as a function of gas jet backing pressure. The reservoir temperature was 100K.

To detect the production of DD fusion neutrons in these experiments we employed various arrays of neutron sensitive scintillator detectors at various distances from the interaction point. Though different detector configurations were employed, the majority of the data presented below were collected with detectors consisting of 10 x 10 cm^2 scintillator slabs of 7mm thickness coupled to PMTs via light pipes. These detectors were located outside the vacuum interaction chamber and were shielded with 6 mm of lead. The time of flight of particles from the time the laser irradiates the plasma is recorded. Each neutron detector was operated in a mode such that a particle was detected every few laser shots, permitting a determination of the flight time of each detected particle with < 1 ns accuracy.

With these neutron detectors we detected the presence of substantial numbers of MeV energy particles when the gas was backed with cooled deuterium at a pressure of 65 atm and a reservoir temperature of -170 C. A characteristic time-of-flight spectrum of particles detected on a neutron detector located 54 cm from the source is illustrated in Figure 3. These time-of-flight spectra were acquired by accumulation of particle hits over ~ 10,000 laser shots. In these data, a distinct peak occurs in the detected TOF spectrum at the position expected for 2.45 MeV neutrons (which have a flight time of 46 ns/m) at 24.5 ns for 54 cm separation. This measurement confirms the presence of DD fusion reactions in the cluster plasma. We also confirmed that all detected particles on the neutron detectors disappear when hydrogen clusters were made in the jet in lieu of the

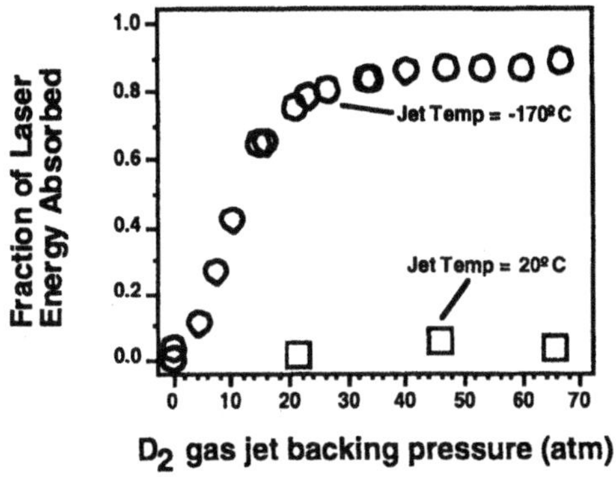

Figure 2: Measured absorption of the 35 fs pulse in the deuterium gas jet as a function of backing pressure for two different jet temperatures.

deuterium clusters (a measurement taken over 50 laser shots assuring better than 1 part in 10^8 that no neutrons were produced when hydrogen replaces deuterium). Furthermore, we found that the fusion neutron production disappeared when deuterium clusters where absent in the D_2 gas jet (when the jet was not cooled). We observed no neutron production in the gas jet with temperatures higher than -120° C, the temperature at which we no longer observe large cluster formation via Rayleigh scattering.

The signal seen in these spectra after the main neutron peak arise from neutron scattering from the gas jet and vacuum chamber walls. We have conducted estimates of the expected scattered neutron signal and have confirmed that these longer time detected particles are consistent with scattered neutrons from material surrounding the fusion plasma, such as the gas jet body and vacuum chamber walls, We also see an occasional hit early in time. It is likely that these are hard x-rays produced from fast electrons released in the laser/cluster interactions. These fast electrons can produce bremsstrahlung photons in the surrounding material, including the stainless-steel gas jet body.
However, these gammas are rare and the vast majority of the signal observed in this experiment are from the fusion neutrons.

Also shown in figure 3 is the neutron TOF distribution very close to the source. This distribution was collected using a very thin (1.5 mm) scintillator directly coupled to the PMT detector. This detector was optimized to yield very high time resolution, which was measured with an x-ray flash to be <250 ps. From this we can derive information on the duration of the neutron pulse width. As can be seen in figure 3, the pulse duration measured with this device at a distance of 9 cm is well under 1 ns. Detailed deconvolution of the neutron pulse with the intrinsic time response of the detector indicated that the neutrons had duration 650 ps here. The 55 cm data indicate that the neutrons have a duration of 1.4 ns. This broadening with distance is a result of Doppler-like broadening of the neutron energy around 2.45 MeV resulting from the ion temperature. These measurements also indicate that the fusion burn time at the plasma is likely < 500 ps.

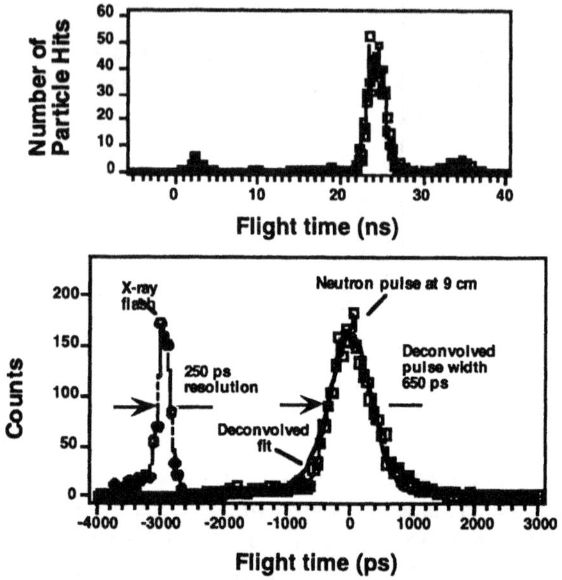

Figure 3: Time-of-flight spectrum of particles detected on an n detector located 54 cm from the source (top) and TOF spectrum 9cm from source (bottom).

To measure the change in neutron yield as different parameters were varied, an additional plastic scintillator detector was placed near the plasma in an re-entrant flange to subtend a large solid angle. It subtended an angle of 0.3 ster and was placed 20 cm from the deuterium plasma. Because the time-of-flight data indicate that there is no noticeable x-ray flash and that virtually all of the detected particles are neutrons, we obtained a shot by shot measure of the fusion yield from the PMT signal. The detection efficiency was calibrated against the other TOF detectors operating in single particle detection mode. We also calibrated the yield by measuring the average pulse height produced when single neutrons struck the detector. Both methods yielded nearly equal values for the yield calibration.

Using this detector, we measured the yield as a function of various parameters. Most importantly, we wish to optimize the fusion yield. Because the fusion cross section is very strongly dependant on the ion energy, small increases in ion kinetic energy from the exploding clusters should yield dramatic increases in the fusion yield. In a Coulomb explosion picture of the fast ion generation, the average ion energy from an exploding sphere of ions that have had their electrons stripped is given by $Qn_c e^2 a^2/5\varepsilon_0$ where Q is the total charge on the cluster, n_c is the initial cluster ion density and a is the cluster radius. Thus, we expect that increasing the cluster size in the jet will (to a certain point) increase the ion energy and hence the fusion yield.

To examine this, we used the well known fact that cluster size from a gas jet expansion is a strong function of the initial reservoir temperature [13]. We varied the temperature of our jet by using liquid nitrogen to cool the jet over a range roughly >100 K and then used liquid helium to further cool the jet. We measured the temperature of the jet body with a thermocouple. Though it did not directly register the temperature of the gas, it acted as good parameter to characterize the gas cooling.

The measured average size of clusters from the jet as a function of measured temperature is shown in figure 4a. We can produce average cluster sizes up to 10 nm (though it is likely that there is a broad distribution of clusters in the jet.) The measured fusion yield as a function of measured temperature is shown in figure 4b. We observe

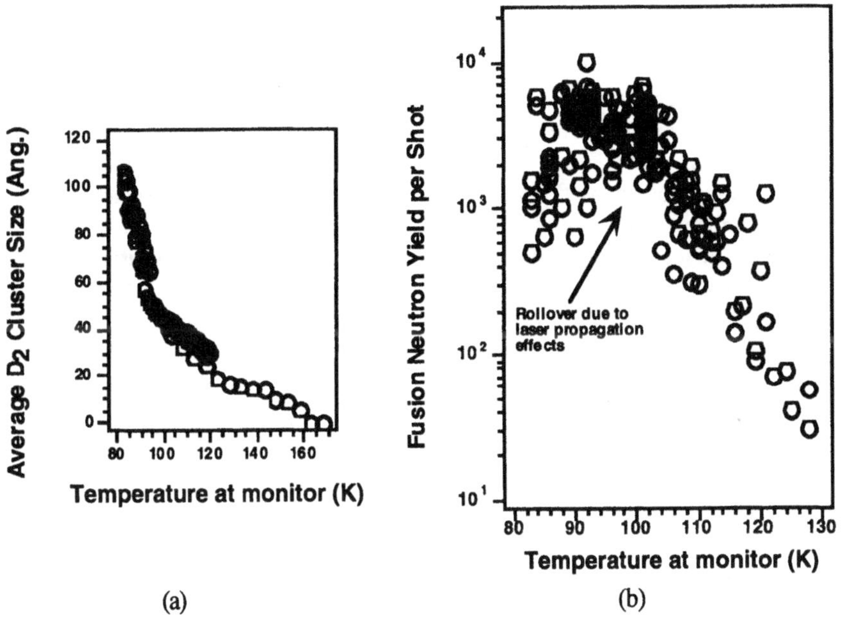

Figure 4: (a) Measured average size of clusters from the jet as a function of measured temperature (b) Measured fusion yield as a function of measured temperature

the expected rapid increase in fusion yield as the cluster size increases with lower temperature down to a temperature of roughly 100K. Detailed analysis of these data indicate that the observed rise in yield is, indeed, consistent with a Coulomb explosion model. The roll over in fusion yield at temperatures below 100 K results from laser propagation effects in the gas jet. Other measurements in this parameter regime indicate that this roll over results from the fact that the laser is strongly absorbed in thedistribution of gas out in front of the center of the plume [14]. As the cluster size is increased, this gas more effectively depletes the laser pulse as it propagates toward the center of the plume preventing further rise in yield.

2. Hot Electron Generation from intense Irradiation of Wavelength Scale Particles

2.1 INTRODUCTION

The interesting phenomena seen in the explosions of nm scale clusters suggest that there might be interesting effects in the intense irradiation of somewhat larger particles. One area of potential interest is in examining hot electron generation from intense irraction of ultrafast laser pulses with particles whose size approaches that of the laser's wavelength. Over the last few years, experimental studies have shown that sub-picosecond to femtosecond lasers focused to intensities of 10^{17} to 10^{20} W/cm^2 on solid targets will, indeed, accelerate substantial numbers of electrons to high energy, and that these fast electrons can produce large fluxes of hard x-rays via bremsstrahlung emission in the target [15]. A number of such experiments have observed photons emitted in the > 1 MeV energy range indicating that such focused laser intensity can accelerate electrons to relativistic temperatures of several hundred keV to well over an MeV [16].

The mechanisms leading to the acceleration of fast electrons responsible for the hard x-rays in laser-solid interactions has been the topic of many experimental and theoretic studies over the past two decades. At intensities approaching 10^{18} W/cm^2 the ponderomotive potential of light with wavelength around 1 μm nears 100 keV and direct acceleration of electrons can occur through electron interactions with the laser field at a plasma surface (Brunel heating [17]) or through **J x B** heating [18]. In addition, when a long scale-length (many wavelength) preformed plasma exists in front of the target, collective electron oscillations driven by resonance absorption or stimulated Raman scattering can also play an important role in accelerating the electrons. Generally, both theoretical and experimental evidence indicates a linear or sub-linear scaling of fast electron temperature with laser ponderomotive potential [19] with calculated hot electron distributions for p-polarized pulses incident on solids exhibiting temperature four to five times the laser ponderomotive potential.

The extension of these studies to small particles is motivated by the fact that linear absorption and scattering of light by particles whose size is comparable to the light wavelength (around 1 μm) can be increased over that of bulk media from enhancements in the light field surrounding the sphere, so called Mie resonances. Thus we might expect that the dynamics of laser driven electron motion will be quite different around such a wavelength scale sphere, and that vacuum heating of hot electrons will take on a different character. To study this effect we measured hard x-ray spectra from a target consisting of micron scale droplets.

2.2 EXPERIMENTAL RESULTS USING A DROPLET SPRAY

This experiment was performed with the same laser system used in the fusion experiments. For the droplet experiments, 0.12 J of laser energy was focused in vacuum with an f/15 lens to a focal spot of roughly 20 μm diameter ($1/e^2$). Far field measurements indicate a peak intensity of 7 x 10^{17} W/cm^2. The pre-pulse intensity contrast was ~ 10^{-6} on the 1 ns time scale before the main pulse and was ~ 10^{-4} on the 10 ps time scale. We estimate that plasma pre-expansion from pre-pulse light was well under 1 μm. These pulses were focused onto either a plane, plastic target in P-polarization at an incidence angle of 10° or into a spray of water droplets. Plastic was chosen as the target material because it exhibits a similar density and has a similar Z to water, the target material in our droplet spray.

The water droplet target was produced by an apparatus capable of creating a high density, polydisperse spray containing micron-scale liquid water droplets in vacuum (similar to the apparatus described in ref. [20]). The droplet source consisted of a gas line backed with high pressure argon (at 67 bar), an inert propellant gas, over a liquid reservoir. The line terminated in a solenoid driven pulsed valve with a 750 μm orifice. The valve was pulsed for 500 μs, causing propellant gas carrying liquid droplets to stream through the orifice.

The composition of the spray produced by the device was characterized in vacuum by measuring the angular scattering and attenuation of a continuous wave 543 nm helium-neon laser polarized perpendicular to the scattering plane. The droplet sizes were measured by fitting theoretical Mie scattering curves of the 543 nm light to the data [21]. As the radius of the sphere increases, the ratio of the intensity of light scattered in the forward direction increases relative to that scattered at larger angles, making the angular distribution of scatter a sensitive test of particle size. Relative Mie scattering data were obtained by integrating the observed signal over the entire 500 μs open time. Angular scatter data for the droplet spray used in our experiment is shown in figure 5.

To ascertain the droplet size, we calculate expected scattering angular distributions using a code calculating the differential Mie scattering cross section. We find that we can fit these scattering data if a two size distribution is assumed. A fit with a single size

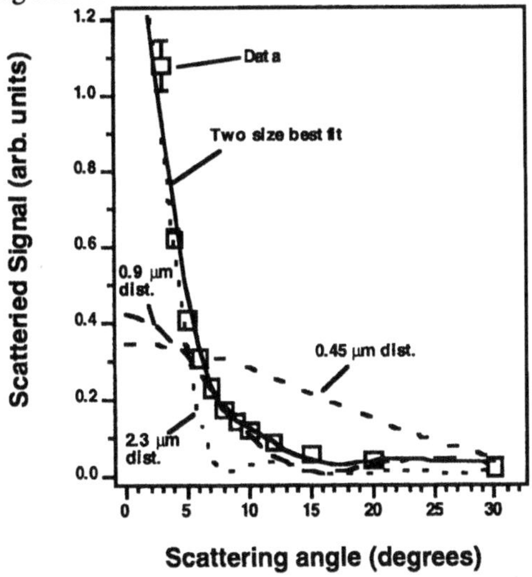

Figure 5: Measured Mie scattering with best fits of droplets in our water spray target

distribution centered at r = 0.9 μm is plotted in the figure. This calculation adequately fits the data at mid-angles (5 – 15 degrees) though it contains oscillations at large angles not seen in the data and does not reproduce the large scattering at small angles. To adequately fit the observed data, we assume a droplet size distribution with a Gaussian profile centered at r = 0.9 μm with a 20% spread (at the 1/e point) with a small additional contribution of droplets with r = 2.3 μm. The large droplets contribute predominately to the small angle scatter signal. The data are reasonably well reproduced by calculation when we assume that 5% of the droplets have an r = 2.3 μm. For comparison, we also show curves for purely r = 2.3 μm droplets as well as small r = 0.45 μm.

The droplet density was determined from light attenuation measurements. With knowledge of the droplet size, and the propagation length of the probe laser through the droplet plume (determined by imaging the scattering onto a CCD), one can use the Mie theory to calculate the average density of particles in the plume. Our measurements indicate that the droplet spray was roughly 2 mm wide and exhibited a density of 6×10^7 droplets/cm^3 (or 7×10^{18} molecules/cm^3). This implies that there are a few tens (10 –100) of droplets in the focal volume per laser shot.

Hard x-rays from the target region, either droplet or planar solid, were observed 90° from the laser propagation axis using detectors employing NaI crystals coupled to photo-multiplier tubes. These detectors were housed in a thick lead casing with an aperture pointed at the target region. X-rays were observed through a 5 mm thick glass window and data were taken with three different filters in front of the detector. We acquired data with an 0.5 cm thick Cu filter and 0.35 cm and 4 cm thick Pb filters. These filters were chosen to yield spectral information in the 100 keV to 1 MeV photon energy range. In addition, data were collected with an unshielded reference detector, filtered by only the 5 mm stainless steel vacuum chamber wall. This detector provided information on x-rays produced with energy above 50 keV and was used to discriminate high and low intensity shots.

Abundant hard x-ray emission was observed from both the solid plastic targets and the water droplet target. The x-ray yield measured from both targets is plotted in figure 6. Here the total x-ray energy deposited in the detector per shot is plotted versus the cut-off energy ($1/e^2$) of each filter. Consequently the data points roughly represent the integrated x-ray yield above this photon energy point. From this illustration it is evident that the total x-ray yield observed from the two kinds of targets is comparable over the 0.1 to 1 MeV energy range. Hard x-rays from the solid target occur due to bremsstrahlung of fast electrons in the bulk material, We further surmise that hard x-ray emission occurs in the droplet target due to bremsstrahlung in the stainless steel droplet jet body from hot electrons produced in the droplets themselves.

While a direct comparison of x-ray yields between these two targets does not yield quantitative information because of the differing mechanisms for producing the bremsstrahlung (plastic as opposed to stainless steel converter for the droplets), it is clear from this figure that the comparable x-ray yields indicate at least roughly comparable hot electron production in the two cases.

To determine the approximate hot electron temperature from these data, we performed simulations of bremsstrahlung production from various exponential electron distributions using the Monte Carlo code, ITS [22]. We calculate hard x-ray spectra from single temperature electron distributions in plastic and in solid iron (to model the droplet x-ray production) and account for the target geometry and direction of detection (90° to the laser). We assume a single temperature Maxwellian electron energy distribution and numerically generated a photon spectrum, $f(E_\gamma)dE_\gamma$ where E_γ is the photon energy. These results were then fit to the data points using a least squares analysis by varying the initial hot electron temperature using the fact that the signal measured in each filtered detector is proportional to $\int E_\gamma T(E_\gamma) f(E_\gamma)\, dE_\gamma$ where $T(E_\gamma)$ is the transmission of each filter as a function of x-ray wavelength.

The best spectrum for each of the targets is shown also in figure 6. We find that the best fit hot electron temperature of the solid plastic target is 400 keV with a single standard deviation of ±100 keV. The droplet target data exhibited a best fit with a 700 keV hot electron temperature and a standard deviation of ±150 keV. The temperature of the solid target is consistent with previously published hot electron temperature measurements in this intensity range (approaching 10^{18} W/cm^2) [16] however, the hot electron temperature inferred from the droplet measurements is nearly twice that inferred from the solid target.

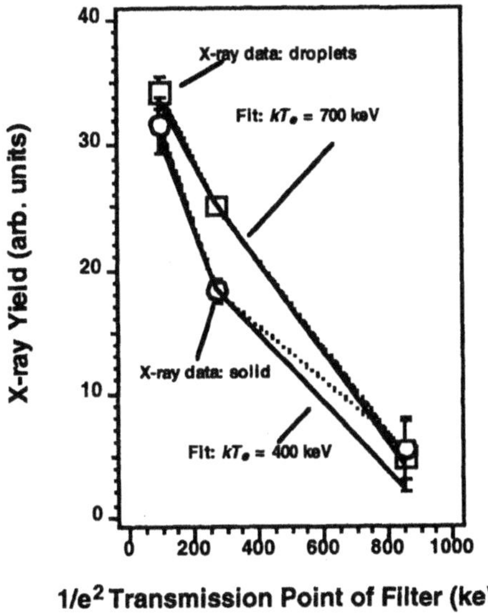

Figure 6: Measured hard x-ray yield from the droplet and solid targets with best fit electron temperatures.

2.3 SIMULATION RESULTS

To understand these observed differences, we have conducted 2D particle-in-cell simulations of a 30 fs pulse irradiating a round solid density plasma and planar solid targets at a peak intensity of 5 x 10^{17} W/cm^2. These simulations were conducted with the code ZOHAR [23] in which solid material was approximated as an overdense, 5 x 10^{22} cm^{-3} plasma. In the planar target calculations, the laser field was p-polarized and was incident at an angle of 19°. An example of the field distribution calculated around a one micron spherical (droplet) plasma is shown as a contour plot in figure 7 near the peak of the laser pulse envelope. As exhibited in figure 7, we observe the presence of enhanced regions of electric field near the droplet surface in a roughly dipole pattern (i.e. both at the surface of incidence as well as the back surface). This arises from the oscillating electrons perpendicular to the laser **k** vector.

The hot electron spectra produced from simulations of the droplet and the planar slab are shown in figure 8. The calculation predicts a hot electron temperature of 125 keV from the flat target and a temperature of 240 keV from the droplet, nearly a factor of two higher than the planar slab, much like the enhancement observed in the experiments. Though these electron temperatures are somewhat lower than observed in the experiment, they do confirm that the electron temperature resulting from the interaction with a small sphere is significantly higher than in the interaction with a planar slab.

In the simulation, the majority of the hot electron production is due to Brunel-type heating of the oscillating electrons at the surface in both the droplet and the slab target simulations. However, the enhanced electron heating observed in the case of the particle appears to arise from the differing electron trajectories possible around the droplet target geometry. In the case of the slab target, electrons acquire energy from the laser field through acceleration in vacuum toward the slab surface. Thus, most electrons acquire substantial velocity in the direction perpendicular to the slab surface (the direction in which the field adiabaticity is broken). In the droplet simulation, however, the laser field diffracts around the droplet and electrons can undergo numerous surface penetrations at various angles through the surface of the plasma sphere. In this case, the particle plasma is large enough to shield the laser field (unlike the case of the much smaller nm scale clusters studied) and the electrons can acquire substantial momentum, both parallel and perpendicular to the laser **k** direction.

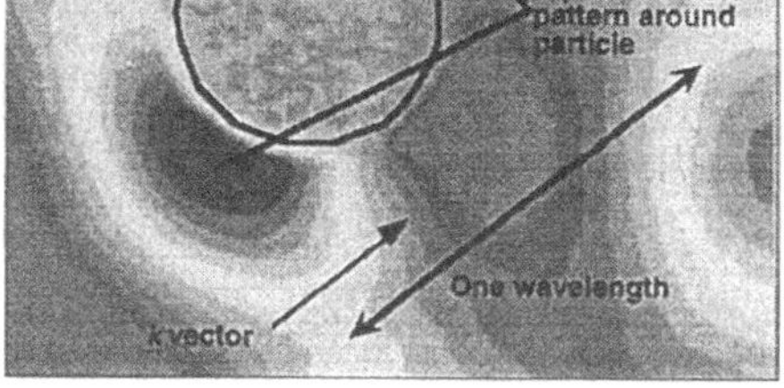

Figure 7: Field distribution calculated around a one micron spherical plasma.

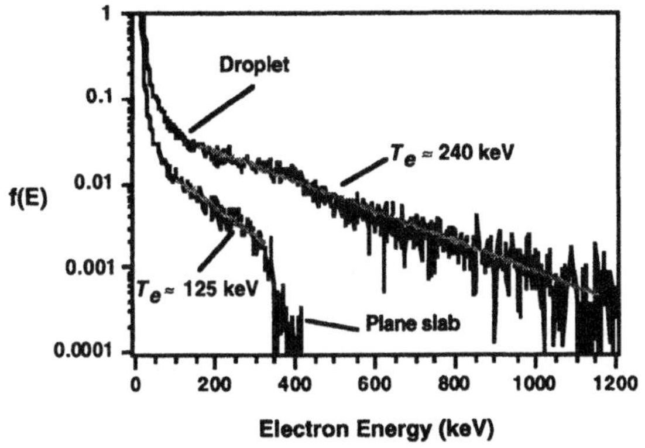

Figure 8: Hot electron spectra produced from simulations of the droplet and the planar slab with best fit temperatures.

3. Conclusion

In conclusion, we have observed the production of 2.45 MeV deuterium fusion neutrons when a gas of deuterium clusters is irradiated with a 120 mJ, 35 fs laser pulse. When the focal position is optimized, we have observed as many as 10^4 neutrons per laser shot. This yield is consistent with some simple estimates for the fusion yield. We also find that the fusion yield is a sensitive function of the deuterium cluster size in the target jet, a consequence of the Coulomb explosion origin of the fast deuterons. We also find that the neutron pulse duration is fast, with a characteristic burn time of well under 1 ns. This experiment may represent a means for producing a compact, table-top source of short pulse fusion neutrons for applications.

Furthermore, we have measured hard x-ray yield from femtosecond laser interactions with both solid and micron scale droplet targets. Strong hard x-ray production is observed from both targets. However, the inferred electron temperature is somewhat higher in the case of irradiation of the droplets. These data are consistent with PIC simulations. This finding indicates that quite unique hot electron dynamics occur during the irradiation of wavelength scale particles by an intense laser field and likely warrants further study

We would like to acknowledge many useful conversations with John Perkins and Howard Powell as well as the technical assistance of Vince Tsai, Gerry Anderson and Rich Shuttlesworth. This work was supported by the Department of Energy under contract W-7405-Eng-48.

4. References

1. McPherson, A., Thompson, B. D., Borisov, A. B., Boyer, K., and Rhodes, C. K. (1994) Multiphoton-induced X-Ray Emission at 4-5 keV from Xe Atoms with Multiple Core Vacancies, *Nature* **370**, 631 .
2. Ditmire, T., Donnelly, T. D., Falcone, R. W., and Perry, M. D. (1995) Strong X-ray Emission from High Temperature Plasmas Produced by Intense Irradiation of Clusters, *Phys. Rev. Lettt.* **75**, 3122.
3. Ditmire, T., Smith, R. A., Tisch, J. W. G., and Hutchinson, M. H. R. (1997) Absorption of High Intensity Laser Pulses in Gases of Clusters, *Phys. Rev. Lett.* **78**, 3121 .
4. Kondo, K., Borisov, A. B., Jordan, C., McPherson, A., Schroeder, W. A., Boyer, K., and Rhodes, C. K. (1997) Wavelength Dependence of Multiphoton-induced Xe(M) and Xe(L) Emissions from Xe Clusters, *J. Phys. B: At. Mol. Opt. Phys.* **30**, 2707 .
5. Ditmire, T., Donnelly, T., Rubenchik, A. M., Falcone, R. W., and Perry, M. D. (1996) The Interaction of Intense Laser Pulses with Atomic Clusters, *Phys. Rev. A* **53**, 3379 .
6. Lezius, M., Dobosz, S., Normand, D., and Schmidt, M. (1997) Hot Nanoplasmas from Intense Laser Irradiation of Argon Clusters, *J. Phys. B: At. Mol. Opt. Phys.* **30**, L251 .
7. Snyder, E. M., Wei, S., Purnell, J., Buzza, S. A., and Castleman. Jr., A. W. (1996) Femtosecond Laser-Induced Coulomb Explosion of Ammonia Clusters, *Chem. Phys. Lett.* **248**, 1 .
8. Ditmire, T., Tisch, J. W. G., Springate, E., Mason, M. B., Hay, N., Smith, R. A., Marangos, J., and Hutchinson, M. H. R. (1997) High-Energy Ions Produced in Explosions of Superheated Atomic Clusters, *Nature (London)* **386**, 54 .
9. Zweiback, J., Ditmire, T., and Perry, M. D. (1999) Femtosecond Time Resolved Studies of the Dynamics of Noble Gas Cluster Explosions, *Phys. Rev. A* **59**, R3166 .
10. Shao, Y. L., Ditmire, T., Tisch, J. W. G., Springate, E., Marangos, J. P., and Hutchinson, M. H. R. (1996) Multi-keV Electron Generation in the Interaction of Intense Laser Pulses with Xe Clusters, *Phys. Rev. Lett.* **77**, 3343.
11. Ditmire, T., Zweiback, J., Yanovsky, V. P., Cowan, T. E., Hays, G., and Wharton, K. B. (1999) Nuclear Fusion from Explosions of Femtosecond-Laser Heated Deuterium Cluster, *Nature* **398**, 489.
12. Smith, R. A., Ditmire, T., and Tisch, J. W. G., (1998) Characterization of a Cryogenically Cooled High-Pressure Gas Jet for Laser/Cluster Interaction Experiments, *Rev. Sci. Instrum.* **69**, 3798.
13. Hagena, O.F. and Obert, W. (1972) Cluster Formation in Expanding Supersonic Jets: Effect of Pressure, Temperature, Nozzle Size, and Test Gas, *J. Chem. Phys.* **56**, 1793.
14. Zweiback, J., Smith, R. A., Yanovsky, V. P., Cowan, T. E., Hays, G., Wharton, K. B. and Ditmire, T. (1999) Nuclear Fusion from Coulomb Explosions of D_2 Clusters Ionized by a Femtosecond Laser, in *Multi-photon Processes* eds. R. Freeman, K. Kulander, L. DiMauro.
15. Kmetec, J. D., et al.(1992) *Phys. Rev. Lett.* **68**, 1527.
16. Chichkov, B. N., et al. (1996) *Applied Physics Letters* **68**, 2804-2806 .
17. Brunel, F. (1987) *Phys. Rev. Lett.* **59**, 52.
18. Wilks, S. C., Kruer, W. L., Tabak, M., and Langdon, A. B. (1992) *Phys. Rev. Lett.* **69**, 1383 .
19. Gibbon, P. (1994) *Phys. Rev. Lett.* **73**, 664.
20. Mountford, L. C., Smith, R. A., and Hutchinson, M. H. R. (1998) *Rev. Sci. Intrs.* **69**, 3780.
21. Kerker, M., (1969) Scattering of Light and Other Electromagnetic Radiation. Academic, New York. .
22. Halbleib, J. A., and Mehlhorn, T. A. (1986) *Nucl. Sci. Eng.* **92**, 338.
23. Langdon, A. B. and Lasinski, B. F. (1976) Methods in Computational Physics. Vol. 16, Academic Press, New York p. 327.

THE ABSORPTION OF ENERGY BY LARGE ATOMIC CLUSTERS FROM SUPER-INTENSE LASER PULSES

V.P. KRAINOV
Moscow Institute of Physics and Technology
141700 Dolgoprudny, Moscow Region, Russia
e-mail: krainov@cyberax.ru
and
M.B. SMIRNOV
Research Scientific Center "Kurchatov Institute"
123182 Moscow, Kurchatov Square 1, Russia

1. Introduction

A highly excited medium is formed as a result of the irradiation of the large clusters containing thousands of atoms or molecules by a super-intense ultra-short laser pulse of a tens, or a hundred of femtoseconds (i.e. of the order of 20 - 30 periods of the laser field). The heating of initially ionized electrons in clusters from rare gas atoms and impossibility of rapid rejection of thermal energy (unlike the usual hot macroscopic plasma) allow one to reach much larger excitation of the matter compared to the case of the isolated atoms and molecules. Atomic ions in the cluster ion are practically not heated during the whole laser pulse. After the quick initial multiple ionization, during the most part of the irradiation time this matter presents an ideal plasma consisting only from conducting electrons and multicharged ions. The evolution of the cluster in the laser field is of great interest.

2. Inner Ionization of Atoms Inside the Cluster

Initial rapid barrier-suppression ionization of valence atomic electrons inside the large cluster ($>10^4$ atoms) is produced by the laser pulse with the intensity $I = 10^{16} - 10^{18}$ W/cm^2. The H. Bethe's condition for the laser field strength F is of the form (Planck constant, electron mass and electron charge are equal to unity here and thereafter):

$$F > E_Z^2/4Z. \tag{1}$$

Here E_Z is the ionization potential of the valence atomic shells.

Initial Q free electrons are escaped by the super-intense laser field from the cluster (the so called "cold" emission). The positive cluster ion with the charge Q is produced. For example, the cluster Xe_{1100} loses its first 1000 electrons by direct electron removal in the laser field with the intensity $I = 10^{16}$ W/cm^2 [1].

The static Coulomb field strength of this cluster ion becomes equal to

$$F_{st} = Q / R^2 > F \tag{2}$$

at the cluster surface (R is the cluster radius).

B. Piraux and K. Rzążewski (eds.), Super-Intense Laser-Atom Physics, 219–226.

Further the inner barrier-suppression ionization of the next atomic shells inside the cluster ion is produced by the static electric field F_{st} at the cluster surface. The modified H. Bethe's condition is

$$F_{st} > E_Z^2/4Z. \tag{3}$$

The charge Q of the cluster ion continues to increase (also due to thermal evaporation of electrons from the cluster surface, see Section 3): previous stages are repeated. This is the so called "ignition model" for inner ionization [2].

Finally, inelastic electron-ion (e-2e) collisions produce the further (some times even total) inner ionization of all atomic electrons at high electron temperature T due to high charges of the produced atomic ions. The degree of ionization depends on the laser intensity.

All above cited processes occur during the leading edge of the ultra-short (30-100 fs) laser pulse.

The rate for inelastic electron – ion collisions A(Z) + e → A(Z+1) + 2e is given by the simple Lotz formula:

$$w = 9{,}76N\sqrt{\frac{T}{E_Z}}\exp\left(-\frac{E_Z}{T}\right) \text{ a.u.} \tag{4}$$

Here Z is (as above) the charge of atomic ion (do not mix with the cluster charge Q), E_Z is the ionization potential of this ion, the quantity N is the concentration of atoms inside the cluster. The Born approximation (i.e. the perturbation theory) is realized when the electron temperature T is higher than the ionization potential. But this limit does not achieved in clusters.

The Eq. (4) is valid when the thermal electron energy is larger than its ponderomotive energy in the external laser field. In the opposite limit we should substitute the thermal energy $3T/2$ by the average ponderomotive energy in Eq. (4), so that the obtained cross-section will be dependent on the laser field strength and on the laser frequency significantly.

The Saha statistical equation gives the values of the electron temperature T (in keV) for the total inner ionization of the cluster by e – 2e –collisions (see Table 1).

TABLE 1. An electron temperature (in keV) for the total inner ionization of various metal clusters

Li	Na	Al	K
0,046	0,34	0,50	0,78
Cu	Mo	Ag	Cs
2,06	3,77	4,54	4,90

It is seen that the values of the electron temperature are sufficiently low. The reason is the large pre-exponential factor in the Saha formula for the thermal equilibrium. The accumulated charge Q on the cluster resides on its surface.

3. Electron Emission from the Cluster

Here we consider the outer ionization, i.e. escape of free electrons from the cluster during the laser pulse. Cold (field) emission is the initial stage of the outer ionization only. Thermal evaporation of heated electrons from the cluster surface is the main process. It is determined by the well-known Richardson – Dushman formula:

$$\frac{dQ}{dt} = \frac{2}{\pi}(TR)^2 \exp\left(-\frac{Q}{RT}\right) \text{ a.u.} \tag{5}$$

Here Q/R is the ionization potential of the cluster ion (the work function can be neglected), and T is the electron temperature. Since $Q/R >> T$, a small part of free electrons leave the cluster, as a rule (though the deuterium clusters lose all their electrons, see Sect. 5).

The charge $Q(t)$ of the cluster ion after the time t of the laser pulse is found from (5) by integration over time t (with logarithmic accuracy):

$$Q(t) = RT \ln\frac{2TRt}{\pi}. \tag{6}$$

The energy conservation law for the cluster ion during the laser pulse is of the form

$$\underset{\substack{\uparrow \\ \text{escape of} \\ \text{electrons} \\ \text{from the cluster}}}{\frac{d}{dt}\left(\frac{Q^2}{2R}\right)} + \underset{\substack{\uparrow \\ \text{heating of} \\ \text{cluster} \\ \text{electrons}}}{\frac{d}{dt}\left(\frac{3}{2}n'T\right)} = \underset{\substack{\uparrow \\ \text{absorption} \\ \text{of laser} \\ \text{radiation by electrons}}}{I\sigma_a(n')n.} \tag{7}$$

The quantity n' is the number of free electrons in the cluster ion, n is the number of atoms in the cluster ion, σ_a is the cross-section for the absorption of the laser field (the mechanism of the absorption is discussed below).

The equation for derivation of the electron temperature $T(t)$ during the laser pulse is obtained from (6) and (7):

$$3n'T + 2RT^2 \ln^2\left(\frac{2TRt}{\pi}\right) = 2\sigma_a It. \tag{8}$$

Free electrons in the cluster plasma are heated up to the temperature of some keV in accordance with the experimental data. Ions are staying cold during the laser pulse.

The cluster presents a small plasma ball since the cluster radius R is significantly larger than a Debye length

$$l_D = \sqrt{T / 4\pi N_e} \,. \tag{9}$$

Here N_e is the free electron concentration inside the cluster ion.

4. Heating of Electrons in the Cluster Ion

Now we consider the various mechanisms of the heating of electrons in the cluster ion during the laser pulse.

1. Elastic electron – ion collisions. The rate (averaged over the Maxwell distribution) is given by the well known Spitzer formula

$$\nu_{ei} = \frac{4\sqrt{2\pi} Z N_e}{3T^{3/2}} \ln\Lambda \,. \tag{10}$$

Here N_e is the free electron concentration, $\ln\Lambda$ is the Coulomb logarithm, Z is the charge of the atomic ion. This expression is valid when the electron thermal energy is large compared to the ponderomotive energy. In the opposite limit of high laser intensity

$$T << U_p \tag{11}$$

The thermal electron energy in Eq. (10) should be substituted by the average ponderomotive energy (12) analogously as we discussed above after Eq. (4).

Average increase of the electron energy at each electron-ion collision due to inverse stimulated bremsstrahlung (the absorption of a photon by an electron is more probable than the emission of a photon) is equal to the ponderomotive energy (in a linearly polarized field):

$$\Delta E = U_p = \frac{F^2}{4\omega^2} \,. \tag{12}$$

Detailed discussion of this problem for rapid and slow electrons can be found in Ref. [3].

Due to electron-ion collisions the laser radiation penetrates inside the cluster to the *small* length compared to the cluster radius [4] (at $\omega << \nu_{ei}$):

$$l = \frac{cZ \ln\Lambda}{8\pi} \left(\frac{\pi}{2T} \right)^{3/2} << R \,. \tag{13}$$

The conclusion can be made that the laser radiation does not penetrate inside the hot cluster ion, though plasma skin depth is large compared to the cluster radius!

At $\omega >> \nu_{ei}$ the penetration length is also small, but it has another estimate:

$$l' = \frac{3c\omega^2 T^{3/2}}{32\pi\sqrt{2\pi}Z^2 N^2 \ln\Lambda} << R \cdot \tag{14}$$

The laser electric field density is shielded from the cluster interior also due to large plasma dielectric constant.

2. Model of Ch. Rhodes group [5]: one or several external electron shells of an atom inside the cluster are ionized simultaneously by the super-intense laser field and produces a "quasiparticle" with large charge. It is not clear how such a quasiparticle can exist during a long time being localized in small volume: it must quickly expanded because of the Coulomb repulsion between unshielded electrons of this quasiparticle.

3. Large positive potential Q/R at the surface of the cluster ion can be modeled as the *impenetrable wall* for inner free electrons. An electron can escape the cluster only if its kinetic energy exceeds the Coulomb potential energy at the cluster surface. An electron acquires (in average) the ponderomotive energy U_p at the elastic reflection from this wall in the presence of the laser field at each collision, since the probability for the absorption of a photon is greater than the probability for its emission. This is valid both at $T > U_p$ and at $T < U_p$ in the assumption of the intensive ee- and ei- collisions inside the cluster.

If the ee- and ei- collisions are rare, an electron acquires the energy which is much less than U_p at each elastic reflection from the wall. The dynamic chaos begins, and diffusion increasing of the electron energy takes place. This difficult numerical problem reduces to the *Sinai billiards* in the presence of the laser field. It is unsolved until now.

4. Alternative model was suggested by F. Brunel (*vacuum heating*) [6]. Free inner electrons having small kinetic energy, are dragged into the vacuum from the cluster and sent back into the cluster in a half of the laser field period acquiring of order of the ponderomotive energy U_p.

The maximum acquired electron energy is $3{,}17U_p$ [7]. But the average acquired electron energy is much less than U_p. If $T > U_p$, this problem is unsolved until now, since it has no analogy in the tunneling ionization of atoms.

5. The absorption cross-section of the laser energy due to collective excitation of dipole oscillations of the electron subsystem in the cluster (at $\nu << \omega$):

$$\sigma_a = \frac{\pi R^3 \nu}{c} \frac{\omega_{\mathrm{Mie}}^2}{(\omega - \omega_{\mathrm{Mie}})^2 + \nu^2/4}. \tag{15}$$

Here the resonance Mie frequency is given by

$$\omega_{\mathrm{Mie}} = \omega_p / \sqrt{3} \tag{16}$$

and

$$\omega_p = \sqrt{4\pi N_e} \tag{17}$$

is the plasma frequency.

The collision frequency ν is determined by one of the above cited mechanisms, so that this is not a new mechanism for absorption of the electromagnetic energy, but it can accelerate the previous mechanisms in the resonance case.

In the opposite limit $\nu \gg \omega$ the small non-resonance cross-section for absorption can be obtained from (15) in the form (n' is the number of the cluster electrons)

$$\sigma_a = \frac{4\pi}{c}\frac{n'}{\nu}. \tag{18}$$

5. Neutron Emission from Deuterium Clusters

Nuclear reaction $d + d \rightarrow {}^3He + n$ requires large kinetic energy of the colliding deuterons which is greater than $1/(4R_d)$ =180 keV in order to overcome the Coulomb repulsion between colliding deuterons. Here R_d = 1,96 fermi is the radius of the deuteron. Below this energy the tunnelling fusion occurs.

The deuterium cluster has sufficiently high concentration of deuterium atoms: it is equal to $5{\cdot}10^{22}$ cm^{-3}. It has the same order of magnitude as the atomic concentration in metal clusters. Though the atoms in deuterium cluster are attracted each to other by weak Van der Waals forces, the dimension of the deuterium core is small, inlike the core of atomic ions in metals, and this is the reason for such a high density.

Deuterium cluster ion with the charge $Q = 3000$ and with the initial cluster radius R = 25 Å can be stripped from all of their electrons before the cluster explodes. The mean kinetic energy of the each deuteron is ~ $Q/2R$ = 1 keV after the Coulomb expansion. The expansion time is longer than about 0,5 ps. The fusion is a relatively rare event.

The second mechanism is the sonic hydrodynamic expansion: the heated electrons expand and pull the cold heavy deuterium nuclei outward with them. The velocity is equal to the plasma sound speed [9]. Then the mean kinetic energy of the each deuteron is equal to the electron temperature T, i.e. of the order of some keV.

The experimental data suggest that the average deuterium ion energy is at least 2,5 keV [10] and even 4,5 keV according to recent data of T. Ditmire (see his paper in this Volume).

The rate for increasing of the electron temperature T due to reflection of the free electrons from cluster surface:

$$\frac{d}{dt}\left(\frac{3}{2}T\right)=\eta\frac{F^2}{4\omega^2}\frac{\sqrt{3T}}{2R}. \tag{19}$$

The phenomenological coefficient η takes into account the non-elasticity of the collision, the quantity $\sqrt{3T}$ is the electron thermal velocity.

The final electron temperature is (pulse length $\tau = 35$ fs, wavelength of 820 nm, laser intensity of 10^{17} W/cm^2, the cluster radius $R = 25$ Å):

$$T=\eta^2\frac{F^2\tau^2}{192\omega^4R^2}\sim 3\text{ keV}. \tag{20}$$

It should be noted that the second nuclear reaction d + d → t + p has the same cross-section with the proton kinetic energy of 3 MeV.

The rate of the tunnelling nuclear fusion averaged over the Maxwell distribution depends on the deuteron temperature T_D as (again in atomic units):

$$<\sigma V>=const\cdot T_D^{-2/3}\exp\left(-3\sqrt[3]{\frac{\pi^2M_D}{4T_D}}\right) \tag{21}$$

Here M_D is the deuteron mass. It decreases very slowly with the decreasing of the deuteron temperature. In particular, at $T_D = 1$ keV we obtain

$$<\sigma V>=10^{-21}\text{ cm}^3/\text{s}.$$

Since the cross-section for d + t nuclear reaction is more than 100 times greater than for d + d reaction, the yield of neutrons can be significantly increased if to add tritium atoms into the region of laser focus.

6. Conclusions

The absorption of the laser energy occurs in the vicinity of the cluster surface only. Elastic collisions of the free electrons with inner surface of the cluster ion are the important mechanism for the absorption of the laser energy by the cluster.

The authors acknowledge financial support from research grant N 01-02-16056 from the Russian Foundation for Fundamental Research.

7. References

1. Last, I. and Jortner, J. (1998) *J. Phys. Chem.* **102**, 9655.
2. Rose-Petruck, C. et al (1997) *Phys. Rev.* A **55**, 1182.
3. Krainov, V.P. (2000) *J. Phys.* B **33**, 1585.
4. Krainov, V.P. and Smirnov M.B. (2000) *Physics – Uspekhi*, **190**, 969.
5. McPherson, A. et al (1993) *Appl. Phys.* B **57**, 337.
6. Brunel F. (1987) *Phys. Rev. Lett.* **59**, 52.
7. Corkum, P. (1993) *Phys. Rev. Lett.* **71**, 1994.
8. Ditmire T. et al (1999) *Nature (London)* **398**, 489.
9. Ditmire T. et al (1996) *Phys. Rev.* A **53**, 3379.
10. Zweiback J. et al (1999) in L. DiMauro, R.R. Freeman and K.C. Kulander (eds.), *Multiphoton Processes - ICOMP VIII*, AIP Conference Proceedings 525, Melville NY, p. 534.

PART IV

PRODUCTION OF VERY INTENSE FEMTOSECOND LASER PULSES

A SEMI-CLASSICAL MODEL FOR HIGH-HARMONIC GENERATION

DEJAN B. MILOŠEVIĆ
Faculty of Science, University of Sarajevo
Zmaja od Bosne 35, 71000 Sarajevo, Bosnia and Herzegovina

1. Introduction

High-harmonic generation (HHG) is one of the major topics of intense-field laser-atom physics [1]–[4]. When an atomic gas (or any nonlinear material) is irradiated by a short intense laser pulse the atoms emit coherent radiation at frequencies that are multiples of the laser field frequency ω. The generation efficiency of these harmonics is characterized by a rapid drop at low orders, followed by a broad plateau with approximately constant efficiency and a sharp cutoff at frequencies $n_{\max}\omega \approx (I_{\mathrm{P}} + 3.17U_{\mathrm{P}})/\hbar$, where $I_{\mathrm{P}} = |E_0|$ is the atomic ionization potential and $U_{\mathrm{P}} = \langle e\mathbf{E}_{\mathrm{L}}\rangle^2/(2m\omega^2)$ is the electron ponderomotive energy in the laser electric $\mathbf{E}_{\mathrm{L}}$. The HHG process has been explained using the three-step quasiclassical model [5, 6]: the atom is ionized and the escaped electron is driven by the field and returns to the atomic core, emitting a harmonic photon in the transition back to the ground state. Both the classical [5, 6] and quantum [7, 8] calculations have confirmed this model.

Theory of HHG is usually considered in two steps: single-atom response and macroscopic response. In order to calculate the macroscopic response of the system, one has to solve Maxwell equations for the fundamental and harmonic fields (the so-called propagation theory [3]). We will concentrate here on the single-atom theory that describes the response of an atom on the fundamental laser field. According to [3], there are four methods that have been used to solve the problem of the single-atom response: (i) numerical methods, (ii) classical phase space averaging method, (iii) strong-field approximation (SFA), and (iv) pseudo-potential method based on the zero-range potential (ZRP) model. We will consider here only the semi-analytical approaches (iii) and (iv). The connection of the SFA [7] and ZRP [8] methods was analyzed in [9]. Both of them are based on

B. Piraux and K. Rzążewski (eds.), Super-Intense Laser-Atom Physics, 229–238.

the calculation of the real time-dependent effective-dipole matrix element $\mathbf{d}_{\text{ed}}(t) = \langle\Phi^{(+)}(t)|e\mathbf{r}|\Phi^{(+)}(t)\rangle$, where $|\Phi^{(+)}(t)\rangle$ is the wave vector which describes the electron in presence of both the laser field and the atomic potential. The harmonic spectrum is then obtained as the Fourier transform of $\mathbf{d}_{\text{ed}}(t)$. In [9] the difference between the S-matrix approach and the effective-dipole method is mentioned. In the S-matrix approach one calculates the complex quantity $\mathbf{d}_{\text{sm}}(t) = \langle\Phi^{(-)}(t)|e\mathbf{r}|\Phi^{(+)}(t)\rangle$, where the wave vector $|\Phi^{(+)}(t)\rangle$ ($|\Phi^{(-)}(t)\rangle$) evolves from the initial (final) state under the action of the total Green's operator $G^{(+)}$ ($G^{(-)}$). It was mentioned that the effective-dipole expectation value is required as the source term for the integration of the Maxwell equations. However, if just the harmonic emission of a single atom is to be considered, then the S-matrix element is the quantity of choice. The difference between these approaches deserves a further analysis.

Recently, some new results were published [10]–[15], which put a new light on the above-mentioned methods (iii) and (iv). There is an indication (N. Manakov, private communication) that the effective-dipole method gives the divergent result if one does not use the SFA, while the S-matrix method does not suffer from this problem. More precisely, it was shown [15] that, within the quasistationary quasienergy approach, instead of the time-dependent effective-dipole matrix element one has to use $\mathbf{d}_{\text{dual}} = \langle\Phi^{(+)}_{\text{dual}}(t)|e\mathbf{r}|\Phi^{(+)}(t)\rangle$, where $\Phi^{(+)}_{\text{dual}}(t)$ is the dual state which corresponds to the state $\Phi^{(+)}(t)$. The dual quasistationary states were introduced in order to have the proper normalization condition. The results obtained using the dual-space method coincide (within some approximations) with those of the S-matrix approach. The S-matrix formulation of the atomic processes in the laser field can be found in [16]. The first attempt to apply the S-matrix theory to HHG was done by Ehlotzky [17]. There are also different approaches according to how one treats the laser field and the harmonic field. Usually, both fields are treated classically (see [7] for the effective-dipole approach and [11] for a method similar to the S-matrix approach which gives the complex time-dependent dipole). In Refs. [13, 14] both fields are treated quantum-electrodynamically (i.e., with the photon creation and annihilation operators), and the HHG problem is analyzed as a time-independent scattering process so that the stationary formalism can be used. However, it is possible, by appropriate transformations, to introduce the time dependence into this approach, which enables an interpretation in terms of the three-step model (D.B. Milošević, unpublished). In the present contribution we will use a mixed representation in which the laser field is classical, while the harmonic field is treated quantum-electrodynamically. This approach is justified by the fact that the quantum-electrodynamical treatment of the electromagnetic field, within the laser approximation (i.e., for a large number of photons in the field mode), is equivalent to the classical treatment [16]. In particular, we will use the method presented in [10], in which the laser-assisted x-ray–atom scattering was considered.

The limitation of the fully numerical approaches is that they are quite time consuming. For more complicated cases such as elliptically polarized and bichromatic laser fields, presence of static fields, relativistically strong laser field, etc., the semi-analytical methods (iii) and (iv) are more useful. These methods can be further simplified using the saddle-point-method approximation (SPA) [18, 12], which we will consider in the last section.

2. *S*-matrix Theory

Our starting point is a general form of the S-matrix [19] (in SI units)

$$S_{\rm fi} = i\hbar \lim_{t'\to\infty} \lim_{t\to-\infty} \langle \Phi_{\rm out}(t')|G^{(+)}(t',t)|\Phi_{\rm in}(t)\rangle. \tag{1}$$

The total Green's operator G corresponds to the Hamiltonian

$$H(t) = H_0(t) + e\mathbf{r}\cdot\mathbf{E}_{\rm H}(t),\ H_0(t) = \frac{\mathbf{p}^2}{2m} + V_{\rm A} + H_{\rm L}(t), \tag{2}$$

where $e\mathbf{r}\cdot\mathbf{E}_{\rm H}(t)$, $e = |e|$, is the interaction of the atom with the high-harmonic field (in the length gauge and in the dipole approximation), and $V_{\rm A}$ is the atomic potential. The laser-atom interaction is $H_{\rm L}(t) = e\mathbf{r}\cdot\mathbf{E}(t)$, $\mathbf{E}(t) = \mathbf{E}_{\rm L}(t) + \mathbf{E}_{\rm S}$, where $\mathbf{E}_{\rm L}(t)$ and $\mathbf{E}_{\rm S}$ are the laser and the static electric field vector, respectively. In order to distinguish the high-harmonic radiation field from the (classical) laser field, we consider it as quantized, i.e., according to [20], we define

$$\mathbf{E}_{\rm H}(t) = \sum_{\mathbf{K}} c_{\mathbf{K}} \left(a^{+}_{\mathbf{K}} e^{i\omega_{\mathbf{K}}t}\hat{\mathbf{e}}^{*}_{\mathbf{K}} - a_{\mathbf{K}} e^{-i\omega_{\mathbf{K}}t}\hat{\mathbf{e}}_{\mathbf{K}}\right),\ c_{\mathbf{K}} = -i\left(\frac{\hbar\omega_{\mathbf{K}}}{2\varepsilon_0\mathcal{V}}\right)^{1/2}, \tag{3}$$

where $a_{\mathbf{K}}$ and $a^{+}_{\mathbf{K}}$ are the annihilation and creation operators of the high-harmonic field photon corresponding to the wave vector $\mathbf{K}$, frequency $\omega_{\mathbf{K}}$, and complex unit polarization vector $\hat{\mathbf{e}}_{\mathbf{K}}$. $\mathcal{V}$ is the quantization volume. The interaction with the laser field is off for the *in* and *out* states, so that

$$|\Phi_{\rm in}(t)\rangle = |\psi_{0\rm i}(t)\rangle|0_{\mathbf{K}}\rangle,\ |\Phi_{\rm out}(t)\rangle = |\psi_{0\rm f}(t)\rangle|1_{\mathbf{K}}\rangle, \tag{4}$$

where $|\psi_{0j}(t)\rangle = |\psi_{0j}\rangle\exp(-iE_{0j}t/\hbar)$, $j = \rm i, f$. The atomic state vectors $|\psi_{0j}\rangle$ and energies E_{0j} satisfy the stationary Schrödinger equation $[\mathbf{p}^2/(2m) + V_{\rm A}]|\psi_{0j}\rangle = E_{0j}|\psi_{0j}\rangle$. Ket vectors $|0_{\mathbf{K}}\rangle$ and $|1_{\mathbf{K}}\rangle$ stand for no high-harmonic photon in the initial state and one high-harmonic photon in the final state, respectively. The total Green's operator satisfies

$$G^{(+)}(t,t') = G_0^{(+)}(t,t') + \int dt'' G^{(+)}(t,t'')\, e\mathbf{r}\cdot\mathbf{E}_{\mathrm{H}}(t'')G_0^{(+)}(t'',t'), \tag{5}$$

where the Green's operator $G_0^{(+)}$ corresponds to the Hamiltonian H_0. Introducing (5) into (1), we obtain

$$S_{\mathrm{fi}} = i\hbar \lim_{t'\to\infty}\lim_{t\to-\infty}\langle\psi_{0\mathrm{f}}(t')|\int dt''\langle 1_{\mathbf{K}}|G^{(+)}(t',t'')|1_{\mathbf{K}}\rangle$$
$$\times c_{\mathbf{K}} e\mathbf{r}\cdot\hat{\mathbf{e}}^*_{\mathbf{K}} e^{i\omega_{\mathbf{K}}t''} G_0^{(+)}(t'',t)|\psi_{0\mathrm{i}}(t)\rangle. \tag{6}$$

Using once more Eq. (5), we get $\langle 1_{\mathbf{K}}|G^{(+)}|1_{\mathbf{K}}\rangle = G_0^{(+)}$. This relation is exact if we consider only the interaction corresponding to the high-harmonic photon emission, otherwise it is valid up to the second order in the interaction with the high-harmonic field. From this and the relations

$$i\hbar \lim_{t'\to\infty}\langle\psi_{0\mathrm{f}}(t')|G_0^{(+)}(t',t'') = \langle\Phi_{\mathrm{f}}^{(-)}(t'')|,$$
$$i\hbar \lim_{t\to-\infty} G_0^{(+)}(t'',t)|\psi_{0\mathrm{i}}(t)\rangle = |\Phi_{\mathrm{i}}^{(+)}(t'')\rangle, \tag{7}$$

where the states $|\Phi_j^{(\pm)}(t)\rangle$, $j=\mathrm{i},\mathrm{f}$, satisfy the time-dependent Schrödinger equation with the Hamiltonian $H_0(t)$, we get

$$S_{\mathrm{fi}} = -\frac{i}{\hbar}c_{\mathbf{K}}\int_{-\infty}^{\infty} dt\, e^{i\omega_{\mathbf{K}}t}\,\hat{\mathbf{e}}^*_{\mathbf{K}}\cdot\mathbf{d}_{\mathrm{fi}}(t),\quad \mathbf{d}_{\mathrm{fi}}(t) = \langle\Phi_{\mathrm{f}}^{(-)}(t)|e\mathbf{r}|\Phi_{\mathrm{i}}^{(+)}(t)\rangle, \tag{8}$$

where $\mathbf{d}_{\mathrm{fi}}(t)$ is the time-dependent dipole matrix element between the initial and final laser-field dressed atomic states. Eq. (8) represents a general form of the S-matrix element for the emission of one high-harmonic photon of frequency $\omega_{\mathbf{K}}$, wave vector $\mathbf{K}$, and polarization $\hat{\mathbf{e}}_{\mathbf{K}}$.

3. Time-dependent Dipole Matrix Element and the SFA

In this section we will analyze the time-dependent dipole matrix element $\mathbf{d}_{\mathrm{fi}}(t)$. The wave vectors $|\Phi_j^{(\pm)}(t)\rangle$, $j=\mathrm{i},\mathrm{f}$, satisfy

$$|\Phi_j^{(\pm)}(t)\rangle = |\psi_{0j}(t)\rangle + \int dt' G_0^{(\pm)}(t,t')H_{\mathrm{L}}(t')|\psi_{0j}(t')\rangle. \tag{9}$$

Introducing this into the second equation in (8), we get

$$\mathbf{d}_{\mathrm{fi}}(t) \equiv \mathbf{d}_{\mathrm{sm}}(t) = \langle\psi_{0\mathrm{f}}(t)|e\mathbf{r}|\psi_{0\mathrm{i}}(t)\rangle + \mathbf{d}_{\mathrm{a}}(t) + \mathbf{d}_{\mathrm{e}}(t) + \mathbf{d}_{\mathrm{c-c}}(t), \tag{10}$$
$$\mathbf{d}_{\mathrm{a}}(t) = \langle\psi_{0\mathrm{f}}(t)|e\mathbf{r}\int dt' G_0^{(+)}(t,t')H_{\mathrm{L}}(t')|\psi_{0\mathrm{i}}(t')\rangle. \tag{11}$$

For equal initial and final states, the term $\langle\psi_{0\mathrm{f}}(t)|e\mathbf{r}|\psi_{0\mathrm{i}}(t)\rangle$ is equal to zero for any spherically symmetric potential. The term $\mathbf{d}_{\mathrm{a}}(t)$ describes the process in which, after the laser-atom interaction at the time t' when photons are absorbed from the laser field (that is why we denoted this term with "a"), the system propagates until the time t, when a high-harmonic photon is emitted during the transition to the final state $|\psi_{0\mathrm{f}}(t)\rangle$. This term corresponds to the three-step model described in the introduction. The term $\mathbf{d}_{\mathrm{e}}(t)$ corresponds to the time-reversed of the "a" process (it is denoted with "e" because a harmonic photon is emitted first). Finally, the term $\mathbf{d}_{\mathrm{c-c}}(t)$ corresponds to the process which consists of five steps. It is denoted by "c–c" because the harmonic photon is emitted between the two propagations that mainly include continuum states. We will further analyze the result (11) by employing the SFA [21], i.e. supposing that, in the intermediate states, the atomic interaction V_{A} can be neglected in comparison with the interaction with the laser field $H_{\mathrm{L}}(t)$, so that the Green's operator G_0 can be approximated by the Volkov Green's operator G_{L} [9, 10]. By expressing G_{L} through the Volkov wave vectors in the length gauge, we get

$$\mathbf{d}_{\mathrm{a}}(t) = -\frac{i}{\hbar}\int d^3\mathbf{q}\,\langle\psi_{0\mathrm{f}}|e\mathbf{r}|\mathbf{q}+\frac{e}{\hbar}\mathbf{A}(t)\rangle$$
$$\times\int_{-\infty}^{t} dt'\exp\left[\frac{i}{\hbar}S(\mathbf{q};t,t')\right]\langle\mathbf{q}+\frac{e}{\hbar}\mathbf{A}(t')|H_{\mathrm{L}}(t')|\psi_{0\mathrm{i}}\rangle, \tag{12}$$

$$S(\mathbf{q};t,t') = E_{0\mathrm{f}}t - \int_{t'}^{t} dt''\frac{\hbar^2}{2m}\left[\mathbf{q}+\frac{e}{\hbar}\mathbf{A}(t'')\right]^2 - E_{0\mathrm{i}}t', \tag{13}$$

with $\mathbf{E}(t) = -\partial\mathbf{A}(t)/\partial t$. The three-dimensional integral over the intermediate electron momenta $\hbar\mathbf{q}$ can be presented as a series [22]

$$\hbar^3\int d^3\mathbf{q}\,M(\mathbf{q};t,t')\exp\left[\frac{i}{\hbar}S(\mathbf{q};t,t')\right] = \left[\frac{2\pi m\hbar}{i(t-t')}\right]^{3/2}$$
$$\times\exp\left[\frac{i}{\hbar}S(\mathbf{q}_{\mathrm{s}};t,t')\right]\left[1 - i\frac{m\hbar}{2(t-t')}\frac{\partial^2}{\hbar^2\partial\mathbf{q}^2}+\cdots\right]M(\mathbf{q};t,t')\,|_{\mathbf{q}=\mathbf{q}_{\mathrm{s}}(t,t')}, \tag{14}$$

where M denotes the product of matrix elements in Eq. (12), and where the stationary momentum $\hbar\mathbf{q}_{\mathrm{s}}$, with $m\boldsymbol{\alpha}(t) = e\int^t dt'\mathbf{A}(t')$,

$$\hbar\mathbf{q}_{\mathrm{s}}(t,t') = -\frac{e}{t-t'}\int_{t'}^{t} dt''\mathbf{A}(t'') = \frac{m}{t-t'}[\boldsymbol{\alpha}(t')-\boldsymbol{\alpha}(t)], \tag{15}$$

is the solution of the equation $\boldsymbol{\nabla}_{\mathbf{q}}S(\mathbf{q};t,t') = \mathbf{0}$. The higher order terms in (14) can be neglected in the *semi-classical approximation* (expansion in $\hbar$). For some models that we will consider later this integral can be solved exactly. Similar expressions can be derived for other terms in (10). In Fig. 1

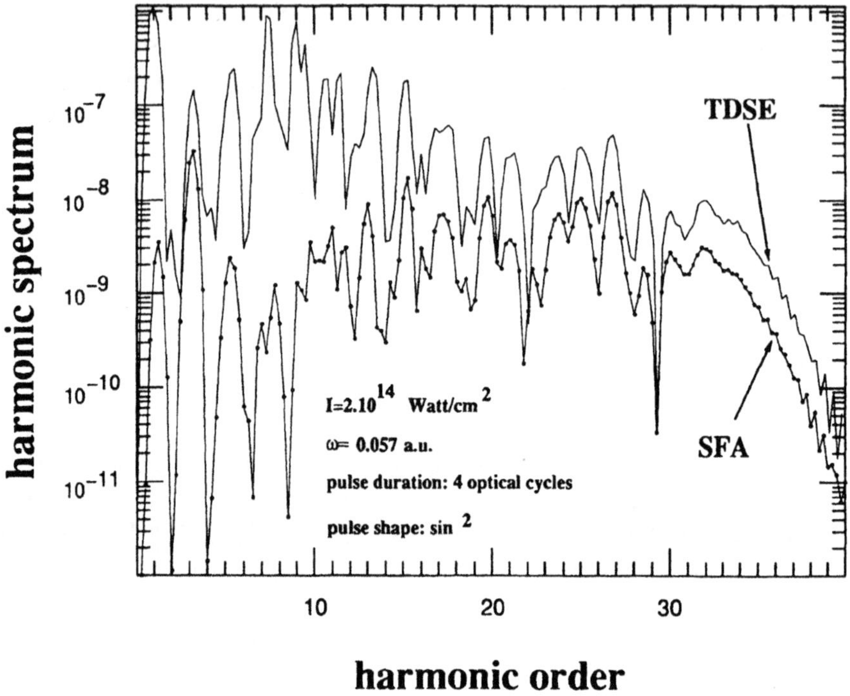

Figure 1. Comparison of the harmonic spectrum for hydrogen atoms, obtained using the SFA and the numerical solution of the time-dependent Schrödinger equation (TDSE, Piraux *et al.*). The laser intensity and frequency are $I = 2\times10^{14}$ W/cm^2 and $\omega = 0.057$ a.u.; the pulse shape is sinsquare having duration of 4 optical cycles.

we compare the results for harmonic spectrum obtained as the Fourier transform of the time-dependent acceleration for "exact" solution of the time-dependent Schrödinger equation with those obtained using the SFA and the above-described method. The agreement is good for the plateau and cutoff harmonics (up to an overall factor about 5), while for low harmonics ($\hbar\omega_{\mathbf{K}} \leq I_{\mathrm{P}}$) the SFA fails [23, 24].

The continuum-continuum term is more difficult to calculate because it includes the Green's function twice. This is a five-step process and, therefore, we expect that its probability is low. In Fig. 2 we show on an example that it can be neglected in comparison with other contributions. The results are obtained using the saddle-point approximation which will be described later. Another way of dealing with this "c–c" term is to combine it with the "a" and "e" terms and to use the ZRP model [9].

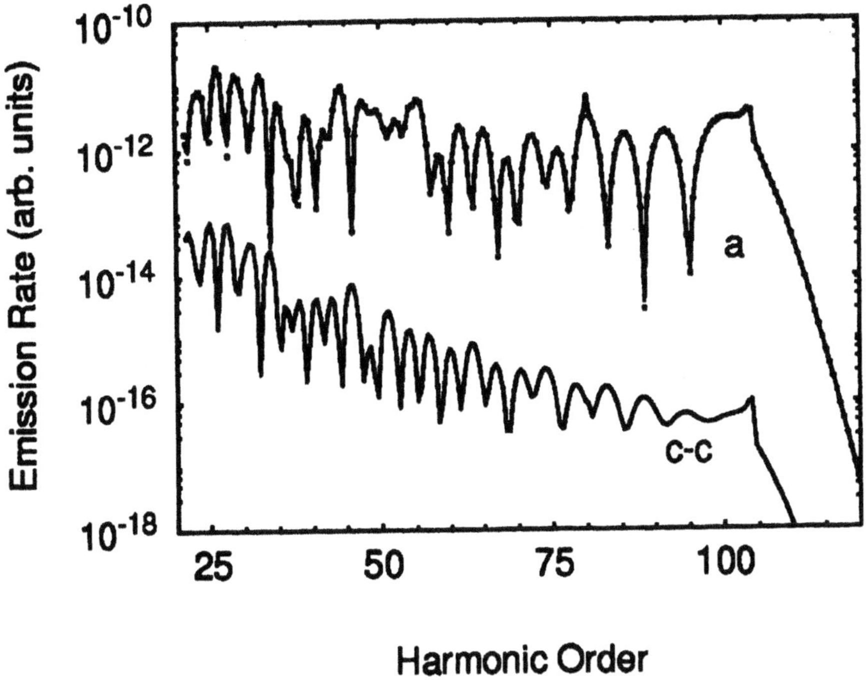

Figure 2. Comparison of the "a" and "c–c" contributions to the harmonic spectrum for $I_P = 21.6$ eV, laser wavelength 800 nm and intensity 7×10^{14} W/cm^2.

4. Comparison of the S-matrix and the Effective-dipole Method

The main contribution to our S-matrix time-dependent dipole matrix element (10) comes from two parts: $\mathbf{d}_{\rm sm}(t) \approx \mathbf{d}_{\rm a}(t) + \mathbf{d}_{\rm e}(t)$. In the effective-dipole approach [7], the corresponding matrix element can be written as: $\mathbf{d}_{\rm ed}(t) = \mathbf{d}_{\rm a}(t) + \mathbf{d}_{\rm a}^*(t) = 2\mathbf{d}_{\rm a}^{\rm R}(t)$, where $\mathbf{d}^{\rm R}$ ($\mathbf{d}^{\rm I}$) denotes the real (imaginary) part of $\mathbf{d}$. We have calculated the relative difference between the harmonic emission rates, obtained using these two approaches, $|w_{n,\rm sm} - w_{n,\rm ed}| / w_{n,\rm sm}$, as a function of the harmonic order n, for various laser field and atomic parameters. For low harmonics the observed differences are less than 40%, while for high harmonics they are negligible. In order to explain why $w_{n,\rm sm} \approx w_{n,\rm ed}$ for $n \gg 1$, we recall that

$$w_n(\hat{\mathbf{e}}_{\mathbf{K}}) = \frac{\omega_{\mathbf{K}}^3}{8\pi^2\varepsilon_0\hbar c^3}\,|T_n|^2\,, \quad \hbar\omega_{\mathbf{K}} + E_{0\rm f} = n\hbar\omega + E_{0\rm i}, \tag{16}$$

where the corresponding T-matrix elements are: $T_{n,\rm sm} = \hat{\mathbf{e}}_{\mathbf{K}}^* \cdot \mathcal{F}(\mathbf{d}_{\rm a} + \mathbf{d}_{\rm e})$ and $T_{n,\rm ed} = \hat{\mathbf{e}}_{\mathbf{K}}^* \cdot \mathcal{F}(2\mathbf{d}_{\rm a}^{\rm R})$, and $\mathcal{F}$ denotes the Fourier transform. Our numerical calculations have shown that: $\mathcal{F}(\mathbf{d}_{\rm a}^{\rm R}) \approx i\mathcal{F}(\mathbf{d}_{\rm a}^{\rm I})$ and $\mathcal{F}(\mathbf{d}_{\rm e}^{\rm R}) \approx -i\mathcal{F}(\mathbf{d}_{\rm e}^{\rm I})$, so that $\mathcal{F}(\mathbf{d}_{\rm a} + \mathbf{d}_{\rm e}) \approx 2\mathcal{F}(\mathbf{d}_{\rm a}^{\rm R})$ and $T_{n,\rm sm} \approx T_{n,\rm ed}$. Another consequence of

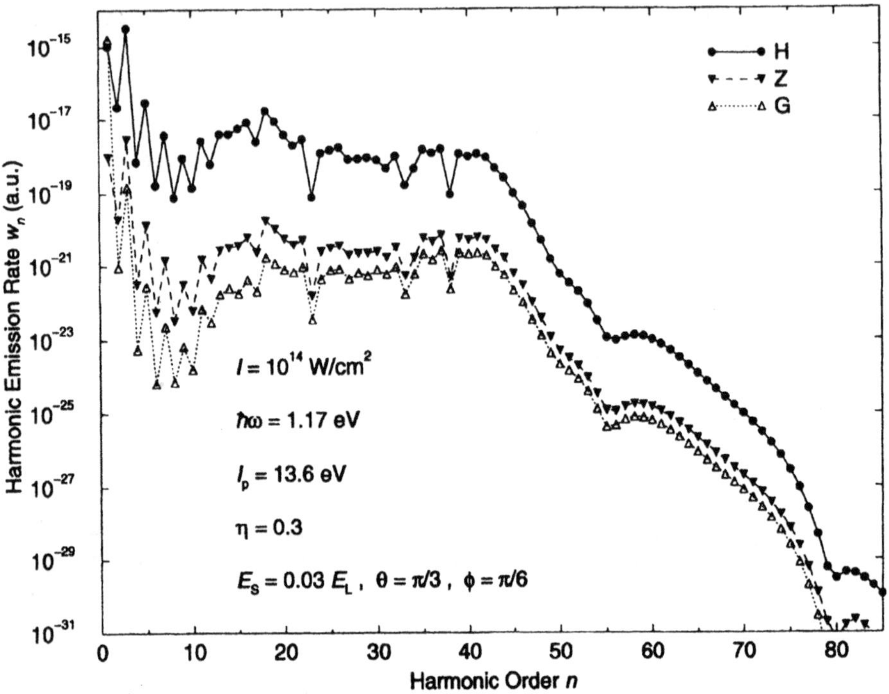

Figure 3. Harmonic emission rates as functions of the harmonic order for three different atomic models: hydrogen-like atoms (H), zero-range potential (Z), and Gaussian (G) model; $I_P = 13.6$ eV. The results correspond to the case of the presence of both an elliptically polarized laser field (with $\varphi_1 = \pi/2$ and $\varphi_2 = 0$) and a static electric field having the strength that is 3% of the laser field strength. The laser field frequency, intensity and ellipticity are $\hbar\omega = 1.17$ eV, $I = 10^{14}$ W/cm^2, and $\eta = 0.3$, respectively. The static field orientation is defined with the angles $\theta = \pi/3$ and $\phi = \pi/6$.

the above results is that the contribution of $\mathbf{d}_e(t)$ can be neglected (for an explicit comparison see Fig. 2 in [10]). A general conclusion is that the S-matrix and the effective-dipole method give almost the same results for the harmonic emission rates, except for low harmonics. However, for low harmonics the SFA is questionable [23].

Let us consider how the emission rates depend on the atomic models. We consider an elliptically polarized laser field $\mathbf{E}_L(t) = \sum_{j=1,2} \hat{\mathbf{e}}_j E_j \sin(\omega t + \varphi_j)$, $E_1 = E_L/\sqrt{1+\eta^2}$, $E_2 = \eta E_1$, where φ_1 and φ_2 are arbitrary phases, η is the ellipticity, $-1 \leq \eta \leq 1$, and $\hat{\mathbf{e}}_1 = \hat{\mathbf{e}}_x$, $\hat{\mathbf{e}}_2 = \hat{\mathbf{e}}_y = \hat{\mathbf{k}} \times \hat{\mathbf{e}}_1$, $\hat{\mathbf{e}}_3 = \hat{\mathbf{e}}_z = \hat{\mathbf{k}}$. We also allow a simultaneous presence of an arbitrary oriented static electric field $\mathbf{E}_S = E_S\,(\sin\theta\cos\phi\,\hat{\mathbf{e}}_1 + \sin\theta\sin\phi\,\hat{\mathbf{e}}_2 + \cos\theta\,\hat{\mathbf{e}}_3)$. We calculate the time-dependent dipole matrix element using Eqs. (10)–(15), addapted to analyzed models [7, 8]. The results presented in Fig. 3 show that the emission rates as functions of the harmonic order have qualitatively the same behaviour for all presented models. The main cutoff appears about $n = 41$. This is close to the cutoff position for a linearly polarized laser field (without $\mathbf{E}_\mathrm{S}$), which is $I_\mathrm{P}+3.17U_\mathrm{p} \approx 40\hbar\omega$. The static field induces a shift of the cutoff position to higher values, while the laser field ellipticity induces a shift to lower values, so that the nett effect is that the cutoff position is almost unchanged. The even harmonics and the additional cutoffs also appear, due to the influence of the static field. The harmonic order which corresponds to these cutoffs is higher, but their emission rates are much lower. The qualitative atomic-model independence of the obtained results indicates that the main behaviour of the emission rates as functions of the harmonic order is determined by the quasiclassical action $S(\mathbf{q};t,t') + \omega_\mathbf{K} t$, that appears in the exponent of the S-matrix element, (8) and (10)–(15).

5. A Semi-classical Method Based on the SPA

In Refs. [18, 12] a semi-classical method that is based on the saddle-point-method approximation was developed. This method is applied to solve the five-dimensional integral in Eqs. (8) and (12). It is particularly useful when the laser field is not linearly polarized or if it is relativistically strong [25]. Details of this method are given in [12, 26] for an elliptically polarized field and in [27, 28] for bichromatic fields. For applications a special case of "bicircular" field is important [27, 29, 30]. We do not have enough space here to give all details of this method. Due to lack of space, we only mention here that, defining the quasiclassical action $S_n(\mathbf{q};t,t') = S(\mathbf{q};t,t') + n\omega t$, one obtains that the T-matrix element can be approximated by

$$T_n \propto \sum_s M_s \left[\det\left(\frac{\partial^2 S_s}{\partial q_k \partial q_l}\right)\right]^{-1/2} \exp\left(\frac{i}{\hbar} S_s\right), \tag{17}$$

where $S_s \equiv S_n(\mathbf{q}_s, t_s, t'_s)$ is evaluated at the saddle points q_i $(i = 1,\ldots,5)$ which comprise the five variables $\mathbf{q}_s \equiv (q_{1s}, q_{2s}, q_{3s})$, t_s, and t'_s. The summation in (17) is over an appropriate subset of the saddle points $(\mathbf{q}_s, t_s, t'_s)$, which are the solutions of equations that express the energy conservation at times t' and t and the condition that the electron returns to its starting point $\mathbf{r}(t) = \mathbf{r}(t')$. This gives a generalization of the three-step model. Due to the tunneling nature of the first step of this process the solutions of these equations are complex. Projecting these solutions into the real plane one obtains the real-space electron trajectories which are very useful for physical understanding of the process [12, 25, 26, 27, 29]. The method described in this section is closely related to the Feynman path integral approach to quantum mechanics. The practical importance of the result (17) is that only few complex solutions (the shortest trajectories) give the main contribution so that the calculations are very fast. Comparison with the

SFA results is given in [12, 25, 27]. In the present work we have used the multi-dimensional SPA (with the boundary stationary points) to calculate the continuum-continuum dipole matrix element (see Fig. 2). A relativistic generalization of the S-matrix method, based on the Klein-Gordon instead of the Schrödinger equation, is given in [25], where numerical results for ultrahigh-harmonic generation by highly charged ions are presented.

References

1. B. Piraux, A. L'Huillier, and K. Rzążewski (eds.) (1993) *Super-Intense Laser-Atom Physics*, NATO Advanced Study Institutes Ser. B, Vol. 316, Plenum, New York.
2. L.F. DiMauro, R.R. Freeman, and K.C. Kulander (eds.) (2000) *Multiphoton Processes*, AIP Conference Proceedings No. 525, Melville, New York.
3. Salières, P., L'Huillier, A., Antoine, Ph., and Lewenstein, M. (1999) *Adv. At. Mol. Opt. Phys.* **41**, 83.
4. Brabec, Th. and Krausz, F. (2000) *Rev. Mod. Phys.* **72**, 545.
5. Krause, J.L., Schafer, K.J., and Kulander, K.C. (1992) *Phys. Rev. Lett.* **68**, 3535; Kulander, K.C., Schafer, K.J., and Krause, J.L., in Ref. [1], p. 95.
6. Corkum, P.B. (1993) *Phys. Rev. Lett.* **71**, 1994.
7. Lewenstein, M., Balcou, Ph., Ivanov M.Yu., L'Huillier, A., and Corkum, P.B. (1994) *Phys. Rev. A* **49**, 2117.
8. Becker, W., Long, S., and McIver, J.K. (1994) *Phys. Rev. A* **50**, 1540.
9. Becker, W., Lohr, A., Kleber, M., and Lewenstein, M. (1997) *Phys. Rev. A* **56**, 645.
10. Milošević, D.B. and Ehlotzky, F. (1998) *Phys. Rev. A* **58**, 2319–2326.
11. Kuchiev, M.Yu. and Ostrovsky, V.N. (1999) *Phys. Rev. A* **60**, 3111.
12. Kopold, R., Milošević, D.B., and Becker, W. (2000) *Phys. Rev. Lett.* **84**, 3831–3834.
13. Gao, J., Shen, F., and Eden, J.G. (1998) *Phys. Rev. Lett.* **81**, 1833; (1999) *J. Phys. B* **32**, 4153–4161; (2000) *Phys. Rev. A* **61**, 043812.
14. Gao, L., Li, X., Fu, P., Freeman, R.R., and Guo, D.-S. (2000) *Phys. Rev. A* **61**, 063407.
15. Borca, B., Flegel, A.V., Frolov, M.V., Manakov, N.L., Milošević, D.B., and Starace, A.F. (2000) *Phys. Rev. Lett.* **85**, 732–735.
16. Mittleman, M.H. (1993) *Introduction to the Theory of Laser-Atom Interactions*, 2nd edn, Plenum, New York.
17. Ehlotzky, F. (1992) *Nuovo Cimento* **14**, 517.
18. Lewenstein M., Salières P., and L'Huillier A. (1995) *Phys. Rev. A* **52**, 4747.
19. Bjorken, J.D. and Drell, S.D. (1964) *Relativistic Quantum Mechanics*, McGraw-Hill, New York.
20. Loudon, R. (1983) *The Quantum Theory of Light*, Clarendon, Oxford.
21. Keldysh, L.V. (1964) *Zh. Eksp. Teor. Fiz.* **47**, 1945; Faisal, F.H.M. (1973) *J. Phys. B* **6**, L89; Reiss, H.R. (1980) *Phys. Rev. A* **22**, 1786; (1990) **42**, 1476.
22. Milošević, D.B. and Piraux, B. (1996) *Phys. Rev. A* **54**, 1522.
23. Antoine, Ph., Piraux, B., Milošević, D.B., and Gajda, M. (1997) *Laser Phys.* **7**, 594.
24. de Bohan, A., Antoine, Ph., Milošević, D.B., and Piraux, B. (1998) *Phys. Rev. Lett.* **81**, 1837–1840.
25. Milošević, D.B., Hu, S.X., and Becker, W. (2000) *Phys. Rev. A* (to be published).
26. Milošević, D.B. (2000) *J. Phys. B* **33**, 2479–2488.
27. Milošević, D.B., Becker, W., and Kopold, R. (2000) *Phys. Rev. A* **61**, 063403.
28. Figueira de Morisson Faria, C., Milošević, D. B., and Paulus, G. G. (2000) *Phys. Rev. A* **61**, 063415.
29. Milošević, D.B. and Becker, W. (2000) *Phys. Rev. A* **62**, 011403(R).
30. Milošević, D.B. and Sandner, W. (2000) *Opt. Lett.* **25**, 1532–1534.

BEYOND THE SIMPLE MAN'S MODEL FOR HIGH HARMONIC GENERATION

MIROSŁAW BREWCZYK
Uniwersytet w Białymstoku
ul. Lipowa 41, 15-424 Białystok, Poland

AND

KAZIMIERZ RZĄZEWSKI
Centrum Fizyki Teoretycznej PAN and College of Science
Al. Lotników 32/46, 02-668 Warsaw, Poland

1. Introduction

Since the first experiments on noble gases exposed to intense laser pulses have been performed and the generation of high harmonics observed [1], there was a hope of getting a source of coherent extreme ultraviolet (XUV) radiation and consequently for a number of applications. Most of this work was done using a low–frequency lasers (Nd:glass, Ti:sapphire) with the photon energies much smaller than the ionization potential. In this case it turned out that the single atom response consists of only odd–integer multiples of the fundamental frequency. The harmonics spectrum falls off for a few first harmonics then a wide frequency region with harmonics of comparable strength (so called plateau) is developed followed by the sharp cutoff. An obvious question connected with the possible applications was the location of the cutoff.

It was shown both experimentally and theoretically that the maximum photon energy emitted in the atomic harmonic response is well described by the simple formula $I_p + 3U_p$, where I_p is the ionization potential, and U_p ($e^2E^2/4m\omega^2$) is the ponderomotive energy, i.e., the mean kinetic energy acquired by a free electron oscillating in the laser field. This cutoff law formula is supported by the simple semiclassical model [2, 3] as well as by rigorous solution of the time–dependent Schrödinger equation within the single active electron approximation [4]. Regarding the first approach (the simple man's model) the electron tunnels through the barrier formed

B. Piraux and K. Rzążewski (eds.), Super-Intense Laser-Atom Physics, 239–248.

by the atomic potential and the electric field of the laser. Once released with the zero velocity it moves classically in the laser field. The maximum energy acquired by the electron from the field when it returns to the nucleus is about $3U_p$. Assuming the electron recombines on return, the photon with the energy $I_p + 3U_p$ is emitted. This simple physical picture was also recovered by a fully quantum mechanical theory developed in Ref. [5]. In several experiments [6] the production of harmonics in excess of 100th order was observed (photon energy above 100 eV).

Recently the attention was focused on the propagation of high harmonics through the ionized medium [7], manipulation of the polarization of the driving field in connection with the production of attosecond pulses [8], and the diagnostics of ultrashort pulses through the properties of the harmonics [9, 10].

2. Model

We investigate the interaction of a multielectron atom with intense radiation at the frequency in vacuum ultraviolet (VUV) range. Such sources of radiation are now being tested at the TESLA test facility at the DESY laboratory. In the first stage the free–electron laser amplifier, based on the self–amplification of spontaneous emission, will allow for a production of photons with energies up to 30 eV and intensities higher than 10^{16} W/cm^2. We have shown in Ref. [11], that the over–the–barrier ionization of valence electrons is suppressed in the regime where the electric field and the photon energy of the laser are comparable to the value of atomic electric field and the binding energy, respectively (see Fig. 1), i.e., in the range of parameters which will be available soon. This suppression can be easily understood in terms of the Keldysh tunneling parameter which depends linearly on the laser frequency. So, increasing the frequency the Keldysh parameter eventually becomes greater than 1 and the ionization is described as a multiphoton process. In such a case one should not expect the generation of harmonics via the simple man's model. Fortunately, the inner–shell electrons are getting sensitive to such a radiation. The core electrons start to vibrate for these frequencies and emit highly energetic (of the order of 1 keV) photons fully replacing the closing of the mechanism so important for optical wavelengths. To find the harmonic spectrum of a multielectron atom we have calculated the electric dipole moment of the electron cloud within the hydrodynamic model. The distribution of the frequencies emitted by the atom is given by the Fourier transform of the dipole moment [12]. Briefly, we describe the electrons as a charged fluid. The Hamiltonian of the system

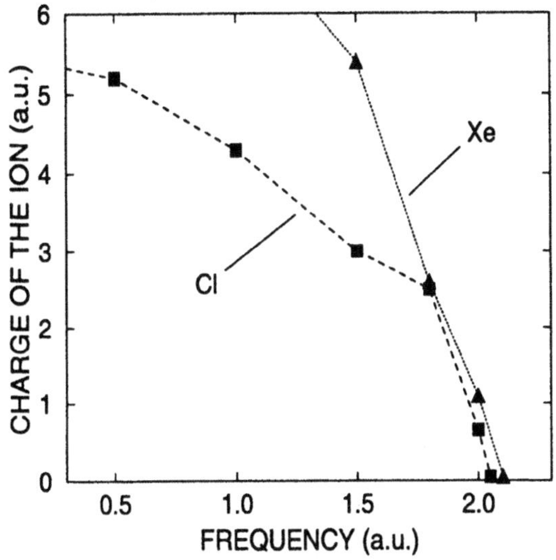

Figure 1. Charge of atomic chlorine and xenon ions as a function of frequency for peak radiation intensity equal to 1 a.u..

is then written in the form (in atomic units):

$$\begin{aligned} \mathrm{H} &= \mathrm{T} + \int V_{ext}(\vec{r},t)\rho(\vec{r})\, d^3r + \frac{1}{2}\int\int \frac{\rho(\vec{r})\rho(\vec{r}\,')}{|\vec{r}-\vec{r}\,'|}\, d^3r\, d^3r' + \\ &\quad \frac{1}{2}\int \rho\,(\nabla\chi)^2 d^3r \ , \end{aligned}$$

where the density $\rho\,(\vec{r},t)$ and the velocity potential $\chi(\vec{r},t)$ are used. The functional T is an intrinsic kinetic energy (the kinetic energy of a locally uniform noninteracting Fermi gas at zero temperature in the simplest approximation), the last term corresponds to the kinetic energy of the macroscopic flow, the second term describes the interaction with the external fields whereas the third the Coulomb repulsion of electrons. Assuming variables ρ and χ are canonically conjugate one can introduce the Lagrangian

$$\mathrm{L} = \int \rho \frac{\partial\chi}{\partial t}\, d^3r - \mathrm{H}$$

and extremizing the action, derive the following set of equations [13]:

$$\begin{aligned} \frac{\partial\rho(\vec{r},t)}{\partial t} + \nabla\cdot[\rho(\vec{r},t)v(\vec{r},t)] &= 0 \\ \frac{\partial\vec{v}(\vec{r},t)}{\partial t} + [\vec{v}(\vec{r},t)\cdot\nabla]\vec{v}(\vec{r},t) &= -\frac{1}{\rho(\vec{r},t)}\nabla\mathrm{P}(\vec{r},t) + \nabla\Phi(\vec{r},t) \\ \nabla\mathrm{P} = \rho\nabla\frac{\delta\mathrm{T}}{\delta\rho} \ , \end{aligned}$$

where the pressure P is already defined and the electric potential Φ is calculated from the following expression:

$$\Phi(\vec{r},t) = \frac{Z}{(b^2+\vec{r}^{\,2})^{1/2}} - \int \frac{\rho(\vec{r}\,',t)\ d^3r\,'}{[c^2+(\vec{r}-\vec{r}\,')^2]^{1/2}} + E(t)\,x\ .$$

One– and two–dimensional versions of the model require smoothing parameters (b,c) which, in fact, are not independent [13]. Smoothing the Coulomb interactions, of course, overestimates the energy of tightly bound electrons.

3. Results

In Fig. 2 we show the maximum number of harmonics produced by the Thomas-Fermi atom as a function of driving frequency; the radiation intensity is 3.5×10^{16} W/cm^2. The cutoff formula predicts no harmonic generation already for the frequency equal to 1 a.u. for such an intensity. However, as we can see in Fig. 2, the production of higher harmonics is quite efficient. For example, the energy of most energetic photons emitted while the driving frequency is equal to 2 a.u. almost attains 2 keV. Certainly, another mechanism of harmonic generation has to be involved for higher than optical driving frequencies while the simple man's model mechanism is getting closed.

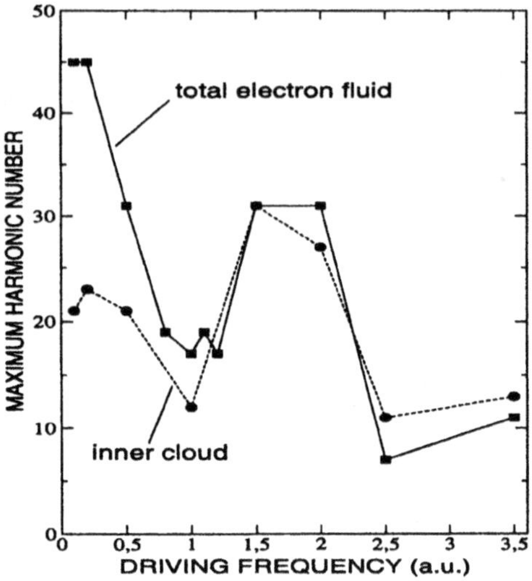

Figure 2. Maximal number of harmonics produced by the total electron fluid (squares) as well as by the inner–cloud electrons (circles). The decrease in the harmonic generation is observed for the driving frequency around 1 a.u.

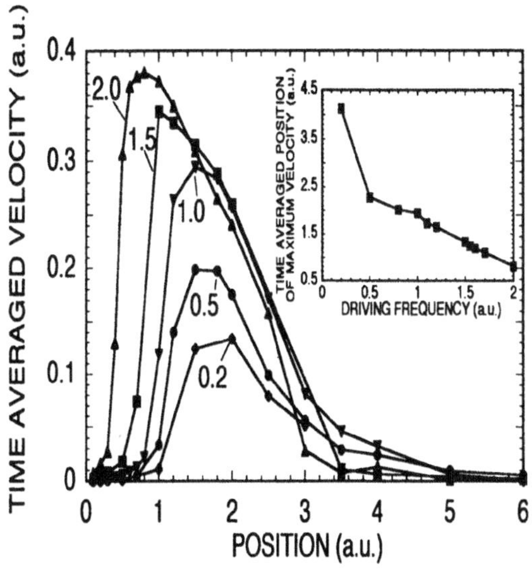

Figure 3. Time–averaged velocity of the electron fluid vs distance from the nucleus. For higher frequencies (marked by the numbers next to the curves) tightly bound electrons start to vibrate leading to the production of photons with the energy up to 1 keV. Inset: Time–averaged position of the maximum velocity of the electron fluid as a function of driving frequency.

To understand the origin of the production of such energetic photons we have plotted some time–averaged properties characterizing the interaction of a multielectron atom with intense radiation field. Fig. 3 (the inset) shows time–averaged (only a "flat" part of the pulse is taken into account) position of the electron fluid element with maximum velocity. When the pumping frequency increases the strongest electron disturbance is shifted towards the nucleus starting from about 4 Bohr radii for the optical wavelengths and reaching the distance below one Bohr radius for the frequency equal to 2 a.u.. The higher frequency the more inner electrons are forced to move rapidly as it is demonstrated in the main frame of Fig. 3 where we plotted the time–averaged velocity of the electron fluid vs the distance from the nucleus. For higher driving frequency the core electrons start to vibrate and produce harmonics. It is illustrated in Fig. 4 where we show the relative intensity of harmonics generated by the electron cloud being within the space interval (-0.2 a.u., 0.2 a.u.) around the nucleus, i.e., their ratios to

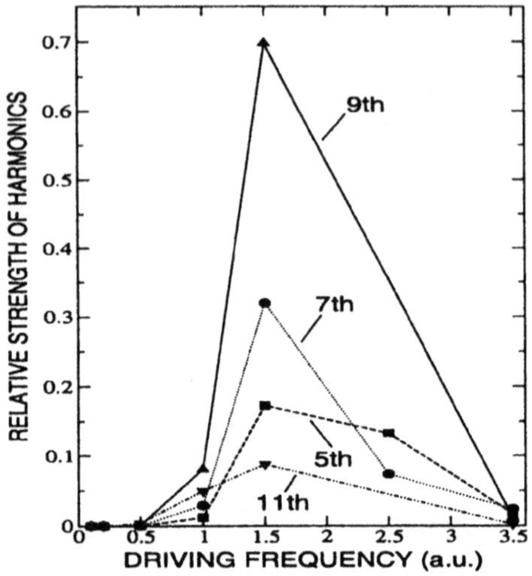

Figure 4. Relative strength of some harmonics produced by the total electron fluid and by the vicinity of the nucleus vs the driving frequency. The chosen harmonics are: 9th (triangles up), 7th (circles), 5th (squares), and 11th (triangles down).

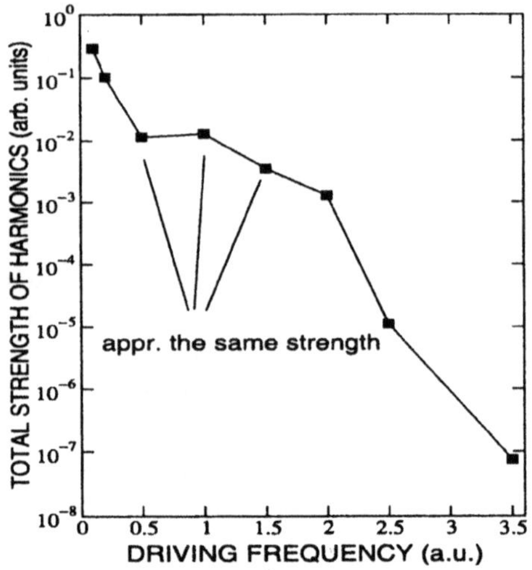

Figure 5. Total strength of the harmonics except the first one as a function of driving frequency. Sudden decrease in harmonics production for the fundamental frequency around 1 a.u. visible in Fig. 2 is compensated by the even harmonics; the total strength stays at the same level.

the intensities of harmonics produced by the whole electron cloud. It is clear that this contribution becomes larger for higher frequencies. The same is also visible in Fig. 2; for optical frequencies the location of the cutoff is determined by the response of the whole electron cloud whereas above 1 a.u. the cutoff position is governed by the core electrons.

In Fig. 5 we have plotted the total strength of harmonics (except the first one) as a function of driving frequency. Although the harmonics production is diminished for the driving frequencies around 1 a.u. (see Fig. 2), it does not influence the total strength. The reason why it happens is explained in Fig. 6. It turns out that for a narrow interval of frequencies around 1 a.u. the even harmonics are generated. So, although the maximum order of harmonic is low, they are "doubled" and the total intensity stays at the same level as for lower and higher frequencies. Interestingly, the even harmonics are in fact in the form of doublets, whose splitting is comparable to the distance between the frequency of neighboring modes. Even harmonics

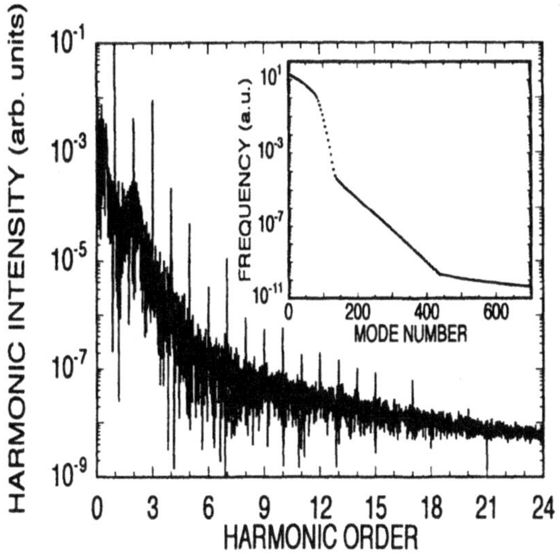

Figure 6. The harmonic spectrum obtained for a chlorine atom at 3.5×10^{16} W/cm^2 driving radiation intensity and 27.2 eV photon energy. Both odd and even harmonics are present. Inset: Normal modes spectrum for a one-dimensional multielectron Thomas-Fermi atom.

were also observed in numerical simulations in Ref. [14] and comprehensively discussed in Ref. [15]. To find the origin of the production of even harmonics we have investigated the response of the Thomas-Fermi atom to an oscillatory perturbation using standard linear–response analysis.

In the inset of Fig. 6 we have plotted the normal modes spectrum of the one–dimensional Thomas-Fermi atom. It shows a rapid transfer of modes from the set with frequencies very close to zero to the set with frequencies 1 a.u. and above. Looking at the shape of the modes which belong to the first group we find that the outer parts of the electron cloud are significantly disturbed (see part (b) in Fig. 7) whereas in the latter case the core electrons oscillate (part (a) in Fig. 7). Moreover, two groups of modes have different phase relation with respect to the disturbance. One is in–phase (the second group) while the other is 90^0 out–of–phase (the first group). For intermediate frequencies (around 1 a.u.) both types of modes, depending on the spectral width of the laser pulse, can be excited leading to the production of even harmonics. We checked that, for example, decreasing laser intensity even harmonics disappear (Fig. 8). Although the Thomas-Fermi model of an atom does not have the shell structure, the analysis of normal modes of the electron cloud clearly shows that increasing the laser frequency inner-cloud electrons are getting disturbed. Therefore we expect to see the even harmonics in the spectrum of a real atom every time the driving frequency excites two consecutive shells.

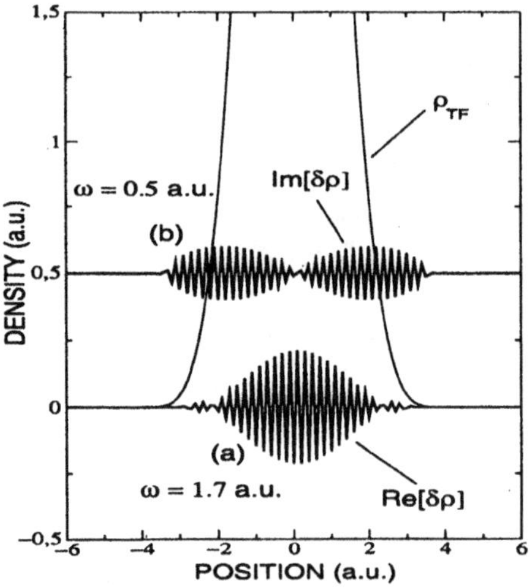

Figure 7. (a) Real part of a high–energy mode. (b) Imaginary part of a low–energy mode (shifted upwards). Solid line marked by ρ_{TF} shows static Thomas-Fermi density.

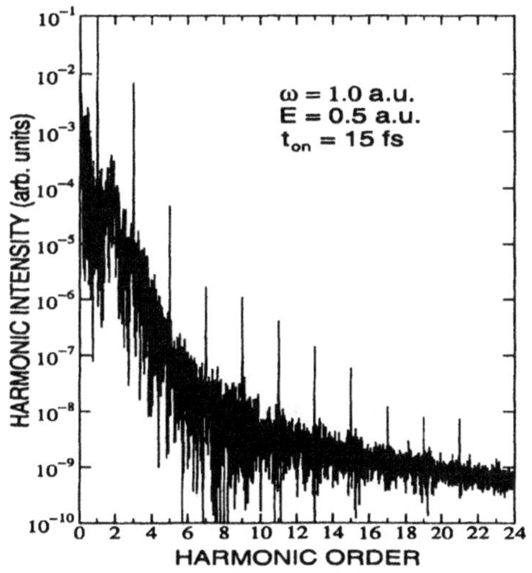

Figure 8. The harmonic spectrum obtained for a chlorine atom at 8.7×10^{15} W/cm^2 driving radiation intensity and 27.2 eV photon energy. Only odd harmonics are present.

4. Conclusions

We have investigated the response of a multielectron atom to the strong (intensity of the order of 1 a.u.) VUV free–electron laser radiation. Our study shows that for such a high driving frequency a new physical mechanism for the generation of high harmonics emerges. The lack of harmonics via the simple man's model mechanism is replaced by the production of harmonics due to the vibration of inner–shell electrons. The highest photon energy found is well in excess of 1 keV. We predict the appearance of even harmonics in the spectrum which could be a sign of "touching" the consecutive electronic shells (more tightly bound assuming the driving frequency increases).

Acknowledgements

This work was supported by the KBN Grant No. 2P03B10313. K.R. acknowledges support of the Subsidy by Foundation for Polish Science. CPU time has been provided by the Interdisciplinary Center for Mathematical and Computational Modeling at Warsaw University.

References

1. McPherson, A., Gibson, G., Jara, H., Johann, U., Luk, T.S., McIntyre, I., Boyer, K., and Rhodes, C.K. (1987), *J. Opt. Soc. Am. B* **4**, 595-601; Li, X.F., L'Huillier, A., Ferray, M., Lompré, L.A., and Mainfray, G. (1989), *Phys. Rev. A* **39**, 5751-5761; Sarukura, N., Hata, K., Adachi, T., Nodomi, R., Watanabe, M., and Watanabe, S. (1991), *Phys. Rev. A* **43**, 1669-1672; Miyazaki, K. and Sakai, H. (1992), *J. Phys. B* **25**, L83-89; Balcou, Ph., Cornaggia, C., Gomes, A.S.L., Lompré, L.A., and L'Huillier, A. (1992) *J. Phys. B* **25**, 4467-4485.
2. Kulander, K.C., Schafer, K.J., and Krause, J.L. (1993), in B. Piraux, A. L'Huillier, and K. Rzążewski (eds.), *Super-Intense Laser-Atom Physics*, NATO ASI Series, Plenum Press, New York and London, pp. 95-110.
3. Corkum, P.B. (1993), *Phys. Rev. Lett.* **71**, 1994-1997.
4. Krause, J.L., Schafer, K.J., and Kulander, K.C. (1992), *Phys. Rev. Lett.* **68**, 3535-3538.
5. Lewenstein, M., Balcou, Ph., Ivanov, M.Yu., L'Huillier, A., and Corkum, P.B. (1994), *Phys. Rev. A* **49**, 2117-2132.
6. Macklin, J.J., Kmetec, J.D., and Gordon III, C.L. (1993), *Phys. Rev. Lett.* **70**, 766-769; L'Huillier, A. and Balcou, Ph. (1993), *Phys. Rev. Lett.* **70**, 774-777.
7. Constant, E., Garzella, D., Breger, P., Mével, E., Dorrer, Ch., Le Blanc, C., Salin, F., and Agostini, P. (1999), *Phys. Rev. Lett.* **82**, 1668-1671.
8. Corkum, P.B., Burnett, N.H., and Ivanov, M.Yu. (1994), *Opt. Lett.* **19**, 1870-1872; Antoine, Ph., L'Huillier, A., and Lewenstein, M. (1996), *Phys. Rev. Lett.* **77**, 1234-1237.
9. Chang, Z., Rundquist, A., Wang, H., Murnane, M.M., and Kapteyn, H.C. (1997), *Phys. Rev. Lett.* **79**, 2967-2970; Spielmann, Ch., Burnett, N.H., Sartania, S., Koppitsch, R., Schnürer, M., Kan, C., Lenzner, M., Wobrauschek, P., and Krausz, F. (1997), *Science* **278**, 661-664.
10. Bohan, A., Antoine, Ph., Milošević, D.B., and Piraux, B. (1998), *Phys. Rev. Lett.* **81**, 1837-1840.
11. Brewczyk, M. and Rzążewski, K. (1999), *J. Phys. B* **32**, L1-4.
12. Kulander, K.C. and Shore, B.W. (1989), *Phys. Rev. Lett.* **62**, 524-526.
13. Brewczyk, M. and Rzążewski, K. (1999), *Phys. Rev. A* **60**, 2285-2295.
14. Kulander, K.C., Schafer, K.J., and Krause, J.L. (1991), *Phys. Rev. Lett.* **66**, 2601-2604.
15. Ivanov, M.Yu. and Corkum, P.B. (1993), in B. Piraux, A. L'Huillier, and K. Rzążewski (eds.), *Super-Intense Laser-Atom Physics*, NATO ASI Series, Plenum Press, New York and London, pp. 63-72.

ANISOTROPY-INDUCED POLARIZATION EFFECTS IN HARMONIC GENERATION BY AN ABSORPTIVE MEDIUM

B. BORCA,* D.B. MILOŠEVIĆ,+ A.F. STARACE
Dept. of Physics and Astronomy, The University of Nebraska
116 Brace Laboratory, Lincoln, NE 68588-0111, USA

AND

A.V. FLEGEL, M.V. FROLOV, N.L. MANAKOV
Voronezh State University, 394693 Voronezh, Russia

1. Introduction

High-order harmonic generation (HHG) has become a main topic of intense laser-atom physics [1]. Both to understand the HHG process better and to enhance its use in applications, many studies have focused on the control of HHG [1]. Among the most recent are those concerned with the polarization characteristics of the harmonics [2-8]. Other studies have focused on the use of a strong static electric field to control the intensities and plateau structure of HHG [9-12]. In this paper we present results of a general formulation of the problem of HHG in the case when the generating medium is anisotropic (see also [13]). Our specific results are for the case of anisotropy introduced by a strong, static electric field. However, we discuss also other ways of introducing anisotropy for which our analysis applies and for which similar results may be expected. We show that a static electric field has striking effects on the polarization of high-order harmonics and on the dependence of harmonic yield on the laser field ellipticity. We demonstrate that: (i) in the presence of a static electric field the harmonics are in general elliptically polarized, even for a linearly polarized driving laser (which contrasts with results obtained in the absence of a static field [5]), and (ii) the static electric field leads to a significant elliptic dichroism effect, i. e., the intensity of a harmonic differs substantially for *right* and *left* helicities of an *elliptically* polarized laser beam. We also discuss elliptic dichroism in the case when the generating medium is isotropic but a polarization sensitive detector is used for harmonic intensity measurements. All of these results

B. Piraux and K. Rzążewski (eds.), Super-Intense Laser-Atom Physics, 249–258.

demonstrate the possibility of significant control of polarization effects in HHG.

2. Symmetry Analysis

We consider a single atom interacting with a laser field of arbitrary polarization, i.e.,

$$\mathbf{F}(\mathbf{r},t) = F\mathrm{Re}\left\{\mathbf{e}\exp\left[i\left(\mathbf{k}\cdot\mathbf{r}-\omega t\right)\right]\right\}, \tag{1}$$

where the complex polarization vector $\mathbf{e}$ is parametrized in a coordinate-frame-invariant way using the laser field wave-vector $\mathbf{k}$ and the unit vector $\hat{\varepsilon}$ along the major axis of the laser polarization ellipse,

$$\mathbf{e} = \frac{\hat{\varepsilon} + i\eta\left(\hat{\mathbf{k}}\times\hat{\varepsilon}\right)}{\sqrt{1+\eta^2}}, \quad -1 \leq \eta \leq +1. \tag{2}$$

Here η is the ellipticity, where $\eta = +1$ (-1) corresponds to right (left) circular polarization and $\eta = 0$ to linear polarization. In terms of these vectors, the degree of linear polarization is given by $l = \mathbf{e}\cdot\mathbf{e} = (1-\eta^2)/(1+\eta^2)$, and the degree of circular polarization by $\xi = i\hat{\mathbf{k}}\cdot(\mathbf{e}\times\mathbf{e}^*) = 2\eta/(1+\eta^2)$, both of which are simply related to the usual Stokes parameters S_1, S_2, S_3 (i. e., $\xi = S_2$, $l = \sqrt{S_1^2+S_3^2}$) [14]. The static electric field $\boldsymbol{\mathcal{F}}$ is oriented along the direction $\hat{\mathbf{e}}_0$, which we assume in the following to lie in the plane of the laser polarization ellipse, perpendicular to $\mathbf{k}$.

Complete information concerning the intensity and polarization properties of the nth harmonic may be extracted from the amplitude $A_{n\omega}(\mathbf{e}')$ describing dipole emission of a harmonic (with frequency $\omega' = n\omega$, *measured* polarization $\mathbf{e}'$, and propagation direction $\hat{\mathbf{k}}' = \hat{\mathbf{k}}$) by an atom in the presence of fields $\boldsymbol{\mathcal{F}}$ and $\mathbf{F}$, where

$$A_{n\omega}(\mathbf{e}') = \mathbf{e}'^* \cdot \tilde{\mathbf{d}}_n, \tag{3}$$

and $\tilde{\mathbf{d}}_n$, defined in Eq. (9), is a (complex) matrix element of the dipole operator. It is useful at this point to specify what can be deduced from symmetry considerations alone, independent of any theoretical model for obtaining numerical results. Taking into account, as in [3], the spatial and temporal symmetry properties of the vectors involved [i.e., $\mathbf{F}(\mathbf{r},t)$, $\boldsymbol{\mathcal{F}}$, $\mathbf{e}$, $\mathbf{e}^*$, and $\tilde{\mathbf{d}}_n$] as well as the fact that only the part of $\tilde{\mathbf{d}}_n$ transverse to $\hat{\mathbf{k}}$ contributes to Eq. (3), one may express $\tilde{\mathbf{d}}_n$ in general as,

$$\tilde{\mathbf{d}}_n = \chi_1\mathbf{e} + \chi_2\mathbf{e}^* + \chi_3\hat{\mathbf{e}}_0. \tag{4}$$

The complex scalars $\chi_{1,2,3}$ are the nonlinear susceptibilities, which depend only on non-vector parameters (e. g., ω, F, $\mathcal{F}$, l) and on the angle φ between

$\hat{\varepsilon}$ and $\hat{\mathbf{e}}_0$. Their values can be obtained by numerical calculation, based on a specific theoretical model. The total intensity of the nth harmonic, summed over its polarization states, is proportional to $|\tilde{\mathbf{d}}_n|^2$, where:

$$|\tilde{\mathbf{d}}_n|^2 = \sum_{i=1}^{3} |\chi_i|^2 + \sqrt{2(1+l)}\,\mathrm{Re}\left[(\chi_1 + \chi_2)\,\chi_3{}^*\right] \cos\varphi$$
$$+ \quad 2l\,\mathrm{Re}(\chi_1\chi_2{}^*) - \xi\sqrt{2/(1+l)}\,\mathrm{Im}\left[(\chi_1 - \chi_2)\,\chi_3{}^*\right]\sin\varphi. \qquad (5)$$

The last term in Eq. (5), involving ξ, exhibits elliptic dichroism: it has opposite signs for right ($\xi > 0$) and left ($\xi < 0$) elliptic polarization of the laser field. Obviously this term vanishes for $\xi = 0$ (linear laser polarization). It vanishes also for $|\xi| = 1$ (circularly polarized laser light), but this is not obvious from Eq. (5) and requires a detailed analysis of the l- and φ-dependence of the susceptibilities χ_i, e. g., in terms of higher order perturbative expansions in F (cf. [3]). Symmetry considerations imply that when $\varphi = \pi/2$ the elliptic dichroism term also vanishes. Therefore, when $0 < |\xi| < 1$ and $0 < \varphi < \pi/2$, *elliptic* dichroism may be observed, caused by an interference between the real and imaginary parts of $\chi_{1,2}$ and the static-electric-field-induced component of the susceptibility tensor, χ_3.

The polarization properties of the nth harmonic are described by its Stokes parameters S_i^n, $1 \le i \le 3$ [14]. These are defined in terms of the intensity of the nth harmonic having a detected polarization $\mathbf{e}'$: $I_{n\omega}(\xi', \theta) \propto |A_{n\omega}(\mathbf{e}')|^2$, where θ is the angle between the directions $\hat{\varepsilon}$ and $\hat{\varepsilon}'$ of the major axes of the polarization ellipses respectively of the laser and the detected harmonic. Thus S_1^n equals the difference between $I_{n\omega}(\xi' = 0, \theta)$ for $\theta = \pm\pi/4$ divided by the sum. Similarly S_2^n involves $I_{n\omega}(\xi' = \pm 1, \theta)$, and S_3^n involves $I_{n\omega}(\xi' = 0, \theta)$ for $\theta = 0,\ \pi/2$. For a linearly polarized laser, Eq. (4) shows that there are only two independent susceptibilities, $\chi_{||} = \chi_1 + \chi_2 + \chi_3 \cos\varphi$ and $\chi_{\perp} = \chi_3 \sin\varphi$. One easily finds that the harmonic's intrinsic degree of circular polarization, ξ_n, and offset angle, θ_n, are:

$$\xi_n \equiv S_2^n = \frac{2\,\mathrm{Im}\left(\chi_{\perp}\chi_{||}^*\right)}{|\chi_{||}|^2 + |\chi_{\perp}|^2}, \qquad (6)$$

$$\tan 2\theta_n \equiv \frac{S_1^n}{S_3^n} = \frac{2\,\mathrm{Re}\left(\chi_{\perp}\chi_{||}^*\right)}{|\chi_{||}|^2 - |\chi_{\perp}|^2}. \qquad (7)$$

As for elliptic dichroism, the ellipticity ξ_n originates from an interference of the real and imaginary parts of the nonlinear susceptibilities and may be observed for $0 < \varphi < \pi/2$. In contrast, the offset angle originates from the obvious anisotropy of the atom in a static field and is nonzero even for real

$\chi_{\|}$ and $\chi_{\perp}$. Note that each of the two dissipation-induced effects [Eqs. (5) and (6)] is described by a different combination of the χ_i's: $\mathrm{Im}\,[(\chi_1 \pm \chi_2)\chi_3^*]$ thus independent measurements of the two effects give complete information on the parameters which govern these effects: $\chi_1\chi_3^*$ and $\chi_2\chi_3^*$.

3. Complex Quasienergy Approach for the Calculation of HHG

The above symmetry analysis is independent of any dynamical model, but quantitative estimations of the magnitude of the effects require numerical calculations of the amplitudes in Eq. (3). An accurate, *ab initio* way to calculate them nonperturbatively is to use the quasistationary, quasienergy states (QQES) or non-Hermitian Floquet states (see, e. g., [15]) for an atom in strong laser and static electric fields, $\Phi_{\mathcal{E}}(\mathbf{r},t)$, where $\mathcal{E}$ is the complex quasienergy. Because of their asymptotically divergent terms in r (in the open ionization channels), the QQES wavefunctions are not normalizable in the standard way; thus matrix elements of the dipole emission operator $\mathbf{r}$ between QQES functions diverge when calculated in the usual way. Instead, properly normalized (dual) functions, $\tilde{\Phi}_{\mathcal{E}}(\mathbf{r},t)$, must be used as bra vectors in a calculation of transition matrix elements [16, 17]. In the presence of two fields, $\boldsymbol{\mathcal{F}}$ and $\mathbf{F}(\mathbf{r},t)$ with an arbitrary elliptical polarization, the proper dual functions are given by

$$\tilde{\Phi}_{\mathcal{E}}(\mathbf{r},t) = [\Phi_{\mathcal{E}}(\mathbf{r},-t;\eta \to -\eta)]^* . \tag{8}$$

For $\eta = 0$, the dual function in Eq. (8) coincides with that introduced by Potvliege and Shakeshaft [16] for a linearly polarized laser field. The time-dependent (complex) dual dipole moment is thus calculated as

$$\tilde{\mathbf{d}}(t) = \langle\tilde{\Phi}_{\mathcal{E}}(\mathbf{r},t)|\mathbf{r}|\Phi_{\mathcal{E}}(\mathbf{r},t)\rangle = \frac{1}{2}\sum_n \tilde{\mathbf{d}}_n e^{-in\omega t}, \tag{9}$$

using $\tilde{\Phi}_{\mathcal{E}}(\mathbf{r},t)$ as the bra vector. The nth Fourier coefficient, $\tilde{\mathbf{d}}_n$, of $\tilde{\mathbf{d}}(t)$ is then used in Eq. (3) for the generation amplitude.

Based on this QQES approach, we analyze the HHG amplitude using a 3-dimensional zero-range potential model for the atom. The QQES-solution for this model [18] has been used in recent HHG-calculations (see, e.g., [19, 9, 10]). We note that if (as done here) the quasienergy $\mathcal{E}$ is approximated by the unperturbed binding energy, E_0, of the model atom and if all but the leading Fourier-coefficients of $\Phi_{\mathcal{E}}(\mathbf{r},t)$, a_n, at the origin ($r \to 0$) are neglected, as in [19], then our approximate QQES amplitude $A_{n\omega}(\mathbf{e}')$ coincides with that obtained in the so-called "S-matrix approach" [7, 10, 20] provided that the latter takes into account the continuum-continuum terms [7]. Ref. [7] discusses the existence of two different definitions for the HHG amplitude, denoted by the terms "S-matrix" and "dipole-moment expectation value". In fact, the authors of [7] argued for the validity of the "dipole-moment expectation value" definition based on Ref. [19], where the expectation value of $\mathbf{r}$ was calculated in the QQES approach, without using $\tilde{\Phi}_{\mathcal{E}}(\mathbf{r}, t)$. As discussed above, this expectation value is divergent. (The divergence of the result in Ref. [19] for $n = 1$ is explicitly shown in that paper, and, for higher n, divergences appear only if one calculates the HHG amplitude beyond the approximation $a_n \sim \delta_{n0}$. However, this latter approximation is a good one for low frequencies ($\omega \ll 1$) and for the intensities which are considered in that paper; therefore their numerical results are consistent with those of the "S-matrix" calculations.) The divergences inherent in the dipole expectation value imply therefore that it cannot be used for *ab initio* calculations of HHG.

4. Results and Discussion for the Zero-Range Potential Model

In the zero-range potential model for calculations of HHG by an elliptically polarized laser in the presence of a static electric field, each susceptibility $\chi_{1,2,3}$ in Eq. (4) involves an infinite sum of one-dimensional time integrals of a product of Bessel functions. These integrals were evaluated numerically. In order to make our numerical results applicable to a variety of atoms and field parameters, we adopt scaled units: our energy unit is the atom binding energy $|E_0|$, and our electric field strength unit is the internal field, $F_0 = \sqrt{2m|E_0|^3}/e\hbar$. The displayed results are calculated for $\hbar\omega = 0.2$, $F = 0.2$, and $\mathcal{F} = 10^{-3}$, all in scaled units, as a particular numerical example.

In Fig. 1 we predict elliptic dichroism for both even and odd harmonics as a function of the angle φ between the directions of the major axis of the laser polarization ellipse, $\hat{\boldsymbol{\varepsilon}}$, and the static electric field, $\hat{\mathbf{e}}_0$. If we denote by I_+ and I_- the intensities obtained for laser ellipticities $\pm|\eta|$ then the dichroic ratio $\delta = (I_+ - I_-)/(I_+ + I_-)$, is a good measure of this effect. Our calculations show that significant values of δ appear as a result of either out-of-phase oscillations of I_+ and I_-, or in-phase oscillations with, e. g., $|I_+| > |I_-|$. One sees from Fig. 1 that δ is significant for both even and odd harmonics, that it is significant throughout the plateau region, and that it is very sensitive to both $|\eta|$ and φ.

Figure 2 shows the circular polarization degree, ξ_n, for low-order even harmonics produced by a linearly polarized laser as a function of the angle φ [cf. Eq. (6)]. For $\mathcal{F} = 0$, a linearly polarized laser field generates only linearly polarized odd harmonics. As our static field strength is very small compared to the laser field strength, it doesn't change the polarization of the odd harmonics significantly (e. g., we found $|\xi_n| < 0.03$ for $n = 3, 5, 7$).

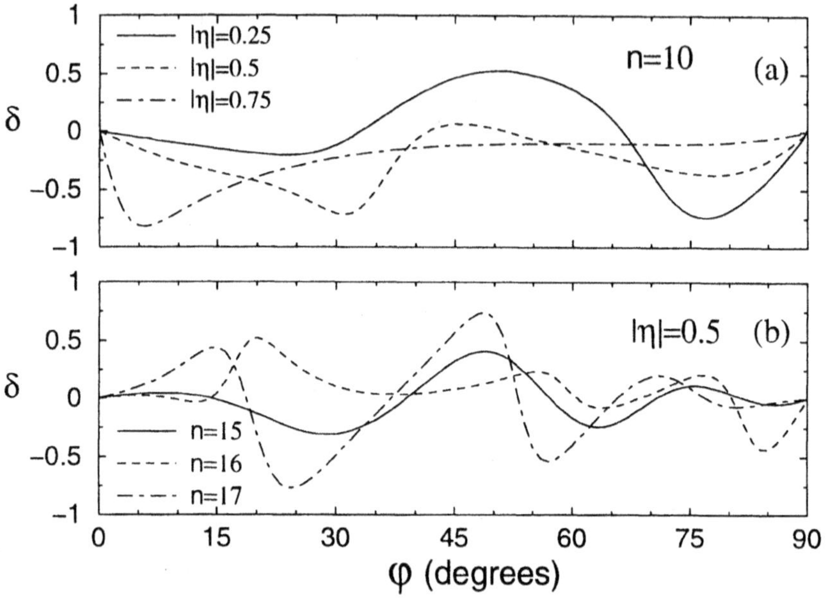

Figure 1. Elliptic dichroism parameter $\delta = (I_+ - I_-)/(I_+ + I_-)$ for (a) n=10 and $|\eta| = 0.25$, 0.5, and 0.75, and (b) $n = 15$, 16, and 17, and $|\eta| = 0.5$. In scaled units (see text), $\omega = 0.2$, $F = 0.2$, and $\mathcal{F} = 10^{-3}$.

However, the even order harmonics, which owe their presence to the static field, are strongly affected by its orientation. It is even possible to produce harmonics with polarizations that are very close to circular over a fairly large range of φ [cf. Fig. 2 for $n = 2$ and 6 and $60° \leq \varphi \leq 70°$]. As the harmonic order increases, the domain of significant non-linear polarization becomes narrower, and is increasingly shifted towards $\varphi = 90°$ (as one can see for the 8th harmonic). When $\varphi = 90°$, all harmonics are emitted with linear polarization but with different orientations: odd harmonics along $\hat{\varepsilon}$ and even ones along $\hat{\mathbf{e}}_0$.

Our analysis above [cf. (5) and (6)] shows that the polarization effects demonstrated in Figs. 1 and 2 stem from the anisotropy induced by the static field [cf. Eq. (4)] in an absorptive medium. We may extend this analysis by noting that the anisotropy can be induced by means other than a static electric field. For example, a similar effect may appear in harmonics generated by reflection of an intense laser beam by a metal surface, where the anisotropy is introduced by the surface normal vector. As another example, closer to the one we have treated, the required anisotropy may be introduced by a second, low-intensity, linearly-polarized laser beam collinear with the first, $\mathbf{F}_\Omega(t) = F_\Omega \hat{\mathbf{e}}_0 \cos \Omega t$. In this case, if one considers only the harmonics of the high-intensity laser, Eqs. (3)-(7) have exactly the

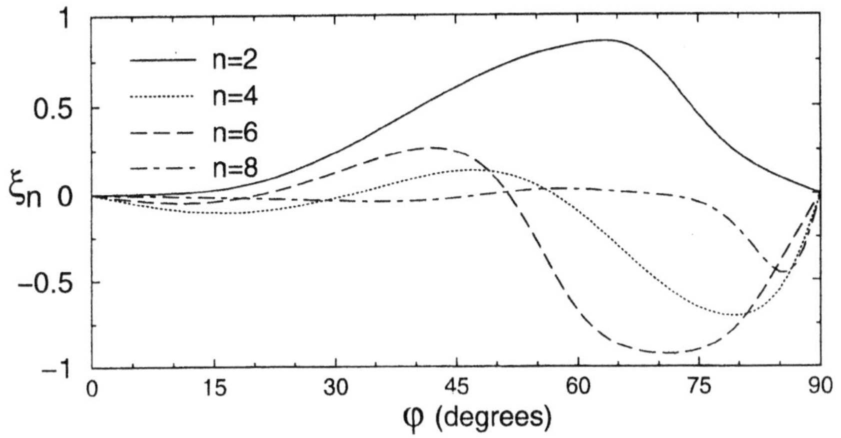

Figure 2. Circular polarization degree, ξ_n, for even harmonics (n = 2, 4, 6, and 8) produced by a linearly polarized laser field for different angles φ. Parameter values ω, F, and $\mathcal{F}$ are as in Fig. 1.

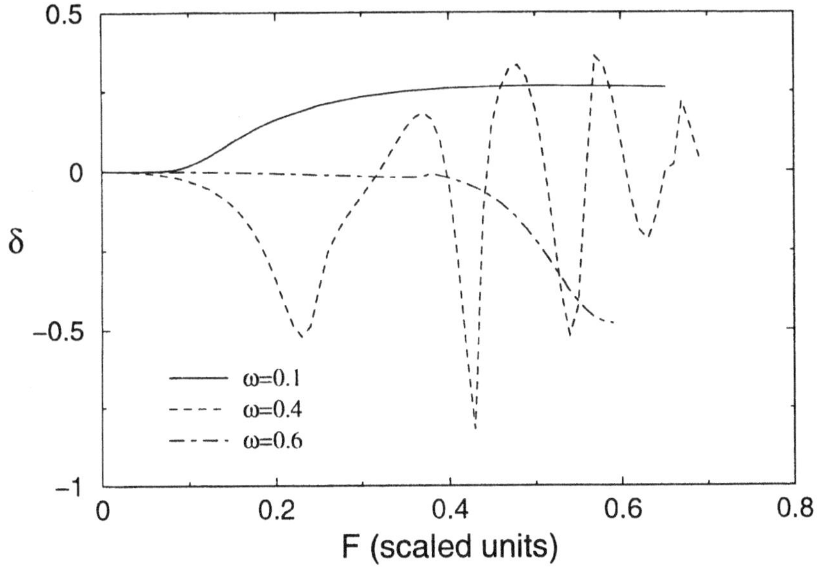

Figure 3. Elliptic dichroism in the case $\mathcal{F} = 0$ for the third harmonic for three values of ω. The laser ellipticity is $|\eta| = 0.5$, the detector detects harmonics with $l' = 1$, and the angle between the laser polarization axis and the detector polarization axis is $\alpha = \pi/4$.

same form and therefore lead to the same polarization effects. For a low frequency, Ω, neglecting (in lowest approximation) the (weak) Ω-dependence of the susceptibilities χ_i, the only effect is to replace the static field amplitude $\mathcal{F}$ by an "effective amplitude" of order F_Ω. Note that the effects predicted are invariant to spatial inversion of $\boldsymbol{\mathcal{F}}$ (or $\mathbf{F}_\Omega$). As $\hat{\mathbf{e}}_0 \to -\hat{\mathbf{e}}_0$, $\varphi \to \varphi + \pi$. This is equivalent in its effect on φ to $\hat{\varepsilon} \to -\hat{\varepsilon}$. From Eqs. (1) and (2), $\hat{\varepsilon} \to -\hat{\varepsilon}$ is in turn equivalent to a phase change, $\omega t \to \omega t + \pi$, which does not affect the predictions (provided the two laser frequencies are incommensurate).

Finally, we note that elliptic dichroism appears also in HHG by an isotropic medium when the detector is polarization sensitive, detecting only photons having polarization $\mathbf{e}'$. If α is the angle between the major polarization axis of the laser photons and the detector polarization axis (*i. e.*, between the major axes of the polarization ellipses defined by $\mathbf{e}$ and $\mathbf{e}'$), the intensity of the detected harmonic is given by [3]:

$$\begin{aligned} I_{n\omega}(\mathbf{e}, \mathbf{e}') = \; & \tfrac{1}{2}\left(|\chi_1|^2 + |\chi_2|^2\right)(1 + ll' \cos 2\alpha) + \tfrac{1}{2}\xi\xi'\left(|\chi_1|^2 - |\chi_2|^2\right) \\ & + \mathrm{Re}(\chi_1\chi_2^*)(l' \cos 2\alpha + l) + \xi l' \mathrm{Im}(\chi_1^*\chi_2) \sin 2\alpha, \end{aligned} \tag{10}$$

which clearly shows that the second and fourth terms are sensitive to the sign of the circular polarization degree, ξ, of the pump laser. The offset angle, θ_n, *i. e.*, the angle between the major axis of the polarization $\mathbf{e}$ of the pump laser and that of the emitted harmonic, is given by:

$$\tan(2\theta_n) = \frac{2\xi \mathrm{Im}(\chi_1^*\chi_2)}{2\mathrm{Re}(\chi_1^*\chi_2) + l\left(|\chi_1|^2 + |\chi_2|^2\right)}. \tag{11}$$

The fact that the polarization of the emitted harmonics may be rotated with respect to that of the pump laser has been measured [2, 5]. Note that, in the case $\mathcal{F} = 0$, only one parameter governs the magnitude of both of these effects: $\mathrm{Im}(\chi_1^*\chi_2)$. This property was used in experimental studies in which the offset angle was extracted from dichroism measurements (see, e. g., Fig. 4 of [2]).

In Fig. 3 we present results for the elliptic dichroism parameter δ (as in Fig. 1) for the third harmonic for the case of a detector of linearly polarized photons oriented at an angle $\pi/4$ with respect to the major axis of the elliptic polarization ($\eta = 0.5$) of the pump laser. One sees that $\delta \approx 0$ for $F < 0.1$, which stems from the fact that $\mathrm{Im}(\chi_1\chi_2^*)$ is negligible until F is of sufficient strength to begin depleting the initial state. One sees also from Fig. 3 that $\delta(F)$ is a very sensitive function of ω.

Though we have presented our analysis and numerical results for the simplest conceptual case, that of HHG in the presence of a static field, we emphasize by the additional examples given above a major goal of our paper: to demonstrate how the introduction of a second polar vector in the problem [in addition to $\mathbf{F}(t)$] leads to interference between the complex susceptibilities χ_i, the results of which are unusual polarization properties of the generated harmonics from an initially isotropic and absorptive medium. (This second polar vector may be a static electric field, the normal to a metal surface, a second linearly polarized laser field, or the orientation of a polarization-sensitive detector.) The predicted effects depend only on the magnitudes of the real and imaginary parts of the susceptibilities χ_i. Physically, the imaginary parts are connected with the ionization of the target atom by the laser field, and hence they are zero if we neglect this ionization. The polarization effects can therefore be called ionization- (or dissipation-) induced effects. Note finally that the dichroic effects discussed above have a different physical origin from the well-known circular dichroism (which appears in chiral systems or magnetic solids), which is not dependent upon the existence of dissipation.

5. Possibility of Experimental Verification

Experimental verification of these harmonic polarization effects depends on the qualitative matching of our scaled parameters with a particular atomic system and a particular set of field strengths and frequencies. For example, for a Xe atom, the scaled parameters employed in Figs. 1 - 3 correspond to a laser with $\lambda = 511.5$ nm and intensity 2.48×10^{14} W/cm^2, which are close to typical values for a frequency-doubled Nd:YAG laser ($\lambda = 532$ nm). The static field strength, however, is 2.16 MV/cm, which is far higher than typical laboratory static fields even though it is weak compared with the laser field ($\mathcal{F}/F = 5\times10^{-3}$). The requisite field strength ratio may nevertheless be achieved in a two color experiment in which the weak laser field, $\mathbf{F}_\Omega(t)$, has a much lower frequency, Ω, than that of the intense laser. For example, the frequency of a CO_2 laser (scaled frequency $\Omega = 0.009$ for Xe) is 22 times smaller than that of a frequency-doubled Nd:YAG laser, while the necessary $\mathcal{F}/F$ ratio could be achieved with an intensity of order 6×10^9 W/cm^2. The weak laser field may be regarded as approximately constant compared with the strong laser field and hence measured polarization results in such an experiment would be qualitatively similar to those predicted here for a laser plus static field configuration as long as the ratio of field strengths is similar.

6. Conclusion

The introduction of anisotropy into a generating medium (e. g., by means of a static electric field) permits significant control over the polarization properties of harmonics. Elliptic dichroism provides a unique case in the analysis of harmonic generation of measuring an effect which depends on the *sign* of the helicity of an elliptically-polarized laser beam. The predicted polarization effects allow the direct determination of the interference between real and imaginary (dissipative) parts of the nonlinear susceptibilities, which is useful for distinguishing between different models of ionization and harmonic generation by atoms in strong fields.

Acknowledgments

This work was supported by NSF Grant PHY-9722110, RFBR Grant 00-02-17843 and INTAS Grant 97-0369 (AVF, MVF, and NLM), and the Alexander von Humboldt Foundation (DBM).

* On leave from Institute for Space Sciences, Bucharest-Măgurele 76900, Romania.
+ Present address: Faculty of Science, University of Sarajevo, Zmaja od Bosne 35, 71000 Sarajevo, Bosnia and Herzegovina

References

1. L. F. DiMauro and P. Agostini, Adv. At. Mol. Opt. Phys. **35**, 79 (1995); P. Salières, A. L'Hullier, Ph. Antoine, and M. Lewenstein, *ibid.* **41**, 83 (1999).
2. F. A. Weihe, S. K. Dutta, G. Korn, D. Du, P. H. Bucksbaum, and P. L. Shkolnikov, Phys. Rev. A **51**, R3433 (1995).
3. N. L. Manakov, Zh. Eksp. Teor. Fiz. **110**, 1244 (1996) [Sov. Phys.-JETP **83**, 685 (1996)].
4. N. H. Burnett, C. Kan, and P. B. Corkum, Phys. Rev. A **51**, R3418 (1995).
5. Ph. Antoine, B. Carré, A. L'Hullier, and M. Lewenstein, Phys. Rev. A **55**, 1314 (1997).
6. M. Kakehata, H. Takada, H. Yumoto, and K. Miyazaki, Phys. Rev. A **55**, R861 (1997); D. Schulze, M. Dörr, G. Sommerer, J. Ludwig, P. V. Nickles, T. Schlegel, and W. Sandner, *ibid.* **57**, 3003 (1998).
7. W. Becker, A. Lohr, M. Kleber, and M. Lewenstein, Phys. Rev. A **56**, 645 (1997).
8. J. Gao, F. Shen, and J. G. Eden, J. Phys. B **32**, 4153 (1999).
9. M. Q. Bao and A. F. Starace, Phys. Rev. A **53**, R3723 (1996).
10. A. Lohr, W. Becker, and M. Kleber, Laser Phys. **7**, 615 (1997).
11. B. Wang, X. Li, and P. Fu, J. Phys. B **31**, 1961 (1998).
12. B. Wang, X. Li, and P. Fu, Phys. Rev. A **59**, 2894 (1999).
13. B. Borca, A.V. Flegel, M. V. Frolov, N. L. Manakov, D. B. Milošević, and A. F. Starace, Phys. Rev. Lett. **85**, 732 (2000).
14. See, e. g., §48 of V. B. Berestetskii, E. M. Lifshitz, and L. P. Pitaievskii, *Quantum Electrodynamics*, 2nd ed. (Pergamon, Oxford, 1982).
15. S. I. Chu, Adv. At. Mol. Opt. Phys. **21**, 197 (1985); N. L. Manakov, V. D. Ovsiannikov, and L. P. Rapoport, Phys. Rep. **141**, 319 (1986).
16. R. M. Potvliege and R. Shakeshaft, in *Atoms in Intense Laser Fields*, ed. M. Gavrila (Academic, New York, 1992), p. 373; Phys. Rev. A **40**, 3061 (1989).
17. N. L. Manakov, M. V. Frolov, A. F. Starace, and I. I. Fabrikant, J. Phys. B **33**, R141 (2000).
18. N. L. Manakov and A. G. Fainshtein, Zh. Eksp. Teor. Fiz. **79**, 751 (1980) [Sov. Phys.-JETP **52**, 382 (1980)].
19. W. Becker, S. Long, and J. K. McIver, Phys. Rev. A **50**, 1540 (1994).
20. M. Yu. Kuchiev and V. N. Ostrovsky, J. Phys. B, **32**, L189 (1999); Phys. Rev. A **60**, 3111 (1999).

ABOUT NEW METHOD OF HIGH HARMONIC AMPLIFICATION

E.A. NERSESOV, S.V. POPRUZHENKO, D.F. ZARETSKY[1]
Moscow State Engineering Physics Institute,
31 Kashirskoe shosse, Moscow, 115409, Russia
[1]Russian Research Center "Kurchatov Institute",
Moscow, 123182, Russia

1. Introduction

High harmonics (HH) produced by atoms in the strong laser field provide a promising source of coherent short-wavelength radiation. Advancements of laser technology have allowed for the generation of a large number of high harmonic quanta per laser pulse $\left(N_s = 10^8 - 10^{10}\right)$ in short laser pulses $(\tau \approx 20-40 fs)$ at moderate $\left(\propto 10^{14} W/cm^2\right)$ intensities [1,2]. However, the high harmonic intensities, that have been reached by now are much less than necessary for many practical applications. As a result, any new method of increasing of high harmonic intensities is of great interest.

The most obvious method of yielding a higher intensity of a high harmonic is the amplification by stimulated emission of HH probe wave passing through the atomic medium in the presence of the strong pump wave. It should be pointed out, however, that for a correct description of the amplification of a high harmonic probe wave the effects of both stimulated emission and absorption of HH photons must be taken into account. It was done for the first time in ref.[3]. It was shown that in the case when both probe and pump pulses are described as a plane monochromatic waves with certain values of wave vectors $\left(\vec{k} \text{ and } \vec{K} \approx s\vec{k}\right)$ and frequencies $(\omega \text{ and } \Omega \approx s\omega)$, the probabilities of stimulated emission W_{em} and absorption W_{abs} almost completely cancel each other with accuracy of the small corrections due to the recoil of atom. As a result, the effect of amplification tends to be very weak. The gain of $s-th$ harmonic was found to be proportional to the product of two small parameters [3]:

$$G(s) \propto \frac{s\hbar\omega}{Mc^2}\left(1-n(\omega)\right), \tag{1}$$

where M is the atomic mass and $n(\omega) \approx 1$ is the refractive index of the atomic medium. The first factor in (1) is due to atomic recoil obtained during the interaction with pump and probe field and the second reflects the fact, that the gain turns to zero in the case of equivalent photons $\left(n((\omega)) = 1, K = sk\right)$. It was shown [4], that the value of the gain (1) is about 10-20% even under optimal experimental conditions [1].

B. Piraux and K. Rzążewski (eds.), Super-Intense Laser-Atom Physics, 259–263.

In the present publication we will consider the problem of high harmonic amplification in the case when both the pump and probe pulses have a finite duration. Besides, we will take into account that pulses may be delayed in time with respect to one another. As will be shown, in this case the probabilities of stimulated emission and absorption of a high harmonic photon no longer coincide even if the recoil of atom is neglected. Moreover, the effect of amplification appears to be essentially larger, than in the case when the recoil mechanism is relevant.

2. Amplitudes of stimulated processes

For the calculation of amplitudes of stimulated emission and absorption of HH photon by a single atom placed at the position $\vec{R}_0$ we will apply well-known strong field approximation [5,6], where the intermediate electron states in the continuum are approximated by plane Volkov waves

$$\Psi_P(t)=\exp\left\{i\vec{P}\vec{r}-i\int_{-\infty}^{t}\varepsilon_P(\tau)d\tau\right\} \tag{2}$$

with drift momentum $\vec{P}$ and time-dependent kinetic energy $\varepsilon_{\vec{P}}(t)=\left(\vec{P}+\vec{A}(t)\right)^2/2$ in the laser field with vector potential $\vec{A}(t)$ (atomic units $e=\hbar=m=1$ are used here and below). In the dipole approximation, which we use, the wave function (2) is an exact solution to the Shrodinger equation in the field of an electromagnetic plane wave with arbitrary temporal envelope.

The amplitude of emission of HH photon with frequency $\Omega\approx s\omega$ has a form [3,4]:

$$A_{em}=-\exp\{i(s\vec{k}-\vec{K})\vec{R}_0\}\sum_P\int_{-\infty}^{+\infty}dt_1\langle\Psi_0|V_2^{(+)}(t_1)|\Psi_P\rangle\int_{-\infty}^{t_1}dt_0\langle\Psi_P|V_1(t_0)|\Psi_0\rangle, \tag{3}$$

where Ψ_0 - is the wave function of the atomic ground state and

$$V_1(t)=\frac{1}{c}\vec{A}_1(t)\hat{\vec{P}}+\frac{1}{2c^2}\vec{A}_1^2(t)\qquad V_2(t)=\frac{1}{c}\vec{A}_2(t)\left(\hat{\vec{P}}+\frac{1}{2c^2}\vec{A}_1(t)\right) \tag{4}$$

are the interaction operators of an electron with pump and probe pulse, respectively.

The amplitude of stimulated emission has a closely related form:

$$A_{em}=-\exp\{-i(s\vec{k}-\vec{K})\vec{R}_0\}\sum_P\int_{-\infty}^{+\infty}dt_0\langle\Psi_0|V_1(t_0)|\Psi_P\rangle\int_{-\infty}^{t_0}dt_1\langle\Psi_P|V_2^{(-)}(t_1)|\Psi_0\rangle \tag{5}$$

To find the total amplitudes of emission (absorption) for the entire system of atoms, we should sum the amplitudes (3,5) over all atomic positions $\vec{R}_0$. In the case of weakly focused pump field the dependence of the amplitudes (3,5) from the atomic position $\vec{R}_0$ reduces to the exponential factors [4]. Therefore, the results of the summing are the same for the cases of emission and absorption and do not influence on the dependence on the Δt of the difference of the probabilities W_{em} and W_{abs}. Because of this, below we will consider the single-atom amplitudes independent from $\vec{R}_0$.

Let us suppose now that both pump and probe pulses are the plane waves of finite duration, and their maxima are shifted by the time Δt. In this case the vector-potentials in (4) have a form:

$$\vec{A}_1(t)=\frac{cF_1}{\omega}\eta_1(t)\vec{e}_x\cos\omega t \quad \text{and} \quad \vec{A}_2(t)=\frac{cF_2}{\Omega}\eta_2(t)\vec{e}_x\cos\Omega t \tag{6}$$

and describe plane-polarized pump and probe waves with temporal envelopes $\eta_1(t)=\eta_1(t/\tau_1)$ and $\eta_2(t)=\eta_2((t-\Delta t)/\tau_2)$ both polarized along the x axis and propagating in the z direction. Here we assume, that each envelope function has a maximum at zero argument $\eta_{1,2}(0)=1$, and $\eta_{1,2}(\pm\infty)=0$. The quantities $F_{1,2}$ denote the maximum amplitudes of the electric-field strengths of the pulses.

It is obvious, that in the general case $(\Delta t\neq 0, \tau_{1,2}\neq\infty)$ the absolute values of the amplitudes (3) and (5) are different, so that the gain has nonzero value. Stimulated emission dominates over absorption for $\Delta t>0$. Therefore, in order to have positive gain, the probe wave should be delayed with respect to the pump one.

3. The physical picture of amplification

The physical nature of the asymmetry between the processes of stimulated emission and absorption for a nonzero temporal delay between pump and probe wave may be understood directly from consideration of the amplitudes (3,5) in terms of simple-man model of HHG proposed by Corkum [7].

In the case of stimulated emission (3) the electron tunnels out from the atom into continuum state $\Psi_{\vec{p}}$ at arbitrary time t_0 under the action of strong pump field. At a later time $t_1>t_0$ it recombines into the ground state of the atom with emission of HH photon. Since the probability of the tunneling effect is a highly nonlinear function of the laser field strength, the transition of the electron into continuum occurs preferably in the small vicinity of pump field maximum, i.e., near $t_0=0$. As a result, the HH emission takes place mainly at positive times (this time interval is shown by the dashed line in Figure 1).

The situation is opposite for the stimulated absorption (see eq.(5)). Now the electron transits into continuum at arbitrary time t_1 with absorption of HH quantum. At the later $t_0>t_1$ the process of stimulated multiphoton recombination under the action of strong pump field takes the electron back into the ground state. As in the case of emission, the interaction of the electron with the pump wave results with highest efficiency near the maximum of the pump field envelope, i.e., at $t_0=0$. Therefore, the absorption of the HH photon occurs predominantly at negative times (the dotted line on Figure 1). Since the probabilities of stimulated emission and absorption are proportional to the probe wave intensity at the times of emission and absorption, respectively, the total probability of stimulated emission should be larger than absorption one in accordance to the temporal delay value Δt.

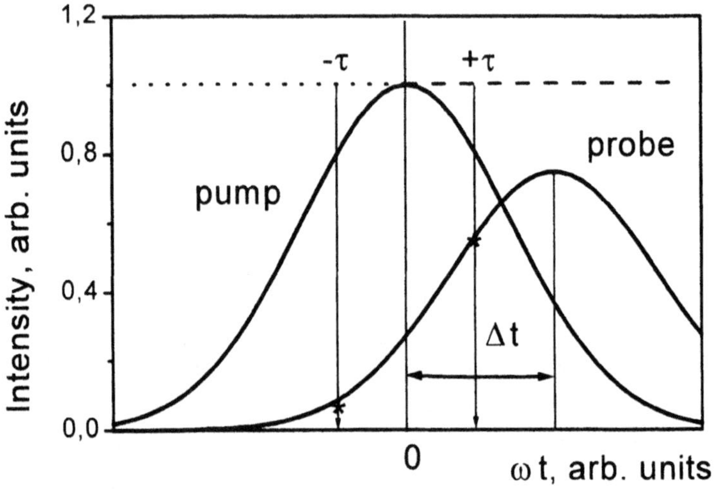

Figure 1. The figure, which qualitatively demonstrates the physical nature of the asymmetry between stimulated emission and absorption for positive Δt. All times are in arbitrary units.

Using the simple-man model of HHG [7] we can obtain an estimate for the gain. Indeed, the model proposes, that the main contribution to the amplitude of the process comes from classical trajectories of the electron in the continuum, that return to the parent ion. Since the typical time of the electron's classical motion in the continuum between ionization and the first return is about half a period of the laser field $\tau \approx \pi / \omega$, the emission occurs mainly at $t_{em} = \tau$ after the pump field maximum, and absorption at the same time before it (see Figure 1). Therefore, the difference between the probabilities of stimulated emission and absorption, as a gain of HH, is proportional to:

$$G(s) \propto W_{em} - W_{abs} \propto V_2^2(\Delta t - \tau) - V_2^2(\Delta t + \tau) \propto \frac{\Delta t \tau}{\tau_2^2} \exp\left\{-\frac{\Delta t^2}{\tau_2^2}\right\} \tag{7}$$

The parameter (7) is very small. For femto- or picosecond laser pulses and temporal delay Δt about a few optical cycles of the pump wave it varies between 10^{-4} and 10^{-6}. However, it is significantly (about 5 - 7 orders in magnitude) larger than the parameter (1), which is due to the recoil effect [3].

The accurate analytical calculation, which confirm the estimate (7) will be presented elsewhere. These calculations show that for a 1ps pump pulse of $Ti:Sa$ laser with a peak intensity $5 \cdot 10^{14} W/cm^2$ the gain of 15-th harmonic should reach unity if the probe wave of the same duration $(\tau_2 \approx \tau_1 \approx 1ps)$ is delayed by $\Delta n = 10$ optical cycles of the pump wave and the yield of spontaneous quanta N_s emitted at the same conditions, but without the probe wave, is about 10^5.

4. Conclusions

We have demonstrated, that the temporal shift between pump and probe waves essentially violates the symmetry between processes of stimulated emission and absorption. This gives rise to the occurrence of the appreciable amplification of the probe wave of HH. It is shown that the gain is positive only if the probe wave is delayed with respect to the pump one and it may be substantial.

Probable schemes for experimental observation of HH amplification will be discussed in details elsewhere. It should be stressed only that the weak probe wave may be obtained in the process of spontaneous HHG in the one atomic jet and directed to the second atomic jet with small temporal delay with respect to the pump pulse. In the second jet the amplification of this probe wave should be observed. Such experimental setup is very close to well-known cross-correlation experiment [8].

5. Acknowledgements

The authors thank P. Agostini and W. Becker for valuable discussions. This work was supported in part by the Deutsche Forschungsgemeinschaft (grant no. 436 RUS 113/536/0), the INTAS (project no. 99-450) and the Competition Center of Basic Natural Science of the Ministry of Education of the Russian Federation (grant no. 97-5.2-117).

6. References

1. Constant, E., Garzella, D., Breger, P., Mevel, E., Dorrer, Ch., Le Blank, C.,Salin, F. and Agostini, P. (1999) Optimizing High Harmonic Generation in Absorbing Gases: Model and Experiment, *Physical Review Letters* **82**, 1668-1671.
2. Durfee III, Ch.G., Rundquist, A.R., Backus, S., Herne, C., Murnane, M. and Kapteyn, H. (1999) Phase Matching og High Harmonics in Hollow Wavequides, *Physical Review Letters* **83**, 21872190.
3. Zaretsky, D.F., Nersesov , E.A. (1999) Gain coefficient of a high harmonic emitted during above-threshold ionization of atoms, *Zhurnal Experimental'noi i Teoreticheskoi Fiziki* **115**, 1630-1641 (Transation: *JETP* **88**, 895-901).
4. Nersesov, E.A., Popruzhenko, S.V., Zaretsky, D.F., Becker, W. (2000) The gain of high harmonics in an atomic jet and in a hollow-core fiber, *Optics Communications* **183**, 289-297.
5. Lewenstain, M., Balcou, Ph., Ivanov, M.Yu., L'Huillier, A. and Corkum, P.B. (1994) Theory of high-harmonic generation by low-frequency laser field, *Physical Review A* **49**, 2117-2132.
6. Becker, W., Long, A. and McIver, J.K. (1994) Modeling harmonic generation by a zero-range potential, *Physical Review A* **50**, 1540-1559.
7. Corkum, P.B. (1993) Plasma Perspective in Strong-Field Multiphoton Ionization, *Physical Review Letters* **71**, 1994-1997.
8. Bouhal, A., Evans, R., Grillion, G., Mysyrowicz, A., Breger, P., Agostini, P., Constantinescu, R.C., Muller, H.G., von der Linde, D. (1997) Cross-correlation measurement of femtosecond noncollinear high-order harmonics, *Journal Optical Society American B* **14**, 950-956.

X-RAY GENERATION VIA STIMULATED RECOMBINATION OF ELECTRONS AND BOHR'S CORRESPONDENCE PRINCIPLE

A. JAROŃ[1], J. Z. KAMIŃSKI[1] and F. EHLOTZKY[2]

[1]*Institute of Theoretical Physics, Warsaw University*
Hoża 69, 00681 Warszawa, Poland
[2]*Institute for Theoretical Physics, University of Innsbruck*
Technikerstrasse 25, A-6020 Innsbruck, Austria

1. Introduction

Knowledge about the structural dynamics of materials on the fundamental time scale of atomic motion is important for a deeper understanding of such processes like chemical and biological reactions, phase transitions and surface effects. These processes are determined by the motion of atoms on the time scale of one vibrational period that is less or of the order of 100 femtoseconds. Until recently, studies of ultrafast structural dynamics were limited to the utilization of femtosecond optical pulses. However, best suited for probing matter is x-ray radiation. Existing synchrotron sources of high-brightness x-rays are two orders of magnitude too slow to resolve atomic motion on the 100 femtosecond time scale. For this reason large efforts are made to generate shorter pulses since the application of ultrashort x-ray pulses to time-resolved x-ray diffraction and absorption would allow direct observation of structural dynamics on the desired time scales. Hence, information on the time-structure of x-ray emission is significant from the point of view of applications. Short x-ray pulses would be of particular interest for ultrafast 'visible-pump–x-ray-probe' experiments in which the structure of a sample, after interaction with a short laser pulse, is subsequently probed by x-rays from a synchrotron [1,2]. The work on designing intense coherent sources at short wavelengths proceeds in different directions and there are different techniques tested for producing x-ray radiation. Among them are high harmonic generation schemes [3–5], recombination pumped lasers, belonging to the group of schemes utilizing features of a laser produced plasma [6], free electron laser schemes, and relativistic Thomson scattering [7,8]. It seems that, compared with synchrotrons, laser-based sources are cheaper and more compact.

B. Piraux and K. Rzążewski (eds.), Super-Intense Laser-Atom Physics, 265–274.

Very promising conditions for x-ray generation can be met in the laser field induced electron-ion recombination, as was proposed many years ago by Farkas and Horváth [9]. In their experimental setup x-ray pulses were generated by shining laser light at grazing incidence on a gold target to create a surface plasma and by applying an additional strong static negative voltage to a metal plate parallel to the gold target at a small distance such that the electrons are pushed back to the gold surface. This scheme presents a simple way of producing strong x-ray fields with rather modest costs of the equipment. The above experiments were repeated with essentially the same conclusions by Dinev et al. [10] and by Hontzopoulos et al. [11].

The laser-assisted recombination process as a mean of generating high frequency radiation was already considered from the theoretical point of view in our previous papers [12,13] (similar considerations were presented in [14]). We showed that the power spectrum of emitted radiation has the shape of a plateau of the width $4\sqrt{2U_pE_p}$, where U_p is the ponderomotive energy of the electron in the laser field and E_p is the electron kinetic energy. We were able to estimate the maximum energy of the emitted x-ray quanta to be $E_p + 2U_p + |E_B| + 2\sqrt{2U_pE_p}$ where E_B denotes the binding energy of an atomic electron. We also considered a classical model of laser-assisted recombination, bearing in mind that classical and quasi-classical models worked very well for treating other multiphoton processes, such as high harmonic generation, above-threshold ionization, or the laser-assisted Auger effect [15–17]. Using Bohr's correspondence principle and the energy conservation law we derived, by means of some elementary classical considerations on the electron motion in a laser field, an expression for the width of the laser-stimulated x-ray spectrum in electron-ion recombination and a classical expression for the probability distribution of x-ray radiation, which both agree surprisingly well with the results of the corresponding quantum-mechanical calculations [18].

2. Differential power spectrum

Let us consider the process of recombination of electrons with ions in the presence of a powerful laser beam, described by the vector potential $\boldsymbol{A}_L(t)$ in dipole approximation. During the action of the laser field the emission of x-ray photons of frequency ω_X, wave vector $(\omega_X/c)\boldsymbol{n}$ and linear polarization $\boldsymbol{\varepsilon}$ takes place. In the length gauge, the Hamiltonian which describes such a process is of the form

$$\hat{H}(t) = \hat{H}_{\rm at} - e\boldsymbol{E}_L(t)\cdot\hat{\boldsymbol{r}} - e\hat{\boldsymbol{E}}_X(\hat{\boldsymbol{r}},t)\cdot\hat{\boldsymbol{r}}, \tag{1}$$

where $\boldsymbol{E}_L(t)$ and $\hat{\boldsymbol{E}}_X(\hat{\boldsymbol{r}},t)$ are respectively the electric field of the laser beam in dipole approximation and the electric field operator of the x-ray radiation. $\hat{H}_{\rm at}$ is the radiation-free atomic Hamiltonian. In our analysis the laser field is treated classically, whereas the spontaneously emitted x-ray radiation is described quantum-mechanically. For a one-mode x-ray radiation the electric field operator is equal to

$$\hat{\boldsymbol{E}}_X(\hat{\boldsymbol{r}},t) = \mathrm{i}\sqrt{\frac{\hbar\omega_X}{2\varepsilon_0 V}}\boldsymbol{\varepsilon}(\hat{a}_X \mathrm{e}^{-\mathrm{i}\omega_X t + \mathrm{i}\frac{\omega_X}{c}\boldsymbol{n}\cdot\hat{\boldsymbol{r}}} - \hat{a}_X^\dagger \mathrm{e}^{\mathrm{i}\omega_X t - \mathrm{i}\frac{\omega_X}{c}\boldsymbol{n}\cdot\hat{\boldsymbol{r}}}), \tag{2}$$

where ε is the polarization vector, $\hat{a}_X$ and $\hat{a}_X^\dagger$ are the annihilation and creation operators, ε_0 is the permittivity of the vacuum, $\boldsymbol{n}$ is the direction of propagation of X-ray radiation of frequency ω_X, and finally V is the quantization volume of the radiation field. For the recombination process the initial state is equal to $|\,\psi_{\boldsymbol{p}}^{(+)}(t);0\rangle$, i.e., it is the vacuum state for x-ray radiation and a scattering state for an electron in a laser field. On the other hand, the final state equals $|\,\psi_B(t);1\rangle$, i.e., it is a one-photon state for x-rays and a laser-dressed bound state for an electron.

From Fermi's golden rule we find that the probability of emitting an x-ray photon by an electron recombining from a scattering state of momentum $\boldsymbol{p}$ into a quasi-bound state is equal to

$$P(\boldsymbol{n},\boldsymbol{p}) = \left|\, \frac{1}{\hbar} \int \mathrm{d}t \langle \psi_B(t);1 \,|\, -e\hat{\boldsymbol{E}}_X(\hat{\boldsymbol{r}},t)\cdot\hat{\boldsymbol{r}} \,|\, \psi_{\boldsymbol{p}}^{(+)}(t);0\rangle \,\right|^2 . \tag{3}$$

After simple algebraic manipulations we arrive at the probability rate ($\alpha = e^2/4\pi\varepsilon_0\hbar c$ is the fine structure constant)

$$\begin{aligned} w(\boldsymbol{n},\boldsymbol{p}) &= (2\pi)^2\alpha\hbar c\frac{\omega_X}{V} \\ &\times \left|\, \frac{\omega_L}{2\pi}\int_0^{2\pi/\omega_L} \mathrm{d}t \langle \psi_B(t) \,|\, \mathrm{e}^{-\mathrm{i}\omega_X t + \mathrm{i}\frac{\omega_X}{c}\boldsymbol{n}\cdot\hat{\boldsymbol{r}}} \boldsymbol{\varepsilon}\cdot\hat{\boldsymbol{r}} \,|\, \psi_{\boldsymbol{p}}^{(+)}(t)\rangle \,\right|^2 \\ &\times \delta(E_B - E_{\boldsymbol{p}} - U_p - n\hbar\omega_L + \hbar\omega_X), \end{aligned} \tag{4}$$

In deriving this formula we have assumed that the imaginary part of the quasienergy for the final quasi-bound state is much smaller than the laser-photon energy $\hbar\omega_L$, and that the laser field is described by a multichromatic plane wave with the fundamental frequency ω_L. Hence the total power of emitted x-rays reads

$$\begin{aligned} P &= \int \frac{\mathrm{d}^3p}{(2\pi\hbar)^3}\frac{\mathrm{d}^3k_X}{(2\pi)^3} V\,\hbar\omega_X w(\boldsymbol{n},\boldsymbol{p}) \\ &= \int \mathrm{d}E_{\boldsymbol{p}}\mathrm{d}\Omega_{\boldsymbol{p}}\mathrm{d}\omega_X\mathrm{d}\Omega_{\boldsymbol{n}} S(\omega_X,\boldsymbol{n},\boldsymbol{p})\delta(E_B - E_{\boldsymbol{p}} - U_p - n\hbar\omega_L + \hbar\omega_X), \end{aligned} \tag{5}$$

in which the so-called differential power spectrum is given by

$$\begin{aligned} S(\omega_X,\boldsymbol{n},\boldsymbol{p}) &= \frac{\alpha\omega_X^4 mp}{(2\pi)^4\hbar c^2} \\ &\times \left|\, \frac{\omega_L}{2\pi}\int_0^{2\pi/\omega_L} \mathrm{d}t \langle \psi_B(t) \,|\, \mathrm{e}^{-\mathrm{i}\omega_X t + \mathrm{i}\frac{\omega_X}{c}\boldsymbol{n}\cdot\hat{\boldsymbol{r}}} \boldsymbol{\varepsilon}\cdot\hat{\boldsymbol{r}} \,|\, \psi_{\boldsymbol{p}}^{(+)}(t)\rangle \,\right|^2 . \end{aligned} \tag{6}$$

This is the quantity we shall analyze below. In our further investigations we shall assume that $\psi_B(\boldsymbol{r},t)$ is the wave function of the ground state of an electron without the laser field, and that $\psi_{\boldsymbol{p}}^{(+)}(\boldsymbol{r},t)$ is represented by a Coulomb-Volkov solution [12]

3. Classical picture

In order to understand qualitatively the structure of the power spectrum of generated x-rays let us begin with a classical model. As we know from classical mechanics, an electron moving in a laser field has an energy equal to (we use atomic units)

$$E(t) = \frac{1}{2}(\boldsymbol{p} - e\boldsymbol{A}_L(t))^2, \tag{7}$$

where $\boldsymbol{p}$ is the momentum of the electron's translational motion whereas

$$\boldsymbol{A}_L(t) = \sum_n \boldsymbol{A}_{0n} \cos(n\omega_L t + \varphi_n) \tag{8}$$

is the vector potential of a (in general) multichromatic laser field in dipole approximation. In all our subsequent investigations, we shall consider recombination of electrons energetic enough to neglect the potential energy of the attracting ions. If at a certain instant t an electron gets captured by an ion in the presence of a laser field, then the energy of the emitted x-ray photon will be equal to $\omega_X(t) = |E_B| + E(t)$. We have assumed that the total kinetic electron energy is transfered into the x-ray photon and that processes in which, apart from this x-ray photon, also some laser photons are emitted to or absorbed from the laser field are much less probable. Moreover, we shall assume in addition that the intensity of emitted x-ray radiation with energy in the interval $[\omega_X, \omega_X + \mathrm{d}\omega_X]$ is proportional to the time during which a 'classical' electron embedded in a laser field possesses energy within the interval $[E(t), E(t) + \mathrm{d}\omega_X]$. This means that the power spectrum of x-ray radiation $S(\omega_X, \boldsymbol{n}, \boldsymbol{p})$ is proportional to the probability density of finding an electron with an instant energy $E(t) = \omega_X(t) - |E_B|$. The classical probability density is equal to

$$\rho_c(\omega_X) = \frac{\omega_L}{2\pi} \sum_{t(\omega_X)} \frac{1}{|\omega_X'(t(\omega_X))|}, \tag{9}$$

where the prime denotes the derivative with respect to time and $t(\omega_X)$ represents the solutions of the equation

$$\omega_X = |E_B| + \frac{1}{2}(\boldsymbol{p} - e\boldsymbol{A}_L(t))^2. \tag{10}$$

The probability density $\rho_c(\omega_X)$ is normalized to 1, i.e.,

$$\int_{\omega_{\min}}^{\omega_{\max}} \rho_c(\omega_X)\mathrm{d}\omega_X = 1, \tag{11}$$

where $\omega_{\min}$ and $\omega_{\max}$ are the classical minimum and maximum frequencies of the generated x-rays. Let us note that the above classical probability distribution resembles the position probability density discussed already by Pauling and Wilson [19] in the context of a one-dimensional harmonic oscillator (see also [20,21]).

One can check that for a monochromatic laser field and for sufficiently large electron energy there are two solutions of Eq. (10) that lead to the same value of $|\omega_X'(t(\omega_X))|$,

$$|\omega_X'(t(\omega_X))| = 2\omega_L\sqrt{(\omega_X - |E_B|)(2U_p - (\sqrt{\omega_X - |E_B|} - \sqrt{E_p})^2)}. \tag{12}$$

In deriving this equation we have already assumed that the electron momentum points parallel to the polarization vector of the laser field, $\boldsymbol{p}||\boldsymbol{A}_0$, i.e., we have considered the most favourable geometry for the generation of x-ray radiation, as was discussed in [12]. For this geometry, the possible energies ω_X of x-ray photons lie within the range $[\omega_{\min}, \omega_{\max}]$, where the minimum and maximum frequencies are correspondingly defined by the expressions,

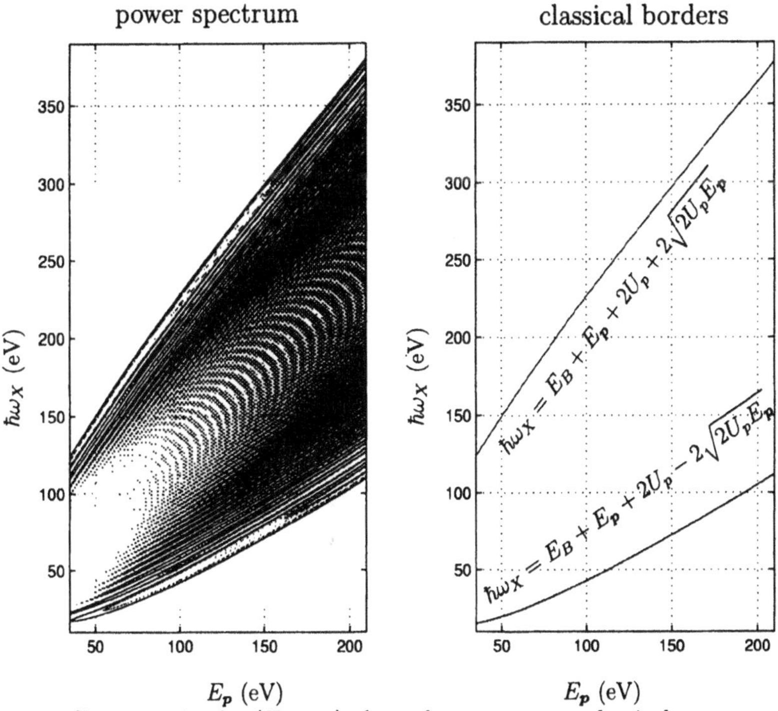

Figure 1. Compares in the (E_p, ω_X)-plane the quantum-mechanical x-ray spectrum and the classical predictions of Eqs. (13) and (14) as a function of the incident electron energy for a fixed intensity of 10^{14}W/cm^2 of a monochromatic Nd:YAG laser field of frequency $\omega_L = 1.17$eV. The borders of the quantum mechanical x-ray spectrum are very well reproduced by the classical considerations.

$$\omega_{\min} = |E_B| + 2U_p + E_p - 2\sqrt{2U_pE_p}, \tag{13}$$

$$\omega_{\max} = |E_B| + 2U_p + E_p + 2\sqrt{2U_pE_p}. \tag{14}$$

In Fig. 1 we compare these classical results with the quantum calculations. As we see, the borders of the contour plot of the x-rays power spectrum are very well described by the above classical formulae.

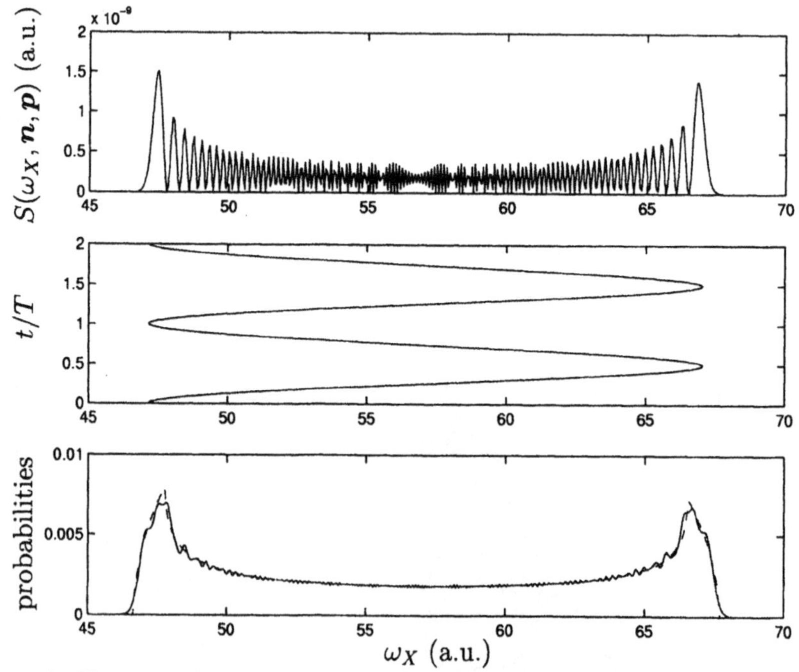

Figure 2. The upper frame presents the differential power spectrum for x-ray generation by electrons of energy $E_p = 56.96$ (in a.u.) recombining with ions in the monochromatic laser field of frequency $\omega_L = 1.5498$ eV and intensity $I = 10^{14}$ W/cm^2 in the geometry in which the electron momentum and the polarization vectors of the laser field and x-ray photons are parallel to each other. In the lower frame we compare the quantum (solid line) with the classical (dashed line) probabilities for a resolution $2\delta = 20\omega_L$. The central frame presents the time-dependence (in units of the laser field period $T = 2\pi/\omega_L$) of the emitted x-ray photon energy (within the classical model).

The classical model predicts that for a given value of the electron momentum the power spectrum of x-ray radiation is limited to the range $[\omega_{\rm min}, \omega_{\rm max}]$. It might seem that the drawback of this model is that the probability density $\rho_c(\omega_X)$ is infinite (although normalizable) at the endpoints of the interval $[\omega_{\rm min}, \omega_{\rm max}]$. In a realistic experiment, however, one measures the intensity of x-ray radiation within a certain vicinity of a given frequency ω_X, which is due to the finite resolution of the measuring apparatus. This means that the above density $\rho_c(\omega_X)$ should be averaged in addition over a frequency range lying in the vicinity of ω_X. As a consequence, this leads to a finite and continuous classical probability density for the generation of high energy photons. For simplicity, we shall assume here that the experimental apparatus counts all photons from the interval $[\omega_X - \delta, \omega_X + \delta]$, that is, its resolution is equal to 2δ. In this case the averaged probability density is equal to

$$\bar{\rho}_c(\omega_X) = \frac{1}{2\delta} \int_{\omega_X-\delta}^{\omega_X+\delta} d\omega \rho_c(\omega). \tag{15}$$

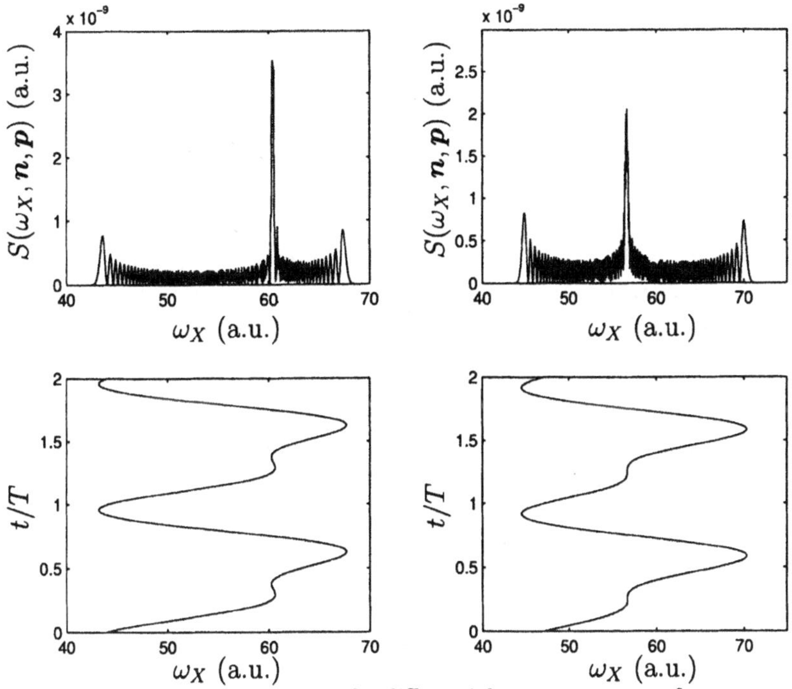

Figure 3. The upper frames present the differential power spectrum for x-ray generation by electrons of energy $E_{\boldsymbol{p}} = 56.18$ (in a.u.) recombining with ions in a bichromatic laser field of frequencies $\omega_L = 1.5498$ eV and $2\omega_L$, of equal intensities 10^{14} W/cm^2 in a geometry in which the electron momentum and the polarization vectors of the laser field and x-ray photons are all parallel to each other. The lower frames present the time-dependence (in units of the period $T = 2\pi/\omega_L$ of the laser field) of the emitted x-ray photon energy (within the classical model). The relative phases of the two laser fields are $\varphi = 45°$ (in the left frames) and $\varphi = 90°$ (in the right frames).

The averaged classical probability density $\bar{\rho}_c(\omega_X)$ is a continuous function of ω_X, while quantum theory predicts that the x-ray frequencies can only differ by multiples of the laser frequency ω_L. Therefore, in order to compare the predictions of the classical and quantum models, we have to discretize the classical probability density. The most natural discretization prescription for this case is the following one,

$$\bar{p}_c(\omega_X) = \int_{\omega_X-\frac{1}{2}\omega_L}^{\omega_X+\frac{1}{2}\omega_L} d\omega \bar{\rho}_c(\omega) \approx \omega_L \bar{\rho}_c(\omega_X), \tag{16}$$

which determines the probability that in the process of recombination a photon of energy from the interval $[\omega_X - \frac{1}{2}\omega_L, \omega_X + \frac{1}{2}\omega_L]$ is emitted. This classical result should be compared with the corresponding quantum probability $\bar{p}_q(\omega_X)$, which is proportional to the power spectrum $S(\omega_X, \boldsymbol{n}, \boldsymbol{p})$ of x-ray radiation calculated by means of quantum theory and normalized in such a way that the total probability of emission of any x-ray photon is equal to one. Moreover, also in this case we have to account for the resolution of the experimental apparatus. We assume that $\bar{p}_q(\omega_X)$ is proportional to the power spectrum averaged in a similar way as the classical probability distribution,

$$\bar{p}_q(\omega_X) = N \sum_{\omega_X - \delta \leqslant \omega \leqslant \omega_X + \delta} S(\omega, \boldsymbol{n}, \boldsymbol{p}), \tag{17}$$

where N is the normalization factor.

In Fig. 2 we compare the predictions of the quantum and classical calculations. We observe that the quantum and classical probabilities agree remarkably well with each other, which is a manifestation of the Bohr's correspondence principle. As expected, the results obtained within the quantum model are smoothed out by the averaging procedure, but the most important feature is that the shapes of the classical and quantum distributions are almost identical. Let us emphasize, however, that the classical model presented here, although it reproduces the shape of the power spectrum and remarkably well determines the borders of the plateau, does not allow to estimate the absolute values of the power spectrum for which a quantum calculation is necessary. It also cannot predict the rapidly changing behaviour of the power spectrum in the inner part of the frequency plateau, that is due to quantum interference effects which result from the summation of probability amplitudes corresponding to different quantum paths, but leading to the same final state.

4. Generation of short x-ray pulses

It is well-known that quantum interference of probability amplitudes can significantly enhance or suppress the corresponding transition probabilities. This effect is also observable in the power spectrum of x-rays considered above. As we have seen, this power spectrum consists of rapid oscillations, especially in the middle of the plateau, which then gradually disappear as the x-ray photon energy approaches the classical borders, given by the formulae (13) and (14), at which we observe its smooth behaviour. Although this is a pure quantum phenomenon, it appears that the main features of these oscillations can be understood within our classical model. Indeed, the lack of oscillations at the borders of the plateau can be attributed to the fact that from the classical point of view the corresponding x-ray photons are emitted coherently (i.e., modulo 2π with the same phases of the probability amplitudes) during the instants at which the electron kinetic energy acquires its minimum or maximum values. This only happens once during an optical cycle of the monochromatic laser field considered above. For the inner part of the plateau, however, there are two instants during one optical cycle that lead to the same energy of the x-ray photons, as is shown in Fig. 2. The probability amplitudes for these two quantum paths have different phases, which result from the instantaneous phase of the electron scattering state. Of course, from the quantum-mechanical point of view, these are not the only paths that lead to the emission of x-ray photons of a given energy. For other quantum paths the electron can virtually emit or absorb a number of laser photons and, after the recombination, another number of laser photons can be virtually absorbed or emitted. Such virtual processes are responsible for the generation of x-ray photons beyond the classical borders. However, the excellent agreement of the quantum results with our classical model for x-ray photons from the plateau region shows that they play a marginal role. Similar conclusions have been drawn by Gribakin and Kuchiev [22] in their analysis of photodetachment.

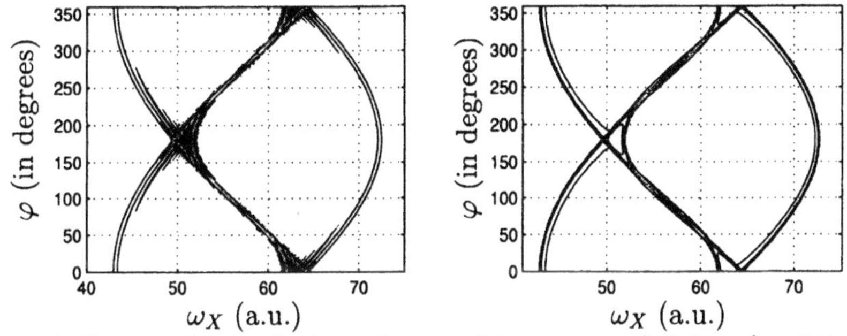

Figure 4. Contour plots in the (ω_X, φ)-plane of the quantum (left frame) and the classical (right frame) probabilities calculated with the resolution $2\delta = \omega_L$ for a bichromatic laser field of frequencies ω_L and $2\omega_L$ and of equal intensities 10^{14}W/cm^2 and for an electron energy of 56.18 (in atomic units). The electron momentum and the polarization vectors of the bichromatic laser field and x-ray photons are all parallel to each other.

Guided by this remarkable agreement between the quantum and classical model calculations one can ask the question: Is it possible to enhance the generation of x-ray photons of a given energy? The classical model suggests that such an enhancement is possible provided that the electron spends a relatively long time in a state with a corresponding instantaneous kinetic energy. Such a situation can be met if we consider the recombination process in the presence of a multichromatic laser field. Varying the intensities and relative phases of the individual harmonics we can always find such parameters for which the electrons are in a state of a given energy for a relatively longer time then in states with other kinetic energies. In Fig. 3 we illustrate such a situation for a bichromatic laser field. We observe that by fixing the intensities and varying the relative phase we can scan the whole plateau region and enhance particular harmonics of x-ray radiation. We see that electrons spend a significant fraction of a laser-field period (for the case considered this fraction is approximately equal to 0.2 of the laser period) in the state of this particular energy that corresponds to the peak in the power spectrum. According to our classical model, these x-ray photons are generated predominantly during this interval of time, suggesting that one should be able to generate in this way a very short pulse of x-ray radiation of the order of one femtosecond.

In Fig. 4 we present the contour plots in the (ω_X, φ)-plane of the quantum and classical probabilities considered above, which again show remarkable similarities and demonstrate the validity of Bohr's correspondence principle for laser-assisted x-ray generation in electron-ion recombination processes.

Acknowledgements

A. J. and J. Z. K. acknowledge support by the Polish Committee for Scientific Research (Grants No. KBN 2 P03B 087 17 and KBN 2 P03B 039 19, respectively). This work was supported in part by the Scientific-Technical Collaboration Agreement between Austria and Poland for 2000/2001 under project number 2/2000.

References

1. Larsson, J. and Sjogren, A. (1999) *Rev. Sci. Instrum.* **70**, 2253.
2. Larsson, J., Heimann, P. A., and Falcone, R. W. (1998) *Appl. Phys.* A **66**, 587.
3. Chang, Z. *et al.* (1997) *Phys. Rev. Lett.* **79**, 2967.
4. Randquist, A. *et al.* (1998) *Science* **280**, 1412.
5. Hentschel, M., Schnürer, M., and Krausz, F. (2000), in L. F. DiMauro, R. R. Freeman and K. C. Kulander (eds.), *Proceedings of the ICOMP VIII*, Melville, New York, p. 361.
6. Rischel, C. *et al.* (1997) *Nature* **390**, 490; Rose-Petruck, C. *et al.* (1997) *Nature* **398**, 310.
7. Schoenlein, R. W. *et al.* (1996) *Science* **274**, 236.
8. Schoenlein, R. W. *et al.* (200) *Appl. Phys.* B **71**, 1.
9. Farkas, Gy. and Horváth, Z. Gy. (1977) *Opt. Commun.* **21**, 408.
10. Dinev, S. G. *et al.* (1979) *Opt. Commun.* **30**, 219; Dinev, S. G. *et al.* (1980) *Opt. Quantum Electron.* **12**, 231.
11. Hontzopoulos, E. *et al.* (1988) *Opt. Commun.* **67**, 124.
12. Jaroń, A., Kamiński, J. Z., and Ehlotzky, F. (2000) *Phys. Rev.* A **61**, 023404.
13. Jaroń, A., Kamiński, J. Z., and Ehlotzky F., submitted to *Laser Physics*.
14. Kuchiev, M. Yu. and Ostrovsky, V. N. (2000) *Phys. Rev.* A**61**, 033414.
15. Corkum, P. B. (1993) *Phys. Rev. Lett.* **71**, 1994.
16. Kulander, K. C., Schafer, K. J., and Krause, J. L. (1993), in B. Piraux, A. l'Huillier and K. Rzążewski (eds.) *Super-Intense Laser-Atom Physics*, Plenum, New York, p. 95.
17. Schins, J. M. *et al.* (1994) *Phys. Rev. Lett.* **73**, 2180.
18. Jaroń, A., Kamiński, J. Z., and Ehlotzky, F., submitted to *Phys. Rev.* A.
19. Pauling, L. and Wilson, E. B. (1935) *Introduction to Quantum Mechanics*, McGraw-Hill, New York.
20. Białynicki-Birula, I., Cieplak, M., and Kamiński, J. (1992) *Theory of Quanta*, Oxford University Press, New York.
21. Robinett, R. W. (1997) *Quantum Mechanics. Classical Results, Modern Systems, and Visualized Examples*, Oxford University Press, New York.
22. Gribakin, G. F. and Kuchiev, M. Yu. (1997) *Phys. Rev.* A**55**, 3760.

PART V

STABILIZATION AND RELATIVISTIC EFFECTS IN SUPER STRONG FIELDS

INTERFERENCE STABILIZATION: Λ-AND V-SCHEMES, DYNAMICS OF IONIZATION, INITIAL COHERENT POPULATION OF RYDBERG LEVELS AND QUANTUM PHASE CONTROL OF THE IONIZATION YIELD

M.V. FEDOROV & N.P. POLUEKTOV
General Physics Institute, Russian Academy of Sciences
38 Vavilov St., 117942 Moscow, Russia

1. Λ- and V-type interference stabilization

As it is well known [1, 2], Λ-type interference stabilization of Rydberg atoms occurs due to Raman–type transitions via the continuum between close Rydberg levels (*Figure 1a*). These Λ-type transitions provide coherent re-population of Rydberg levels

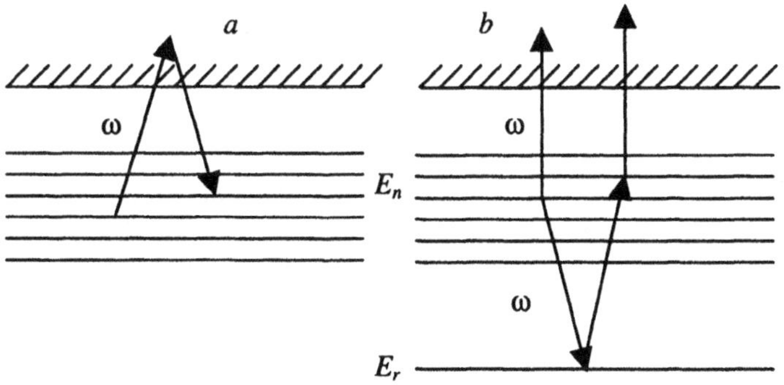

Figure 1. Schemes of transitions for Λ- and V-type (*a* and *b*) interference stabilization

in the process of photoionization by a strong light field. Subsequent transitions to the continuum from the re-populated levels interfere with and partially cancel each other slowing down ionization and stabilizing the atom. Criterion of appearance of the Λ-type interference stabilization is given by the condition that the Fermi-golden-rule ionization width Γ of a Rydberg level E_n becomes of the order of spacing Δ between neighboring levels, $\Gamma \sim E_{n+1} - E_n \approx 1/n^3 \equiv \Delta$ (in atomic units). In the semiclassical approximation (valid at $n >> 1$ and $\omega << 1$) the ionization width is estimated as [3, 4] $\Gamma \sim \varepsilon_0^2 / n^3 \omega^{10/3}$, where ε_0 and ω are the field strength amplitude and frequency of light. This estimate of

B. Piraux and K. Rzążewski (eds.), Super-Intense Laser-Atom Physics, 277–284.

Γ and the condition $\Gamma \sim \Delta$ determine the threshold field of the Λ-type stabilization $\varepsilon_{thr}^{(\Lambda)} = \omega^{5/3}$.

In the V-type stabilization [5-10], in addition to the field-induced transitions to the continuum from Rydberg levels E_n, the same field is assumed to provide resonance transitions to and from a lower-energy bound atomic level E_r (*Figure 1b*). These transitions give rise to a new parameter characterizing the behavior of the system: the Rabi frequency Ω_R estimated as $\Omega_R \sim \varepsilon_0/(n n_r)^{3/2} \omega^{5/3}$, where n_r is the principal quantum number of the level E_r. Owing to resonance transitions via the level E_r, deviations from perturbation theory and interference stabilization arise in the V-type scheme at $\Omega_R \sim \Delta$. Correspondingly, the threshold field of the V-type stabilization is given by $\varepsilon_{thr}^{(V)} = (n_r/n)^{3/2}\, \omega^{5/3}$, and $\varepsilon_{thr}^{(V)} \ll \varepsilon_{thr}^{(\Lambda)}$ if only $n_r \ll n$. Hence, typically, the resonance V-type interference stabilization begins at fields which are significantly weaker than fields required for the non-resonance Λ-type stabilization. For example, if $n = 25$, $n_r = 5$, and $\omega \approx 1/50 \approx 0.8\times10^{15}\text{sec}^{-1}$, the threshold intensities for the Λ- and V-type interference stabilization are of the order of 3×10^{10} and $3\times10^{8}\text{W/cm}^2$, correspondingly. Moreover, in a wide region of intensities and values of the field strength, $(n_r/n)^{3/2}\, \omega^{5/3} < \varepsilon_0 < (n_r/n)^{3/2}\, \omega^{5/3}$, ionization width remains smaller than Rabi frequency, $\Gamma \ll \Omega_R$, and nonperturbative reconstruction of the atomic spectrum is determined by resonance mixing of Rydberg levels. The difference between $\varepsilon_{thr}^{(V)}$ and $\varepsilon_{thr}^{(\Lambda)}$ is the main quantitative difference between Λ- and V-type stabilization. Another, qualitative, difference between these two phenomena concerns structure of quasienergy levels, which is described in the following Section.

2. Complex quasieneries

To describe theoretically dynamics of strong-field photoionization from Rydberg levels, it appears to be useful to expand the wave function Ψ of a field-driven atom in terms of the field-free wave functions of bound Rydberg and resonant states φ_n and φ_r

$$\Psi(t) = a_r(t)\exp(i\omega t)\,\varphi_r + \sum_n a_n(t)\,\varphi_n + \Psi_{cont}\,, \tag{1}$$

where $a_n(t)$ and $a_n(t)$ are the corresponding Time-Depenedent Probability Amplitudes (TDPA) and Ψ_{cont} is the projection of $\Psi(t)$ upon the field-free atomic continuum. Typical approximations used for solving the Schrödinger equation are the rotating-wave approximation and the procedure of adiabatic elimination of the continuum [2]. As the result, it appears to be possible to reduce the non-stationary Schrödinger equation to a set of usual first-order differential equations for TDPA $a_n(t)$ and $a_n(t)$:

$$i\dot{a}_r = (E_r + \omega)\,a_r + \Omega_R \sum_{n'} a_{n'}\,,\quad i\dot{a}_n = \Omega_R\, a_r + E_n a_n - \frac{i}{2}\Gamma \sum_{n'} a_{n'}\,, \tag{2}$$

where the matrix elements determining Ω_R and Γ are assumed to be n-independent.

Equations (2) have stationary solutions $a_r(t) = A_r \exp(-i\gamma t)$, $a_n(t) = A_n \exp(-i\gamma t)$, where A_r and A_n are constants and γ are complex quasienergies of the field-driven atom. An equation for γ follows from (2) and has the form [11]

$$(\gamma-\delta)\left\{1+i\frac{\Gamma}{2}\sum_{n'}\frac{1}{\gamma-(n'-n_0)\Delta}\right\}-\Omega_R^2\sum_{n'}\frac{1}{\gamma-(n'-n_0)\Delta}=0, \tag{3}$$

where δ is the resonance detuning, $\delta = E_r + \omega - E_{n_0}$, and E_{n_0} is the Rydberg level closest to the resonance but such that $E_{n_0} < E_r + \omega$ (at such a definition $0 < \delta \le \Delta$). An additional approximation used for derivation of Equation (3) is the approximation of equidistant levels $E_n \approx E_{n_0} + (n - n_0)\,\Delta$. Under the condition $\Omega_R >> \Gamma, \Delta$ solutions of Equation (3) are given by

$$\gamma_j=\left(j+\frac{1}{2}\right)\Delta-i\frac{\Gamma}{2}\left(j+\frac{1}{2}-\frac{\delta}{\Delta}\right)^2\left(\frac{\Delta^2}{\pi\Omega_R^2}\right)^2, \tag{4}$$

where $j = 0, \pm1, \pm2, \ldots$ Strong-field widths of quasienergy levels are given by $\Gamma_{str}^{(j)} = -\mathrm{Im}(\gamma_j)$. Owing to the strong-field condition $\Omega_R >> \Delta$ the widths $\Gamma_{str}^{(j)}$ are very small ($<< \Gamma$). The widths $\Gamma_{str}^{(j)}$ significantly differ from each other: at large j they grow proportionally j^2. At $\delta = \Delta/2$ one of the widths ($\Gamma_{str}^{(0)}$) turns zero, i.e., the quasienergy level γ_0 does nor decay at all. All these features differ strongly the V-type interference stabilization from the Λ-type one. The latter is characterized by a single strong field width $\Gamma_{str}^{(\Lambda)} \sim \Gamma\times\left(\varepsilon_{thr}^{(\Lambda)}/\varepsilon_0\right)^2$. Structures of quasienergy zones are shown in Figure 2 for Λ- and V-type schemes (*a* and *b*) and they are seen to be significantly different.

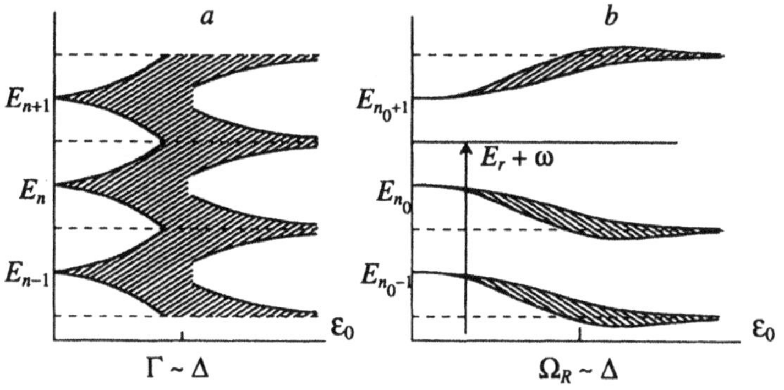

Figure 2. Quasienergy zones of a field-perturbed Rydberg atom in L- and V-schemes.

In the Λ-type scheme of transitions quasienergy levels broaden at weak fields ($\Gamma < \Delta$), overlap and form a quasicontinuum at intermediate fields ($\Gamma \sim \Delta$), and then narrow again around $\tilde{E}_n = ½\,(E_{n+1} + E_n)$ at strong fields ($\Gamma > \Delta$). In the case of V-type scheme the behavior of quasienergy levels is both similar to and different from that of the Λ-scheme. Again, strong-field quasienergy levels are localized near $\tilde{E}_n = ½ (E_{n+1} + E_n)$. But this effect occurs at much weaker fields ($\Omega_R \sim \Delta$ rather than $\Gamma \sim \Delta$). For this reason, quasienergy levels broaden only very little and, then, having no chance to overlap and form a quasicontinuum they narrow.

As mentioned above, in the V-type scheme, at $\delta - \Delta/2$, one of the quasienergy levels (γ_0) has a zero width. Existence of such a non-decaying quasienergy level results in a possibility of absolute stabilization, i.e., nonzero population of bound states in the asymptotic limit of very long time of interaction. This is a specific feature of the V-type stabilization which is missing in the case of Λ-type scheme. The quasienergy wave function of the stable state, $\psi_0 \equiv \psi_{st}$, was found [11] to have the form by

$$\psi_{st} = \frac{1}{\pi} \sum_n \frac{\varphi_n}{n - n_0 - 1/2} . \tag{5}$$

3. Initial coherent population on Rydberg levels

Let us assume that initially (at $t = 0$) and atom is prepared in a state haveing the form of superposition of Rydberg wave functions

$$\Psi(t=0) = \sum_n a_n(0)\, \varphi_n , \tag{6}$$

where $a_n(0)$ are the initial Rydberg probability amplitudes. The initial wave function (6) can be expanded in a series of quasienergy wave functions $\psi_j(t = 0)$, and this expansion can be used to describe evolution of the wave function in time

$$\Psi(t) = \sum_j b_j \psi_j \exp(-i\gamma_j t) , \tag{7}$$

where

$$b_j = \sum_n a_n(0) \langle \psi_j(0) | \varphi_n \rangle . \tag{8}$$

At $\delta = \Delta/2$ and sufficiently long time t ($t > (\Omega_R/\Delta)^4 t_K$, where $t_K = 2\pi n_0^3$ is the Kepler period) all the terms on the right-hand side of equation (7) vanish except the term with $j = 0$ which gives [11]

$$w_i^{as} = 1 - |b_0|^2 = 1 - \frac{1}{\pi^2} \left| \sum_n \frac{a_n(0)}{(n - n_0) - 1/2} \right|^2 . \tag{9}$$

In dependence of a choice of the initial probability amplitudes $a_n(0)$, the asymptotic long-time probability of ionization can be either small ($w_i^{as} \ll 1$) or close to one. For example, $w_i^{as} \equiv 1$ if $a_n(0)$ are given by any even function of $n - n_0 - ½$ ($a_{n_0+1}(0) = a_{n_0}(0), a_{n_0+2}(0) = a_{n_0-1}(0)$, etc.). This is the case when an atom is completely ionized during the time t and there is no asymptotic stabilization. On the other hand, evidently, Equation (9) gives $w_i^{as} = 0$, if its initial probability amplitudes $a_n(0)$ coincide with those of the above-described stable state ψ_{st} (5)

$$a_n(0) = \frac{1}{\pi} \frac{1}{(n-n_0)-1/2}. \tag{10}$$

This is a direct confirmation that he state ψ_{st} (5) is absolutely stable. Evidently, any other cases, intermediate between $w_i^{as} = 1$ and $w_i^{as} = 0$, can be realized with a proper choice of the initial probability amplitudes $a_n(0)$. Creation and control of specific initial superpositions of Rydberg states is a problem discussed in the following Section. The suggested solution is based on the use of the pump-probe two-pulse scheme of an assumed experiment.

4. Pump-probe scheme

Let us consider ionization of an atom by two identical rectangular laser pulses of a duration τ_p with a delay time between them τ_d (*Figure 3*). Let an atom be initially in an

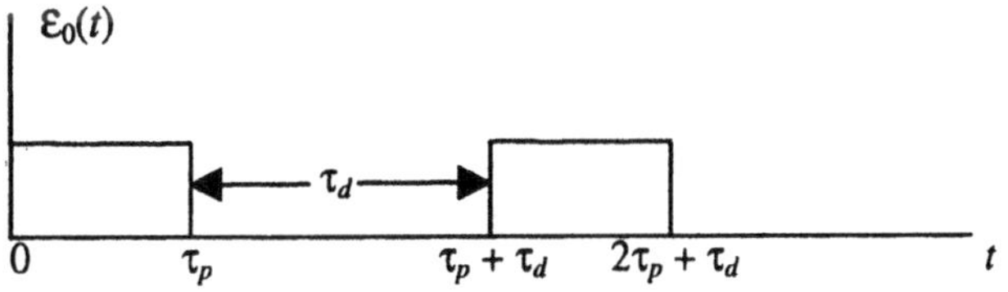

Figure 3. Field strength envelope for a pump-probe scheme

arbitrary state of the form (6) and le $\delta = \Delta/2$. The first pulse ionizes (burns out) all the terms of the superposition (6) except $\psi_{\gamma_0} \equiv \psi_{st}$, and to the wave function $\Psi(t)$ at $t = \tau_p$ turns into

$$\Psi_{bound}(t=\tau_p) = \exp(-i\tau_p\Delta/2)\; b_0\; \psi_{st} = \exp(-i\tau_p\Delta/2)\,\frac{b_0}{\pi}\sum_n \frac{\varphi_n}{n-n_0-1/2}. \tag{11}$$

During the period of time between the two pulses each of the field-free Rydberg states φ_n evolves as $\exp(-itE_n)$ to give at the time $\tau_p + \tau_d$ when the second pulse begins

$$\Psi_{bound}(t=\tau_p+\tau_d) = \exp(-i\tau_p\Delta/2 - i\tau_d E_{n_0})$$
$$\times \frac{b_0}{\pi}\sum_n \frac{\varphi_n \exp[-i\tau_p\Delta/2 - i\tau_d\Delta(n-n_0)]}{n-n_0-1/2}. \tag{12}$$

The coefficients in front of φ_n,

$$\tilde{a}_n = \exp(-i\tau_p\Delta/2 - i\tau_d E_{n_0})\,\frac{b_0}{\pi}\,\frac{\exp[-i\tau_d(n-n_0)\Delta]}{n-n_0-1/2}, \tag{13}$$

play the role of initial probability amplitudes for the second pulse. If $\tau_d\Delta = 2\pi k$ or $\tau_d = k\, t_K$, where k is an integer, the amplitudes $\tilde{a}_n$ (13) are proportional to those of Equation (10), and the function $\Psi_{bound}(\tau_p + \tau_d)$ is proportional to ψ_{st} (5),

$$\Psi_{bound}(\tau_p + \tau_d) = \exp(-i\tau_p\Delta/2 - i\tau_d E_{n_0})\; b_0\, \psi_{st}\,. \tag{14}$$

This function is stable and, hence the second pulse does not provide any additional ionization of an atom. If, however, $\tau_d \neq k\, t_K$, the wave function $\Psi_{bound}(\tau_p + \tau_d)$ is not proportional to ψ_{st} and is given by a superposition of all the quasienergy functions ψ_{γ_j}. All the terms of this superposition, except ψ_{γ_0}, decay under the action of the second pulse. Hence, in these cases the second pulse gives an additional contribution to the ionization yield, and the total probability of ionization appears to be much larger than in the case $\tau_d = k\, t_K$. In particular, in the case $\tau_d = (k + ½)\, t_K$ the probability amplitudes $\tilde{a}_n$ (13) appear to be even functions of $n - n_0 - ½$, and the function $\Psi_{bound}(\tau_p + \tau_d)$ (14) appears to be orthogonal to ψ_{st}:

$$\Psi_{bound}(\tau_p + \tau_d) = \exp(-i\tau_p\Delta/2 - i\tau_d E_{n_0})\, \frac{b_0}{\pi}\sum_n \frac{(-1)^{n-n_0}}{n - n_0 - 1/2}\,\varphi_n\,. \tag{15}$$

In accordance with the discussions of the previous Section, the second pulse ionizes such a state completely and the final probability of ionization equals one. As a whole, the dependence of the total ionization probability w_i on the delay time between the pulses τ_d is expected to be given by an oscillating sinusoidal-like curve. This conclusion is confirmed by the results of a numerical solution of Equations (3) shown in Fig. 4.

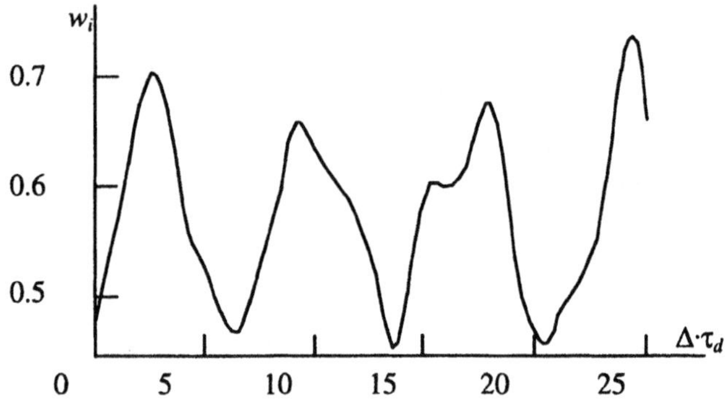

Figure 4. Ionization probability w_i vs. the delay time τ_d between the pump and probe pulses

Specifically, these calculations were carried out for an atom initially excited to the 25s state, under he conditions of a resonance with the 10p-level and for light frequency, intensity, and pulse duration equal to $\omega = 1{,}7\times10^{14}\ \text{sec}^{-1}$, $I = 10^9\ \text{W/cm}^2$, and $\tau = 50{\cdot}t_K = 120$ ps. The calculations were free of many approximations used above for simplicity in a qualitative analysis. In particular, in numerical calculations the approximation of equidistant levels was not used, i.e., the energies of Rydberg levels were taken in the form $E_n = -1/2n^2$ without approximate expansion near $n = n_0$. Similarly, matrix elements determining Rabi frequencies and tensor of ionization widths were not assumed to be n- and n'-independent. This generalizations are rather important. In principle, non-equidistant structure of levels can destroy perfectly periodical variations of phases of Rydberg probability amplitudes during the period of time between the two pulses. Nevertheless, as it follows from the numerical example, for optimally chosen parameters of the field and Rydberg states, the function $w_i(\tau_d)$ remains periodical (though not purely sinusoidal) and shows well pronounced large-amplitude oscillations even if the energy spectrum of an atom is not approximated by the equidistant one.

5. Conclusion

So, in the V-type interference stabilization the yield of photoelectrons depends significantly on phases of the initial population of Rydberg states. With properly chosen phases of the initial Rydberg probability amplitudes, an atom can be made either completely ionized by a sufficiently long laser pulse or almost not ionized at all. Phase distributions of Rydberg probability amplitudes can be created and efficiently controlled in the two-pulse pump-probe scheme of photoionization. In such a scheme the first pulse creates a superposition of Rydberg states stable with respect to photoionization. During the period of time between the two pulses phases of this superposition change in a controllable way. In dependence on the delay time between the pulses, different types (stable or unstable) of Rydberg superposition states arise at the instant of time when the second pulse begins. For this reason, the probability of ionization of an atom by the second pulse depends periodically on the delay time between the pulses and varies in a wide range (approximately from zero to one).

Acknowledgement

The work is carried out under partial support of the Russian Fund of Basic Research, grants ## 99-02-18034 and 97-02-71023 (IR-97-673) and CRDF (grant # RP1-2259).

References

1. Fedorov, M.V. and Movsesian, A.M. (1988) Field-induced effects of narrowing of photoelectron spectra and stabilization of Rydberg atoms, *J. Phys. B* **21**, L155-L158.

2. Fedorov, M.V. (1977) *Atomic and Free Electrons in a Strong Light Field,* World Scientific, Singapore.

3. Delone, N.B., Goreslavsky, S.P., and Krainov, V.P. (1989) Quasiclassical dipole matrix elements for atomic continuum states, *J. Phys. B* **22**, 2941-2945.

4. Bersons, I. Ja. (1981), Multiphoton ionization of high Rydberg states, *Phys. Lett. A* **84**, 364-366.

5. Ivanov, M.Yu. (1994) Suppression of resonant multiphoton ionization via Rydberg states, *Phys. Rev. A* **49**, 1165-1170.

6. Wojcik, A. and Parzinski, R. (1994) Rydberg-atom stabilization against photoionization: an analitically solvable model with resonance, *Phys. Rev. A* **50**, 2475-2489.

7. Wojcik, A. and Parzinski, R. (1995) Dark-state effect in Rydberg-atom stabilization, *J. Opt. Soc. Am. B* **12**, 369.

8. Fedorov, M.V. and Poluektov, N.P. (1997) Λ- and V-type transitions and their role in the interference stabilization of Rydberg atoms, *Laser Physics* **7**, 299-304.

9. Fedorov, M.V. and Poluektov, N.P. (1998) Competition between Λ- and V-type transitions in interference stabilization of Rydberg atoms, *Opt. Express* **2**, 51-57, http://www.opticsexpress.org/oearchive/source/2982.htm.

10. Poluektov, N.P. and Fedorov, M.V. (1998) Stabilization of a Rydberg atom and competition between the Λ and V transition channels, *JETP* **87**, 445-453

11. Fedorov, M.V. and Poluektov, N.P. (2000) Coherent control of strong-field two-pulse ionization of Rydberg atoms, *Opt. Express* **6**, 117-123, http://www.opticsexpress.org/oearchive/source/19077.htm.

COMPUTER EXPERIMENTS ON ATOMIC STABILIZATION IN A STRONG LASER FIELD

A.M. POPOV, O.V. TIKHONOVA and E.A. VOLKOVA
Institute of Nuclear Physics, Moscow State University,
119899, Moscow, Russia, e-mail: popov@mics.msu.su

In this paper, we apply the method of direct numerical integration of the time-dependent Schroedinger equation [1-6] for the analysis of the ionization dynamics of the hydrogen atom in a state $|nlm>$ by a strong linearly polarized laser field. Based on the results of these calculations, we discuss the general scenario of stabilization of an atom in a strong electromagnetic field due to the nonlinear dependence of the matrix element governing the ionization process on the strength of the wave field. We will demonstrate that such a nonlinearity may arise under certain conditions due to different kinds of the transitions between stationary states of the quantum system being reconstructed by the intense laser radiation. In this sence, we can speak of the competition and replacement of stabilization mechanisms, depending on the laser intensity and the set of the quantum numbers characterizing the initial atomic state.

We employ a model of a hydrogen-like system [5,7] with a smoothed Coulomb potential (in atomic units)

$$V_C = -1\Big/\sqrt{a^2 + r^2}\,, \tag{1}$$

where r is the modulus of the radius vector and $a = 0.5$ is the smoothing parameter. The structure of the energy spectrum and the wave functions of stationary states in potential (1) are discussed in detail in [5] and describe with a reasonable accuracy the energy spectrum and the wave functions of a real hydrogen atom.

For a linearly polarized laser field and the wave function $\Phi_0(\vec{r}, t=0)$ of the initial state with the definite angular momentum projection on the direction of the electric field vector the electron wave function in the cylindrical frame at an arbitrary instant of time can be represented as

$$\Psi(\rho, z, \phi, t) = \psi(\rho, z, t)\cdot exp(im\phi)\,, \tag{2}$$

where m is the angular projection momentum determined by the initial condition. The function $\psi(\rho, z, t)$ can be derived from the two-dimensional non-stationary Schroedinger equation

$$i\frac{\partial\psi}{\partial t} = -\frac{1}{2}\left(\frac{1}{\rho}\frac{\partial}{\partial\rho}\rho\frac{\partial\psi}{\partial\rho} + \frac{\partial^2\psi}{\partial z^2}\right) + V_{eff}^{(m)}(\rho, z)\psi(\rho, z, t) + z\varepsilon(t)\cos\omega t\cdot\psi(\rho, z, t) \tag{3}$$

with an effective potential

$$V_{eff}^{(m)}(\rho, z) = V_C(\rho, z) + \frac{m^2}{2\rho^2}\,, \tag{4}$$

B. Piraux and K. Rzążewski (eds.), Super-Intense Laser-Atom Physics, 285–294.

which includes the cetrifugal barrier for an electron with a momentum projection $L_z = m$. Here $\varepsilon(t)$ is the envelope of the laser pulse, and ω is the radiation frequency.

The majority of calculations were performed for the states with the principal quantum number $n = 3$ ($|3s>$, $|3p>$, $|3d>$) for different values of magnetic quantum number, $m = 0,...,l$.

Ionization dynamics was studied for radiation with quantum energy $\hbar\omega = 5.0$ eV in the range of laser intensities $P = 5 \cdot 10^{11} - 2 \cdot 10^{15}$ W/cm^2. All the states of the atom except $|1s>$ were coupled with the continuum by the one-photon transitions. The pulse envelope was assumed to have a trapezoidal shape with smoothed fronts with a duration $\tau_f = 5T$ and a plateau with a duration $\tau_{pl} = 10T$ ($T = 2\pi/\omega$ is the duration of the optical cycle)

$$\varepsilon(t) = \begin{cases} \varepsilon_0 \sin^2 \dfrac{\pi t}{2t_f}, & t \le t_f, \\ \varepsilon_0, & t_f \le t \le t_f + t_{pl}, \\ \varepsilon_0 \sin^2 \dfrac{\pi[t-(2t_f + t_{pl})]}{2t_f}, & t_f + t_{pl} \le t \le 2t_f + t_{pl}. \end{cases} \qquad (5)$$

The main result of our simulations is the dependence of the ionization probability of the hydrogen-like atom on the laser intensity for a set of initial atomic states (Fig.1). For weak fields, $P \le 10^{14}$ W/cm^2, the ionization probability $W_i(P)$ is linear function of the intensity. As the laser intensity grows up to the critical value P^*, (P^* is specific for each initial state) the ionization probability reaches its maximum and then starts to decrease ($|3d, m = 2>$ and $|3p, m = 1>$ states) or becomes a nonmonotonic function of laser intensity. In all the cases the ionization probability saturates at the level much lower than unity.

It is known that the stabilization effect is related to a considerable reconstruction of the atomic spectrum in the radiation field and the decrease of ionization width of states dressed by the field with the growth of the laser intensity. Therefore, the choice of a basis that would provide an adequate description of an atom in the presence of an electromagnetic field is of importance for the analysis of the dependencies obtained and the physical mechanisms of the stabilization.

Analysis of ionization dynamics of a model one-dimensional quantum system [8] has shown that, when the condition $\omega > I$ is fulfilled (I is the ionization potential), the ionization dynamics of an atom should be described in terms of Kramers-Henneberger (KH) method, which implies that the states of a dressed atom are represented as the states in the KH potential, and ionization is determined by the harmonics of the KH potential. In the opposite case ($\omega < I$) the KH approach is valid only for superatomic laser intensities. The same results for three-dimensional quantum systems were found to be valid in [9,10].

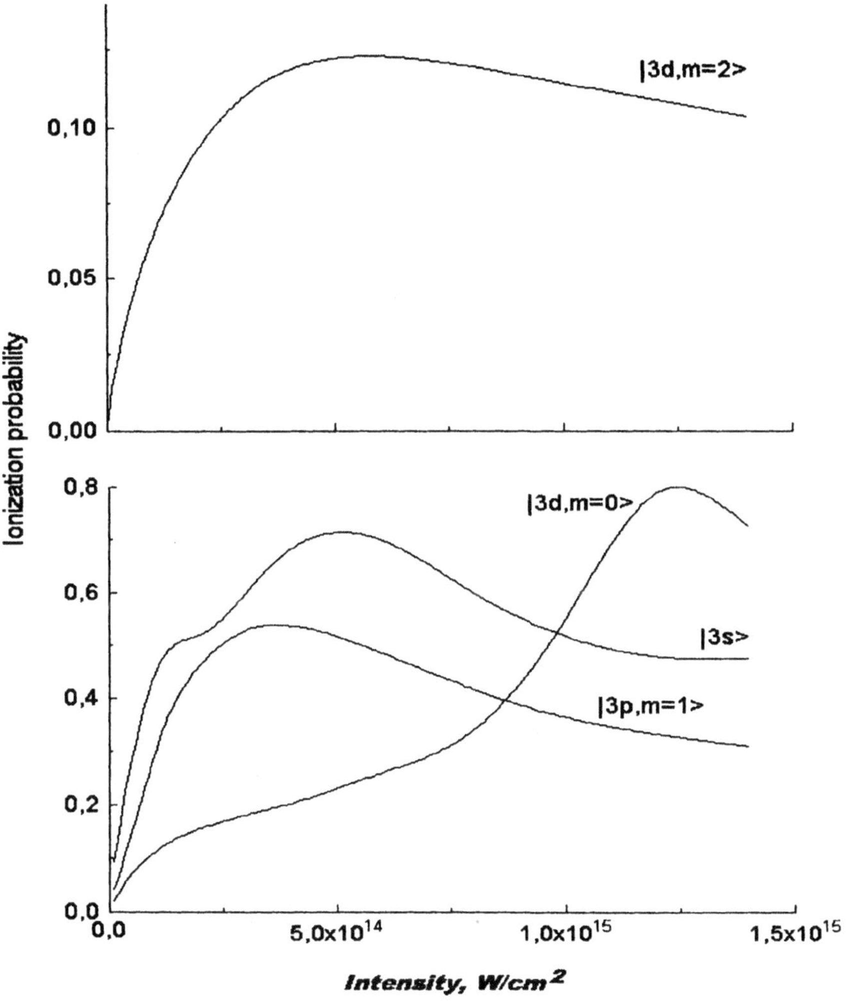

Fig.1. Ionization probability of a hydrogen atom initially prepared in the $|\,3d,m=2>$, $|\,3d,m=0>$, $|\,3p,m=1>$, $|\,3s>$ states as a function of the radiation intensity.

It should be mentioned, that the stabilization threshold (see Fig.1) corresponds to the intensity at which quiver amplitude of a free electron, α_e, is less than the characteristic size of the initial wave function $|\,nlm>$ localization, i.e.,

$$\alpha_e = \varepsilon_0 / \omega^2 < n^2 , \tag{6}$$

and the wave function of stationary states in the KH potential are close to the wave functions of a field-free atom. In such a situation in [9,11] it was demonstrated that the energy spectrum of an electron in the KH potential coincides with the spectrum of an atom with allowance for the Stark shift of levels.

Within the framework of the KH approach, electron transitions from bound states to the continuum are caused by the harmonics of the KH potential:

$$V_{KH}^{(j)}(\rho,z)=\frac{1}{2\pi}\int_{0}^{2\pi}V_C(\rho,z-\alpha_e\cos\xi)\exp(-ij\xi)d\xi . \tag{7}$$

The jth harmonic in this model gives rise to transitions where the energy is changed by $j\hbar\omega$, $j=\pm 1,\pm 2,...$ Stabilization under these conditions is the result of the decrease in the probability of transition from the relevant bound KH state to the KH continuum under the action of harmonics of the KH potential $V_{KH}^{(j)}(\rho,z)$. This transition probability is described by matrix elements of the following form: $<V_{KH}^{(j)}>=<(nlm)_{KH}\,|\,V_{KH}^{(j)}\,|\,E_{KH}>$, where $|(nlm)_{KH}>$ and $|\,E_{KH}>$ are the wave functions of the discrete and continuous spectra of the KH potential.

Since stabilization occurs in our calculations for $\alpha_e<n^2$, we neglect the difference of the wave functions in the free atomic and KH bases and represent $V_{KH}^{(j)}$ in the following form:

$$V_{KH}^{(1)}(\rho,z)=-\frac{1}{2}\cdot\frac{\alpha_e}{r^2}P_1(\cos\theta)-\frac{3}{8}\cdot\frac{\alpha_e^3}{r^4}P_3(\cos\theta)+...,\tag{8}$$

$$V_{KH}^{(2)}(\rho,z)=-\frac{1}{2}\cdot\frac{\alpha_e^2}{r^3}P_2(\cos\theta)+...\tag{9}$$

For weak fields, the first term of the harmonic $V_{KH}^{(1)}(\rho,z)$ is of most importance. This term is linear in the field and governs transitions changing the orbital momentum by $\Delta l=\pm 1$. In fact, this term describes one-photon transitions in the atomic basis. Similarly, the first term in the expansion $V_{KH}^{(2)}(\rho,z)$ gives $\Delta l=0,\pm 2$ and corresponds to two-photon absorption (emission) processes. However, the harmonic $V_{KH}^{(1)}(\rho,z)$ describes not only one-photon absorption (emission) processes: the second term in (8) gives the selection rules in the momentum $\Delta l=\pm 1,\pm 3$, corresponding to three-photon transitions with absorption of two photons and emission of one photon. In the atomic basis, where an atomic electron is involved in nearly free oscillations under these conditions, transitions from bound states to the continuum may occur through various channels:

$$\begin{aligned}&|\,n>\to|\,E>,\\&|\,n>\to|\,E'>\to|\,E''>\to|\,E>.\end{aligned}\tag{10}$$

Therefore, the nonlinear dependence of the matrix element $<V_{KH}^{(1)}>$ on the electric field strength in the radiation wave, which (as will be shown below) is responsible for the stabilization effect, has the interference nature, similar to [12,13]. Stabilization in such a situation is a result of the coherent repopulation of the continuum states by the free-free transitions and the interference of the amplitudes of one- and three-photon transitions to the continuum.

Now let us discuss the dependencies of ionization probability on laser intensity, $W_i(P)$, for different initial atomic states $|\,3lm>$.

1. *Ionization of* $|\,3d, m=2>$ *circular state.*

The circular state $|\,3d, m=2>$ is the lowest state among the states with $m=2$. We define the probability amplitude of the transition to the continuum $|\,Ef, m=2>$ as

$$C_{3d\to Ef} \sim <Ef, m=2\,|\,V_{KH}^{(1)}\,|\,3d, m=2> \sim \frac{\alpha_e}{2\sqrt{7}}\left\{<Ef\,|\,\frac{1}{r^2}\,|\,3d> - \frac{\alpha_e^2}{2}<Ef\,|\,\frac{1}{r^4}\,|\,3d>\right\}. \tag{11}$$

As can be seen from this expression, it is the nonlinearity of the operator $V_{KH}^{(1)}$ as a function of the wave filed that saturates the ionization. Let us estimate the stabilization threshold as

$$\alpha_e^2 \approx \frac{<Ef\,|\,1/r^2\,|\,3d>}{<Ef\,|\,1/r^4\,|\,3d>}. \tag{12}$$

Assuming that

$$<Ef\,|\,1/r^4\,|\,3d> \cong \sum_n <Ef\,|\,1/r^2\,|\,nd><nd\,|\,1/r^2\,|\,3d> \approx 2<Ef\,|\,1/r^2\,|\,3d><3d\,|\,1/r^2\,|\,3d>,$$

we derive $\alpha_e^2 \cong \left(2\cdot <3d\,|\,1/r^2\,|\,3d>\right)^{-1} \cong 30$.

Hence, for $\hbar\omega = 5$ eV, we have $P^* \cong 10^{15}$ W/cm^2. This estimation is only slightly higher than the value obtained in computer simulations, $P^* \cong 6\cdot 10^{14}$ W/cm^2.

Closing this section, we should note that, physically, the scenario of ionization of the $|\,3p, m=1>$ state and mechanism of the stabilization of this state are completely analogous to the scenario of ionization and mechanism of ionization considered above. The most significant difference of ionization and stabilization processes in the case of $|\,3p, m=1>$ state is associated with the fact that, due to the weaker centrifugal potential, this state is localized closer to a nucleus, which increases the matrix element $<V_{KH}^{(1)}>$ and the corresponding ionization probability. The stabilization threshold in this case can be estimated as $\alpha_e^2 \cong (2<3p\,|\,1/r^2\,|\,3p>)^{-1}$ which gives $P^* = 6\cdot 10^{14}$ W/cm^2.

2. *Ionization of* $|3d, m=0>$ *and* $|3s>$ *states.*

Let us first consider ionization of the $|3d, m=0>$ state. This case is seemingly analogous to the case of the $|3d, m=2>$ circular state, if to neglect the difference of the integral over angular part of the wave functions. Indeed, the dependence $W_i(P)$ demonstrates saturation of the ionization probability in the range $P = 3-7\cdot 10^{14}$ W/cm^2. However, the ionization probability then increases again with growth in laser intensity, followed by the stabilization in the range $P \geq 1.5\cdot 10^{15}$ W/cm^2. Such a behavior of $W_i(P)$ can be understood from the structure of the energy spectrum of a hydrogen atom with $m=0$ (see Fig.2). Besides the direct transitions to the continuum, described by the matrix element $<V_{KH}^{(1)}>$, transitions to the continuum through the low-lying $|2p, m=0>$ state can also occur in this case. The influence of transitions of this type (V-type transitions) on the ionization of atoms and stabilization effect was pointed out in [14-16]. Taking into account such transition through the $|2p, m=0>$ state, we derive the following expression for the amplitude of ionization:

$$C_{3d\to Ep} \sim <Ep|V_{KH}^{(1)}|3d> + \frac{<Ep|V|3s><3s|V|2p><2p|V|3d>}{(E_{2p}-E_{3d}+\omega)(E_{3s}-E_{3d})} + \ldots \qquad (13)$$

Here, $V = -\frac{\alpha_e}{2r^2}\cdot\cos\theta$ is the first term in the power-series expansion of the harmonic $V_{KH}^{(1)}$ in the field and the energy levels E_{nl} correspond to the levels of "dressed" atom. We should take into account that

$$E_{3s} - E_{3d} = (\beta_{3s} - \beta_{3d})\varepsilon_0^2/4 > 0, \qquad (14)$$

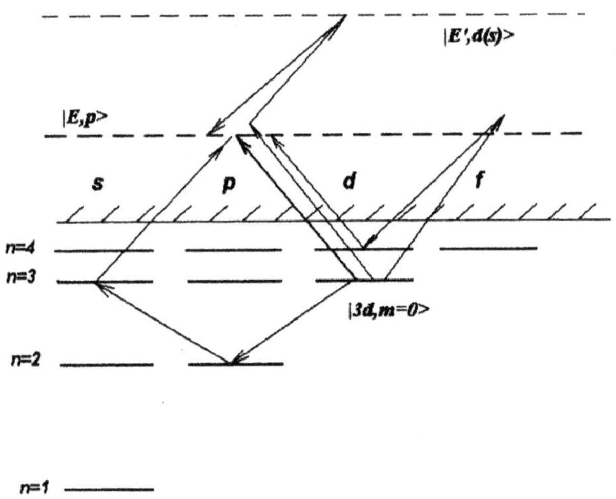

Fig.2. Diagram of levels of a hydrogen atom for $m=0$ and different channels of transitions to the continuum.

(here β is the atomic susceptibility) and the quantity $E_{2p} - E_{3d} + \omega$ remains positive, but decreases with the growth of the intensity, as $\beta_{2p} > \beta_{3d}$ [11]. Therefore, the contribution of the second term in (13) increases with intensity growth, which leads to destabilization of the system. The data presented in Fig.3 and 4 provide the additional confirmation of the statements made above. Specifically, Fig.3 presents the distribution of the electron density at the instant of time corresponding to the termination of the laser pulse for two values of the laser intensity. Fig.3a illustrates the case of a weak field, when only the direct transition $|3d, m=0> \rightarrow \begin{cases} |Ep, m=0> \\ |Ef, m=0> \end{cases}$ plays a significant role. Fig.3b represents the stabilization area, where the amplitudes of direct and three-photon transitions interfere through intermediate continuum states. Due to this interference the distribution of the electron density becomes strongly nonuniform in space. The data presented in Fig.4 provide a convincing proof of the reality of V-transitions in the system under study: the $|2p, m=0>$ state is populated in the process of laser action, while the population of $|3s>$ state monotonically increases in time.

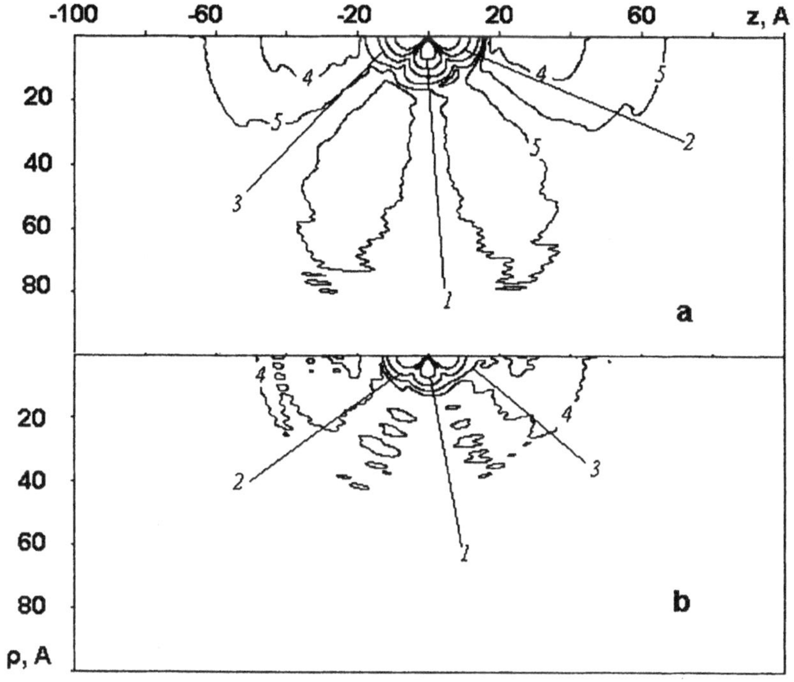

Fig.3. The distribution of the electron density at the instant of time corresponding to the termination of the laser pulse for the ionization $|3d, m=0>$ state for laser intensities equal to (a) $5 \cdot 10^{13}$ and (б) $3 \cdot 10^{14}$ W/cm^2. The level lines correspond to 10^{-2} (1), 10^{-3} (2), 10^{-4} (3), 10^{-5} (4), and 10^{-6} (5).

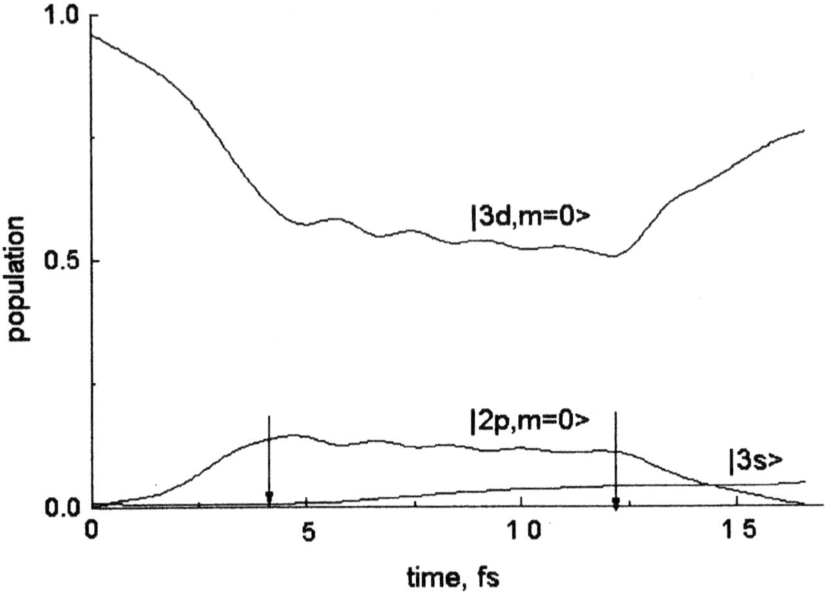

Fig.4. Dynamics of population of the $|3d, m=0>$, $|2p, m=0>$, and $|3s>$ states modified by the field for the radiation intensity equal to $P = 2\cdot 10^{14}$ W/cm^2. The arrows indicate the plateau of the pulse.

However, the stability of the system with respect to the ionization increases again above $P^* \cong 1.4\cdot 10^{15}$ W/cm^2. Ionization suppression in this case is due to the coherent repopulation of $|nd>$ $(n \geq 4)$ Rydberg states through Raman two-photon Λ-type transitions and interference of the amplitudes of direct $|3d> \to |E>$ and three-photon $|3d> \to |E'> \to |nd> \to |E>$ transitions to the continuum. These transitions are also shown at Fig.2. This statement is confirmed by the distribution of populations for $|nl, m=0>$ discrete atomic states at the instant of time corresponding to the laser pulse with peak intensity $P^* \cong 1.4\cdot 10^{15}$ W/cm^2 (Fig.5), which reveals the influence of Λ-type transitions during the laser pulse, resulting in the predominant population of states with $\Delta l = 0,2$. Estimations of the threshold of the interference stabilization performed within the framework of the model described in [12,13] demonstrate that the overlapping of the ionization widths for $|3d, m=0>$ state appears for $P \approx 2\cdot 10^{15}$ W/cm^2, which agrees well with the results of numerical simulations.

Now, let us analyze the dependence $W_i(P)$ for the $|3s>$ state. Besides three-photon transitions through continuum states $|3s> \to |E', p> \to \begin{cases} |E'', s> \\ |E'', d> \end{cases} \to |E, p>$ we should also take into account both V-type $|3s> \to |2p, m=0> \to |3d, m=0>$ and Λ-type Raman transitions $|3s\rangle \to |E',p\rangle \to \begin{cases} |ns\rangle, \\ |nd,m=0\rangle \end{cases}$ with subsequent transitions to the $|Ep\rangle$ state in the continuum. With allowance for V-type transitions through the intermediate $|2p,m=0\rangle$ state, we have

$$C_{3s\to Ep} \sim \langle Ep|V_{KH}^{(1)}|3s\rangle + \frac{\langle Ep|V|3d\rangle\langle 3d|V|2p\rangle\langle 2p|V|3s\rangle}{(E_{2p}-E_{3s}+\omega)(E_{3d}-E_{3s})} + \dots \quad (15)$$

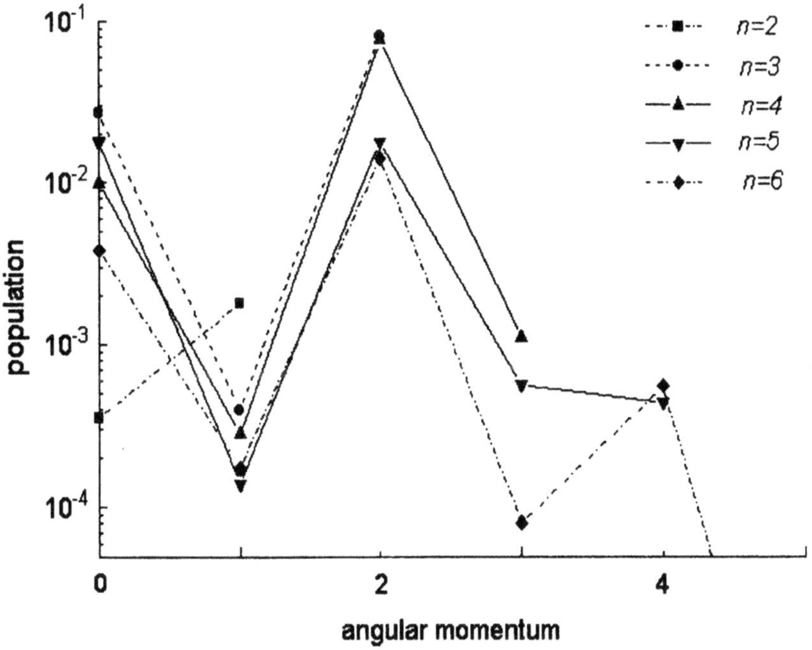

Fig.5. Population of different states $|nl,m=0\rangle$ of the discrete spectrum in the after-pulse regime for the ionization of $|3d,m=0\rangle$ state with the intensity equal to $P = 1.4\cdot 10^{15}$ W/cm^2.

In the case of weak fields, the amplitude of transitions to the continuum through intermediate states of the discrete spectrum is opposite in sign to the analogous term in (14). Hence, this term decreases the ionization amplitude. As a result, the ionization probability saturates at the level of about 0.5 at the intensity $P \cong 2\cdot 10^{14}$ W/cm^2. This value of the intensity corresponds to a zero resonant denominator in (15) $E_{2p}-E_{3s}+\omega=0$. With a further increase of laser intensity, the amplitude of three-photon transition reveals its sign, which leads to destabilization effect: the ionization probability grows again with the increase in the intensity for $P \geq 2\cdot 10^{14}$ W/cm^2. As for the increase in the stability of the system with respect to ionization for $P > 5\cdot 10^{14}$ W/cm^2, it is related also to the interference mechanism [12,13]. Numerical estimates show, that for such intensities ionization widths overlap, giving rise to the interference of amplitudes of transitions to the continuum from $|ns>$ states coherently repopulated by Λ-type transitions:

$$|3s> \to |Ep>,$$
$$|3s> \to |E', p> \to |ns> \to |Ep>.$$

Summarizing the aforesaid, we should emphasize once again that the explanation of the dependence of the ionization on laser intensity requires the analysis of different type of the transitions coupling the initial electronic state with the final state in the continuum. The contribution of different channels changes depending on radiation intensity. In this sense, it is possible to speak on the competition of different stabilization mechanisms and the replacement of one mechanism by another. Nevertheless, all this mechanisms have the interference nature.

This study was supported by Russian Foundation for Basic Research, grant 00-02-16046 and INTAS, grant 99-1495.

References

1. Kulander, K.C., (1987) Multiphoton ionization of hydrogen: A time-dependent theory, *Phys.Rev.A*, **35**, 445-447
2. Kulander, K.J.Shafer, K.J., and Krause, J.I. (1991) Dynamic stabilization of hydrogen in an intense high-frequency pulsed laser field, *Phys.Rev.Lett.* **20**, 2601-2604
3. Im, K., Grobe, R, and Eberly, J.H, (1994) Photoionization of the hydrogen 4s state by a strong laser pulse: Bare-state dynamics and extended-charge-cloud oscillations, *Phys.Rev.A* **49**, 2853-2860
4. Gajda, M., Piraux, B., and Rzazewski, K. (1994) Ionization of an excited hydrogen atom by a high-frequency circulary polarized pulsed field, *Phys.Rev.A* **50**, 2528-2539
5. Volkova, E.A., Popov, A.M., and Tikhonova, O.V. (1998) Numerical modeling of the photoionization of Rydberg atoms by the field of an electromagnetic wave, *JETP* **86**, 328-334
6. Potvliege, R.M., and Smith, P.H.G. (1993) Adiabatic stabilization of excited states of H in an intense linearly polarized laser field, *Phys.Rev.A*, **48**, R46-R49
7. Tikhonova, O.V., Volkova E.A., Popov A.M., and Fedorov, M.V., (1999) Interference stabilization of Rydberg atoms: quasiclassical analytical theory and exact three-dimensional numerical simulations, *Phys.Rev.A* **60**, R749-R752
8. Popov A.M., Tikhonova, O.V., and Volkova E.A. (1999) Applicability of the Kramers-Henneberger approximation in the theory of strong-field ionization, *J.Phys.B* **32**, 3331-3345
9. Popov A.M., Tikhonova, O.V., and Volkova E.A. (2000) A Hydrogen atom in a strong laser field, *Laser Phys.* **10**, 188-197
10. Popov A.M., Tikhonova, O.V., and Volkova E.A. (2000) Stabilization of circular states of a hydrogen atom in the dichotomous regime, *Laser Phys.* **10**, 779- 784
11. Krylovetsky, A.A, Manakov, N.L., and Marmo, S.I. (1997) Quadratic Stark effect and dipole dynamic polarizabilities of hydrogen-like levels, *Laser.Phys.* **7**, 781-796
12. Fedorov, M.V., and Movsesian A.M. (1988) Field-induced effects of narrowing of photoelectron spectra and stabilization of Rydberg atoms, *J.Phys.B* **21**, L155-L158
13. Fedorov, M.V. (1999) Analytical models and exact numerical calculations in the theory of interference stabilization of Rydberg atoms, *Laser Phys.* **9**, 209-214
14. Noordam, L.D., Stapelfeldt, H., Duncan, D.I., and Gallagher, T.F. (1992) Redistribution of Rydberg states by intense picosecond pulses, *Phys.Rev.Lett.* **68**, 1496-1499
15. Wojcik, A., and Parzinski, R. (1994) Rydberg-atom stabilization against photoionization: An analytically solvable model with resonance, *Phys.Rev.A* **50**, 2475-2489
16. Fedorov, M.V., and Poluektov, N.P. (1997) Λ- and V-type transitions and their role in the interference stabilization of Rydberg atoms, *Laser Phys.* **7**, 299

QUASISTATIONARY STABILIZATION OF THE DECAY OF A WEAKLY-BOUND LEVEL AND ITS BREAKDOWN IN A STRONG LASER FIELD

N. L. MANAKOV, M. V. FROLOV
Voronezh State University, 394693 Voronezh, Russia

AND

BOGDAN BORCA,* ANTHONY F. STARACE
Department of Physics and Astronomy, The University of Nebraska 116 Brace Laboratory, Lincoln, NE 68588-0111, USA

1. Introduction

Although it was pointed out about 10 years ago [1, 2] that an atomic decay rate might decrease as the intensity of a high-frequency laser field increases, there still does not exist any complete understanding of either the physical origin of this interesting nonlinear phenomenon or its dependence on the atomic and field parameters. Essentially, the problem consists in that the phenomenon requires a major modification of the standard picture of photoeffect in a strong laser field. In Ref. [1] the origin of this stabilization is related to a particular distortion of an atomic potential by an intense monochromatic high-frequency field. This phenomenon is called adiabatic or quasistationary stabilization (QS). For the case of Rydberg levels, another (interference) mechanism of QS was suggested [2]. Both theories predict an unlimited decrease of the decay rate (or of the width Γ of an atomic level, i.e., of the imaginary part of the complex quasienergy, $\epsilon = \mathrm{Re}\,\epsilon - i\Gamma/2$)) as the laser field amplitude increases. In recent years the idea of "dynamic stabilization" (DS) [3] has become popular. It originates from the pulse form of a laser field rather than from any intrinsic property of the atom in a strong monochromatic field. Within this model the numerous simulations point also to the possibility of a breakdown of stabilization for the case of superintense short laser pulses [4]. However, a recent paper [5], using the quasistationary quasienergy states (QQES) as an adiabatic basis for the laser pulse has shown that DS has the same (quasistationary) origin

B. Piraux and K. Rzążewski (eds.), Super-Intense Laser-Atom Physics, 295–304.

as QS. Finally, a number of authors deny the existence of stabilization, in particular, of QS for ionization from a short-range potential [6] (see also [7]) and of DS in pulsed fields [8]. Obviously, these controversies and ambiguities are caused by the complexity of the numerical solution of the Cauchy problem for the time-dependent Schrödinger equation in a strong field and by the absence of analyses for exactly solvable analytical models. We analyze below (see also [9]) the exactly solvable problem of an electron in a three-dimensional, short-range potential and consider the two questions: does a QS-like behavior of the decay rate exist for this model, and, if so, is there an upper intensity limit of the QS regime?

2. Basic equations and numerical results

We consider the quasistationary decay of a bound state in a monochromatic laser field with electric vector

$$\mathbf{F}(t) = \frac{F}{\sqrt{1+\eta^2}}\{\cos\omega t, \eta\sin\omega t, 0\}, \quad -1 \le \eta \le +1,$$

and intensity $I = cF^2/8\pi$ based on the QQES approach [10] for the exactly solvable model of a 3-dimensional, short-range (δ -) potential having one bound state with energy E_0. In this model, the complex quasienergy ϵ is the eigenvalue of a one-dimensional integral equation for the periodic (in time) function $\varphi_\epsilon(\tau = \omega t)$, which determines the asymptotic behavior of $\Phi_\epsilon(\mathbf{r}, t)$ at origin ($r \to 0$) [11]:

$$(\sqrt{\mathcal{E}_F - \epsilon} - 1)\varphi_\epsilon(\tau) = \sqrt{\frac{\omega}{4\pi i}} \int_0^\infty \frac{d\tau'}{\tau'^{3/2}} e^{-i(\mathcal{E}_F - \epsilon)\tau'/\omega} \times$$
$$\times \left\{\varphi_\epsilon(\tau - \tau') e^{4i\Delta\frac{\sin^2\tau'/2}{\tau'}[1 - l\cos(2\tau - \tau')]} - \varphi_\epsilon(\tau)\right\}. \quad (1)$$

This equation involves the linear polarization degree of the laser field, $l = (1-\eta^2)/(1+\eta^2)$, as well as the characteristic dimensionless parameters of the problem: $\hbar\omega/|E_0|$ and the ratio of the ponderomotive energy of a free electron, $(eF)^2/(4m\omega^2)$, to either the photon energy, $\Delta = (eF)^2/(4m\hbar\omega^3)$, or to $|E_0|$, $\mathcal{E}_F = (eF)^2/(4m\omega^2|E_0|) = \Delta\,(\hbar\omega/|E_0|)$. In scaled units, $\mathcal{E}_F = F^2/\omega^2$ and $\Delta = F^2/\omega^3$. (The Keldysh parameter, $\gamma = \sqrt{2m|E_0|}\omega/eF$, in our units is: $\gamma = \omega/\sqrt{2}F$.)

For the case of circular polarization of $\mathbf{F}(t)$ (i.e., $\eta = \pm 1$), $\varphi_\epsilon(\tau) \sim const$, in which case Eq. (1) reduces to the transcendental equation [12]:

$$\sqrt{\mathcal{E}_F - \epsilon} - 1 = \sqrt{\frac{\omega}{4\pi i}} \int_0^\infty \frac{d\tau}{\tau^{3/2}} e^{-i(\mathcal{E}_F - \epsilon)\tau/\omega} \left\{e^{4i\Delta\frac{\sin^2\tau/2}{\tau}} - 1\right\}. \quad (2)$$

Since $\mathrm{Im}\,\epsilon < 0$, the integrals in Eqs. (1) and (2) are formally divergent at their upper limit and hence must be understood as an analytical continuation of complex ϵ from the upper half-plane. For this analytical continuation we use the following relation,

$$\int_0^\infty \frac{d\tau}{\tau^{1/2}} e^{-i\alpha\tau} f(\tau) = \frac{1}{\sqrt{4\pi i}} \int_{-\infty}^{\infty} \frac{dk}{\sqrt{\alpha + k}} \int_{-\infty}^{\infty} dx e^{ikx} f(x), \tag{3}$$

where the double integral is convergent for any $\alpha = (\mathcal{E}_F - \epsilon)/\omega$. Eq. (3) is a **key result** for numerical calculations involving strong fields. Note that a number of authors dealing with the general QQES formalism for the δ - model potential [11] (see, e.g., [13]) remove the divergence of integrals of the kind (3) by the (approximate) substitution $\epsilon \approx E_0 = -1$, which is obviously incorrect for strong fields. In particular, it was what led to the incorrect results in Fig. 5 of Ref. [6] and, as a result, to the incorrect statement in Ref. [6] regarding the absence of the QS regime for short-range potentials. Although a number of other authors have performed acurate numerical calculations of Γ for the $\delta-$potential (for predominantly circular (e.g., [14]), but also linear [15] polarization), they do not analyze the results in the context of stabilization.

In Fig. 1 we present our numerical results for $\Gamma(F)$ at a scaled frequency $\omega = 1.55$, which corresponds to the Nd:YAG laser frequency for the case of the H^--ion (with $|E_0| \approx 0.752$ eV). One sees that for not too high (but nonperturbative) F up to $F = F_{cr}$, $\Gamma(F)$ demonstrates the typical QS-behavior (cf. Fig. 4 for the H-atom in [16]). The decrease of $\Gamma(F)$ in the QS domain is more pronounced for the case of linear polarization. In contrast with QS-calculations for the Coulomb potential, Fig. 1 shows a sharp breakdown of QS (i.e., a sharp rise of $\Gamma(F)$) beginning at a critical field, $F = F_{cr}$, which does not depend on the polarization (see Eq. (4) below). Fig. 2 gives the $\omega-$dependence of $\Gamma(F)$ for frequencies $\omega > 1$ for the case of circular polarization. Both the size of the interval of QS (in F) and the magnitude of the decay rate Γ in the QS regime increase with increasing ω. Moreover, we find that at fixed F, for $F \lesssim F_{cr}$, $\Gamma(\omega)$ is approximated by the power law, $\Gamma(\omega) \sim \omega^\alpha$ with $\alpha \approx 3.92$. This result is in reasonable agreement with the value $\alpha = 4$ given by the high-frequency theory of QS for a circularly-polarized field [1].

Concerning the breakdown of QS we note firstly that near the point $F = F_{cr}$ there is a narrow interval of fields, $F \leq F_{cr}$, in which the width Γ exhibits an irregular behavior (see Fig. 2). An exact numerical calculation of Γ in this interval is difficult. For a more thorough analysis of Γ at $F \sim F_{cr}$ we performed nonperturbative calculations of partial widths, $\Gamma^{(n)}(F)$, corresponding to the absorption of a fixed number, n, of photons. Since

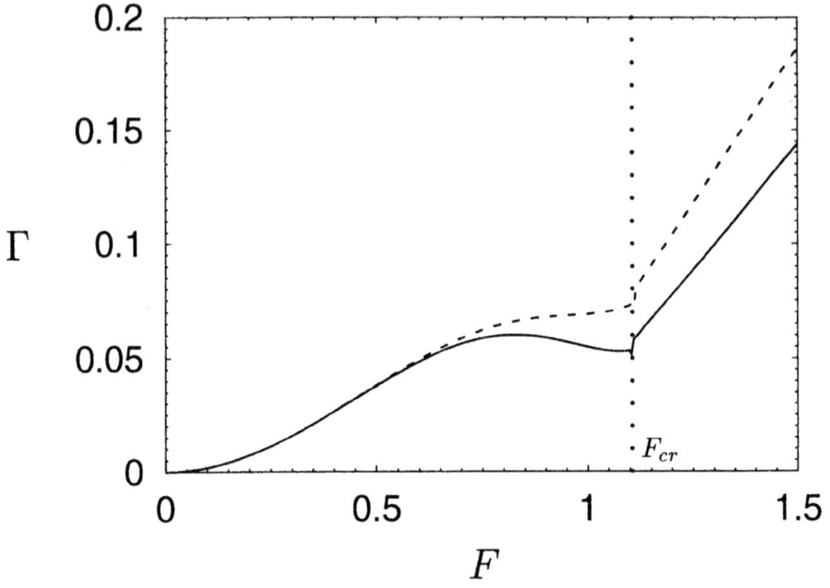

Figure 1. Dependence of Γ on F for linear (solid curve) and circular (dashed curve) polarizations of a laser field with $\omega = 1.55$. The dotted vertical line denotes F_{cr}.

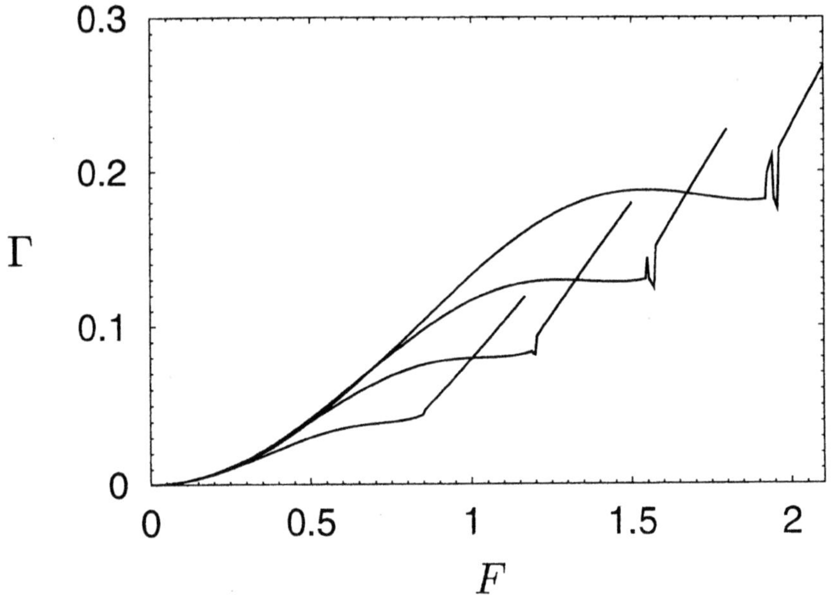

Figure 2. Dependence of Γ on F for a circularly polarized laser field having $\omega > 1$. The curves (bottom to top) correspond to $\omega = 1.4$, 1.6, 1.8, and 2.0.

the available space does not permit a detailed discussion, we note only that in the vicinity of F_{cr} the one-photon width, $\Gamma^{(1)}$ (which corresponds to the photoelectron energy $E_p = \mathrm{Re}\,\epsilon - \mathcal{E}_F + \omega$ and almost completely determines the total width Γ up to the midpoint of the QS interval), has a deep minimum. Thus the major contribution to Γ in the breakdown domain stems from (interfering) channels of the direct n-photon above-threshold decay. Assuming that the breakdown of QS originates from the closeness of the direct photoionization channel, the estimation of F_{cr} may be obtained from the relation $\mathrm{Re}\,\epsilon - \mathcal{E}_F + \omega = 0$. Neglecting the Stark-shift (which is small for $\omega \sim 1$) one obtains

$$F_{cr.} \approx \omega\sqrt{\omega - 1}. \tag{4}$$

This estimation is in excellent agreement with values of F_{cr} generated from the numerical calculations. A slight lowering of the calculated F_{cr} with increasing ω compared to that in Eq. (4) is caused by the Stark-shift.

3. Analytical results for circular polarization

These results on the breakdown of stabilization and on the magnitude of the critical field F_{cr} have important physical consequences, implying, in particular, the non-existence of the so-called Death Valley [1] (i.e., of a global deep minimum in the F-dependence of the lifetime of a quasistationary level) for a weakly-bound system in a strong monochromatic field. Hence to justify our numerical results and their physical interpretation, we present below some analytical considerations. It is well-known that for the case of circular polarization, ϵ may be obtained as the (complex) eigenvalue of the stationary QQES-Hamiltonian (see, e.g., [10]) in a coordinate frame rotating with the frequency ω,

$$\mathcal{H}_{\mathrm{rot.}}(\mathbf{r}) = -\nabla_{\mathbf{r}}^2 + U(\mathbf{r}) + Fx \pm \omega \hat{L}_z\,,$$

where $\hat{\mathbf{L}}$ is the orbital angular momentum operator. We take into account the operator $\omega\hat{L}_z$ using perturbation theory (PT) in a basis of quasistationary states of an electron in the δ-potential, $U(\mathbf{r}) = 4\pi\delta(\mathbf{r})(\partial/\partial r)r$, and in a static field $\mathbf{F}$. (For PT for quasistationary states see, e.g., [17], and for a convenient form for the Green's function of our problem in terms of regular and irregular Airy functions, $\mathrm{Ai}(x)$ and $\mathrm{Bi}(x)$, as well as for techniques for PT calculations to second order, see [18].) The result for ϵ (taking into account corrections of order ω^2) is:

$$\epsilon = E - \frac{\omega^2}{360F^{2/3}}\frac{I^{(4)}(\xi)}{I(\xi)}, \quad \xi = -\frac{E}{F^{2/3}} \tag{5}$$

where E is the exact (complex) energy of the quasistationary state in a static field $\mathbf{F}$. E is determined as the root of the transcendental equation:

$$1+\pi F^{1/3}J(-EF^{-2/3})=0, \tag{6}$$

where $J(x)=\mathrm{Ai}'(x)\mathrm{Ci}'(x)-x\mathrm{Ai}(x)\mathrm{Ci}(x)$; $\mathrm{Ci}(x)=\mathrm{Bi}(x)+\mathrm{iAi}(\mathrm{x})$; $I(x)=\mathrm{Ai}(x)\mathrm{Ci}(x)$; $I^{(4)}(x)=\mathrm{d}^4I(x)/\mathrm{d}x^4$. (For an analysis of the function $E=E(F)$, see [18].) Eqs. (5) and (6) do not assume that F is small. Thus the known properties of the Airy functions allow a simple analysis of the applicability of PT in the operator $\omega\hat{L}_z$ (i.e., for the conditions under which the correction of order ω^2 in Eq. (5) is small compared to E) for the limiting cases of weak and strong F.

In the weak-field limit, $F<<1$, we obtain from Eqs. (5) and (6):

$$\epsilon=-1-\frac{1}{16}F^2\left[1+\frac{3}{2}F^2+\frac{7}{24}(1+13F^2)\omega^2\right]-\frac{\mathrm{i}}{4}F\left[1+\frac{4}{45}\frac{\omega^2}{F^3}\right]\mathrm{e}^{-\frac{4}{3F}}. \tag{7}$$

As may be seen from Eq. (7), a PT-account of $\omega\hat{L}_z$ in weak fields is possible only for small ω ($\omega^2<F^3<<1$), so that the frequency correction to the tunnelling pre-exponential factor is small. The Stark-shift, $\mathrm{Re}\,\epsilon+1$, given by Eq. (7) coincides exactly with the first two terms of the ω-expansion for the known dynamic polarizability and hyperpolarizability of a weakly-bound particle [19]. The inapplicability of Eq. (7) for $\omega>1$ is evident already from the known fact that for weak fields and above-threshold frequencies the level width has the form of a power series in F. In particular, in the lowest PT order in F (for an arbitrary polarization),

$$\Gamma_0^{(1)}=(8F^2/3\omega^4)(\omega-1)^{3/2}. \tag{8}$$

Finally, we note that $\mathrm{Im}\,\epsilon$ in Eq. (7) coincides with the first two terms of the expansion (in ω^2) of the known result for the level width Γ^{ad} for a circularly-polarized field in the adiabatic (Keldysh) approach [12]:

$$\Gamma^{ad}=\frac{F}{2}\exp\left[-\frac{4}{3F}\left(1-\frac{\omega^2}{15F^2}\right)\right]. \tag{9}$$

In the limit $F>>1$, Eqs. (5) and (6) give the following result:

$$\begin{aligned}\epsilon\; &=\; F^{2/3}\Big\{0.44133122\mathrm{e}^{-\mathrm{i}\frac{\pi}{3}}-\left[0.86328690\mathrm{e}^{-\mathrm{i}\frac{\pi}{6}}-0.02548960\mathrm{e}^{\mathrm{i}\frac{\pi}{3}}\frac{\omega^2}{F^3}\right]F^{-1/3}\\ &-\; 0.3763512F^{-2/3}+\dots\Big\},\end{aligned} \tag{10}$$

showing that in ultrastrong fields a perturbative account of the term $\omega\hat{L}_z$ is possible for any frequencies satisfying the condition $\omega^2<F^3$. (It is

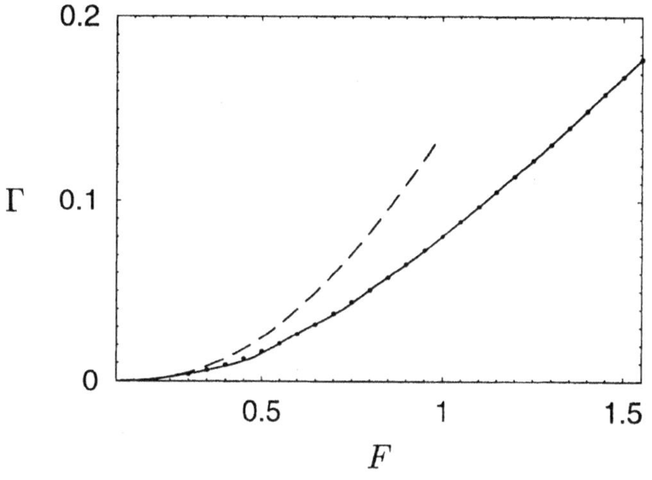

Figure 3. Dependence of Γ on F for a circularly polarized laser field having $\omega = 0.74$. Solid curve: exact (numerical) result from Eq. (2); dashed curve: the weak-field approximation, Eq. (7); solid circles: the PT result in ω, Eq. (5).

interesting that this condition is the reverse of that for the convergence of the standard PT in F for the complex quasienergy [20]: $\Delta \equiv F^2/\omega^3 < 1$.) Although the asymptotic expansion of the energy E in a static field, which was used in the derivation of (10), is valid only for ultrastrong fields ($F >> 10$) [18], the result (10) is noteworthy since it shows analytically the breakdown of QS in ultrastrong fields.

Concerning the applicability of the approximate result (5), we note that for $\omega < 1$ it agrees well with the exact result (cf. Eq. (2)) both for moderate and strong fields. Fig. 3 shows that the exact curve for $\Gamma(F)$ at $\omega = 0.74$ and that calculated using Eq. (5) are almost indistinguishable for $F \geq 0.5$. Moreover, the weak-field (tunnelling) approximation, Eq. (7), is not applicable in this region of F, which corresponds to "overbarrier" ionization for the case of an atomic potential. It is interesting that, beginning from $F \approx 1$, the width Γ exhibits actually a linear dependence on F similar to the so-called "intermediate" asymptotic for $\Gamma(F)$ for overbarrier decay in a Coulomb potential [21]. In other words, for $\omega < 1$ the action of a strong circularly-polarized field is equivalent to the action of a strong static field **F**, and it is reasonably described by the analytic result (5).

For $\omega > 1$ PT in $\omega \hat{L}_z$ is applicable only for ultrastrong fields. Thus results for the domain of stabilization and for $F \geq F_{cr}$ may be obtained only numerically. In Fig. 4 we compare the strong-field asymptote, from Eq. (10), with the exact (numerical) result for $\omega = 1.55$. Although the range of F considered is not enough to achieve close coincidence of the solid and dashed curves, the agreement improves as F increases. Fig. 4 demonstrates also

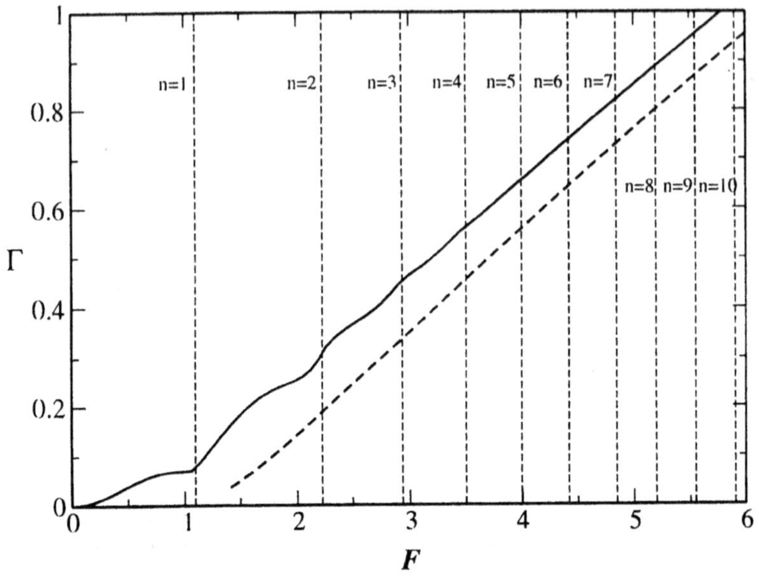

Figure 4. Comparison of exact (solid line) and approximate (dotted line) results for Γ [from Eqs. (2) and (10)]for strong, circularly polarized fields with $\omega = 1.55$. Vertical dashed lines denote threshold values of F, at which $\mathrm{Re}\,\epsilon - \mathcal{E}_F + n\,\omega = 0$.

threshold peculiarities for higher ATI-channels with $n > 1$, although these peculiarities are smoothed out compared with $n = 1$ (see also [13, 14]). Our detailed calculations show that, for linear polarization, high-order threshold peculiarities are more pronounced than for circular polarization (see also [15]). Moreover, the stabilization-like regime (up to the closure of the direct photoionization channel) presented in Fig. 1 for $l = 1$ exists also for $\omega < 1$, although both the interval of QS (in F) and the variation of Γ in this interval decrease rapidly with decreasing ω.

4. Discussion and conclusions

Numerical and analytical analyses allow one to understand the major peculiarities of the quasistationary decay of a weakly-bound level for various relations among the key parameters, $|E_0|$, ω, and F. Since (in contrast to the case of a static field) $F = 0$ is the point of analyticity for the function $\epsilon = \epsilon(F)$, for $\Delta << 1$ the multiphoton (or perturbative) regime for decay is realized for any F and ω including the case of $\omega << 1$, i.e., $\Gamma \sim F^{2N}$ with $N = [|E_0|/\omega]$. For increasing F the situation depends qualitatively on the frequency: for small ω the relation $\Delta \sim 1$ is achieved at relatively weak values of F, and the PT-regime merges smoothly into the tunnelling regime of Keldysh (see Eq. (9) for the case of circular polarization). This mechanism is realized for fields that are weak compared to interatomic ones (see Eq. (7) and the dashed curve in Fig.1). It has been analyzed in detail for arbitrary γ [22]. For higher F (and small ω) the tunnelling regime merges into the "overbarrier" regime (the solid curve in Fig. 3), which exists up to the passing of the quasistationary level into continuum.

For $\omega > 1$ the decay mechanism is multiphotonic even for fields at which the first nonvanishing PT-result (see Eq. (8) for the case of a short-range potential) is already inapplicable. Hence it is necessary to take into account high-order corrections, which correspond to the re-emission of absorbed photons and to direct n-photon, above-threshold absorption. However, in this domain the linear dependence of Γ on the intensity [see Eq. (8)] is changed: the correction of order F^2 to $\Gamma_0^{(1)}$ is negative [19]. Thus, the QS regime (in which Γ no longer increases or even decreases with F) arises instead of the tunnelling regime. Its beginning corresponds to fields $F < \omega$, at which higher-order PT-corrections become important. Obviously in the domain of an 'established' stabilization, PT in F is inapplicable. In contrast to the low-frequency case, where Coulomb effects change only the pre-exponent factor in the tunnelling regime or the slope of the linear in F dependence of Γ for overbarrier decay, the QS regime is much more sensitive to the details of the potential. For a short-range potential, the Stark-shift is negative in the QS interval and the breakdown of QS is connected with the closure of the direct photoionization channel (see Figs. 1 and 4) and later with the passing of the quasibound state into the continuum for superstrong F. For a Coulomb potential, the Stark-shift is positive for $\omega > 1$ and moderate F. Thus the closure of the photoionization channel is impossible here, and only the passing of the bound level into the continuum may lead to the breakdown of QS. But detailed analysis of this interesting question on the breakdown of QS for the Coulomb potential has not yet been performed.

In conclusion, our analysis demonstrates the existence of a QS regime for a weakly-bound particle at above-threshold frequencies, but only for a limited interval of laser intensities, whose upper limit, F_{cr}, is estimated analytically. The δ-model potential describes photoprocesses in negative ions reasonably, especially for H^-. Fig. 1 shows that (using the Nd:YAG laser) the observation of QS for H^- is possible at $F \sim 1$ ($I \approx 3\times10^{12}$ W/cm^2). Since the lifetime of H^- in such a field is small, $\tau \sim 1/\Gamma \sim 10$ fs, stabilization may be observable using femtosecond pulses.

Acknowledgments

The authors are grateful to M.V.Fedorov for useful discussions. This research is supported in part by Grants INTAS-RFBR (No. 97-673), RFBR (No. 00-02-17843), and NSF (No. PHY-0070980).

* On leave from Institute for Space Sciences, Bucharest-Măgurele 76900, Romania.

References

1. M.Pont and M.Gavrila, Phys. Rev. Lett. **65**, 2362 (1990); M.Gavrila, in *AIP Conference Proceedings 525, Multiphoton Processes: ICOMP VIII, 8th Interantional Conference*, ed. by L.F.DiMauro, R.R.Freeman, and K.C.Kulander (Melveille, New York, 2000) p.103.
2. M.V.Fedorov and A.M.Movsesian, J. Phys. B **21**, L155 (1988); M.V.Fedorov, Laser Phys. **9**, 209 (1999).
3. Q.Su, Laser Phys. **2**, 241 (1993); Q.Su, B.P.Irving, C.W.Johnson, and J.H.Eberly, J. Phys. B **29**, 5755 (1996).
4. Q.Su, B.P.Irving, and J.H.Eberly, Laser Phys. **7**, 1 (1997); A.Patel, N.J.Kylstra, and P.L.Knight, J. Phys. B **32**, 5759 (1999); A.M.Popov, O.V.Tikhonova, and E.A.Volkova, Laser Phys. **10**, 779 (2000).
5. M.Dorr and R.M.Potvliege, J. Phys. B **33**, L233 (2000).
6. V.P.Krainov and M.A.Preobragenskii, Zh. Exp. Teor. Fiz. **103**, 1142 (1993) [Sov. Phys. JETP **76**, 559 (1993)].
7. J.Z.Kamiński, Phys. Rev. A **52**, 4976 (1995).
8. S.Geltman, J. Phys. B **32**, 853 (1999); T.Mercouris and C.Nicolaides, *ibid.* **32**, 2371 (1999); C.Figueira, A.Fring, and R.Schrader, Laser Phys. **9**, 379 (1999).
9. N.L.Manakov, M.V.Frolov, B.Borca, and A.F.Starace, Pis'ma ZETF, **72**, 426 (2000) [JETP Lett. **72**, 294 (2000)].
10. N.L.Manakov, V.D.Ovsiannikov, and L.P.Rapoport, Phys. Rep. **141**, 319 (1986).
11. N.L.Manakov and A.G.Fainshtein, Dokl. Akad. Nauk SSSR, **244**, 567 (1979) [Sov. Phys. Dokl. **24**, 41 (1979)]; Zh. Exp. Teor. Fiz. **79**, 751 (1980) [Sov. Phys. JETP **52**, 382 (1980)].
12. N.L.Manakov and L.P.Rapoport, Zh. Exp. Teor. Fiz. **69**, 842 (1975) [Sov. Phys. JETP **42**, 430 (1975)]; I.J.Berson, J. Phys. B **8**, 3078 (1975).
13. W.Becker, J.K.McIver, and M.Confer, Phys. Rev. A, **40** 6904 (1989); W.Becker, S.Long, and J.K.McIver, *ibid.* **42** 4416 (1990);
14. F.H.M.Faisal, P.Filipowicz, and K.Rzazewski, Phys. Rev. A **41**, 6176 (1990); P.S. Krstić, D. B. Milošević, and R. K. Janev, Phys. Rev. A **44**, 3089 (1991).
15. P. Filipowicz, F.H.M.Faisal, and K.Rzazewski, Phys. Rev. A **44**, 2210 (1991).
16. M.Dorr, R.Potvliege and R.Shakeshaft, Phys. Rev. Lett. **64**, 2003 (1990).
17. Ya.B.Zel'dovich, Zh. Exp. Teor. Fiz. **39**, 776 (1961) [Sov. Phys. JETP **12**, 542 (1961)].
18. N.L.Manakov, M.V.Frolov, A.F.Starace, and I.I.Fabrikant, J. Phys. B **33**, R141 (2000).
19. N.L.Manakov, L.P.Rapoport, M.A.Preobragenskii, and A.G.Fainshtein, Zh. Exp. Teor. Fiz.**75**, 1243 (1978) [Sov. Phys. JETP **48**, 626 (1978)].
20. N.L.Manakov and A.G.Fainshtein, Teor. Mat. Fiz. **48**, 375 (1981) [Theor. Math. Phys. **48**, 815 (1982)]; R.M.Potvliege and R.Shakeshaft, Phys. Rev. A **41**, 1609 (1990).
21. V.S.Popov, Zh. Exp. Teor. Fiz.**118**, 56 (2000) [Sov. Phys. JETP **91**, 48 (2000)].
22. L.V.Keldysh, Zh. Exp. Teor. Fiz. **47**, 1945 (1964) [Sov. Phys. JETP **20**, 1307 (1964)]; A.I.Nikishov and V.I.Ritus, *ibid.* **50**, 255 (1966) [**23**, 168 (1966)]; A.M.Perelomov, V.S.Popov, and M.V.Terent'ev, *ibid.* **50**, 1393 (1966) [**23**, 1924 (1966)]; **51**, 309 (1966) [**24**, 207 (1967)].

THE STRONG FIELD LIMIT OF ATOMIC STABILIZATION IN ULTRASHORT PULSES

M. DÖRR
Max-Born-Institut, D-12489 Berlin, Germany

AND

N.J. KYLSTRA AND R.M. POTVLIEGE
Department of Physics, University of Durham, Durham DH1 3LE, UK

Generating controlled few-cycle Ti:Sapphire laser pulses with peak intensities well above 10^{15} W/cm^2 has recently become possible [1, 2]. If the electric field amplitude increases sufficiently fast to sufficiently high values, the main effect of such pulses on a weakly bound electron is to rapidly displace it from the vicinity of the nucleus into regions of space far from the core, where the atomic potential is much weaker than in the initial state. The electron is periodically brought back near the nucleus by the oscillations of the field, but at a high velocity and only for a small fraction of each cycle. During the pulse, the electron's dynamics is therefore less and less governed by the atomic potential and more and more by simple wave packet spreading. In the limit of high peak intensities, for a fixed, finite pulse duration, the oscillating electron essentially undergoes free wave packet spreading during the entire pulse duration. In this case, the state of the atom at the end of the pulse only depends on the initial state, on the pulse duration, and on the final position and mean velocity of the electronic wave packet. If this final position is at the nucleus and the final mean velocity is zero, the final state of the atom does not depend on the temporal profile of the laser pulse intensity, on the phase of the field, or on its wavelength: the atom is stabilized, in the sense that increasing the peak intensity further, for a constant pulse duration, does not result in a higher ionization probability. Wave packet spreading is relatively slow on the time scale of one period of the field when the photon energy is larger than the binding energy of the initial state, in which case one can expect a small ionization probability if the pulse encompasses but a few optical cycles.

B. Piraux and K. Rzążewski (eds.), Super-Intense Laser-Atom Physics, 305–312.

As we will see below, it is unlikely that this regime of stabilization will ever be reached in practice, although it can be observed in model calculations assuming the dipole approximation. The peak intensities required for having *free* wave packet spreading during most of the pulse seem indeed to be prohibitively high, and radiation pressure will suppress stabilization in strong fields. However, the peak intensities that can now be achieved for few-cycle Ti:Sapphire pulses are sufficiently high for the evolution of an atom prepared in a well chosen excited state to be dominated by spreading and transfer of population between bound states, as will be shown below.

The relevance of wave packet spreading at high laser intensities has been known for some time [3]. In particular, its role in adiabatic stabilization has been emphasized by Grobe and Fedorov [4, 5]. (Atoms initially in a state whose binding energy is less than the photon energy may adiabatically follow the dressed initial state and undergo adiabatic stabilization in short high-frequency pulses [6, 7]. Here, however, we consider the case where the temporal variation of the intensity is too fast, and the peak intensity is too high, for this to occur.) An illustration of drift and spreading in the low frequency regime is given in Ref. [8]. The approximation whereby the effect of the atomic potential is neglected during the action of the incident field is usually called the strong field approximation (SFA). Its validity in photodetachment from a one-dimensional short range potential by a short laser pulse has been examined recently by Patel, Kylstra and Knight [9]. These authors showed how the residual bound state populations converge at high peak intensities to the values predicted by the SFA. Further numerical examples can be found in Ref. [10], and a mathematical analysis in Ref. [11]. The limit in which the electronic wave packet undergoes free spreading during the whole pulse duration will be called the SFA limit in the following.

The convergence towards this limit in one-dimensional model systems is illustrated by Figure 1 for photodetachment from the ground state of the soft-core Coulomb potential $V(x) = -V_0/(a^2+x^2)^{1/2}$. The parameters of the potential, $V_0 = 0.432$ a.u. and $a = 2.315$ a.u., are such that the binding energy of this state is 0.125 a.u. We consider a vector potential of the form $\mathcal{A}(t) = \mathcal{A}_0 f(t)\cos\omega t$ with $\omega = 1$ a.u. The peak intensity is $I_0 = c\epsilon_0\mathcal{A}_0^2/2\omega^2$. Three types of pulses are defined by the functional form of the pulse envelope $f(t)$. The solid curve in the figure represents the variation with I_0 of the probability that the system is still in the ground state at the end of a square pulse, that is $f(t) = 1$ during 14 cycles and 0 otherwise. The dashed curve shows the corresponding result for $f(t)$ increasing from 0 to 1 like a $\sin^2$ function during the first three cycles, then remaining equal to 1 for eight cycles, and finally decreasing back to 0 in three cycles, again like a $\sin^2$ function. The chain curve refers to an even smoother temporal profile, with $\sin^2$, six-cycle long turn on and turn off periods. The results were

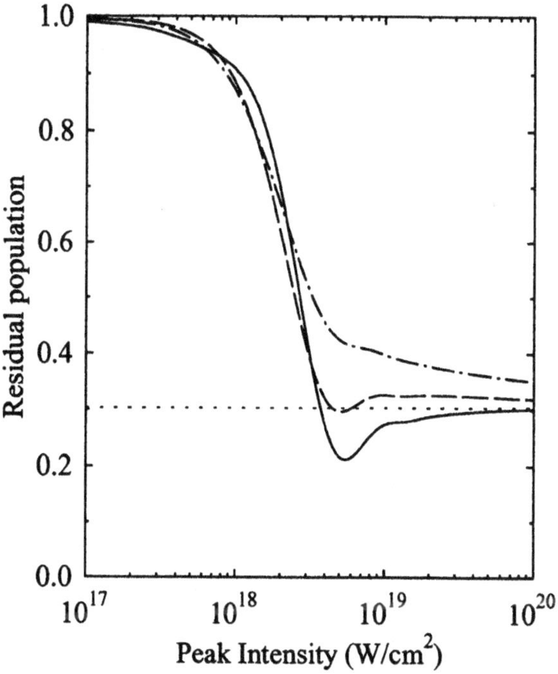

Figure 1. Population left in the ground state of a 1D soft-core Coulomb potential at the end of a 14-cycle pulse, vs. the pulse's peak intensity. Solid curve: square pulse; dashed curve: (3,8,3) pulse; chain curve: (6,2,6) pulse. The dotted curve represents the residual ground state population after 14 optical cycles of free wave packet spreading.

calculated by a standard Crank-Nicolson integration of the time-dependent Schrödinger equation in the velocity gauge. A free electron initially at rest at the origin is at rest at the origin at the end of each of the three pulses considered. Within the strong field approximation, the population left in the ground state at the end of the pulse is thus the same as that left in the ground state after 14 optical cycles of free wave packet spreading. It is represented by a dotted line. As seen from the figure, the exact results depend somewhat on the pulse shape at intermediate intensity and always tend towards the SFA limit in very strong fields. Convergence is faster for the square pulse than for the smooth pulses, because the electron is exposed to relatively weak fields for a longer time in the latter. Note that the residual population is substantial, even for the square pulse, despite the pulse's large bandwidth.

One may wonder whether a similar asymptotic regime of dynamic stabilization exists for real atoms. At first sight, convergence towards the SFA limit may be hindered by a number of effects absent from the model calcu-

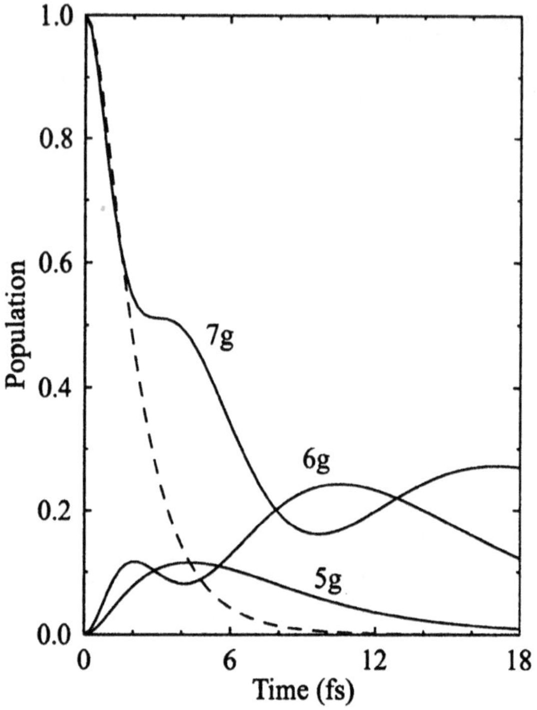

Figure 2. Temporal evolution of the populations of the $5g_{m=4}$, $6g_{m=4}$ and $7g_{m=4}$ states during free wave packet spreading of the initial $7g_{m=4}$ state (solid curves) and temporal evolution of the $5g_{m=4}$ state during free wave packet spreading of the initial $5g_{m=4}$ state (dashed curve). All results are for atomic hydrogen.

lation, such as population transfer at Stark-shift induced resonances, core excitation, radiation pressure, and ionization at relatively low intensity in the pulse pedestal (i.e., during the relatively long periods of non-zero field on either side of the central spike). Radiation pressure is unavoidable and must be taken into account in strong fields [5, 12, 13, 14, 15, 16, 17]. There is obviously no core excitation in the case of atomic hydrogen. Ionization in the pedestal part of the laser pulse can be alleviated by starting with an atom initially in an excited circular state that undergoes adiabatic stabilization at low intensity [7, 18, 19]. This also eliminates the possibility of multiphoton resonances. Taking the initial state to be a sufficiently excited circular state, rather than the ground state, has the additional advantage of avoiding too rapid wave packet spreading in the SFA regime. For example, the probability that a hydrogen atom initially in the ground state is still in this state after 5.3 fs of free wave packet spreading is only 1.5×10^{-5}. (This result is easily obtained by working in momentum space [9]; 5.3 fs is equivalent to 2 optical cycles at the typical wavelength, 800 nm, of a

TABLE 1. Probability that an hydrogen atom is still in the initial state at the end of a 5-cycle pulse, as obtained in the dipole approximation. The carrier wavelength is 800 nm and the field is polarized parallel to the axis of quantization of the angular momenta.

Pulse peak intensity	Initial state $7g_{m=4}$	Initial state $5g_{m=4}$
0.125×10^{15} W/cm^2	0.78	0.80
0.25	0.60	0.62
0.5	0.46	0.42
1.0	0.47	0.24
2.0	0.55	0.13
4.0	0.54	0.07
8.0	0.44	0.04
∞ (SFA limit)	0.23	0.001

Ti:Sapphire laser beam.) As shown in Figure 2, the $7g_{m=4}$ state, and to a lesser degree the $5g_{m=4}$ state, spread much more slowly.

We have calculated how these two states evolve, in dipole approximation, when irradiated by a linearly polarized ultrashort pulse of exactly 5 optical cycles. This was done by solving the time-dependent Schrödinger equation in the Kramers-Henneberger frame using a split-operator algorithm, working in cylindrical coordinates. The results are presented in Table 1, together with the the SFA limits of the residual populations. The carrier wavelength is 800 nm and the field is defined by the vector potential $\mathcal{A}(t) = \mathcal{A}_0 \sin^2(\Omega t) \sin(\omega t)$ with $\Omega = \omega/10$. The probability that the atom remains in the initial state at the end of the pulse is about the same for the two states below 3×10^{14} W/cm^2 peak intensity. The probability is still close to 0.6 at that intensity, but it decreases and approaches the SFA prediction in stronger fields. The faster decrease for the $5g_{m=4}$ state is consistent with Figure 2. As will be discussed in more detail elsewhere, and although the probability is still somewhat higher than the SFA values at 8×10^{15} W/cm^2, the evolution of the wave function at the largest intensities considered in the table is best understood as being dominated by wave packet spreading.

The drift of the electron induced by the magnetic field component of the laser pulse, i.e., radiation pressure, was neglected when obtaining the results of Table 1. The importance of this drift can only be evaluated by calculations beyond the dipole approximation. As a rule of thumb, recently verified for a two-dimensional model [17], one can expect that the Lorentz force significantly affects the dynamics of an atom when I_0 exceeds about $c\omega^3$, in atomic units — namely at or above 2×10^{15} W/cm^2 at 620 nm wavelength or 9×10^{14} W/cm^2 at 800 nm. (Other criteria have been proposed, too; see, e.g., Refs. [15, 16, 20].) So far, calculations for a highly excited circular state that take this effect into account have been reported only for the $5g_{m=4}$ state of hydrogen at 620 nm: in the range of intensities for which the computation could be carried out, i.e., up to 8×10^{14} W/cm^2, the magnetic field of the incident wave was found to have no significant influence on the Floquet rate of ionization [21].

In order to clarify the role of the Lorentz force in this context, we have investigated the short pulse dynamics of an excited state of a model two-dimensional atom without making the dipole approximation. The electron is confined to the $x-z$ plane. Its interaction with the core is represented, in polar co-ordinates (ρ, θ), by the effective potential $V(\rho) = -V_0/(a^2+\rho^2)^{1/2}$, with $V_0 = 1$ a.u. and $a = 0.8$ a.u. The method of calculation is the same as in Ref. [17]. However, the laser pulses considered here have a wavelength of 800 nm and a duration of only 2 optical cycles. The field is linearly polarized in the x-direction; its vector potential, $\mathcal{A} = A_0 f(\omega t - kz)\sin(\omega t - kz)\hat{x}$, is ramped on and off linearly without any plateau. The initial state adopted in the calculation is described by a wave function of the form $\psi(\rho,\theta) = R(\rho)\sin\theta$, the polar angle θ being measured with respect to the x-axis. Its zero-field binding energy, 0.023 a.u., is less than half the photon energy, 0.057 a.u. Its extension in the polarization direction is comparable to that of the $7g_{m=4}$ state of hydrogen. Since this state is of odd parity for $\theta \to -\theta$ and the dipole operator does not depend on the sign of θ, it is not coupled, in dipole approximation, to the states of even parity under this transformation. Hence, the probability that the electron encounters the nucleus when oscillating in the field is zero in dipole approximation, just as in the case of the three-dimensional circular states of hydrogen that undergo adiabatic stabilization.

The results are presented in Figure 3. The probability that the atom is still in the initial state at the end of the pulse is predicted fairly well by the SFA at the highest peak intensities. However, the probability does not approach a constant value, which shows that strong field stabilization is not yet achieved. The ionization varies little with the peak intensity and is also correctly predicted by the SFA. The dipole approximation rapidly becomes inaccurate above 1×10^{15} W/cm^2, in agreement with the rule of thumb mentioned above. At the highest intensities considered, radiation pressure produces a considerable redistribution of population among bound states but hardly any increase in ionization. Its effect is similar to that found by Bugacov, Pont and Shakeshaft [16] in their study of photoïonization of low excited states in an intense high-frequency field.

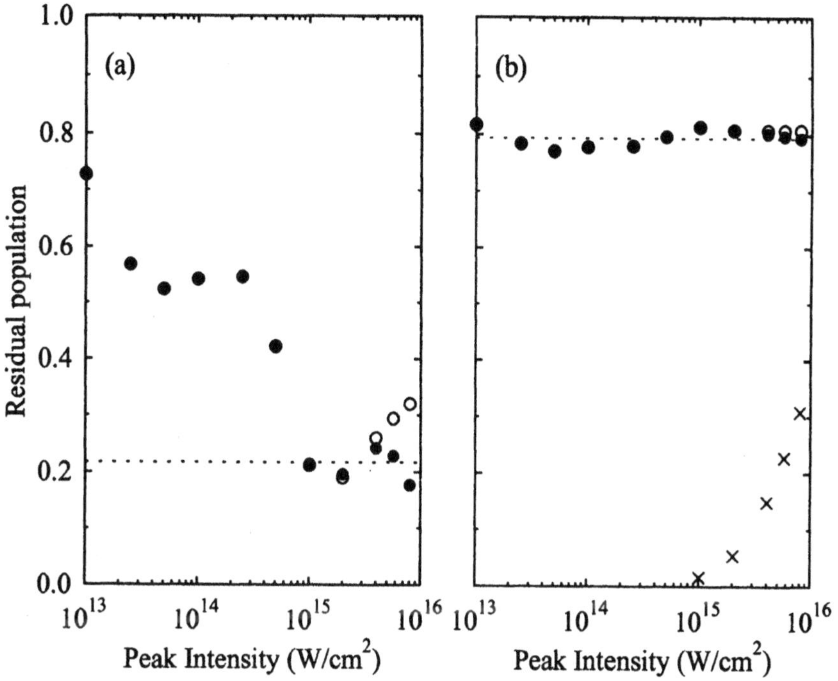

Figure 3. Photodetachment from a 2d soft Coulomb potential by a 800 nm, 2-cycle pulse. (a) Population left in the initial state at the end of the pulse, as calculated without (closed circles) or with (open circles) making the dipole approximation; the SFA limit, in dipole approximation, is indicated by a dotted line. (b) Same as (a), but for the population transferred to the continuum. The crosses represent the population left in the bound states of parity opposite to that of the initial state under $\theta \to -\theta$, as obtained in the non-dipole calculation. (There is no transfer of population to these states in dipole approximation.)

These results, and those presented in Table 1, suggest that circular states such as those considered here remain stable against ionization when exposed to an intense ultrashort laser pulse, in a regime where their dynamics are dominated by wave packet spreading. For Ti:Sapphire pulses with intensities of about 10^{16} W/cm^2, the magnetic field component of the pulse induces a considerable redistribution of population between bound states with little enhancement of ionization.

The computations were carried out on equipment financed by the UK EPSRC.

References

1. Apolonski, A., Poppe, A., Tempea, A., Spielmann, Ch., Udem, Th., Holzwarth, R., Hänsch, T.W., and Krausz, F. (2000) *Phys. Rev. Lett.* **85**, 740.
2. Brabec, T. and Krausz, F. (2000) *Rev. Mod. Phys.* **72**, 545.

3. Antunes Neto, H.S. and Davidovich, L. (1984) *Phys. Rev. Lett.* **53**, 2238.
4. Grobe, R. and Fedorov, M.V. (1992) *Phys. Rev. Lett.* **68**, 2592; *ibid.* **69**, 3591(E).
5. Grobe, R. and Fedorov, M.V. (1993) *Phys. Rev. Lett.* **70**, 1562.
6. Dörr, M. and Potvliege, R.M. (2000) *J. Phys. B* **33**, L233.
7. Popov, A.M., Tikhonova, O.V. and Volkova, E.A. (2000) *Laser Phys.* **10**, 779.
8. Kim, A.V., Ryabikin, M.Yu, Sergeev, A.M., Farina, D. and Lontano, M. (1999) *SPIE* **3735**, 158; Kim, A.V., Ryabikin, M.Yu and Sergeev, A.M. (1999) *Usp. Fiz. Nauk* **169** 58 [*Phys. Usp.* **42**, 54].
9. Patel, A., Kylstra, N.J. and Knight, P.L. (1999) *J. Phys. B* **32**, 5759.
10. Dörr, M., Kylstra, N.J. and Potvliege, R.M. (2001) *Laser Phys.* **11**, to be published.
11. Fring, A., Kostrykin, V. and Schrader, R. (1997) *J. Phys. A* **30**, 8599.
12. Kyrala G.A. (1987) *J. Opt. Soc. Am. B* **4**, 731.
13. Grochmalicki, J., Lewenstein, M. and Rzążewski, K. (1991) *Phys. Rev. Lett.* **66**, 1041.
14. Horbatsch, M. (1993) *Z. Phys. D* **25**, 305.
15. Katsouleas, T. and Mori, W.B. (1993) *Phys. Rev. Lett.* **70**, 1561.
16. Bugacov, A., Pont, M. and Shakeshaft, R. (1993) *Phys. Rev. A* **48**, R4027.
17. Kylstra, N.J., Worthington, R.A., Patel, A., Knight, P.L., Vázquez de Aldana, J.R. and Roso, L. (2000) *Phys. Rev. Lett.* **85**, 1835.
18. van Druten, N.J., Constantinescu, R.C., Schins, J.M., Nieuwenhuize, H. and Muller, H.G. (1997) *Phys. Rev. A* **55**, 622.
19. Piraux, B. and Potvliege, R.M. (1998) *Phys. Rev. A* **57**, 5009.
20. Reiss, H.R. (1992) *Progr. Quant. Electron.* **16**, 1.
21. Potvliege, R.M. (2000) *Laser Phys.* **10**, 143.

3–D NUMERICAL CALCULATIONS OF LASER ATOM INTERACTIONS

Subrelativistic and weakly relativistic regime

A. SCRINZI AND T. BRABEC
Institut für Photonik, Vienna University of Technology
Gusshausstr. 27/387, A-1040 Vienna, Austria
e-mail: scrinzi@tuwien.ac.at

AND

M. WALSER
Institut für Theoretische Quantendynamik, U. of Freiburg
Hermann Herder-Straße 3, D-79104 Freiburg, Germany

1. Introduction

Laser–atom interactions at routinely used high laser intensity have defied a complete numerical description because of the very large quiver momenta and excursion amplitudes of the electron in the laser field. For nuclear charge Z, laser intensity I, and frequency ω the dimensions of the bound states are $\sim 1/Z$ while the quiver motion covers distances of $\propto \sqrt{I}/\omega^2$. When I approaches one atomic unit in a Ti:Sapphire laser of $\lambda = 800\,nm \approx 0.057\,a.u.$, the quiver amplitude exceeds atomic dimensions by $300Z$, making the simultaneous description of electron–ion rescattering and non–bound electronic motion exceedingly difficult.

Models have stepped in and provided essential physical insight. Analytic and semi–analytic calculations based on the strong field approximation (SFA) [1, 2, 4, 3] were able to explain experimental data in great detail [5], in spite of the fact that several important approximations are made in these models. Numerical calculations were performed for one– and two–dimensional models [6, 7, 8, 9]. From all these calculations a generally accepted picture of strong laser–atom interactions has emerged: an electron is set free from the atom with low initial velocity, it is accelerated in the laser field with hardly any influence of the ionic potential, where the free motion is only disturbed, when the electron happens to return to the nucleus one or several times. If electron velocities become comparable to the

B. Piraux and K. Rzążewski (eds.), Super-Intense Laser-Atom Physics, 313–322.

velocity of light, the magnetic field component of the laser field makes itself felt by pushing the electron into the direction of laser propagation.

At present theory cannot give definite answers to the following questions: What is the initial momentum spectrum after release of the electron ? What is the influence of the Coulomb potential on the non–bound electronic motion, is there "Coulomb focusing" [10] of the electronic motion ? How are harmonic generation and non–sequential ionization affected by the suppression of rescattering, when the electron drifts away due to magnetic field effects ? These questions motivate our effort to integrate the successful ideas of the SFA into a complete numerical description of intense laser–atom interaction. In its present form, the method is capable of solving the time–dependent Schrödinger equation for an atom in fields up to $10^{17} W/cm^2$ at the Ti:Sapphire wave length of $800\,nm$. The necessary extensions to include lowest order non–dipole effects will be discussed in detail.

2. Method

Our computational method is based on three main ideas:

(A) The main difficulty in numerically solving the time dependent Schrödinger equation for an atom in a strong laser field are the very large excursions of the electron compared atomic scale. Since that motion is nearly the motion of a free electron in the laser field, the exact solution is expanded into Volkov solutions for the unbound electron.

(B) Since the ground state, when represented in terms of Volkov solutions, moves very violently, the ground state is included explicitly in the expansion and its time evolution is treated separately.

(C) The smoothness of the atomic potential operator in the Volkov representation, which has its origin in the spatially limited action of the potential, is exploited for an efficient implementation of that integral operator.

To be specific, we assume that the Hamiltonian of the system can be written as

$$H(t) = T(t) + V + B(t) \tag{1}$$

with the kinetic energy operator (generalized for velocity gauge)

$$T(t) = \frac{1}{2m}\left[\vec{p} - \vec{A}(t)\right]^2 \tag{2}$$

and the atomic potential $V(\vec{r})$. $B(t)$ is a first order non–dipole correction that will be specified below. The complete solution of the time dependent

Schrödinger equation is expanded in a time–dependent basis as

$$\Psi(\vec{r},t) = c(t)|0,t\rangle + \int d^3k\, b(\vec{k},t)|\vec{k},t\rangle \tag{3}$$

with the (gauge transformed) ground state function

$$[T(t)+V]\,|0,t\rangle = E_0|0,t\rangle \tag{4}$$

for ground state energy E_0 and the Volkov solutions

$$[T(t)+B(t)]\,|\vec{k},t\rangle = i\frac{\partial}{\partial t}|\vec{k},t\rangle. \tag{5}$$

We first restrict ourselves to the Hydrogen atom $V(\vec{r}) = -1/r$ and the laser field in dipole approximation $B(t) = 0$. In that case the explicit form of the expansion functions is

$$\begin{aligned} |0,t\rangle &= \pi^{-1/2}\exp(i\vec{r}\cdot\vec{A}(t) - r) & (6)\\ |\vec{k},t\rangle &= (2\pi)^{-3/2}\exp\left[-\frac{i}{2m}\Phi\left(\vec{k},t\right)\right]\exp\left[i\vec{k}\cdot\vec{r}\right] & (7)\end{aligned}$$

with the Volkov phase

$$\Phi(\vec{k},t) = \int_{t_0}^{t} dt'\left[\vec{k}-\vec{A}(t')\right]^2. \tag{8}$$

Note that the expansion is not uniquely defined by Eq. (3), since the Volkov solutions $|\vec{k},t\rangle$ by themselves form a complete basis. Uniqueness is attained by imposing the orthogonality condition

$$\int d\vec{k}\langle 0,t|\vec{k},t\rangle b(\vec{k},t) = 0. \tag{9}$$

By substituting Eq. (3) into the time-dependent Schrödinger equation and enforcing the orthogonality condition (9) one obtains the equations for the expansion coefficients

$$\begin{aligned} i\frac{\partial}{\partial t}c(t) &= -(g|W^*|b) & (10)\\ i\frac{\partial}{\partial t}|b) &= -W|g)c(t) + |g)(g|W^*|b) - |g)(g|T|b) + V_\perp|b). & (11)\end{aligned}$$

Here we have adopted the notation $|b)$ for the coefficient vector $b(\vec{k},t)$ with the corresponding scalar product denoted by $(g|b) = \int d^3k\, g^*(\vec{k},t)b(\vec{k},t)$.

The vector $|g\rangle$ denotes the Volkov expansion of the ground state (with a trivial extra phase)

$$|g\rangle = e^{-itE_0}\langle\vec{k},t|0,t\rangle = \frac{\sqrt{8}}{\pi}\exp\left[\frac{i}{2m}\Phi(\vec{k},t) - itE_0\right]\left[1+\left(\vec{k}-\vec{A}\right)^2\right]^{-2} \tag{12}$$

The kinetic energy operator in Volkov representation is $T = (\vec{k}-\vec{A})^2$ and $W(t)|g\rangle$ is defined by

$$W(t)|g\rangle := \left[-T(t) - i\frac{\partial}{\partial t}\right]|g\rangle = i4\left(\frac{\partial}{\partial t}\vec{A}\right)\cdot\frac{\vec{k}-\vec{A}}{1+\left(\vec{k}-\vec{A}\right)^2}|g\rangle \tag{13}$$

Finally $V_\perp$ is the atomic potential in Volkov representation restricted to act only on the non–ground state part of the wave function

$$V_\perp = [1-|g\rangle\langle g|]\,V\,[1-|g\rangle\langle g|]\,; \qquad V_\perp|g\rangle = 0. \tag{14}$$

The physical meaning of the individual terms in Eqs. (10) and (11) is easily identified. Terms involving W describe transitions to and from the continuum (more precisely: non–ground states), i.e. ionization and recombination. These terms vanish for field $\partial\vec{A}/\partial t = 0$. The term with T has its origin in the orthogonality condition (9) and describes the disturbance of the free electron motion by the removal of the ground state content from the Volkov solutions. Finally $V_\perp$ describes the action of the atomic potential on the electron outside the ground state. No approximations have been made in obtaining Eqs. (10) and (11), but by putting into the expansion functions $|0,t\rangle$ and $|\vec{k},t\rangle$ information that can be obtained by analytic means, the numerical calculation only involves truly unknown parts of the time evolution.

This formulation resembles the SFA, where however orthogonality is violated by neglecting $|g\rangle\langle g|T$, where the non–bound atomic potential $V_\perp$ is neglected, and $c(t)$ is set $= 1$ or replaced with a suitable approximation. In the SFA, one is left with only the first two terms on the right hand side of Eq. (11), which can be further treated by analytic methods.

2.1. IMPLEMENTATION

Eqs. (10) and (11) were solved on a grid in $\vec{k}$-space. To exploit cylindrical symmetry, we use the coordinates k_x, $k_\rho = \sqrt{k_y^2+k_z^2}$, and $\sin\theta = k_z/k_\rho$. For linear polarization and in dipole approximation cylindrical symmetry is exact and the θ–coordinate drops out. At high intensities with very large excursions along one axis, expansion in cylindrical rather than spherical coordinates is mandatory, since convergence of the spherical expansion becomes very slow.

The required grid sizes depend on the parameters used. The minimal grid must faithfully represent the field free initial state. We obtain three digits accurate ground state energies with maximal momenta $|\vec{k}| \leq 4a.u.$ and a grid spacing of $\Delta k \approx 0.05\,a.u.$. Many more points are required in polarization direction, since the full range of quiver momenta must be covered by the grid. As an example, at a laser peak intensity of $10^{16} W/cm^2$ and wave length $\lambda = 800\,nm$, the electron reaches a maximum momentum of about 10 a.u. and moves to distances as large as 160 a.u., which requires a grid spacing of $\Delta k \lesssim 2\pi/160 \approx 0.04a.u.$, or about 500 points in the range $k_x = [-10, 10]$. The total number of points in the k_x-k_ρ plane will then be about 40000. The number of grid points significantly increases, when cylindrical symmetry becomes broken by non–dipole effects. However, since electrons that drift away perpendicular to the polarization direction have little likelihood to ever return to the nucleus they can be absorbed without altering the laser–atom interaction dynamics, which allows to keep the expansion in the θ–coordinate short by using a few 10s of expansion functions. In that way the total number of expansion coefficients becomes of the order of 10^6.

For the time–propagation of such a large number of expansion coefficients, an efficient implementation of the operators W, T and $V_\perp$ is needed. Application of W and T requires little more than a single multiplication per point and operator, but a straight forward application of $V_\perp$, even when using fast Fourier transform techniques, would involve at least $\sim 6N \log_2 N_1$ operations, when N is the total number of grid points and N_1 is the maximum number of points in one spatial dimension. Therefore $V_\perp$ is approximated on a coarser grid. This is justified by observing that the integration kernel of the full Coulomb operator

$$V(\vec{k} - \vec{k}') \sim \frac{1}{(\vec{k} - \vec{k}')^2} \tag{15}$$

is smooth except around $\vec{k} = \vec{k}'$, i.e. except for small momentum transfer, or, large impact parameters in $\vec{r}$–space. Since rapid oscillations of the wave function in $\vec{k}$–space will be averaged out by applying V, we reduce the $\vec{k}$–space wave function to a coarser grid by averaging over neighboring points by a procedure which in one dimension is given by

$$b_{\text{coarse}}(k_{i\cdot m}) = \sum_{j=-m}^{m} w_j b(k_{i\cdot m} + j\Delta k), \qquad i = 1, N_1/m - 1, \tag{16}$$

where we used linear tent functions for the weights w_j. Only every m'th point $k_{i\cdot m}$ of the original grid points k_j is used for the coarse grid rep-

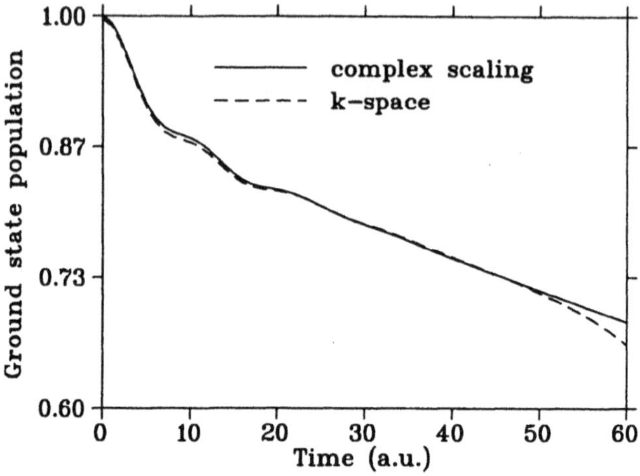

Figure 1. Depletion of the hydrogen ground state by a static electric field of 0.08 a.u. The results of complex scaling [12] and the present method coincide until times 50 a.u., when the wave functions moves off the $\vec{k}$–grid.

resentation b_{coarse}. It was found that sufficiently large grid reductions by factors 2 to 4 in each dimension can be used without compromising the accuracy of the results, such that the computational effort for application of $V_\perp$ becomes similar to the time spent with the rest of the operator.

3. Results

3.1. STATIC FIELD IONIZATION

The calculation of static field ionization rates, in spite of its apparent simplicity, is far from trivial. Even for hydrogen quantitatively accurate results require involved numerical calculations. To obtain electron spectra for static field ionization was not possible so far. While such spectra by themselves may be of little practical interest, they provide important input in SFA calculations, where the standard assumption of initial momentum =0 "after tunneling of the electron" is inadequate for certain considerations [11]. Apart from that, the problem provides a good testing ground for the procedures introduced above.

Figure 1 compares the ground state depletion of hydrogen by a static field of 0.08 a.u. as obtained by a complex scaling calculation [12] with the present method. Results coincide up to times $50a.u.$, when in the present calculation the limits of the $\vec{k}$–grid where reached. Figure 2 shows the electron spectrum at $t = 30\,a.u.$, where a stationary state with exponential ground state depletion has been reached. The exact spectrum is compared to the spectrum obtained, when the atomic potential outside the ground

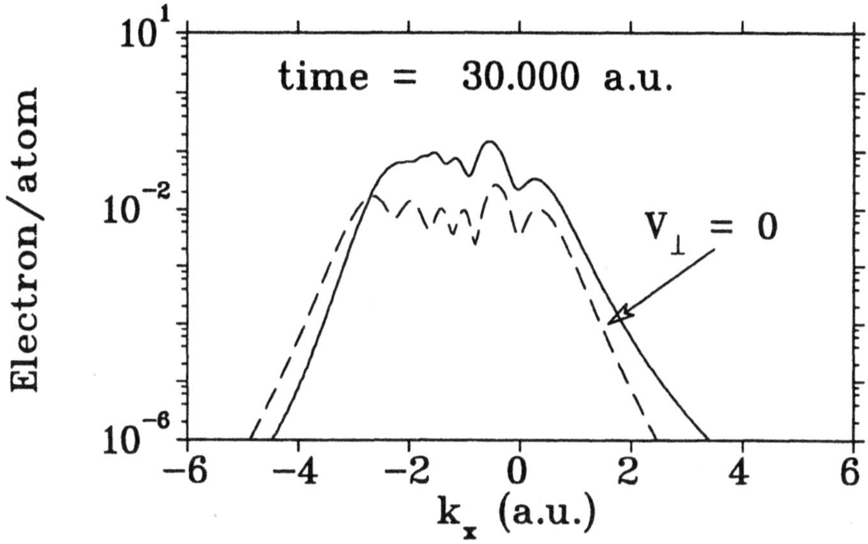

Figure 2. Electron spectrum for static field ionization of hydrogen at time 30 a.u. after sudden turn on of a field of 0.08 a.u. When the action of the atomic potential on the continuum $V_\perp$ is neglected, ionization is strongly underestimated.

state $V_\perp$ is neglected, which is the main approximation made in the SFA. One sees, that qualitatively the spectra agree, but SFA severely underestimates total ionization.

3.2. HARMONIC SPECTRA

The SFA is known to produce qualitatively correct, but quantitatively flawed results. The present code allows to systematically study the source of error in the SFA. As in SFA, the action of the Coulomb potential on the non–ground states is neglected by omitting $V_\perp$ in Eq. (11). The resulting harmonic spectra are compared with the SFA in Fig. 3. Clearly, neglecting the Coulomb potential introduces the main error in the SFA as far as the highest harmonics are concerned. More specifically, ionization and therefore the electron flux available for harmonic generation is underestimated due to the incorrect density of Volkov states in the absence of the Coulomb potential. Discrepancies for the lower harmonics are only partially due to the absence of excited bound states in SFA, apparently also the stationary phase approximation used to perform the $\vec{k}$–integration in SFA fails, as is indicated by the discrepancy between the purely numerical result without $V_\perp$ and the semi–analytic SFA result.

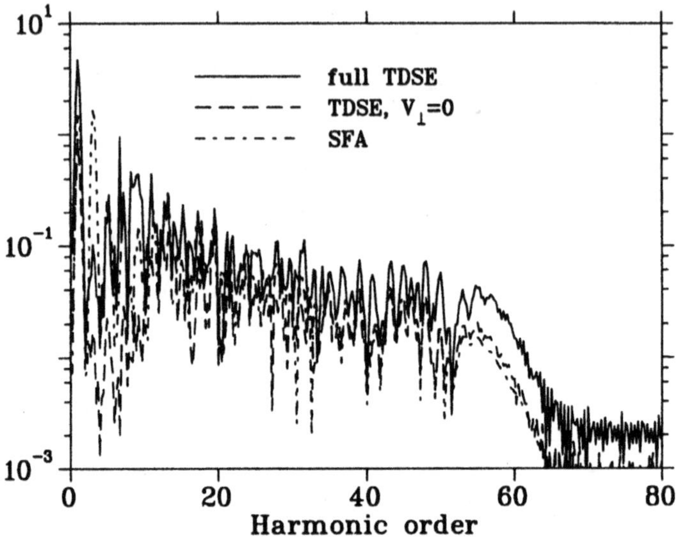

Figure 3. Harmonics produced by a 5 fs laser pulse on hydrogen. Results from the complete solution of the time–dependent Schrödinger equation are compared to a calculation with the potential $V_\perp$ switched off and with the SFA.

3.3. CARRIER PHASE DEPENDENCE OF E⁻–SPECTRA

With very short laser pulses, the phase φ of the carrier relative to the pulse envelope $A_0(t)$

$$\vec{A}(t) = \vec{A}_0(t)\sin(\omega t + \varphi) \tag{17}$$

becomes increasingly important. In recent experiments [13] the phase has been stabilized without obtaining its absolute value. At higher intensities, electron spectra provide a signature of the absolute phase by the forward–backward asymmetry of electron emission. Fig. 4 shows electron spectra in polarization direction for a 5 fs pulse with peak intensity of $10^{16} W/cm^2$ and wavelength $\lambda = 800\,nm$ at phases $\varphi = 0, \pi/2$ and π. As expected, a very pronounced dependence of the electron spectra on the carrier phase is observed, which can serve as signature of the carrier phase.

4. Beyond the dipole approximation

Equations (10) and (11) are not restricted by the dipole approximation. Their derivation relies only on the general form of the Hamiltonian Eq. (1) and on finding solutions for Eqs. (4) and (5). It is further assumed that $\langle 0, t|B(t)|0, t\rangle = 0$. The non–dipole term in lowest order of $1/c$ is

$$B(t) = -\frac{z}{mc}\left(\frac{\partial}{\partial t}\vec{A}\right)\cdot\left(\vec{p} - \vec{A}\right), \tag{18}$$

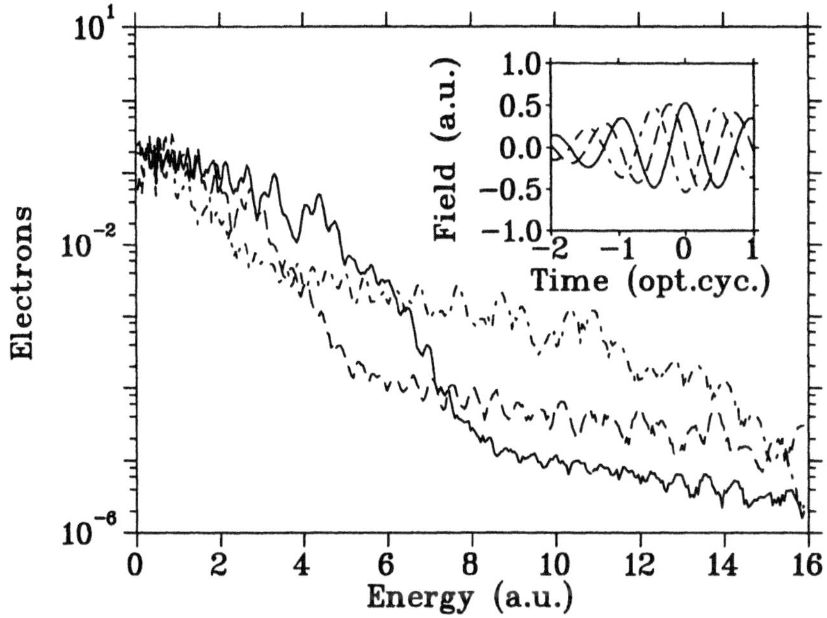

Figure 4. Energy spectrum of electrons emitted in polarization direction for different carrier phases. The inset shows the electric field of the three different pulses.

in which case the Volkov solutions have exactly the form of Eq. (7), if for the Volkov phase Φ one uses

$$\Phi(\vec{k},t) = \int_{t_0}^{t} dt' \left[\vec{k} - \vec{A}(t') + \frac{1}{mc}\left(\vec{a}(t) - \vec{a}(t')\right)\right]^2 \tag{19}$$

with $a_x = a_y = 0$ and

$$a_z(t) = k_x A_x(t) - A_x^2(t)/2. \tag{20}$$

The expressions for $|g)$ and $W(t)$ must be modified accordingly. The main new feature in lowest non–dipole order is the breaking of cylindrical symmetry due to the extra terms $\propto k_z$ appearing in $\Phi(\vec{k},t)$. As long as the deviation from cylindrical symmetry is not too pronounced, an expansion of $|g)$ and $|b)$ with respect to the θ coordinate in the following form is suitable:

$$|g) = \frac{1}{\sqrt{2\pi}} \sum_{m=-N_\theta}^{N_\theta} g(k_x, k_\rho) e^{-in\theta}. \tag{21}$$

Acknowledgement

This work was supported by the Austrian Science Fund, special research program F016 and project Y142-TPH.

References

1. Keldysh, L.V. (1964), *Sov. J. Phys. JETP* **20**, 1307; Faisal, F.H.M., (1973), *J. Phys. B* **6**, L89; Reiss, H.R. (1980), *Phys. Rev. A* **22**, 1786.
2. Lewenstein, M. *et al.* (1995) Rings in above-threshold ionization: A quasiclassical analysis *Phys. Rev. A* **51**, 1495.
3. Becker, W., Lohr, A., and Kleber, M. (1994) Effects of rescattering on above-threshold ionization *J. Phys. B* **27**, L325.
4. Milosevic, D.B. and Ehlotzky, F. (1998) Coulomb and rescattering effects in above–threshold ionization *Phys. Rev. A* **58**, 3124.
5. Lohr, A., Kleber, M., Kopold., R., and Becker W. (1997) Above–threshold ionization in the tunneling regime, *Phys. Rev. A* **55**, R4003.
6. Su, Q., Eberly, J.H., and Javanainen, J. (1990) Dynamics of atomic ionization suppression... *Phys. Rev. Letters* **64**, 862.
7. Rathe, U.W., Keitel, C.H., Protopapas, M., and Knight, P.L. (1997) Intense laser-atom dynamics with the two-dimensional Dirac equation *J. Physics B* **30**, L531.
8. Potvliege, R.M., Kylstra, N.J., and Joachain, C.J. (2000) Photoemission by He^+ in intense ultrashort laser pulses *J. Phys. B* **33**, L743.
9. Vazquez de Aldana, J.R. and Roso, L. (2000) Nonrelativistic numerical study... *Phys. Rev. A***61**, 043403.
10. Brabec, T., Ivanov, M. Yu., and Corkum, P.B (1996) Coulomb focusing in intense field atomic processes *Phys. Rev. A* **54**, R2552.
11. R. Bharwaj et al. (2000) Few cycle dynamics of correlated multi–photon double ionization... presented at this conference (SILAP 2000).
12. Scrinzi, A. (2000) Ionization of multielectron atoms by strong static electric fields *Phys. Rev. A* **61**, 041402.
13. Jones, D.J. *et al.* (2000) Carrier-envelope phase control... *Science* **288**, 5466; Apolonski, A. *et al.* (2000) Controlling the phase evolution of few-cycle light pulses *Phys. Rev. Lett.* **85**, 740.

MOMENTUM SPACE DESCRIPTION OF HYDROGEN ATOM INTERACTING WITH A LOW FREQUENCY, STRONG LASER FIELD

A. DE BOHAN AND B. PIRAUX
Laboratoire de Physique atomique et moléculaire,
Université catholique de Louvain, chemin du cyclotron, 2,
B-1348 Louvain-la-Neuve, Belgium

AND

L. PONCE AND A. MAQUET
Laboratoire de Chimie Physique,
Université Pierre et Marie Curie,
11, rue Pierre et Marie Curie,
F-75231 Paris CEDEX 05, France

1. Introduction

Ab-initio calculations aimed at studying the dynamics of atoms driven by short and intense laser pulses are now becoming an essential step towards a deeper understanding of such interactions.

Considerable efforts have been put into the development of numerical methods to solve the corresponding time-dependent Schrödinger equation (TDSE). Except for two recent calculations [1, 2], the time propagation of the atom wavefunction has always been treated in the configuration space. In this contribution, we consider the momentum space. Despite the non-locality of the atomic potential, the momentum representation is actually more natural since by contrast to the configuration representation, the wavefunction stays localized during the whole interaction of the atom with the laser field. Besides, working in momentum space allows a direct control of the density of (discretized) continuum states taken into account in the calculations. It has also the merit to shed a different light and therefore to gain insight into the dynamics of the laser-atom interaction. For the sake of illustration, we consider a hydrogenic system exposed to an intense, low fre-

B. Piraux and K. Rzążewski (eds.), Super-Intense Laser-Atom Physics, 323–328.

quency laser field and analyse in detail the wavefunction in the momentum space.

2. TDSE in momentum space

In momentum space, the TDSE for the wavefunction of a hydrogenic system exposed to a laser field reads:

$$i\frac{\partial\psi(\mathbf{p},t)}{\partial t} = \frac{\mathbf{P^2}}{2}\psi(\mathbf{p},t) - \frac{Z}{2\pi^2}\int \mathbf{dp}'\frac{\psi(\mathbf{p}',t)}{(\mathbf{p}-\mathbf{p}')^2} + H_I(t)\psi(\mathbf{p},t) \qquad (1)$$

In the above equation, the first term of the right hand side denotes the kinetic energy operator, whereas the second one corresponds to the Coulomb potential which, in momentum space, becomes non-local. Finally, $H_I(t)$ is the laser-atom interaction Hamiltonian expressed in its velocity form: $H_I(t) = \mathbf{A(t)}\cdot\mathbf{P}$.

In order to solve eq(1), we expand the total wavefunction $\psi(\mathbf{p},t)$ over a complete set of B-splines [3] for the "radial" part and over a basis of spherical harmonics for the angular part. The time propagation of the total wavefunction is performed within the interaction picture and in the atomic basis with an usual explicit Runge-Kutta type of method. The non-locality of the Coulomb potential has required a special attention because the corresponding kernel is singular. The way this problem is solved, will be discussed in a forthcoming publication. Besides, and before presenting our results, it is worth stressing the main advantages of the present method. In the strong low frequency field regime where ionisation plays a dominant role, reflection of the wavefunction on the boundaries of the space covered by a finite basis becomes the major drawback of the configuration representation. By contrast, in "**p**" space, the wavefunction remains localized mainly between $\pm\mathbf{A}(t)$ at the beginning of the interaction and within a relatively small value p_{max} when ionisation takes place. Moreover, the use of an expansion over B-splines enables us to impose the density of continuum states. In fact, in order to avoid a continuum state to be viewed as a pseudo-bound state (which may lead to serious errors), it is important to make sure that the energy gap between two continuum states is much smaller than the pulse frequency bandwidth.

In the following, we consider a Ti:Sapphire linearly polarized laser ($\omega = 0.057$ a.u. which defines the optical period T) within a range of intensity from $3.10^{14} W/cm^2$ to $10^{15} W/cm^2$. In our calculations the electric field is given by $\mathbf{E}(t) = E_0 f(t) cos(\omega t + \phi)$ where the pulse envelope takes two different forms: $f_1(t) = sin^2(\frac{\pi}{\Delta t}t + \frac{\pi}{2})$ and $f_2(t) = (1 - cos^8(\frac{\pi}{\Delta t}t + \frac{\pi}{2}))$.

3. Results

Let us first start by looking at the probability density as a function of time and p_z, the projection of the canonical momentum along the polarization axis. The result for $I = 3.10^{14} W/cm^2$, which in the case of hydrogen slightly exceeds the suppression barrier intensity is presented in fig(1). As predicted

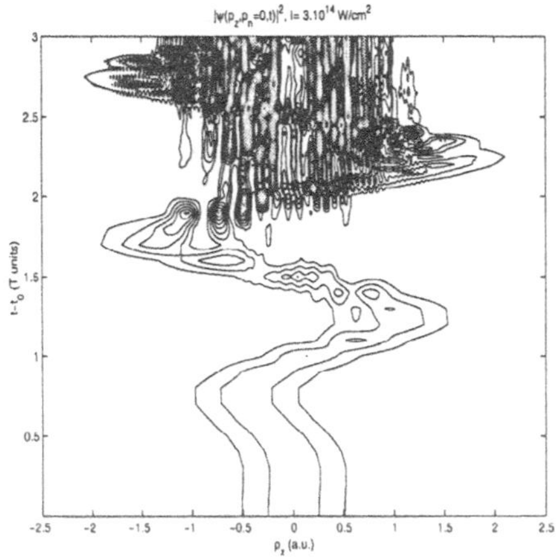

Figure 1. Density profile along the polarization axis as a function of time, for $I = 3.10^{14} W/cm^2$, $\Delta t = 8T$, the pulse shape is given by $f_2(t)$ with $\Delta t = 8T$

by the strong field approximation [4], $\langle v_z \rangle = 0$ as long as the ionisation is negligible. This means that the canonical momentum $\langle p_z \rangle = -A_z(t)$ which manifests itself by oscillations of the density profile whose amplitudes are given in good approximation by E_0/ω. As soon as the ionisation becomes significant, the density profile exhibits a series of "stripes" emerging from the ground state and then remaining constant as time evolves. We believe these stripes as being the signature of ionisation. Indeed, referring to Ehrenfest's theorem and regarding the "ionised" part of the wavefunction only, one has: $\frac{d}{dt}\langle p_z \rangle_{ion} = -\langle \nabla_z V \rangle_{ion} \approx 0$. It is interesting to mention that the same argument has been already invoked for electron trajectories which lead to high order harmonic generation (HOHG) in the semi-classical theory of Lewenstein [4]. Therefore, this analysis in "**p**" space can provide a nice way of probing the Coulomb potential effects on the dynamics.

Regarding the tranverse direction, we display in fig(2) the density profile as a function of time. By contrast to "**r**" space, the wavefunction along the transverse coordinate does not spread but instead is "squeezed", as shown on fig(2a). This is in fair agreement with Kroll and Watson theory [5] which suggests that when the intensity increases the probability of scattering in the transverse direction dramatically decreases. Fig(2b) displays half-cycle bunches of scattering probability in the transverse direction. Note that this structure is clearly visible a few cycles after the beginning of the pulse and for a lower intensity.

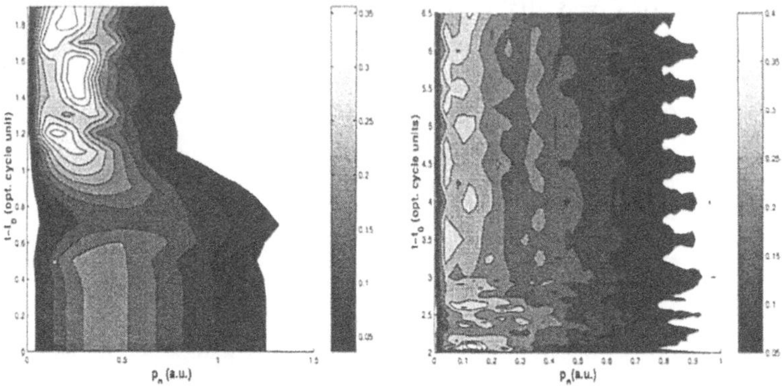

a. $I = 10^{15} W/cm^2$, $\Delta t = 4T$ pulse shape $f_1(t)$

b. $I = 5.10^{14} W/cm^2$, $\Delta t = 8T$ pulse shape $f_2(t)$

Figure 2. contour plot of the time evolution of the density probability along the transverse coordinate.

In the frame of semi-classical studies, we now turn to a different analysis, namely the computation of the Wigner distribution [6] which has notably been used in the context of atomic stabilization [7]. It provides, in phase space, a "quasi-probability" of finding the electron at coordinates $(\mathbf{r}, \mathbf{p})$, but cannot be defined as a real probability since it can experience negative values. Anyway, they can only be attributed to "quantum interferences" whereas the positive ones bring into relief the quasi-classical behaviour of the wavefunction.

Let us recall the general expression of the Wigner distribution:

$$W_{\psi}(\mathbf{r}, \mathbf{p}) = \frac{1}{\pi^3} \int d\mathbf{p}' \psi^*(\mathbf{p} - \mathbf{p}') \psi(\mathbf{p} + \mathbf{p}') e^{2i\mathbf{p}' \cdot \mathbf{r}} \tag{2}$$

and stress the fact that we focus our attention on the polarization direction restricting the integration of $W_{\psi}(\mathbf{r}, \mathbf{p})$ to the p_z coordinate. We perform this calculation using an intensity of $3.10^{14} W/cm^2$, at a time when the vector potential is zero. The result is displayed on fig(3a). Three main quasi-classical zones are present: the trace of the ground state centered at the origin, a branch escaping from it towards negative z and a second one extending between $z = 0$ and $z = 2\alpha_0$ (where α_0 is the maximum excursion of a free electron oscillating in a laser field). These two latter branches are presumably featuring ongoing ionization but no further quantum argument

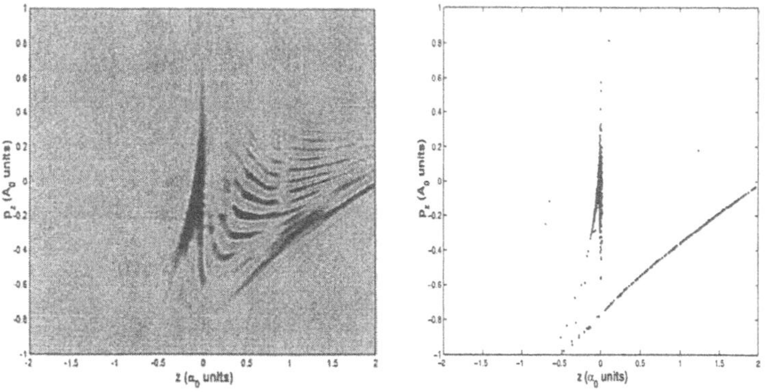

a. Wigner distribution

b. Classical distribution

Figure 3. a) $I = 3.10^{14} W/cm^2$, $\Delta t = 8T$, pulse shape $f_2(t)$, the dark zones corresponds to positive values of the distribution and the light ones to negative values. b) $I = 10^{15} W/cm^2$, $\Delta t = 8T$, pulse shape $f_2(t)$. In both cases, $t = t_0 + 2T$ and z and p_z are normalized in terms of, respectively, the maximum excursion of the electron in the laser field α_0 and the amplitude of the vector potential A_0.

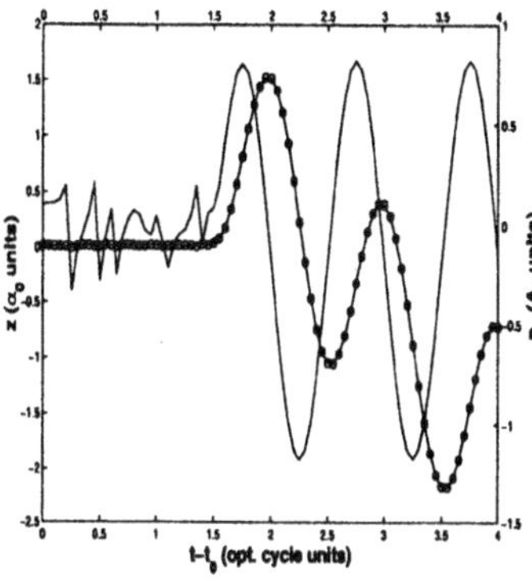

Figure 4. Position (line with open circles) and velocity (solid line) as a function of time of a single trajectory selected from fig(3b) among a thousand trajectory simulation. It is located nearby $z = 1.5\alpha_0$ on the latter figure.

to prove it is available. Therefore, we haved perfomed classical calculations, solved Newton's equations and defined a phase space distribution by means of Monte-Carlo simulations. These classical techniques have already brought a lot of insights about the dynamics of strong field ionisation [8], atomic stabilization [7, 9] and of HOHG [10, 11, 12]. The result on fig(3b) is amazingly similar to the Wigner distribution. By selecting a single classical trajectory, we can analyse its time profile on fig(4). It brings into relief that the same part of the Wigner distribution corresponds to non-direct ionisation since it leaves the nucleus after multiple returns.

4. Conclusion

This short contribution illustrates that the momentum representation is probably the most natural one to describe laser-atom interaction. Together with a Wigner function type of analysis, it provides a powerful tool to analyse in detail both the ionisation and harmonic generation dynamics. In particular, this approach looks promising in probing the role of the Coulomb potential during these processes.

References

1. Pen U., Jiang T., *Phys. Rev. A* **53** 623 (1996)
2. Collins L.A., Merts A.L. *Phys. Rev. A* **45** 6615 (1992)
3. De Boor C., *A practical guide to splines*, Springer-Verlag (1978)
4. Lewenstein M., Balcou Ph., Ivanov M.Y., L'Huillier A., Corkum P.B. *Phys. Rev. A* **49** 2117 (1994)
5. Kroll N.M., Watson K.M., *Phys. Rev. A* **8** 804 (1973)
6. Wigner E. *Phys. Rev.* **40** 729 (1932)
7. Watson J.B., Keitel C.H., Knight P.L., Burnett K. *Phys. Rev. A* **52** 4023 (1995)
8. Chu S.I., Yin R.Y. *J. Opt. Soc. Am B* **4** 720 (1987)
9. Grochmalicki J., Lewenstein M., Rzążewski K. *Phys. Rev. Lett.* **66** 1038 (1981)
10. Bandarage G., Maquet A., Cooper J. *Phys. Rev. A* **41** 1744 (1990)
11. Bandarage G., Maquet A., Menis T., Taïeb R., Véniard V., Cooper J. *Phys. Rev. A* **46** 380 (1992)
12. Gajda M., Grochmalicki J., Lewenstein M., Rzążewski K. *Phys. Rev. A* **46** 1638 (1992)

RELATIVISTIC EFFECTS IN THE ATOMIC RESPONSE TO SUPER-INTENSE LASER PULSES

A. MAQUET, R. TAÏEB AND V. VÉNIARD
Laboratoire de Chimie Physique-Matière et Rayonnement
Université Pierre et Marie Curie
11, Rue Pierre et Marie Curie
75231 Paris Cedex 05 FRANCE.

1. Relativistic effects in multiphoton transitions in the XUV and X-ray ranges

The advent of new powerful sources of radiation in the XUV and soft X-ray ranges, either from Free-Electron Laser (FEL) devices [1], or from high harmonic sources [2], will make possible in the near future, the observation of multiphoton transitions involving the inner shells of high-Z atoms. Although it is well known that relativity plays a determinant role in the fine structure of the atomic levels, it is not so easy to determine to what extent it affects the radiative transition probabilities. For instance it has been observed in standard X-ray spectroscopy that, *for one-photon transitions*, the non-relativistic dipole (NRD) approximation leads to correct estimates well beyond its expected range of validity, [3]. In order to account for this observation, it has been often said that there exist a partial cancellation between different higher order corrections (retardation, spin effects and other relativistic effects) beyond the dipole approximation. The open question is to determine if whether or not such cancellations would still hold in the strong field context, i.e. for multiphoton transitions. If the answer was positive that would greatly simplify the analysis of multiphoton processes.

Let us first outline how relativity and Quantum Electrodynamics (QED) modify the level energies, as compared to the non-relativistic Schrödinger value, in the archetypal case of hydrogenic systems. For a nucleus with charge Z, the energy of a state (n, j) is derived from the Dirac equation, resulting in:

B. Piraux and K. Rzążewski (eds.), Super-Intense Laser-Atom Physics, 329–338.

$$E_{n,j} = \frac{mc^2}{\sqrt{1+(\frac{\alpha Z}{n-j-\frac{1}{2}+\sqrt{(j+\frac{1}{2})^2-\alpha^2 Z^2}})^2}} \tag{1}$$

where $\alpha \approx 1/137$ is the fine structure constant. To compare with the non-relativistic result E_n, it is convenient to expand this expression in powers of $Z\alpha$ which is the natural expansion parameter of QED. Discarding the recoil corrections due to the finite mass of the nucleus and adding the QED contributions, the result can be given as follows, [4]:

$$\tilde{E}_{n,j} = mc^2 + E_n(1 + \frac{\alpha^2 Z^2}{n^2}(\frac{n}{j+\frac{1}{2}} - \frac{3}{4})) + \Delta E_{n,j}. \tag{2}$$

Here, the term in $\alpha^2 Z^2$ accounts for the lowest-order relativistic correction to the kinetic energy together with the contributions of the spin-orbit coupling and of the Darwin term, [5], and E_n is the non relativistic hydrogenic energy. On the other hand, the $\Delta E_{n,j}$ term is the so-called Lamb-shift, which contains higher order terms and the QED corrections. It can be expressed also as an expansion in terms of αZ, [4]:

$$\Delta E_{n,j} = \frac{4\alpha mc^2}{3\pi n^3} \sum_{k=0}^{\infty} \sum_{i=4}^{\infty} A_{i,k} (\alpha Z)^i ln^k[(\alpha Z)^{-2}]. \tag{3}$$

The $A_{i,k}$ coefficients, corresponding to the contributions of one-loop diagrams, are known from $A_{4,0}$ up to $A_{6,2}$. The calculation of higher-order contributions is still under active investigation, [4].

The noticeable point in these results is the fact that the dominant relativistic corrections originating from the spin-orbit coupling and from the corrections to the kinetic energy, in the presence of the strong field of the nucleus, are of the same order of magnitude. As we shall see below, this is no longer the case in the presence of a strong laser field.

Regarding inner-shell multiphoton transitions, it appears that, in spite of the remarkable brightness of the currently designed sources in the X-ray range, a perturbative treatment of the processes is adequate. This is because the relevant atomic unit of intensity for hydrogenic systems with charge Z, scales like $I_Z = Z^6\, 3.5 \times 10^{16}\, W/cm^2$. It appears that the expected peak intensities of the planned X-ray FEL sources at DESY and Livermore (SLCS) are announced to stay below the critical intensities at which non-perturbative effects are of importance in high-Z systems, [1]. The same remark holds for the presently developed high harmonic sources, [2]. In this context, it is important to have at one's disposal reliable data on multiphoton transition amplitudes involving inner shells in high-Z elements.

As a typical example, we present next the results of an "exact" computation of the second-order perturbative transition amplitude for a typical two-photon absorption process, namely the two-photon $| 1^2S_{1/2}\rangle \rightarrow | 2^2S_{1/2}\rangle$ transition. By "exact" computation, we mean that the infinite sums (running over the discrete and continuous Coulomb spectrum) entering the second order amplitude, have been performed exactly with the help of a sturmian expansion of the Dirac-Coulomb Green's function, [6], [7].

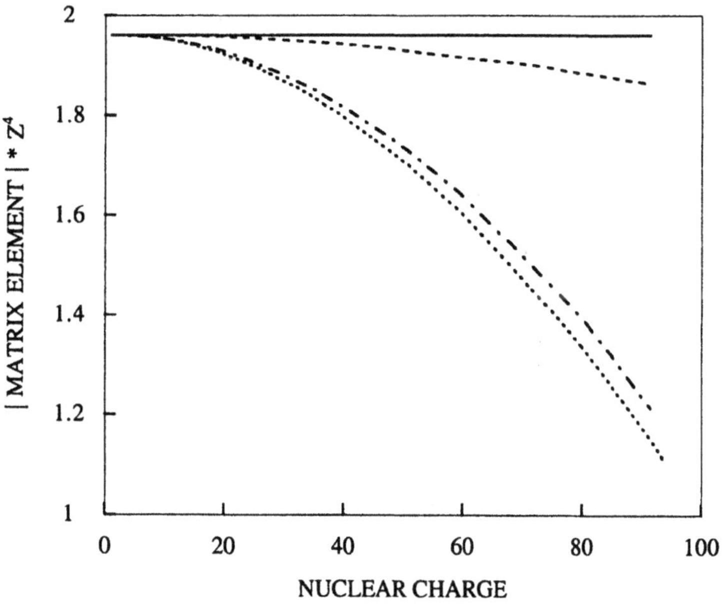

Figure 1. Modulus of the matrix elements M for the transition $|1\,^2S_{1/2}\rangle \rightarrow |2\,^2S_{1/2}\rangle$ in atomic units multiplied by a factor Z^4 as a function of nuclear charge Z. The solid line represents the nonrelativistic dipole (NRD), the dashed line the nonrelativistic retarded (NRR), the dot-dashed line the relativistic dipole (RD), and the dotted line the relativistic retarded (RR) result.

When comparing with less sophisticated computations, it appears clearly that relativistic corrections become significant already for $Z = 20$, at frequencies $\omega = 2.04\,keV$ and that they can amount to more than 10% for $Z = 50$, at frequencies $\omega = 12.75\,keV$. This is shown in the figure 1, where the corresponding scaled two-photon amplitudes are displayed, as computed at different levels of approximation. This is in sharp contrast with the case of *one-photon* transitions in the X-ray range where, for instance at $Z = 50$, the difference between the transition amplitudes for ionization as computed either from a non-relativistic dipole treatment or from the fully relativistic computation does not exceed 5% for photon energies at least twice higher,

[3]. These results clearly demonstrate that there is a need for more sophisticated calculations, beyond the non-relativistic dipole approximation, in order to determine the multiphoton rates involving inner-shell states in high-Z elements. We turn now to a discussion of typical phenomena which are expected to occur in the presence of strong *infrared* fields.

2. Signatures of mass-shift effects in super-intense infrared laser fields

Atoms submitted to infrared pulses with intensities exceeding the so-called atomic unit of intensity ($I_0 = 3.5 \times 10^{16}\ W/cm^2$), are readily stripped of their outer electrons. Recently, rare gas ionic species, with charges up to 18^+ have been created in this way, [8, 9, 10, 11]. The subsequent dynamics of the ejected electrons within the laser focal spot is then dominated by relativistic effects. This arises because the classical averaged quiver energy (also termed as the ponderomotive energy) $U_p = q^2 F_0{}^2/(4m\omega^2)$ can become comparable to the electron's rest mass energy. Here F_0 is the field strength of the laser with frequency ω. The question then arises of the relative importance of different relativistic effects and more precisely of delineating the respective roles of spin and other corrections. As we shall show, the magnitudes of the effects resulting from the so-called "mass-shift" are, by far, dominant.

In the presence of an electromagnetic field, the electron's mass is renormalized. Within a classical picture, the mass-shift can be directly associated to the quiver motion energy and, more precisely to the ratio of U_p to the rest mass energy of the electron. In the case of a constant amplitude (plane wave) field, one has:

$$m^* = m\sqrt{1 + \frac{2U_p}{mc^2}}. \tag{4}$$

On the other hand, within a fully quantum (QED) description, the same mass-shift appears naturally as resulting from the occurrence of inelastic (Compton) scattering of the laser photons, accompanied by the transfer of forward momentum in the propagation direction of the laser, [12]. The effect has been seen in experiments, when colliding a beam of relativistic electrons from a LINAC with an infrared laser, [13]. The interesting point is that the correction term does not contain $\hbar$, and that the classical and quantum approaches lead to the same expression, [12]. This partially explains the success of classical mechanics for describing the dynamics of electrons within the field.

A typical example is provided by the dependence in terms of the field strength of the angular distribution of the photoelectrons stripped from the outer shells of atoms within the laser field. Indeed, as a direct consequence of the forward momentum transfer, the polar angle of ejection θ of a photoelectron with respect to the laser wave vector is related to its quiver energy U_p through the relation:

$$\theta = tan^{-1}\sqrt{\frac{2mc^2}{U_p}}, \tag{5}$$

which is exact in the limits of zero initial velocity and in the long pulse regime, [14, 15]. For a treatment valid under more general conditions, see [16]. This relation has been used successfully for analyzing the results of recent experiments on the photoionisation of rare gas atoms submitted to ultra intense infrared radiation pulses from a Nd: Yag laser ($\omega = .043\, a.u.$), [8]. In particular, it provides the basis for a convincing description of the mechanisms governing the multiple ionization of atoms. It turns out that the process is dominantly sequential, each ionic stage being created when, during the rise of the pulse, the field strength reaches a critical value corresponding to the suppression of the Coulomb barrier associated to the previous ionization stage. When this happens, accurate numerical simulations, based on the resolution of the Time-Dependent Schrödinger Equation (TDSE), show that ionization is completed within a few half laser cycles, [17]. As an illustration, the typical variations in time of the populations of different ionic stages for a model Kr atom submitted to an ultra-intense laser pulse, is shown in Figure 2.

This scenario, together with the numerical simulations, enable ones to account for the polar angle distributions of the photoelectrons recorded in the experiments, [8, 9]. This demonstrates that, *for infrared lasers with frequencies* $\omega \approx 1\, eV$, mass-shift corrections are already most important at laser intensities $I_L \approx 10^{18}\, W/cm^2$.

Still regarding the angular dependence of the differential cross sections, it should be noted however, that the question of the azimuthal distribution of the photoelectrons, in the plane perpendicular to the propagation direction of the field, is still matter of discussion, as the results reported in the references [8] and [9] do not agree with each other. For recent discussions, see [17] and [18].

Besides these signatures seen in photoionization experiments, several theoretical studies have shown that mass-shift effects should play an important role in other laser-induced processes. Here follows a non-exhaustive list of related topics of interest:

i) *laser-assisted electron-atom scattering.* During the course of the collision in the presence of the laser field, it is known that the projectile can absorb (or emit) photons. In the limit of super-intense fields, very large numbers of photons can be exchanged: up to several thousands for "hard" collisions accompanied by large transfer of momentum. In the test case of

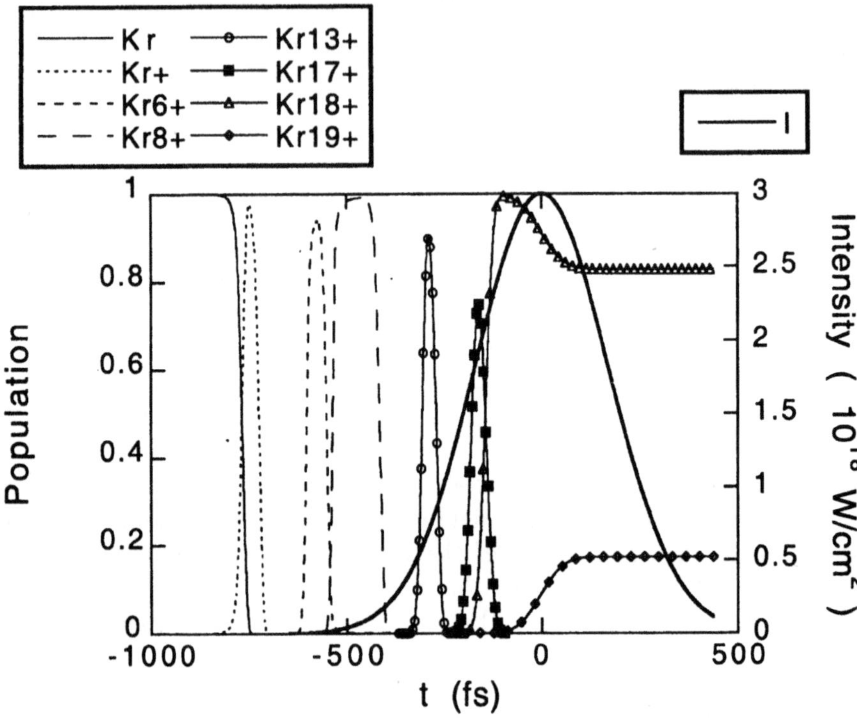

Figure 2. Populations of different ionic stages for a model Kr atom as a function of time. The intensity of the laser field is $3 \times 10^{18}\, W/cm^2$ and the frequency is $\omega = 0.043\ a.u.$ The intensity envelop is a gaussian of $400fs$ FWHM, [9]

laser-assisted Mott scattering for relativistic electrons, one can show that mass-shift effects manifest themselves in the width of the energy span for the scattered electron, see [19]. In the same work it has been shown, by comparing the results derived from a Klein-Gordon or from a Dirac treatment, that spin effects remain significantly smaller.

ii) *atomic stabilization.* In the presence of a super-intense *high frequency* field, theory predicts that an atom could be "stabilized" in the sense that the atomic state lifetime could increase when the field strength grows, [20]. We note that this *a priori* counter-intuitive dependence holds in adiabatic conditions, i.e. for relatively slow turn-on. Earlier numerical simulations were done in the non-relativistic dipole approximation. Now, the question at stake is whether or not the effect will survive in the relativistic regime, i.e. when taking into account mass-shift and other corrections. Regarding the effect of mass-shift, it turns out that it tends to reduce the excursion length of the bound electron, thus reducing the width of the potential well in which it is trapped and thus reinforcing its binding energy, [21, 22]. The question of the signature of such an effect in photoelectron spectra has been discussed in Ref. [23]. Broadly speaking, the global influence of the mass-shift is to stabilize further the atom. Then comes the question of the roles of the forward momentum transfer and of the coupling with the magnetic field. Both processes induce an irreversible drift of the atomic electron, the resulting effect being to hinder stabilization. This has been confirmed in very recent simulations, see for instance, [24]. However, it should be noted that, although it appears that the range of intensities in which stabilization does occur is reduced, the question of the respective importance of mass-shift and other effects is not settled yet.

iii) *occurrence of mass-shift effects in the solutions of the Dirac equation.* Solving the time-dependent Dirac equation for a bound particle in the presence of a super-intense laser field is a formidable task. Besides the above-mentioned works, several numerical simulations have been reported, for 1- and 2-dimensional model systems, which clearly display significant changes in the probability density distributions, as compared to the non-relativistic limit, see for instance [25, 26]. A thorough discussion of the technical difficulties encountered in solving the Dirac equation, as well as the question of the peculiarities of the relativistic dynamics (see below), can be found in [27]. It is not clear, however, to decide to which extent these changes do affect the dynamical response of the atom.

iv) *harmonic generation spectra.* Since the early studies on the importance of relativistic effects in super-intense laser physics, it has been surmised that their signature might be found in harmonic generation spectra, [28]. For a more recent account, see [29]. In fact, in the case of a sample of atoms in the presence of a super-intense infrared field, a significant proportion of electrons are freed. They can subsequently interact non-linearly with the field and generate harmonics. The open problem is to determine how this contribution would combine with the atomic ones, thus making the analysis more involved.

We shall turn now to the question of the relative importance of other relativistic effects, including spin and more exotic effects, such as zitterbewegung and radiation reaction.

3. Spin and more exotic effects in super-intense laser fields

3.1. SPIN EFFECTS

Up to now, no experimental evidence of effects arising from the coupling of the spin with the magnetic component of the laser field has been reported. The reason is most likely due to the fact that these effects are significantly smaller than those discussed in the preceding section. This contrasts with standard relativistic corrections in atoms where the spin-orbit coupling is important, if not dominant.

Nevertheless, the question of the relative importance of spin-laser interactions has been addressed in several theoretical analysis and also in numerical simulations. The signature of spin coupling appears clearly in the computed energy distribution of the scattered electrons in laser-assisted Mott scattering, [19]. It turns out however, that the results derived from a full Dirac treatment differ only slightly from those obtained by using the Klein-Gordon equation. The same situation prevails when considering the influence of spin on harmonic generation spectra, [30], or when evaluating spin-induced forces, [31]. In fact, significant spin effects show up under the combined effects of the strong laser and of the static Coulomb field of a high-Z nucleus, the signature being notable changes in the spin-orbit splitting in X-ray emission lines, [32].

3.2. ZITTERBEWEGUNG

The elusive effects of zitterbewegung, i.e. of the virtual oscillations on a time scale of $(2mc^2)^{-1}$ of a relativistic particle between negative and positive energy states, pervade the literature on relativistic quantum dynamics. In the context of super-intense laser-atom physics, the question at stake is if, whether or not, signatures of the effect could be detected. The reason is that it is expected that stimulated multiphoton transitions between negative and positive energy states could be induced in the presence of a super-intense laser. This would certainly modify the dynamics of an electron embedded within the radiation field. In fact, the existence of such high frequency oscillations, induced by the zitterbewegung, in the time dependence of the wave function is a nuisance, in the sense that it imposes using extremely small time steps, when computing the solutions of a time-dependent relativistic wave equation. This point has been addressed notably in the paper by Braun *et al.*, [27].

Assuming that the problem of the numerical stability of the solution of the time-dependent wave equation has been solved, remains the question of the signature of zitterbewegung in atomic phenomena. It turns out that, although noticeable effects are predicted in harmonic generation spectra, their magnitudes remain extremely small, [33].

3.3. RADIATION REACTION

An electron embedded within a super-intense radiation field, can experience considerable time-varying accelerations and, consequently, can emit radiations. Then the question arises of the influence of this damping, i.e. of the so-called radiation reaction, on its motion. Another question is that of the signatures of the effect. It has been surmised that it could induce changes in Compton scattering or in harmonic generation spectra. A fully quantum (QED) treatment being out of reach, a classical approach, based on the so-called Lorentz-Dirac equation, has been advocated, [34]. The difficulty is however that the solutions are plagued by the so-called runaway solutions, the divergence of which being linked to the point-like character of the electron, [35]. Nevertheless, a clever integration scheme, running backwards in time, has been proposed which permits to extract finites quantities from the equation. Then, via a standard Classical Trajectory Monte-Carlo treatment, it is possible to compare the dynamics of an ensemble of electrons depending on whether the radiation reaction terms are included or not in the equations of motion, [36]. The simulations show noticeable, though very small, changes in the emission radiation spectra, already at intensities around $I_L \approx 10^{20}\ W/cm^2$.

As a conclusion, it appears that at the intensities currently achieved in experiments on atom-laser phenomena, the dominant relativistic corrections originate from the mass-shift effect. Other, more exotic, phenomena would be observed at significantly higher intensities. Another possibility has been suggested, which is to add a strong external magnetic or electric field. Interesting results of several simulations performed along these lines have been reported during the Conference and are presented in this volume.

ACKNOWLEDGEMENTS

The Laboratoire de Chimie Physique-Matière et Rayonnement is a Unité Mixte de Recherche, Associée au CNRS, UMR 7614, and is Laboratoire de Recherche Correspondant du CEA, LRC # DSM-98-16. Parts of the computations have been performed at the Centre de Calcul pour la Recherche (CCR, Jussieu, Paris) and at the Institut du Développement et des Ressources en Informatique Scientifique (IDRIS).

References

1. Krzywinski J., this volume.
2. Brabec T. and Krausz F.,*Rev. Mod. Phys.* **72**, 545 (2000); see also Brabec T., this volume.
3. Ron A. *et al.*,*Phys. Rev. A* **50**, 1312 (1964).
4. A recent review is: Pachucki K., Leibfried D., Weitz M., Huber A., König W. and Hänsch T.W. , *J. Phys. B* **29**, 177 (1996).
5. A recent text-book reference is: Bransden B. and Joachain C.J., *"Quantum Mechanics of Atoms and Molecules"* (Longman, London, 1983).
6. Maquet A., Véniard V. and Marian T.A., *J. Phys. B* **31**, 3743 (1998).
7. Szymanowski C., Véniard V., Taïeb R. and Maquet A.,*Phys. Rev. A* **56**, 700 (1997).
8. Moore C.I., Knauer J.P. and Meyerhofer D., *Phys. Rev. Lett.* **74**, 2439 (1995); Meyerhofer D. *et al.*, *J. Opt. Soc. Am. B* **13**, 113 (1996); McNaught S.J., Knauer

J.P. and Meyerhofer D.,*Phys. Rev. Lett.* **78**, 626 (1997); McNaught S.J., Knauer J.P. and Meyerhofer D.,*Phys. Rev. A* **58**, 1399 (1998).
9. Moore C.I. *et al.*,*Phys. Rev. Lett.* **82**, 1688 (1999).
10. Maeda H. *et al. Phys. Rev. A* **62**, 035402 (2000).
11. Moshammer R. *et al.*, *Phys. Rev. Lett.* **84**, 447 (2000).
12. Brown L.S. and Kibble T.W.B., *Phys. Rev.* **133**, A705 (1964).
13. Bula C.*et al. Phys. Rev. Lett.* **76**, 3116 (1996).
14. Reiss H., *J. Opt. Soc. Am. B* **7**, 574 (1990).
15. Corkum P.B., Burnett N.H. and Brunel F., in *Atoms in Intense Laser Fields*, Gavrila M., ed. (Academic, New York, 1992) p. 109.
16. Salamin Y.I. and Faisal F.H.M., *Phys. Rev. A* **55**, 3678 (1997); more recent references include: Salamin Y.I. and Faisal F.H.M., *J. Phys. A* **31**, 1319 (1998); Salamin Y.I. and Faisal F.H.M.,*Phys. Rev. A* **61**, 043801 (2000).
17. Taïeb R., Véniard V. and Maquet A., unpublished.
18. Maquet A., Taïeb R. and Véniard V., in *Atoms, Molecules and Quantum Dots in Laser Fields: Fundamental Processes*, edited by Rizzo A. and Rahman N., (2000)
19. Szymanowski C., Véniard V., Taïeb R., Maquet A. and Keitel C.H., *Phys. Rev. A* **56**, 3846 (1997).
20. Gavrila M., in *Atoms in Intense Laser Fields*, Gavrila M., ed. (Academic, New York, 1992) p. 435.
21. Protopapas M., Keitel C.H. and Knight P.L., *J. Phys. B* **29**, L591 (1996); see also the review: Protopapas M.,Keitel C.H. and Knight P.L., *Rep. Prog. Phys.* **60**, 389 (1997).
22. Ermolaev A.M., *J. Phys. B* **31**, L65 (1998).
23. Taïeb R., Véniard V and Maquet A., *Phys. Rev. Lett.* **81**, 2882 (1998).
24. Kylstra N.J. *et al.*, *Phys. Rev. Lett.* **85**, 1835 (2000).
25. Kylstra N.J., Ermolaev A.M. and Joachain C.J., *J. Phys. B* **30**, L449 (1997).
26. Rathe U.W., Keitel C.H., Protopapas M. and Knight P.L., *J. Phys. B* **30**, L531 (1997).
27. Braun J.W., Su Q. and Grobe R., *Phys.Rev. A* **59**, 604 (1999).
28. Keitel C.H., Knight P.L. and Burnett K., *Europhys. Lett.* **24**, 539 (1993); Keitel C.H. and Knight P.L., *Phys. Rev. A* **51**, 1420 (1995).
29. Milosevic D.B., this Conference.
30. Walser M.W., Szymanowski C. and Keitel C.H., *Europhys. Lett.* **48**, 533 (1999).
31. Walser M.W. and Keitel C.H., *J. Phys. B*, **33**, L221 (2000).
32. Hu S.X. and Keitel C.H., *Phys. Rev. Lett.* **83**, 4709 (1999).
33. Szymanowski C., Maquet A. and Keitel C.H., *Laser Phys.* **9**, 133 (1999).
34. Hartemann F.V. and Kerman A.K., *Phys. Rev. Lett.* **76**, 624 (1996).
35. a recent discussion is: Efremov G.F., *Sov. Phys. JETP* **87**, 899 (1998).
36. Keitel C.H., Szymanowski C., Knight P.L. and Maquet A., *J. Phys. B*, **31**, L75 (1998).

WEAKLY RELATIVISTIC STABILIZATION

The effect of the magnetic field

H.G. MULLER
FOM-Institute for Atomic and Molecular Physics
Kruislaan 407, 1098 SJ Amsterdam, The Netherlands

1. Introduction

This paper adresses a problem that lately attracted a lot of interest, namely the effect of the magnetic field component of light on stabilization. In particular, it derives an analytical estimate for the maximum allowable turn-on rate of the light pulse, and the resulting optimal turn-on shape. We then present ab-initition numerical simulations in three dimensions using this pulse shape.

2. Breakdown of the dipole approximation

The lowest-order effect of the finite speed of light is the breakdown of the dipole approximation. Taking into account the position dependence of the electromagnetic wave (polarized along the z-axis) in the direction of its propagation x leads to the expansion

$$A_z(t + x/c) = A_z(t) + x/c\partial_t A_z(t) + O(1/c^2). \tag{1}$$

The corrections of order $1/c$ sometimes are not considered relativistic corrections at all, but just retardation effects. Indeed, all truly relativistic behavior of the *electron*, such as mass correction or spin-orbit coupling, is of $O(1/c^2)$. The treatment upto $O(1/c)$ thus only makes sense in the weakly relativistic regime, where the speed of the electron stays significantly below c (which in atomic units is approximately 137). For velocities v from 10 to 30 a.u., the $1/c^2$ corrections range from 0.3% to 3%, and although they lead to small changes in the quantitative behavior of the electron, they do not introduce essentially new effects.

This is quite different for the retardation, which shows up as a magnetic field that can deflect a fast electron significantly through its Lorentz force, $B \times v/c$. Plugging eq.1 into the Schrödinger equation for a single-electron

B. Piraux and K. Rzążewski (eds.), Super-Intense Laser-Atom Physics, 339–344.

atom leads to a hamiltonian (upto $O(1/c)$)

$$H \doteq \frac{1}{2}(\mathbf{p} - \mathbf{A}(t + x/c))^2 + V \tag{2}$$
$$= \frac{1}{2}p^2 - \mathbf{A}(t)\cdot\mathbf{p} + \frac{1}{2}A^2(t) + x/c\mathbf{A}(t)\partial_t\mathbf{A}(t) - x/c\partial_t\mathbf{A}(t)\cdot\mathbf{p}.$$

The last term is negligible for weakly bound electrons that quiver with the field (which have $\mathbf{p} \approx 0$). Due to its non-zero cycle average, however, the forelast term causes a drift that would sweep a free electron away with the propagating pulse:

$$\langle x/c\, A_z(t)\partial_t A_z(t)\rangle = \frac{x}{2c}\partial_t\left\langle A_z^2(t)\right\rangle = \frac{x}{4c\omega^2}\partial_t I = \frac{x}{c}\partial_t U_p. \tag{3}$$

From the last form it can be seen that this effect is basically just the ponderomotive force due to gradients in the leading and trailing edge of the pulse. Eq.3 has the form of a homogeneous electric field, and it can be cast into velocity form

$$1/2c\, A_z^2(t)\, p_x \tag{4}$$

through the introduction of an extra vector potential $A_x = 1/2c\, A_z^2(t)$, and the associated gauge transform $\exp(iA_x x)$ on the wave function. Since for a free electron $\mathbf{p}$ is conserved in the velocity gauge, it can immediately be seen that the presence of this term leads to a drift velocity in the x-direction equal to the vector potential, and that the x-component of this motion is solely determined by the instantaneous magnitude of A_x.

In half an optical cycle, the displacement due to this drift is

$$\frac{\pi}{4\omega c}A_z^2 = \frac{\pi}{4c}\,\alpha_0^2\omega. \tag{5}$$

This shows that the drift grows with respect to the quiver amplitude α_0 in the polarization direction at higher frequencies ω. This effect could potentially be very detrimental for high-frequency phenomena like adiabatic stabilization[2]. Indeed it has been claimed that adiabatic stabilization in the weakly relativistic regime becomes impossible [3, 4]. Even at $\omega = 1$ the a wave packet pulled off an atom would not return after half a cycle but miss the nucleus by 1.5 Bohr if $\alpha_0 = 15$.

3. Barrier suppression in the KH frame

What holds for a free electron is of course not necesaryl true for a bound one. The ponderomotive turn-on force of eq.3 can be made as small as we want by limiting the rate of intensity ramp-up. To reach the same intensity then will take a longer time, so that a *free* electron would pick up the same

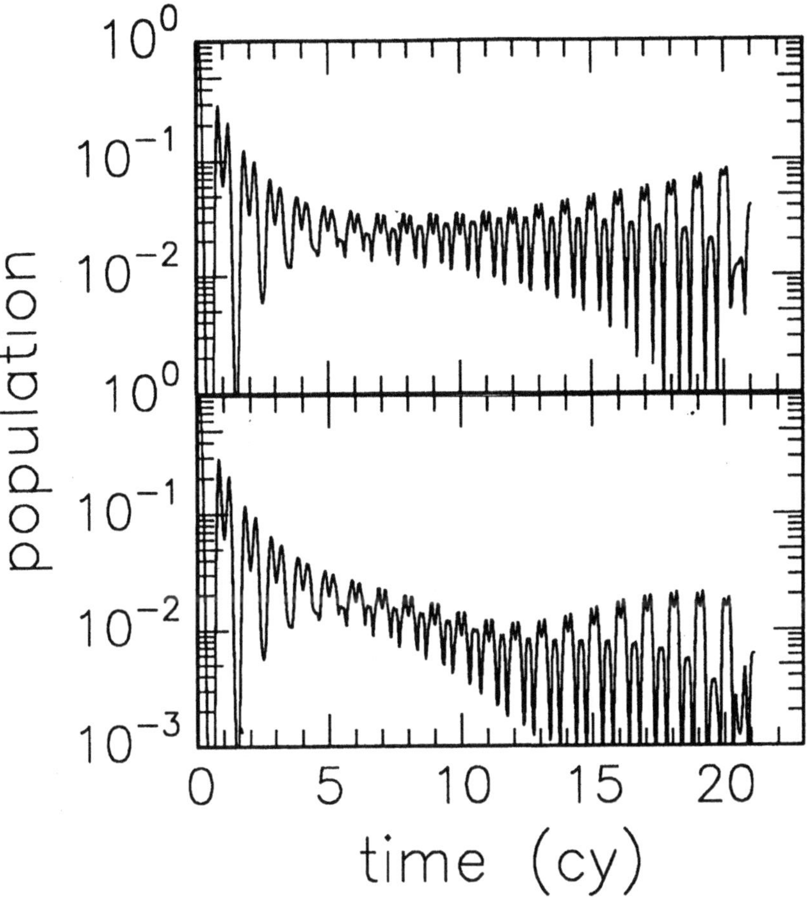

Figure 1. **Overlap with the unperturbed atomic ground state as a function of time in a 21-cycle laser pulse with $t^{1/3}$ turn-on and turn-off, obtained from a full 3D solution to the TDSE. The frequency is 1 Hartree, the peak field amplitude 10 a.u.. The upper trace does not take the effect of the magnetic field into account. The lower trace does, and shows about five times lower survival.**

drift velocity anyway. But a *bound* electron can be hold against a force for an exponentially long time if its energy level is located below the the edges of the potential well that binds it. These edges are depressed by the addition of the ponderomotive potential, and the interplay between ponderomotive force and binding force will determine at which field the bound electron will be pulled from its atom.

The situation is completely analogous to ordinary barrier suppression by the electric component E of the light. The main difference is that the binding potential for an adiabatically stabilized atom is the dressed potential V_0. At large α_0 the electron is mainly bound to the endpoints of this potential, which is given by the asymptotic form

$$V_{EP}(\mathbf{r}) = \frac{P_{1/2}(\cos\vartheta)}{\sqrt{2\alpha_0 r}}. \tag{6}$$

Making use of this scaling, we can determine the height of the barrier in the usual way, by looking for the saddle point of $V_{EP}(r) + \mathbf{E}\cdot\mathbf{r}$. Equating the derivative to zero leads to a barrier at

$$r \simeq E^{-2/3}\alpha_0^{-1/3}, \tag{7}$$

(we are only interested here in the powers, not the pre-factors) of magnitude

$$V_{barrier} \simeq E^{1/3}\alpha_0^{-1/3}. \tag{8}$$

Finally, we equate this to the energy level of the KH ground state, which in the asymptotic region is at $U_0 = W\alpha_0^{-2/3}$, to obtain

$$E_{BS} \simeq \alpha_0^{-1}. \tag{9}$$

As long as the ponderomotive force stays well below this threshold, it will have no significant effect on the stabilized atom. This leads to the condition

$$\partial_t I = \omega^2 \partial_t A^2 \simeq 4c\omega^3 A^{-1}, \tag{10}$$

which for the fastest possible ramp-on (the equality) can be solved as

$$I \simeq t^{1/3}. \tag{11}$$

This shows that any intensity can still be reached in a finite time without violating the stability condition.

4. Simulations in 3D

The extra field component in the x-direction breaks the cylindrical symmetry one usually has in linear polarization, and thus forces a truly three-dimensional simulation. This situation is very similar to the case of elliptical (or circular) polarization. In fact it is so similar, that the computer code designed for solving the TDSE in elliptically polarized fields [1] had to be changed in only one statement to solve the weakly-relativistic linear-polarization case: the x-component of the vector potential, which normally is a sine that laggs the z-field by 90o has to be replaced by expression 4.

This was easy enough to give it a try, and the results are shown im Fig. 1. The pulse used consisted of a $t^{1/3}$ turn-on, immediately followed

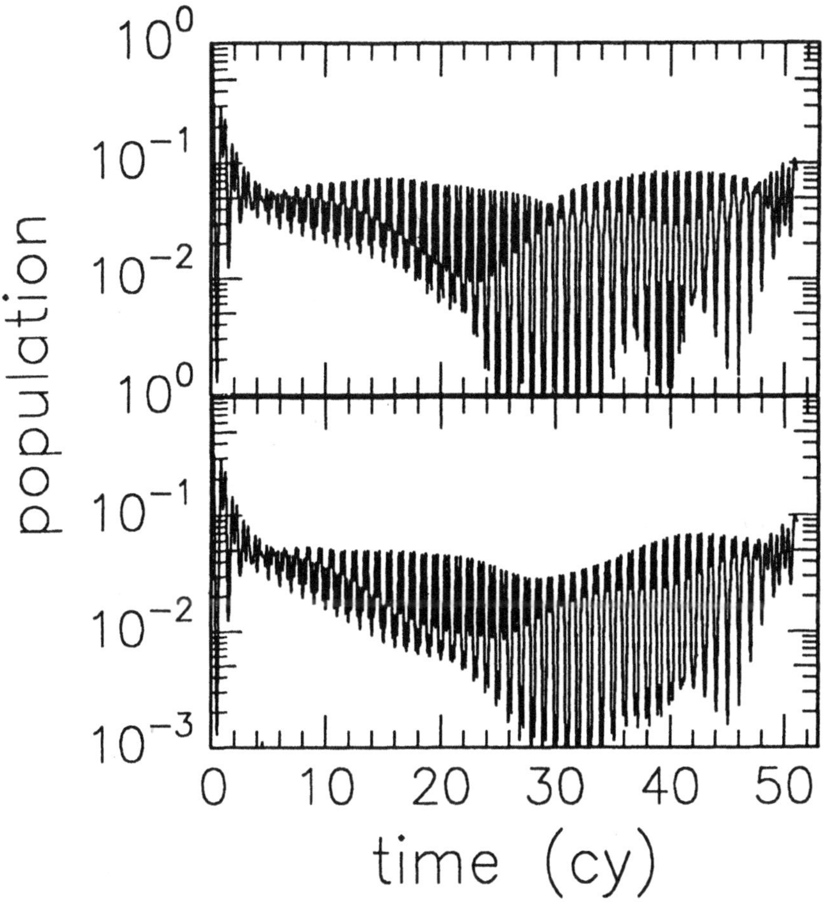

Figure 2. **The same as Fig. 1, but now for a 51-cycle pulse.**

by a similar, equally long turn-off for the vector potential. For a pulse duration of 21 cycles the simulation including the $B \times v$ drift produces significantly less survival than the (2-d) simulation neglecting it. In the non-relativistic treatment most loss of ground state occurs in the initial (non-relativistic!) part of the pulse, mostly due to shake-up to other bound states (the $t^{1/3}$ turn-on is not very favorable in this region). This explains the assymmetry of the curves; at the very end there is no significant loss of ground-state population, since the shake-up now simply reshuffles the population between all the surviving bound states.

The difference caused by the magnetic term largely disappears on increasing the pulse duration to 51 cycles, however (Fig. 2). The overlap at the peak of the pulse is slightly smaller for the relativistic treatment, due to the polarization of the wave function by the magnetic term. After turn-off it simply snaps back to the axis, and the survival probability is the same in the relativistic and non-relativistic case. The conclusion thus seems that adiabatic stabilization can persist in the relativistic regime; it is just more difficult to satisfy the adiabaticity condition.

Acknowledgement - This work is part of the research program of Fundamental Research on Matter (FOM), which is supported by the Netherlands Organization for the Advancement of Research (NWO).

References

1. H.G. Muller (1999) *Laser Phys.* **9**, 138.
2. M. Pont and M. Gavrila (1990) *Phys. Rev. Lett.* **65**, 2362.
3. C.H. Keitel and P.L. Knight (1995) *Phys. Rev. A* **51**, 1420.
4. N. J. Kylstra, R. A. Worthington, A. Patel, P. L. Knight, J. R. Vázquez de Aldana, and L. Roso (2000) *Phys. Rev. Lett.* **85**, 1835.

INTENSE LASER-ATOM INTERACTIONS: BEYOND THE DIPOLE APPROXIMATION

N. J. KYLSTRA, R. M. POTVLIEGE, R. A. WORTHINGTON
Physics Department
University of Durham
Durham DH1 3LE, United Kingdom

A. S. PATEL, P. L. KNIGHT
Optics Section, Blackett Laboratory,
Imperial College,
London SW7 2BZ, United Kingdom

J. R. VAZQUEZ DE ALDANA, L. ROSO
Departamento de Fisica Aplicada
Universidad de Salamanca
E-37008 Salamanca, Spain

AND

C. J. JOACHAIN
Physique Théorique, CP 227
Université Libre de Bruxelles
Boulevard du Triomphe
B-1050 Brussels, Belgium

1. Introduction

High harmonic generation by atoms [1,2] and the stabilization of atoms [3] are two intense-field phenomena that have been studied extensively in recent years. Harmonic generation can be understood qualitatively in terms of a semi-classical, three-step model [4]. In this so-called "simpleman's model", the active electron of the atom or ion is first detached by tunneling through an effective potential barrier formed by the instantaneous electric field of the laser pulse and the potential that binds the electron. The ejected electron is then assumed to move like a free particle in the electric field of the incident laser pulse. As the field reverses, the electron can re-collide with the core and recombine radiatively. A fully quantum mechanical description

B. Piraux and K. Rzążewski (eds.), Super-Intense Laser-Atom Physics, 345–354.

of harmonic generation that incorporates the essential ideas of the semi-classical model has been developed by Lewenstein and co-workers [1,5].

Stabilization in the high-intensity, high-frequency regime can, on the other hand, be understood with the help of the high-frequency Floquet theory (HFFT) developed by Gavrila and co-workers [6]. First, a transformation is made to the rest frame of a classical electron in the laser field. In this Kramers-Henneberger (K-H) frame, the electron experiences an effective time-dependent Coulomb potential that oscillates in space. Gavrila and co-workers have shown that in a stationary field, to first approximation, the dynamics of the atom in the field are governed by the time-average of this potential. This theory can be modified to accommodate the non-adiabatic evolution of the atom in a short laser pulse and, if the turn-on of the laser pulse is short enough, the electron wavepacket dynamics is essentially those of a free electron in the pulse [7].

At high laser intensities, these relatively simple descriptions of harmonic generation and stabilization, respectively, can break down since non-dipole and relativistic effects must be accounted for. Indeed, for only moderately high intensities, the magnetic field component of the laser pulse cannot be *a priori* neglected. This magnetic field induces a drift in the laser propagation direction that, for a free classical electron, can be quite large. If taken literally, the semi-classical model of harmonic generation predicts no harmonic generation beyond the dipole approximation since an ejected electron would never return to the core. Similarly, in the high-frequency regime, a K-H transformation to the rest frame of an oscillating and drifting electron in a laser pulse would imply that the Coulomb potential moves away from the electron during the pulse. This, in turn, would affect the stabilization of the atom in the laser pulse. Obviously, these observations are oversimplified. However, we will show that the magnetic field component of an intense laser pulse can indeed strongly influence both high harmonic generation and stabilization.

2. The Classical Equations of Motion

We start by considering the influence of the Lorentz force, due the magnetic field component of the laser, on a classical electron. The Lorentz equation for the electron in a laser pulse is (in atomic units, a.u.)

$$\frac{d}{dt}\mathbf{p} \quad = \quad -\left[\mathbf{E}(\eta) + \frac{\mathbf{v}}{c} \times \mathbf{B}(\eta)\right] , \tag{1}$$

where $\mathbf{p} = \gamma\mathbf{v}$ and $\gamma = 1/\sqrt{1-\beta^2}$, with $\beta = |\mathbf{v}|/c$. The electric and magnetic fields, respectively, depend only on space and time via the quantity $\eta = \omega(t - z/c)$. The laser pulse propagates along $\hat{\mathbf{z}}$.

By expanding in powers of $1/c$, one obtains the Lorentz equations correct to order $1/c$:

$$\frac{d}{dt}v_x = -E(\omega t)\,, \quad \frac{d}{dt}v_z = -\frac{v_x}{c}B(\omega t)\,. \tag{2}$$

Here we have set $\mathbf{k} = k\hat{\mathbf{z}}$, $\mathbf{E}(\eta) = \hat{\mathbf{x}}E(\eta)$ and $\mathbf{B}(\eta) = \hat{\mathbf{z}}B(\eta)$. The electron motion along the polarization direction is unaffected by the magnetic field. For an electron initially at rest at the origin, its displacement is

$$\alpha(t) = \int^t dt' A(\omega t')\,, \quad z(t) = \frac{1}{2c}\int^t dt' A^2(\omega t')\,, \tag{3}$$

where we have used $E(\omega t) = -dA(\omega t)/dt$. Since the velocity in the propagation direction is never negative, the drift in the propagation direction increases monotonically during the pulse. In particular, for a step pulse

$$\alpha(t) = \frac{\mathcal{E}_0}{\omega^2}\sin\omega t\,, \quad z(t) = \frac{\mathcal{E}_0^2}{2c\omega^2}\left(\frac{t}{2} + \frac{\sin 2\omega t}{4\omega}\right)\,. \tag{4}$$

The drift per cycle in the propagation direction can then be expressed in terms of $k = \omega/c$ as

$$\chi_0 \simeq \frac{\mathcal{E}_0^2}{c\omega^3} = \frac{1}{k}\beta^2\,. \tag{5}$$

This result is valid when i) the ratio of the ponderomotive energy of the electron to its rest mass is small, i.e. the electron dynamics are essentially non-relativistic, and ii) the drift of the electron in the propagation direction becomes comparable to the width of the electron wave packet. Setting $\beta = \mathcal{E}_0/\omega c$, we must therefore have

$$\frac{1}{4}\beta^2 \ll 1 < \frac{1}{ka}\beta^2\,, \tag{6}$$

where a is the width of the wavepacket along the propagation direction. This, in turn, implies that $ka \ll 1$. One recognizes this as the usual requirement for the applicability of the dipole approximation. However, for intense laser pulses this criterion leads to a long wavelength approximation in which both the electric field and the magnetic field of the laser are spatially homogeneous.

Taking $\omega = 0.057$ a.u., corresponding to the angular frequency of a Ti:Sapphire laser beam, and the width of the wavepacket to be one Bohr radius, magnetic field effects can become important when $I \simeq 1 \times 10^{15}$ W cm^{-2}, while for $\omega = 1$ a.u. intensities of approximately $I \simeq 5 \times 10^{18}$

W cm^{-2} are required. On the other hand, for $\omega = 0.057$ a.u. we see that $\beta^2 = 1$ when the intensity is $I = 2 \times 10^{18}$ W cm^{-2} while for $\omega = 1$ a.u., we have $\beta^2 = 1$ when $I = 7 \times 10^{20}$ W cm^{-2}.

3. Non-relativistic Volkov wave functions: beyond the dipole approximation

Let us now discuss a quantum mechanical description of an electron in a laser field using the time-dependent Schrödinger equation (TDSE). The goal is to obtain, in analogy with the classical case considered above, a non-dipole solution that accounts for to leading order in $1/c$ the magnetic field component of the laser pulse. Our starting point is the TDSE

$$i\frac{\partial}{\partial t}\Psi(\mathbf{r},t) = \frac{1}{2}\left[-\nabla^2 - 2i\mathbf{A}(\eta)\cdot\boldsymbol{\nabla} + A^2(\eta)\right]\Psi(\mathbf{r},t)\,, \tag{7}$$

containing the fully spatially dependent vector potential. The electron spin is neglected. Exact solutions to equation (7) exist [8]. However, we proceed by expanding the vector potential to lowest order in $1/c$:

$$\mathbf{A}(\eta) \simeq \mathbf{A}(\omega t) + \frac{z}{c}\mathbf{E}(\omega t)\,. \tag{8}$$

To leading order in $1/c$, the TDSE is now

$$i\frac{\partial}{\partial t}\Psi(\mathbf{r},t) = \left(\frac{1}{2}\left[-i\boldsymbol{\nabla} + \mathbf{A}(\omega t)\right]^2 + \frac{z}{c}\left[-i\boldsymbol{\nabla} + \mathbf{A}(\omega t)\right]\cdot\mathbf{E}(\omega t)\right)\Psi(\mathbf{r},t)\,. \tag{9}$$

Transforming to the length gauge, which is more appropriate for the low frequency regime considered in the next section, yields [9]

$$i\frac{\partial}{\partial t}\Psi^{\mathrm{L}}(\mathbf{r},t) = \left(-\frac{1}{2}\nabla^2 + \left[\mathbf{r} - i\frac{z}{c}\boldsymbol{\nabla}\right]\cdot\mathbf{E}(\omega t)\right)\Psi^{\mathrm{L}}(\mathbf{r},t)\,, \tag{10}$$

where $\Psi^{\mathrm{L}}(\mathbf{r},t) = \exp[i\mathbf{A}(\omega t)\cdot\mathbf{r}]\Psi(\mathbf{r},t)$. An exact solution to this equation can be obtained. This non-relativitic, non-dipole Volkov wave function reads

$$\Psi^{\mathrm{L}}_{\mathbf{p}}(\mathbf{r},t) = \frac{1}{(2\pi)^{\frac{3}{2}}}\exp\left(i\boldsymbol{\Pi}(\mathbf{p},t)\cdot\mathbf{r} - \frac{i}{2}\int^{t} dt''\left[\boldsymbol{\Pi}(\mathbf{p},t'')\right]^2\right)\,, \tag{11}$$

where

$$\boldsymbol{\Pi}(\mathbf{p},t) = \mathbf{p} + \mathbf{A}(\omega t) + \frac{1}{c}\left[\mathbf{p}\cdot\boldsymbol{A}(\omega t) + \frac{1}{2}A^2(\omega t)\right]\hat{\mathbf{z}}. \tag{12}$$

The wave function (11) reduces to the familiar dipole Volkov wave function for $1/c \to 0$.

The displacement of a wavepacket of Volkov waves is governed by Hamilton's equations, with the Hamiltonian taken to be classical analogue of the the Hamiltonian operator in equation (9). The solutions to these equations are

$$\alpha(t) = \int^t dt' \left[A(\omega t') + \frac{1}{c} E(\omega t') z(t') \right], \quad z(t) = \frac{1}{2c} \int^t dt' A^2(\omega t') \quad (13)$$

Note that here, as in equation (11), terms of order $1/c^2$ appear. To order $1/c$ these results agree with equations (3) obtained above.

4. Harmonic Generation Beyond the Dipole Approximation

In this section we discuss a non-relativistic, non-dipole version of the Lewenstein model of harmonic generation [1,5]. This model is valid in the regime where the inequalities of equation (6) are satisfied. The model is used to study photon emission by light hydrogenic ions driven by Ti:Sapphire pulses [10]. Positive ions can survive higher peak intensities than neutrals, and can therefore emit more energetic photons, due to their larger ionization potentials [11,12]. At high intensities, as will be demonstrated below, the Lorentz force can strongly influence the photon emission process [10,13,14].

In the non-relativistic model of Lewenstein and co-workers, the dipole moment of the atom can be expressed as

$$\mathbf{d}(t) \simeq -\int_{-\infty}^{t} dt' \int d\mathbf{r} d\mathbf{r}' \phi_0^*(\mathbf{r}, t) \mathbf{r} G_{\mathrm{V}}^{(+)}(\mathbf{r}, t; \mathbf{r}', t') \mathbf{E}(\omega t) \cdot \mathbf{r} \phi_0(\mathbf{r}', t') + \text{c.c.} \quad (14)$$

Here $G_{\mathrm{V}}^{(+)}(\mathbf{r}, t; \mathbf{r}', t')$ is the Volkov Green's function associated with the Hamiltonian describing a free electron in the laser field. The atom or ion is initially in an eigenstate of the field-free Hamiltonian, given by the wave function $\phi_0(\mathbf{r}, t) = \phi_0(\mathbf{r}) \exp(I_p t)$, with I_p being the ionization potential of the state. Continuum-continuum transitions are neglected. Equation (14) has been analyzed in detail within the dipole approximation [1,5].

This model can be readily generalized to include the influence of the magnetic field component of the laser pulse. Using the non-dipole Green's function, constructed from the Volkov wave functions (11), and the non-dipole laser-atom interaction Hamiltonian $H_{\mathrm{int}}(t)$ (see equation 10) in expression (14) above, one obtains

$$\mathbf{d}(t) \simeq -\int_0^t dt' \int d\mathbf{p}\, \mathbf{d}_{\mathrm{rec}}[\mathbf{\Pi}(\mathbf{p}, t)] \exp[-iS(\mathbf{p}, t, t')]\, d_{\mathrm{ion}}[\mathbf{\Pi}(\mathbf{p}, t'), t'] + \text{c.c.} \; (15)$$

The dipole moments $\mathbf{d}_{\mathrm{rec}}(\mathbf{q})$ and $d_{\mathrm{ion}}(\mathbf{q}, t)$ are

$$\mathbf{d}_{\mathrm{rec}}(\mathbf{q}) = \int \frac{d\mathbf{r}}{(2\pi)^{\frac{3}{2}}} e^{-i\mathbf{q}\cdot\mathbf{r}} \mathbf{r} \phi_0(\mathbf{r}), \quad d_{\mathrm{ion}}(\mathbf{q}, t) = \int \frac{d\mathbf{r}}{(2\pi)^{\frac{3}{2}}} e^{-i\mathbf{q}\cdot\mathbf{r}} H_{\mathrm{int}}(t) \phi_0(\mathbf{r}), \; (16)$$

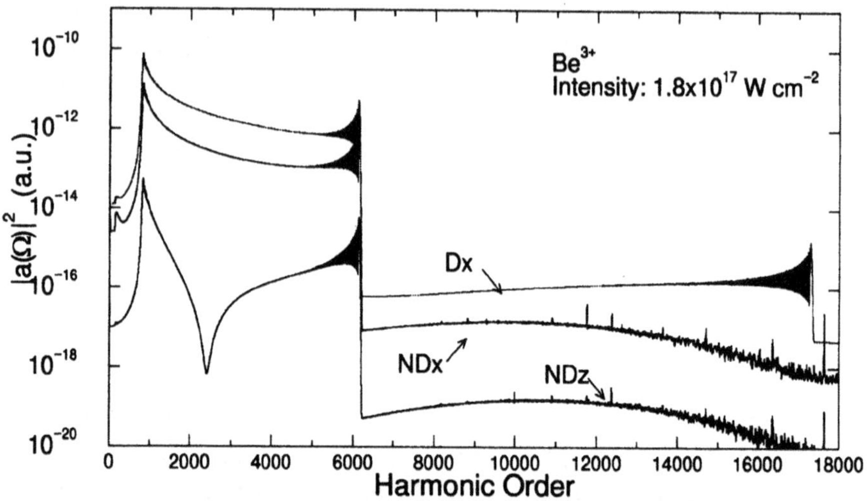

Figure 1. The magnitude squared of the Fourier transform of the dipole acceleration (in a.u.) of Be^{+3} as a function of the photon energy (in units of $\hbar\omega$). The peak intensity is $\mathcal{E}_0^2 = 5.1$ a.u. Dipole results (Dx) are displayed, as well as non-dipole results for photons polarized along $\hat{\mathbf{x}}$ (NDx) and $\hat{\mathbf{z}}$ (NDz), respectively.

and the quasi-classical action $S(\mathbf{p}, t, t')$ is

$$S(\mathbf{p}, t, t') = \frac{1}{2}\int_{t'}^{t} dt'' \left[\mathbf{\Pi}(\mathbf{p}, t'')\right]^2 + I_p(t - t'). \tag{17}$$

The integration over $\mathbf{p}$ can be carried out using the saddle point method. The integral over t' is performed numerically, and the photon emission spectrum then obtained by calculating $|\hat{\mathbf{x}} \cdot \mathbf{a}(\Omega)|^2$ and $|\hat{\mathbf{z}} \cdot \mathbf{a}(\Omega)|^2$, where $\mathbf{a}(\Omega)$ is the Fourier transform of $\ddot{\mathbf{d}}(t)$. Our results have been obtained for a two cycle pulse with carrier wavelength $\lambda = 2\pi c/\omega = 800$ nm that is described by the vector potential

$$\mathbf{A}(\eta) = \hat{\mathbf{x}}\,(\mathcal{E}_0/\omega)\sin^2(\eta/4)\sin(\eta)\,. \tag{18}$$

In figure 1 results are shown for Be^{3+} with $\mathcal{E}_0^2 = 5.1$ a.u. In order to illustrate more clearly the differences between the dipole and non-dipole results, the strong oscillations in the spectra have been removed by convoluting with a Gaussian window function. As is seen in the figure, a marked decrease in the emission efficiency of photons polarized along $\hat{\mathbf{x}}$ occurs. This is due to the Lorentz force pushing the returning electron wavepacket in the propagation direction so that it "misses" the ionic core. (This situation has analogies with the dramatic decrease in photon emission that occurs when the ellipticity of the driving laser field is increased [15].) Differences between the dipole and non-dipole results are largest at the highest photon energies, where a pronounced "bending over" of the spectra is seen. This effect has also been found by Milošević et al. [14] who have calculated harmonic generation rates using a model based on the Klein-Gordon equation. Our results show that this is not a relativistic effect. Indeed, for $\mathcal{E}_0^2 = 5.1$ a.u., the relativistic parameter $\beta^2 = 8 \times 10^{-2}$ is still relatively small, and corrections of order $1/c^2$ that are not accounted for in this model should not be significant. We have also found that photon emission saturates as the peak intensity is increased from 1.8×10^{17} W cm^2 to 3.6×10^{17} W cm^2 [10]. This effectively sets a practical limit on XUV photon generation using ions.

5. Non-dipole effects in the stablization regime

We now discuss the influence of the magnetic field component of an intense laser pulse on stabilization in the high frequency regime. As mentioned above, non-dipole effects can strongly modify the stabilization dynamics [16–18]. We study a model 2D atomic system [19] (see also Ref. [20]) governed by the TDSE

$$i\frac{\partial}{\partial t}\Psi(\mathbf{r},t) = \left(\frac{1}{2}\left[\mathbf{p}+\mathbf{A}(\eta)\right]^2 + V(r)\right)\Psi(\mathbf{r},t)\,, \tag{19}$$

with $\mathbf{r} = (x, z)$. A soft-core Coulomb potential, $V(r) = -1/\sqrt{r^2+a^2}$, is used, with $a = 0.80$ a.u. so that the ground state has the same binding energy as hydrogen. The laser field is described classically by the fully spatially dependent vector potential

$$\mathbf{A}(\eta) = \frac{\mathcal{E}_0}{\omega}\hat{\mathbf{x}} f(\eta)\sin\eta\,. \tag{20}$$

The function $f(\eta)$ is the laser pulse envelope, which is taken to be trapezoidal. Equation (19) is then numerically integrated on an uniform grid. The probability that the atom does not ionize during the laser pulse is determined by projecting the wave function at the end of the laser pulse onto the bound field-free states of the atom. We also performed calculations in the dipole approximation by setting $\eta = \omega t$ in equation (20) for the vector potential.

In the upper plots of figure 2, the ionization probability is shown as a function of the peak electric field strength of the laser, $\mathcal{E}_0$, for angular frequencies of $\omega = 1$ a.u. (left) and $\omega = 2$ a.u. (right). As indicated on the plots, the solid lines are non-dipole results (ND) and broken lines are results in the dipole approximation (D). A trapezoidal pulse with three cycles turn-on (turn-off) and six cycles constant intensity was used. From the results, it is seen that the dipole approximation breaks down when $\mathcal{E}_0 > 10$ a.u. for

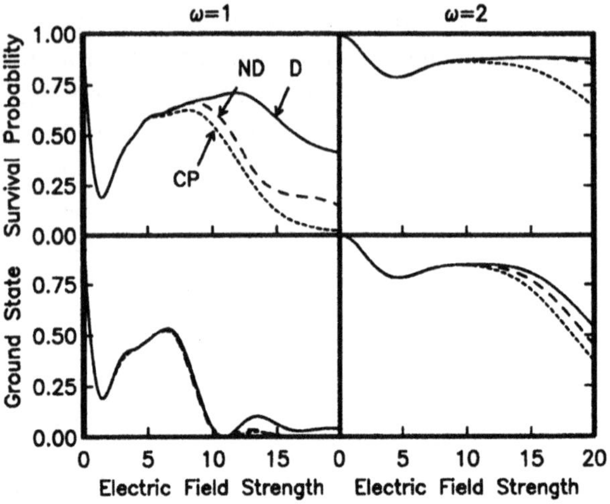

Figure 2. The upper plot shows the survival probability as a function of the maximum electric field strength $\mathcal{E}_0$ (in a.u.) of the laser. On the left, results are shown for an angular frequency of $\omega = 1$ a.u. while on the right $\omega = 2$ a.u. A trapezoidal pulse with a 3 cycle turn-on (turn-off) and 6 cycles constant intensity was used. In the lower plot, the population in the ground state is shown. The dipole (D) results are given by the solid curves, non-dipole (ND) results are given by the broken curves while the dotted curves (CP) correspond to results for two counter propagating pulses.

$\omega = 1$ a.u. However, for $\omega = 2$ a.u. it remains a good approximation up to $\mathcal{E}_0 = 20$ a.u., the maximum field strength considered here.

In the lower plots of figure 2, the populations in the ground state at the end of the pulse are shown. For a 12 cycle pulse and $\omega = 1$ a.u., nearly all of the surviving population returns to the ground state when $\mathcal{E}_0 < 5$ a.u. This is an indication of the so-called "adiabatic stabilization" of the atom. When $\mathcal{E}_0 > 10$ a.u., the evolution of the system is highly non-adiabatic and in the dipole approximation there is significant population transfer ("shake up") to the excited states. When $\omega = 2$ a.u., adiabaticity is maintained up to $\mathcal{E}_0 \simeq 10$ a.u.

The laser-atom interaction dynamics in the high-frequency, high-intensity regime depends strongly on the duration of the pulse. If the pulse is ramped on sufficiently slowly, the atom can be described, in the first approximation within the framework of the HFFT, by a single Floquet state [6]. However, when the pulse duration is very short and $\mathcal{E}_0$ large, a strong field approximation (SFA) can be used to describe the evolution of the system [7]. In this regime, ionization is governed primarily by wave packet spreading. (For a further discussion of the SFA stabilization regime, see the contribution of Doerr, Kylstra and Potvliege in this volume.) Between these two regimes, ionization and excitation occur via "shake-up" and "shake-off" processes that can be understood within the HFFT in terms

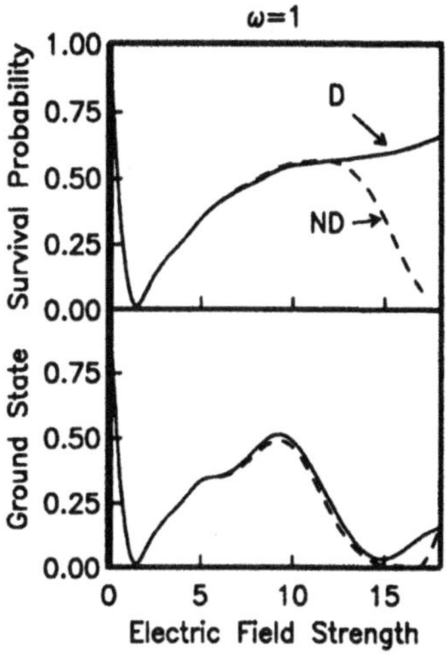

Figure 3. As in figure 2, however a trapezoidal pulse with a 6 cycle turn-on (turn-off) and 10 cycles constant intensity was used. The angular frequency is $\omega = 1$ a.u.

of non-adiabatic transitions occurring between bound and continuum Floquet eigenstates. In this intermediate regime, the initial evolution of the wavepacket is essentially that of a free wavepacket as long as the width of the electron wavepacket along the polarization direction is much smaller than the classical electron excursion amplitude [21]. The breakdown of stabilization seen in figure 2 occurs in this intermediate non-adiabatic regime. Therefore, the breakdown of the dipole approximation at relatively moderate intensities, and the improvement of the dipole approximation with increasing frequency can be understood by considering the motion of a classical electron in the laser field. From equation (5) above, the dipole approximation is expected to fail when the drift of the electron per cycle in the propagation direction, $\chi_0 \simeq \mathcal{E}_0^2/c\omega^3$, becomes comparable to the width of the initial wavepacket. This is borne out by our calculations. We have also performed calculations for a longer pulse consisting of a 6 cycle turn-on (turn-off) and 10 cycles constant intensity. These results are shown in figure 3. For this pulse, the atom evolves adiabatically when $\mathcal{E}_0$ is less than about 8 a.u., and dramatic differences between the dipole and non-dipole results now appear for $\mathcal{E}_0$ larger than 12 a.u. We stress that for the laser frequencies and intensities considered here, the dynamics of the electron in the laser field is essentially non-relativistic. Indeed, for the highest field strength considered, $\mathcal{E}_0 = 20$ a.u., the relativistic parameter β^2 is 2×10^{-2}.

To lowest order in $1/c$, the detrimental effect of the magnetic field can be eliminated by having the atom interact with two counter propagating laser pulses. We have calculated the time evolution of the initial ground state of our two-dimensional model atom interacting with two such counter-propagating pulses. In this case, magnetic field effects still strongly influence the stabilization dynamics. From figure 2, it is seen that in fact the counter propagating pulses (labeled CP) result in survival probabilities that are smaller than when a single pulse is used. Classical Monte Carlo simulations give similar results [19].

6. Acknowledgements

This work has been supported by the UK Engineering and Physical Sciences Research Council (EPSRC) and the Spanish Dirección General de Enseñanza Superior e Investigación Científica (grant PB98-0268).

References

1. Salières, P., L'Huillier, A., Antoine, P. and Lewenstein, M. (1998) *Adv. At. Mol. Phys.* **41**, 83
2. Brabec, T. and Krausz, F. (2000) *Rev. Mod. Phys.* **72**, 545
3. Protopapas, M., Keitel, C. H. and Knight, P. L. (1997) *Rep. Progr. Phys.* **60**, 389; Joachain, C. J., Dörr, M. and Kylstra, N. J. (2000) *Adv. At. Mol. Opt. Phys.* **42**, 225
4. Corkum, P. B. (1993) *Phys. Rev. Lett.* **71**; Kulander, K. C., Schafer, K. J. and Krause, J. L. (1993) *Super-Intense Laser-Atom Physics*, ed B. Piraux, A. L'Huillier and K. Rzążewski (New York: Plenum Press) p. 95
5. Lewenstein, M., Balcou, Ph., Ivanov, M. Yu., L'Huillier, A. and Corkum, P. B. (1994) *Phys. Rev.* A **49**, 2117
6. Pont, M. and Gavrila, M. (1990) *Phys. Rev. Lett.* **65**, 2362; Gavrila, M. (1997) *Photon and Electron Collisions with Atoms and Molecules*, ed. P. G. Burke and C. J. Joachain (New York: Plenum Press) p. 147
7. Patel, A. S., Kylstra, N. J. and Knight, P. L. (1999) *J. Phys. B* **32**, 5759
8. Bergou, J. and Varro, S. (1981) *J. Phys. A* **14**, 2281; Druhl, K. and McIver, J. K. (1983) *J. Math. Phys.* **24**, 705
9. Reiss, H. R. (2000) unpublished
10. Kylstra, N. J., Potvliege, R. M. and Joachain, C. J. (2000) unpublished
11. Kulander, K. C., Schafer, K. J. and Krause, J. L. (1992) Phys. Rev. Lett. **68**, 3535
12. Potvliege, R. M., Kylstra, N. J. and Joachain, C. J. (2000) *J. Phys. B* **33**, L743
13. Kim, A. V., Ryabikin, M. Yu. and Sergeev, A. M. (1999) *Usp. Fiz. Nauk.* **169**, 58 [(1999) *Phys. Usp.* **42**, 54]; Taranukhin, V. D. (2000) *Laser Phys.* **10**, 330
14. Milosević, D. B., Hu, S. X. and Becker, W. (2000) unpublished
15. Corkum, P. B., Burnett, N. H. and Ivanov, M. Yu. (1994) *Opt. Lett.* **19**, 1870
16. Grochmalicki, J., Lewenstein, M., and Rzążewski, K. (1991) *Phys. Rev. Lett.* **66**, 1038; Katsouleas, T. and Mori, W. B. (1993) *Phys. Rev. Lett.* **70**, 1561
17. Keitel, C. H. and Knight, P. L. (1995) *Phys. Rev. A* **51**, 1420
18. Bugacov, A., Pont, M. and Shakeshaft, R. (1993) *Phys. Rev. A* **48**, R4027; Latinne, O., Joachain, C. J. and Dörr, M. (1994) *Europhys. Lett.* **26**, 333; Potvliege, R. M. (2000) *Laser Phys.* **10**, 143
19. Kylstra, N. J., Worthington, R. A., Patel, A., Knight, P. L., Vázquez de Aldana, J. R. and Roso, L. (2000) *Phys. Rev. Lett.* **85**, 1835
20. Vázquez de Aldana, J.R., and Roso, L., (1999) *Opt. Express* **5**, 144; (2000) *Phys. Rev. A* **61**, 043403
21. Grobe, R. and Fedorov, M. (1992) *Phys. Rev. Lett.* **68**, 2592; *ibid.* **69**, 3591(E)

A PICTUREBOOK OF RELATIVISTICALLY DRIVEN WAVEPACKETS

Ernst Lenz, Martin Dörr and Wolfgang Sandner

Max-Born-Institut, D-12489 Berlin, and
Technische Universität Berlin
mailto: doerr@mbi-berlin.de
http://mitarbeiter.mbi-berlin.de/doerr

1. INTRODUCTION AND THEORY

We consider the motion of wavepackets in the one-dimensional Dirac equation (D), the relativistic Schrödinger equation (R), and the non-relativistic Schrödinger equation (N). We consider wavepackets being driven by an extremely strong and short field pulse and interacting with a deep potential well. Such field intensities are not totally out of reach when one considers deeply bound ionic states interacting (possibly colliding) with an ultrashort and ultrastrong laser pulse. The dynamics of the free wavepacket in the relativistic regime is itself interesting: since relativity constrains the speed to be less than the speed of light c, spreading is inhibited by field driving [2, 3]. At turning points of the motion, however, spreading sets in. The non-relativistic Schrödinger equation exhibits superluminal spreading. For free wavepackets, the Dirac equation is equivalent to the R equation (Appendix B). Both can of course be solved "analytically", using the plane wave eigenfunctions [12]. We highlight the numerical pitfalls to be avoided when studying the negative energy states of the Dirac equation.

During interaction with a potential, we consider on the one hand the scattering of a wavepacket from a potential (well or barrier) and on the other hand the rapid ionization of a bound electron by a strong laser pulse. We also discuss scattering of such wavepackets from a potential, both a deep well (an atomic core) and a high barrier, allowing for "Klein tunneling" [4]. We do not include multi particle effects such as the trident process occurring during ultrarelativistic electron scattering from a potential. In the ionization problem, the potential is a model for a very deeply bound ion. The laser pulse, although very short, is longer than the natural orbit time of

B. Piraux and K. Rzążewski (eds.), Super-Intense Laser-Atom Physics, 355–368.

the particle bound in the potential. We consider very narrow wavepackets, both for the free wavepacket propagation and for the initial bound states.

Several groups have studied relativistic wave equations in the context of strong laser fields and we do not intend to give an overview of the field here. The relevant references up to 1999 can be found in the review [5]. See also the paper [6], concerned with non-dipole effects which are very important when more than one spatial dimension is considered.

The non-relativistic form of the Schrödinger equation (N) is [7]

$$i\partial_t \Psi(x,t) = \left[(p - A(t))^2/2m + V(x)\right] \Psi(x,t). \tag{1}$$

Its relativistic counterpart, the relativistic Schrödinger equation (R) is obtained by inserting the relativistic energy term instead of the kinetic energy term in the N equation

$$i\partial_t \Psi(x,t) = \left[\{(p - A(t))^2 c^2 + m^2 c^4\}^{1/2} + V(x)\right] \Psi(x,t), \tag{2}$$

where $V(x)$ is the local binding potential and $A(t)$ is the time dependent vector potential of the external, strong field pulse. Since we restrict ourselves to one dimension, the dipole approximation is implied, and thus $A(t)$ is independent of x. Alternatively to the form given above, equations (1) or (2) can be transformed to the length gauge, removing the $A(t)$ and adding a term $F(t)\cdot x$, where the electric field is $F(t) = -\partial_t A(t)$.

The numerical solution of the N and the R equations is performed by a standard grid method, with equidistant spatial points, possibly using an absorbing potential at the grid boundaries. The time propagation for the N equation is performed by the standard Crank-Nicolson scheme, using the three-point discretization of the Laplacian. The R equation, on the other hand, due to its kinetic energy term, demands a treatment commonly termed split-operator, involving Fourier transformation to momentum space (see [8] and references therein). We have also implemented the split-operator propagation of the N equation.

The other relativistic wave equation we consider in the present work is the Dirac equation (D), which in its one-dimensional form reads

$$i\partial_t \Psi(x,t) = \left[\alpha c(p - A(t)) + \beta mc^2 + \mathbf{1}V(x)\right] \Psi(x,t). \tag{3}$$

We use the representation of the matrices

$$\alpha = \begin{pmatrix} 0 & -i \\ i & 0 \end{pmatrix}, \quad \beta = \begin{pmatrix} 1 & 0 \\ 0 & -1 \end{pmatrix}. \tag{4}$$

In one dimension only a two-component spinor must be considered (see Appendix A: The 1D Dirac equation). Its two components correspond to the two energy solutions of the equation, since spin plays no role in the 1D equation. The positive energy solutions are the particles (electrons) and the negative energy solutions are the antiparticles (positrons). The two components of the spinor wavefunction are called the upper and the lower component, respectively. Instead of the form (3), we use the length gauge, reached by a gauge transformation equivalent to the one for the Schrödinger equation [9].

The numerical solution of the Dirac equation is performed in a way similar to the solution of the N equation; we propagate the wavefunction with the Crank-Nicolson method on a staggered spatial grid for the two components. The spatial derivative is again approximated by the lowest order finite difference formula. This naturally and straightforwardly avoids the problem of fermion doubling [10, 11].

2. FREE WAVEPACKETS

2.1 Free wavepackets without field

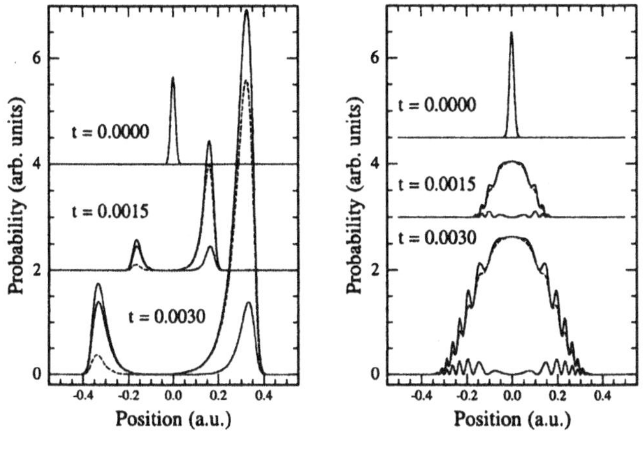

Figure 1. Free wavepackets in the Dirac equation. Solid upper lines: total probability density (sum of the upper and lower component probabilities), dashed lines: upper component, solid lower lines: lower component. The initial width of the wavepackets for both figures is 0.02 a.u. Each figure displays three frames from the time evolution, shifted in the vertical direction from each other. The initial wavepacket has only a Gaussian upper and no lower component. The mean kinetic energy is 10000 a.u. for the left figure and 0 for the right figure. The initial wavefunction is multiplied by a factor of 0.05 and the wavefunction at time 0.0015 is multiplied by a factor of 0.2. The timestep required is dt= 5×10^{-7}. The spatial discretization is dx= 6×10^{-4} for the right figure and dx= 6×10^{-5} for the left figure at larger $E_{\rm kin}$.

We solve the time dependent Dirac equation (D) numerically, using a particular wavepacket as an initial state. In figure 1, we illustrate some numerical pitfalls. The left portion depicts the solution for an initial wavepacket at kinetic energy $E_{\text{kin}} = E - mc^2 = 10^4$ a.u. that is just a Gaussian in the upper component [8]. We see that such a wavepacket contains a larger positive energy component (moving to the right) and a smaller (by the factor $p/(E + mc^2)$) negative energy component moving to the left. When the same initial wavepacket is chosen for a particle at rest (right portion of figure 1), then an oscillatory structure appears on the spreading wavepacket. This can be termed a manifestation of the Zitterbewegung. However, we simply view it as a consequence of an improperly chosen initial wavepacket. The wavepacket must instead be constructed according to [16, 12]

$$\psi(x) = \int dk\, G(k, k_0, x_0, w)\, \tilde{\psi}(k). \tag{5}$$

Here $\tilde{\psi}(k)$ is the eigenfunction corresponding to momentum k and the weight function $G(k, k_0, x_0, w)$ is Gaussian, centered on momentum k_0 and position x_0, with width w in position. However, even when the initial wavepacket is constructed according to the exact equation 5, on a log scale (figure 2) one can still see a negative energy wavepacket moving off to the left. This is caused by discretization error due to the finite step size used in the numerical calculation and the probability in this negative energy part decreases with x-stepsize dx as dx^2.

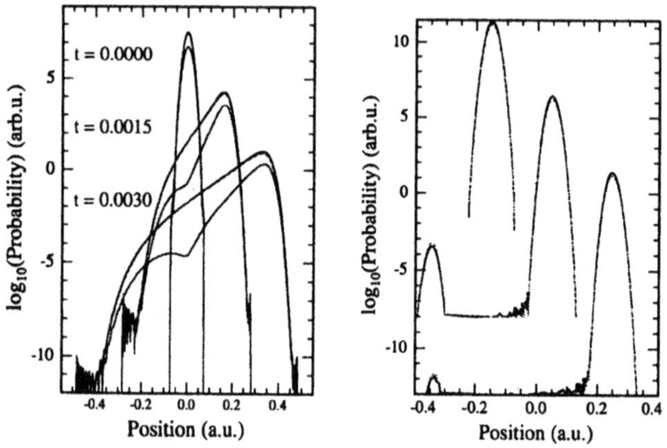

Figure 2. Probability density as a function of position for the propagation of a free Dirac wavepacket with mean kinetic energy $E = 10^4$ a.u. for the left graph and $E = 6 \times 10^4$ a.u. for the right graph. The initial wavepackets have been computed from equation 5. Top graphs: initial wavepacket at $t = 0$ of width $w = 0.06$; middle graphs: at $t = 0.0015$ a.u.; bottom graphs: at $t = 0.003$ a.u. The graphs at different times have been shifted in the y-direction. Solid line: upper component; dashed line: lower component.

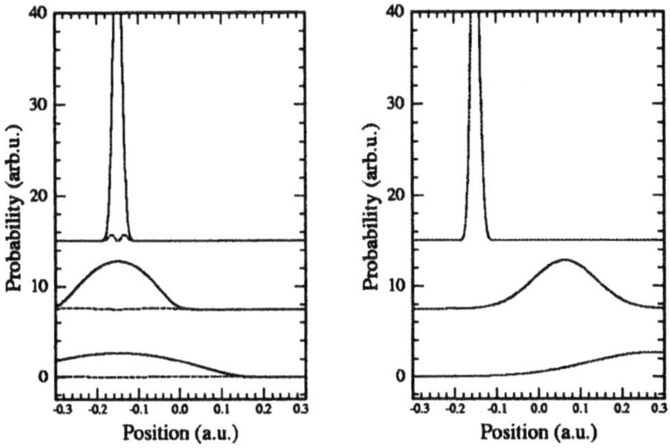

Figure 3. Probability density as a function of position for the propagation of a free wavepacket. Top graphs: initial Gaussian at $t = 0$ of width $w = 0.02$; middle graphs: at $t = t_{max}/2$; bottom graphs: at $t = t_{max}$. The graphs at different times have been shifted in the y-direction. Final time $t = t_{max} = 0.003$ a.u. Left graphs: Dirac equation, mean kinetic energy $E = 0$. Solid line: upper component; dashed line: lower component. Right graphs: Schrödinger equation, mean kinetic energy $E = 10000$ a.u.

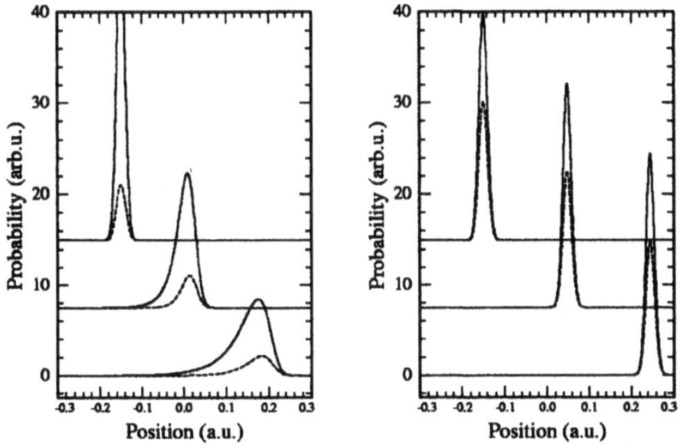

Figure 4. Probability density as a function of position for the propagation of a free Dirac wavepacket with mean kinetic energy $E = 10^4$ a.u. for the left graph and $E = 6 \times 10^4$ a.u. for the right graph. Top graphs: initial Gaussian at $t = 0$ of width $w = 0.02$; middle graphs: at $t = t_{max}/2$; bottom graphs: at $t = t_{max}$. The graphs at different times have been shifted in the y-direction. Solid line: upper component; dashed line: lower component. Final time $t = t_{max} = 0.003$ a.u.

In figures 3 and 4 we explore the effect of relativistic suppression of wavepacket spreading [2, 3]. The explanation of this effect is simply classical: if the speed of the wavepacket approaches the speed of light, then all of its energy components move at the same speed and the spreading is inhibited. Figure 3, left hand panel, shows the spreading for a D wavepacket that is initially at rest. The initial wavepacket width is only 0.02 a.u. When compared to the N solution (not shown) one notes that the agreement between D and N is rather good. When a wavepacket width of 0.01 is chosen (not shown in the figure, but in a figure in [13]) then the difference between the D and N solutions is much more apparent since the energy tails then extend to relativistic energies. The right hand panel of figure 3 shows the time evolution for a N wavepacket for an initial kinetic energy of 10^4 a.u. The left hand panel of figure 4 gives the corresponding D results. There is a significant difference between the D and N solution since the spreading of the N solution is not limited by the speed of light c. The right panel of figure 4 shows that for an ultrarelativistic kinetic energy of 6×10^4 a.u. the spreading for the D solution has ceased to be visible on the graph.

It must be noted that in all cases discussed above, the R solution gives identical results to the D solution as far as the total probability density is concerned, as can be inferred from the discussion in Appendix A. It is therefore unnecessary to solve the Dirac equation for the propagation of a free wavepacket that does not contain negative energy components (or, alternatively, for a wavepacket that *only* contains negative energy components).

2.2 Free wavepackets with an external field pulse

In this subsection we consider a free wavepacket, with some definite initial kinetic energy, subject to a single-cycle field pulse, defined as

$$F(t) = F_0 \frac{1}{\tau} \sin\left(\frac{2\pi t}{\tau}\right) \sin^2\left(\frac{\pi t}{\tau}\right). \tag{6}$$

Here $F(t)$ denotes the electric field, related to the vector potential $A(t)$ by $F(t) = -\partial_t A(t)$ (we have omitted the extra factor c in our definition of A).

Figure 5 compare the R and D solutions for a wavepacket driven by a field pulse (in dipole approximation). The wavepacket is accelerated very quickly to relativistic energies, but the field is not strong enough to generate negative energy wavepackets in the D equation visible on the linear scale. Thus, the R solution agrees perfectly with the D solution (the sum of the upper and lower component probability densities). The arrows on the middle plots indicate the direction of motion of the wavepackets after the first half of the pulse. It must be noted that for the present propagation

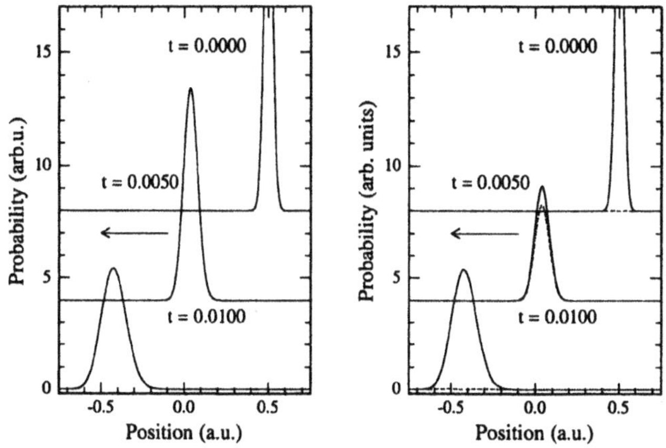

Figure 5. Time evolution of a wavepacket driven by a single-cycle laser pulse. The initial wavepacket has a width of 0.05 a.u. and a mean kinetic energy 0. The field pulse duration is $\tau = 0.01$ a.u. and the vector potential peak amplitude $A_0 = 2 \times 10^3$ a.u. Left graphs: R, right graphs: D. Right graph: solid line: upper component, dashed line: lower component. The discretization used is dx= 10^{-4} and dt= 2×10^{-7}.

method a tiny stepsize in time is necessary. For a too large dt, negative energy wavepackets propagate at an incorrect speed, since in the numerical implementation we subtract mc^2 from the diagonal of the Hamiltonian.

3. INTERACTION WITH A POTENTIAL

3.1 Potential scattering without field

In figure 6 we show the results from scattering a wavepacket off a deep well or a high barrier. In both cases, we see that at the high energies considered, the major part of the wavepacket is either simply transmitted (over a well, left) or reflected (from a barrier, right). However, for the well (left), there is a reflected part, with a double hump structure due to scattering off the two sides of the potential. One notes the difference between the R and D solutions: in the R case, the reflected probability, although small, is a factor 10^2 larger than in the D case.

For the tunneling under the high barrier (right), the D solution exhibits a much larger tunneling probability than the R solution, since Klein tunneling occurs through (virtual) negative energy states inside the barrier in the D equation [4, 14]. The negative energy states inside the barrier are evident since the lower component (dashed) is larger than the upper component (solid). The corresponding tunneling probability in the R equation is about a factor 10^{-3} lower than in the D equation.

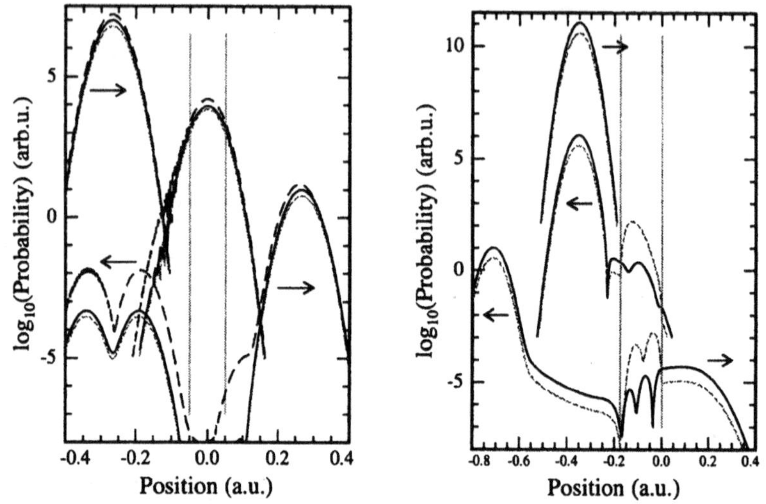

Figure 6. Scattering of a wavepacket by a static potential. The position of the potential is indicated by the vertical lines. Left part: initial kinetic energy $E = 5 \times 10^4$ a.u., initial wave packet width $w = 0.05$ a.u., potential width 0.1 a.u. and depth $d = 6 \times 10^4$ a.u. Upper curves at $t = 0$, middle at $t = 0.002$, bottom at $t = 0.004$ a.u. Solid lines: upper component of Dirac solution, short-dashed lines: lower component of Dirac solution, long-dashed lines: relativistic Schrödinger solution. The arrows indicate the direction of motion. Right part: initial kinetic energy $E = 2 \times 10^4$ a.u., initial wave packet width $w = 0.05$ a.u., barrier width 0.17 a.u. and height $d = 5 \times 10^4$ a.u. Upper curves at $t = 0$, middle at $t = 0.003$, bottom at $t = 0.006$ a.u. In the right figure only the upper (solid) and lower (dashed) components of the D solution are shown.

3.2 Ionization from a deep well

Let us now consider the physically important case of an electron that is initially bound in a deep potential well (as for example the ground state of a highly-charged ion) that is being ionized by a short field pulse. Since we are only interested in the ionization dynamics, we can restrict ourselves to the same single-cycle pulse as we have used above. Furthermore, any laser frequency will be much lower than the binding frequency for a highly charged ion. Although we shall be using an ultrashort field pulse of total duration 0.01 a.u., it must be stressed that the pulse duration is much longer than the orbit time of the particle bound in the deep well, the latter time being of the order of 0.002 a.u. for the potential we are considering, which is given by a smooth "square" well

$$W(x) = -\left[\left(\frac{2x}{w}\right)^{40} + \frac{1}{d}\right]^{-1}. \qquad (7)$$

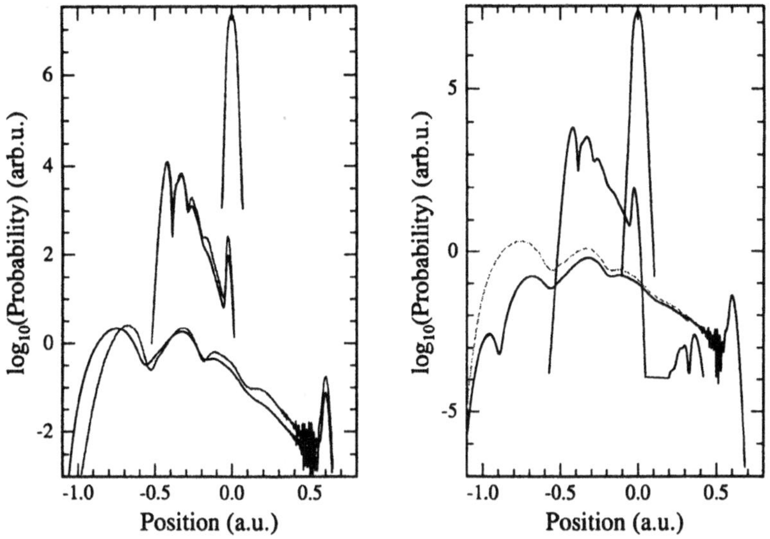

Figure 7. Ionization from a deep well. Potential width $w = 0.01$ and depth $d = 10^4$. Field pulse duration $\tau = 0.01$ and peak strength $A_0 = 2 \times 10^3$. Left part: comparison of D and R solutions, total probabilities are shown. Top: bound state at $t = 0$. Middle: wavepacket after the first half-cycle pulse, at $t = 0.005$ a.u. Bottom: wavepacket at end of single-cycle pulse, at $t = 0.01$ a.u. Right part: D solution, as in left part, on an expanded log scale; dashed line: upper component, solid line: lower component.

We are therefore in the *low-frequency* regime for our choice of pulse duration, and the ionization occurs nearly adiabatically, as in a slowly turned on static field. In figure 5, after the first half-pulse the atom is mostly ionized and the structure visible in the ionized wavepacket tells of the interplay between ionization and bound-state dynamics: the ionized probability is modulated by the oscillations of the wavepacket in the potential. At the end of the pulse, the ionized portion around $x = -0.5$ has been decelerated towards the right, and the small residual population that had not been ionized during the first half-pulse is ejected (visible at $x = 0.6$) [13]. During ionization, also a small negative-energy wavepacket is formed, visible in the right-hand plot at $t = 0.005$ and $x = 0.3$.

4. CONCLUSIONS AND INTERPRETATION

We have presented the results from the numerical solution of the time dependent Dirac equation in one dimension. For a free wavepacket, the relativistic Schrödinger equation is equivalent to the Dirac equation. For ionization by a laser pulse, the R equation is also in very close agreement with the D equation. Small differences are visible due to the slight difference in the bound state in the two equations. With proper care, negative energy wavepackets can be identified that are created by the ultrastrong field during the driving process. Such wavepackets occur when the field intensity becomes close to (but still significantly below) the critical vacuum polarization field strength.

Pair creation is a multiparticle process. Thus, in order to describe this process, in principle a second quantization or field theoretical formalism must be used. The equation we are using in the present work, however, is the single particle ("first quantization") Dirac equation. The time evolution of this equation is strictly unitary (apart from numerical tricks such as absorbing boundaries, which are not relevant for the present discussion). Nevertheless, the single particle Dirac equation can be invoked in certain cases to compute pair creation probabilities. In particular, there is a large literature on the solution of the Klein paradox using the Dirac equation [4]. There is also substantial amount of work on the pair creation probability in the collision of highly charged ions [10]. In all cases, a reinterpretation of the negative energy currents is required. Sometimes, the "Dirac sea" picture is invoked: positrons – the electrons' antiparticles – are nothing but vacancies in the "filled sea" of negative energy levels. This explanation has its use for Fermions; in the present one-dimensional case, however, spin plays no role, fermionic statistics are not enforced by our single particle Dirac Hamiltonian. And in fact the Klein paradox appears also in the solution of the Klein-Gordon equation (for bosons). Thus the Dirac sea picture is of no help here. What does shed light, however, is the Feynman picture that the antiparticles move backwards in time. Thus a negative energy wavepacket which we follow during its time evolution in the single particle Dirac equation can be reinterpreted as an antiparticle (positron) moving in the opposite time direction.

Many open questions remain, however: if the Klein paradox is correctly described by the D results, interpreted as pair emission by the barrier, then how could this pair emission be restricted to the time the wavepacket is at the barrier, as happens in time dependent wavepacket studies [15, 8] ? These issues and the ones related to the pair creation during ionization by a time-dependent laser pulse must be addressed in future work.

A. Appendix: The 1D Dirac Equation

When only one-dimensional motion is allowed and the momenta in the other two spatial dimensions are set to 0, then the Dirac equation can be represented by a two-component spinor. The two spinor components are called the upper and the lower component. Consequently, the spin degree of freedom of the Dirac particle plays no role.

The Dirac equation in three dimensions for a free particle reads:

$$E\Psi(x) = H\Psi(x) = \left\{ c\sum_{i=1}^{3} \alpha_i \, \mathbf{p}_i + \beta mc^2 \right\} \Psi(x) \tag{8}$$

It is well known [16] that in this equation α_i and β are 4×4 matrices and $\Psi(x)$ is a 4-dimensional vector valued function. The matrices α_i and β obey the following relations:

$$\alpha_i \, \alpha_k + \alpha_k \, \alpha_i = 2\delta_{ik}\mathbf{1} \tag{9}$$

$$\alpha_i \, \beta + \beta \, \alpha_i = \mathbf{0} \tag{10}$$

$$\beta^2 = \mathbf{1} \tag{11}$$

In one dimension the Dirac equation for a free particle reads :

$$E\Psi(x) = H\Psi(x) = \left\{ c\alpha_1 \, p_1 + \beta mc^2 \right\} \Psi(x) \tag{12}$$

The conditions the matrices have to fulfill are now only

$$\alpha_1 \, \alpha_1 + \alpha_1 \, \alpha_1 = 2 \cdot \mathbf{1} \tag{13}$$

$$\alpha_1 \, \beta + \beta \, \alpha_1 = \mathbf{0} \tag{14}$$

$$\beta^2 = \mathbf{1} \tag{15}$$

These conditions are fulfilled for every pair $\{\alpha_1, \beta\}$ out of

$$\sigma_1 = \begin{pmatrix} 0 & 1 \\ 1 & 0 \end{pmatrix}, \; \sigma_2 = \begin{pmatrix} 0 & -i \\ i & 0 \end{pmatrix}, \; \sigma_3 = \begin{pmatrix} 1 & 0 \\ 0 & -1 \end{pmatrix}, \tag{16}$$

which are the well-known Pauli matrices. Thus the free Dirac equation in one dimension can be fulfilled by a two-component formalism (see e.g. [15] for a similar detailed derivation).

B. Appendix: Equivalence of relativistic Schrödinger and Dirac equations

In this appendix, we show that the Dirac equation for a free particle is equivalent to the relativistic Schrödinger equation, following [9]. This is no longer true when the particle interacts with a potential. Also, when the particle is driven by a spatially homogeneous time-dependent vector potential field, this can no longer be true, since the field can become so strong to produce pairs, mixing the positive and negative energy subspaces of the Dirac equation. However, as long as the field is not too strong (and its turn-on time slow enough, on the relativistic scale) the positive and negative energy subspaces remain effectively uncoupled and the equivalence between D and R equations still holds adiabatically.

The free Dirac Hamiltonian in position space representation reads :

$$H_0 = c\alpha \cdot \mathbf{p} + \beta mc^2 = \begin{pmatrix} mc^2\mathbf{1} & -ic\hbar\sigma\cdot\nabla \\ -ic\hbar\sigma\cdot\nabla & -mc^2\mathbf{1} \end{pmatrix}. \tag{17}$$

By the principle of minimal coupling and assuming the dipole approximation the Hamiltonian can be rewritten by replacing p by p$-A(t)$ and thus $-i\hbar\nabla$ by $-i\hbar\nabla - A(t)$

$$H = c\alpha\cdot(\mathbf{p} - \mathbf{A(t)}) + \beta mc^2 = \begin{pmatrix} mc^2\mathbf{1} & c\sigma\cdot((-i\hbar)\nabla - \mathbf{A(t)}) \\ c\sigma\cdot((-i\hbar)\nabla - \mathbf{A(t)}) & -mc^2\mathbf{1} \end{pmatrix}. \tag{18}$$

Here σ denotes the triplet of Pauli matrices

$$\sigma = (\ \sigma_1,\ \ \sigma_2,\ \ \sigma_3\ ). \tag{19}$$

If we take the Dirac equation in the one dimensional form we obtain:

$$H(t) = c\sigma_1((-i\hbar)\partial_x - A(t)) + \sigma_3 mc^2 = \begin{pmatrix} mc^2 & c(-i\hbar\partial_x - A(t)) \\ c(-i\hbar\partial_x - A(t)) & -mc^2 \end{pmatrix} \tag{20}$$

By a double Fourier transformation $\mathcal{F}_{x,t}$ from position to momentum space $\mathcal{F}_x$ and from time to energy $\mathcal{F}_t$, due to the linearity of the Fourier transformation and the dipole approximation one obtains:

$$H(p) = \begin{pmatrix} mc^2 & c(p - \mathcal{F}_t\{A(t)\}) \\ c(p - \mathcal{F}_t\{A(t)\}) & -mc^2 \end{pmatrix}, \tag{21}$$

This is a Hermitian 2×2 matrix with eigenvalues (defining $\tilde{p} \equiv p - \mathcal{F}_t\{A(t)\}$)

$$\lambda_1 = -\lambda_2 = \sqrt{(c\tilde{p})^2 + m^2c^4} = \lambda(\tilde{p}) \tag{22}$$

The corresponding unitary transformation which brings the Hamiltonian to diagonal form is given by the 2×2 matrix

$$u(\tilde{p}) = \frac{(mc^2 + \lambda(\tilde{p}))\mathbf{1} + \sigma_3 c\sigma_1\tilde{p}}{\sqrt{2\lambda(\tilde{p})(mc^2 + \lambda(\tilde{p}))}} = a_+(\tilde{p})\mathbf{1} + a_-(\tilde{p})\sigma_3\frac{\sigma_1\tilde{p}}{|\tilde{p}|}, \tag{23}$$

with

$$a_\pm(\tilde{p}) \equiv \frac{1}{\sqrt{2}}\sqrt{1 \pm \frac{mc^2}{\lambda(\tilde{p})}} \tag{24}$$

and with the inverse matrix

$$u(\tilde{p})^{-1} = a_+(\tilde{p})\mathbf{1} - a_-(\tilde{p})\sigma_3 \frac{\sigma_1 \tilde{p}}{|\tilde{p}|}. \tag{25}$$

Thus, the two dimensional Hilbert space, in which the Dirac Hamiltonian acts, can be rewritten as an orthogonal direct sum of two one dimensional subspaces. One represents the positive energy subspace and the other one represents the negative energy subspace.

$$U_{FW} H U_{FW}^{-1} = \begin{pmatrix} \sqrt{-c^2\tilde{\Delta} + m^2c^4} & \mathbf{0} \\ \mathbf{0} & -\sqrt{-c^2\tilde{\Delta} + m^2c^4} \end{pmatrix} \tag{26}$$

with

$$U_{FW} \equiv \mathcal{F}_{x,t}{}^{-1} u \mathcal{F}_{x,t} \tag{27}$$

and

$$\tilde{\Delta} \equiv (-i\hbar\partial_x - A(t))^2. \tag{28}$$

So it turns out that the free Dirac Hamiltonian is unitarily equivalent to a pair of relativistic Schrödinger (square root Klein-Gordon) equations.

These equations remain valid if one turns back to four dimensions simply by turning the σ_1 into α and σ_3 into β:

$$\alpha_i = \begin{pmatrix} 0 & \sigma_i \\ \sigma_i & 0 \end{pmatrix}, \ \beta = \begin{pmatrix} \mathbf{1} & 0 \\ 0 & -\mathbf{1} \end{pmatrix} \quad i = 1, 2, 3. \tag{29}$$

The Fourier operators are as usual

$$\mathcal{F}_x\{f\} = \frac{1}{\sqrt{2\pi}} \int_{-\infty}^{\infty} e^{-ipx} f \, dx \ , \ \mathcal{F}_t\{f\} = \frac{1}{\sqrt{2\pi}} \int_{-\infty}^{\infty} e^{-iEt} f \, dt. \tag{30}$$

Acknowledgments

We are thankful for many useful remarks from and discussions with Wilhelm Becker. EL and MD are supported by the Deutsche Forschungs-Gemeinschaft.

References

1. S. Brandt and H. D. Dahmen *The picture book of quantum mechanics*, Springer (1995, 2nd edition)
2. Q. Su, B. A. Smetanko and R. Grobe, Laser Phys. **8**, 93 (1998)
3. Q. Su, B. A. Smetanko and R. Grobe, Opt. Expr. **2**, 277 (1998)
4. N. Dombey and A. Calogeracos, Phys. Rep. **315**, 41 (1999); A. Calogeracos and N. Dombey, Physica B **255**, 126 (1998)
5. C. J. Joachain, M. Dörr and N. J. Kylstra, Adv. At. Mol. Opt. Phys. **42**, 225-286 (2000)
6. N. J. Kylstra, R. A. Worthington, A. Patel, P. L. Knight, J. R. Vazquez de Aldana and L. Roso, Phys Rev Lett **85**, 1835 (2000)
7. Atomic units ($e = m_e = \hbar = 1$) are used throughout the text unless otherwise indicated. This implies that $c = 137.0$ a.u.
8. J. W. Braun, Q. Su and R. Grobe, Phys. Rev. A **59**, 604 (1999)
9. B. Thaller *The Dirac Equation*, Springer (1992)
10. C. Bottcher and M. R. Strayer, Ann. Phys. **175**, 64, (1986)
11. Details of the numerical approach shall be published by M. Dörr, E. Lenz and W. Sandner elsewhere.
12. J. San Roman, L. Roso and H. R. Reiss, J. Phys. B **33**, 1869 (2000)
13. E. Lenz, M. Dörr and W. Sandner, Laser Phys., to be published (2001)
14. P. Krekora, Q. Su and R. Grobe, preprint (2000)
15. H. Nitta,T. Kudo and H. Minowa Am. J. Phys. **67**, 966 (1999)
16. J. D. Bjorken, S. D. Drell *Relativistic Quantum Mechanics*, Mc Graw-Hill (1964)

POSITION DEPENDENT TUNNELING SPEED FOR PARTICLES UNDER A BARRIER

P. KREKORA, Q. SU AND R. GROBE
Intense Laser Physics Theory Unit and Department of Physics
Illinois State University
Normal, IL 61790-4560 USA

1. Introduction

The phenomenon of tunneling in which a quantum mechanical particle can penetrate a repulsive potential barrier with a height that exceeds the total energy of the particle is counterintuitive. Intuitions based on classical mechanics fail to explain this process. At the same time, tunneling is extremely important and has been studied widely. The Josephson effect in semiconductors, β-decay in nuclear physics and instantons in high-energy physics are just a few examples. It was already recognized in 1932 [1] that there was no appreciable temporal delay in the transmission of wave packets through barriers. Wigner [2] discussed the possibility for a particle to effectively travel faster than the speed of light under the barrier.

In this note we intend to address the following questions: Can one trust the predictions of a non-relativistic theory at all if superluminal effects are being investigated? Does the relativistic quantum theory predict superluminal speeds? Does a fully relativistic treatment of tunneling increase or reduce the tunneling probability? Can one define a position dependent quantity that describes the time evolution of a wave packet inside the barrier directly without relying on the average quantities available to date?

Quite remarkably, despite the large amount of literature on non-relativistic tunneling [3], we are aware of only a few works which addressed some of these questions. [4] Leavens and Aers used the stationary state approach to analyze Larmor-clock transmission times for single and double rectangular barriers. In special cases the quantum mechanical

B. Piraux and K. Rzążewski (eds.), Super-Intense Laser-Atom Physics, 369–378.

tunneling can be mapped to the electrodynamics of evanescent waves studied theoretically and experimentally. [5-8] For non-relativistic tunneling, the question of how much time it takes a particle to pass the barrier has triggered considerable debates to the present day. Even though by 1993 the community has largely accepted that a time scale can be associated with the duration of tunneling, there is still a lack of consensus on the existence and implications of a unique expression for this time scale.

We introduce a new concept of an instantaneous tunneling speed that can be associated with each location under the barrier and corresponds to the peak velocity when computed after the scattering. We evaluate this speed analytically and compare it directly with exact numerical wave function solutions to the time-dependent Dirac equation. It can differ from previously proposed average speeds by more than a factor of two. It also provides a microscopic picture of the tunneling process in full temporal and spatial resolution.

2. The model system

We study the effects of relativity on the tunneling process for the simplest possible model system, a particle that tunnels through a one-dimensional repulsive barrier. The time-evolution of this system is given by the solution of the Dirac equation [in atomic units]:

$$i\,\partial\Psi/\partial T = -i\,c\,\alpha_x\,\partial\Psi/\partial x + c^2\beta\,\Psi + W(x)\Psi \qquad (1)$$

where the repulsive potential $W(x)$ is centered around $x=0$ and has an effective width of w and a height W_0. To check the generality of our results we have used a variety of different tunneling potentials $W(x)=W_0\exp[-(2x/w)^n]$. For large even integers n we recover the rectangular barrier for which the energy eigenstates can be found and approximate analytical estimates can be derived. Here α_x and β denote the 4×4 Dirac matrices. The time-dependent solution of the spinor wave function $\Psi(x,T)=[\Psi_1,\Psi_2,\Psi_3,\Psi_4]$ can be obtained numerically on a space-time grid on a massively parallel computing platform. [9]

For the incoming wave packet we use the state

$$\Psi(x,T=0)=N\,\exp[-(x-x_0)^2/(4\Delta x^2)]\,\exp(ik_0x)\,\psi(k_0) \qquad (2)$$

where spinor $\psi(k_0) \equiv [1,0,0,ck_0/(E_0+2c^2)]$ and $N \equiv [(E_0+2c^2)/(2(E_0+c^2)\Delta x\sqrt{2\pi})]^{1/2}$. The central canonical momentum k_0 is related to the initial speed v_0 via $v_0=ck_0/\sqrt{[c^2+k_0^2]}$. The initial location of the wave packet x_0 is far enough to left of the barrier, that the total spatial probability for positive x values is negligible at time T=0. The (shifted) total energy $E_0=\sqrt{[c^4+c^2k_0^2]}-c^2$ in our case is consistently chosen smaller than W_0. The potential height W_0 is smaller than $2c^2$ to avoid the effects of the negative energy continuum as characteristic of the Klein-paradox. [9] This restricts our initial velocities to $v_0 < 0.94c$ (=129 a.u.). Predictions of the corresponding Schrödinger equation can be obtained conveniently in our simulations by adjusting the "parameter" c to "infinity". [10]

With the exception of an overall amplitude reduction the transmitted wave packet does not get significantly distorted for spatially broad wave packets $\Delta x > w$. In this regime the center of mass for the transmitted wave packet (denoted in the following by $\langle x \rangle_t$) agrees approximately with the peak value of the probability density. For the non-relativistic case this regime has been studied extensively as the stationary phase approximation is reliable permitting some analytical investigations. Below we will test this approximation by comparing it with exact numerical solution of the time-dependent Dirac and Schrödinger equation.

3. The effective and instantaneous tunneling speeds

In the limit $n \to \infty$ the model potential becomes rectangular and an *effective* average tunneling velocity can be defined as

$$v_e \equiv \frac{w}{T-(T_1+T_2)} = \frac{w}{T+\dfrac{w/2+x_0}{v_0}+\dfrac{w/2-\langle x(T)\rangle_t}{<v(T)>_t}} \tag{3}$$

where T_1 and T_2 correspond to the time intervals spent outside the potential region $|x|<w/2$ and where $\langle v \rangle_t \equiv \langle \Psi_t | c\alpha_x | \Psi_t \rangle$ denotes the central velocity of the transmitted wave packet $|\Psi_t\rangle$. Even though Eq. (3) is usally used to define a tunneling time, we discuss below that this "speed" v_e does not necessarily provide a microscopically correct physical picture in describing the center of mass motion during the tunneling.

The small momentum components in the wave packet are attenuated more during the tunneling because the transmission $|t(k)|$ increases monotonically with momentum k, and the emerging "truncated" wave packet has higher average momentum. As a consequence, the propagation velocities on the two sides of the barrier can be different ($<v>_t > v_0$) as noted earlier for the non-relativistic limit. [3,5]

If $1/(2\Delta x) = \Delta k$ is smaller than the momentum scale on which t(k) varies, then $<v>_t$ agrees with that of the initial wave packet, $<v>_t \approx v_0$, and we can approximate $<x(T)>_t = <x^{(0)}(T)> - d\alpha(k)/dk|_{k_0}$. Here $<x^{(0)}(T)>$ is the position associated with a free electron and $\alpha(k)$ is the phase of the complex transmission amplitude $|t(k,w,W_0)| \exp[i\alpha(k,w,W_0)]$. Using the stationary phase approximation $\alpha(k)$ can be derived analytically .

The "speed" v_e defined in (3) is plotted in Fig. 1 as a function of the barrier width w. The markers correspond to the exact numerical wave-packet solutions to the time-dependent Schrödinger (circles) and Dirac (crosses) equation computed from the spatial

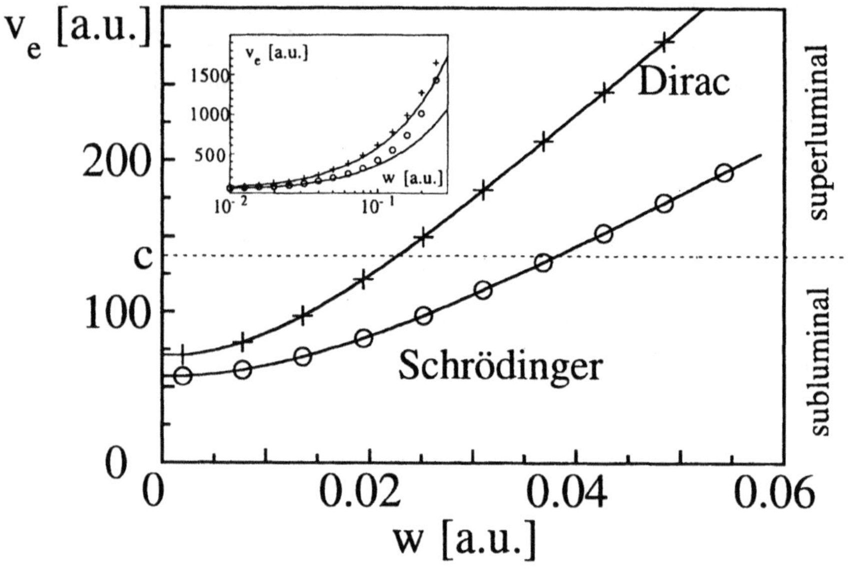

Figure 1. Comparison of the relativistic and non-relativistic predictions for the average tunneling speed defined in Eq. (3). The solid lines are the predictions from the approximate analytical theory. The open circles and crosses are obtained from the exact spatial probability densities calculated from the Schrödinger and Dirac equation. The inset shows same graph as in Fig. 1 but on a larger scale of width w to demonstrate the breakdown of the analytical predictions.

[The parameters were x_0=-100 a.u., Δx=20 a.u., v_0=100 a.u., T=2 a.u., $W_0=1.5E_0$]

probability density $P(x,T)= \sum_{i=1}^{4} |\Psi_i(x,T)|^2$. For each barrier width w we have evolved the initial wave packet in time and computed the effective average speed v_e according to Eq. (3). The agreement between the exact numerical data and the analytical prediction (formula omitted) indicates that the stationary phase approximation is reliable for these parameters in both the relativistic and non-relativistic theory.

Having established the validity of the stationary phase approximation we point out the difference between the relativistic and non-relativistic tunneling speeds. For our initial velocity of v_0=100 a.u. the non-relativistic tunneling speed v_e turns out to be 20% *smaller* than the relativistic prediction. This result is a bit surprising as one expects that smaller velocities are associated with relativistic corrections such as the nonlinear mass increase. In fact our result is in contrast to another work on relativistic tunneling [4] which reported a reduction of the tunneling speed due to relativity for the case of a double-rectangular potential. Despite the different predictions at small barrier width, both Dirac and Schrödinger equations predict superluminal effective average tunneling speeds for large barrier widths (w>0.023a.u. and w>0.038a.u., respectively). In this limit v_e increases linearly with the width: $v_e \rightarrow v_0 w \sqrt{(W_0 - E_0)/2}$. The Dirac theory predicts also a lower tunneling probability than the Schrödinger equation.

A breakdown of the stationary phase approximation is associated with a non-zero third-order momentum derivative of the scattering phase $\alpha(k)$ and an increase of the barrier width w relative to Δx. As shown in the inset of Fig.1 the predictions of the stationary phase approximation become less reliable when barrier width w approaches the wave packet width Δx. In this limit $\langle x(T)\rangle_t$ becomes less meaningful to identify as the spatial density of the transmitted pulse may evolve into multi-peaked structure that varies in time.

To the best of our knowledge most previous theories investigated the tunneling process by *extrapolating* backward the information contained in the final transmitted and reflected state. We will now analyze the process in full microscopic time-resolution *during* the tunneling. As the initial wave packet was prepared to the left of the barrier, the probability density in the barrier region $|x|<w/2$ vanishes until the front edge of the incoming wave packet enters the barrier. Then the density grows and after the tunneling process it shrinks back to zero (Fig. 2). This stresses the point we made earlier that is not trivial to trace directly the peak movement under the barrier. The near parallel lines in the logarithmic plot under the barrier indicate the exponential spatial decay. If these lines were precisely parallel, then the tunneling process could take place instantaneously in

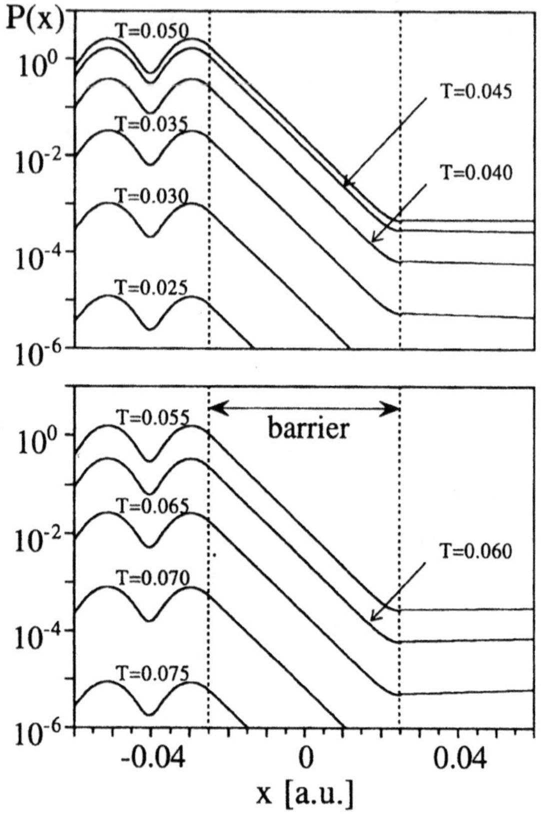

Figure 2. The spatial probability density P(x) during the tunneling displayed in the spatial region near and inside the barrier at various times T in a.u.. [The parameters were x_0=-5 a.u., Δx=0.5 a.u., v_o=100 a.u., w=0.05 a.u., W_0=1.5E_0]

principle. The details of the incoming wave packet at x=-w/2 could then be instantaneously transmitted and copied over to the right edge at x=w/2. The analysis below shows that the tunneling process requires a non-zero time duration associated with a sub- or super-luminal speed.

To study the tunneling under the barrier, we define a new physical quantity. For a given location x, we search for the specific time (denoted by T_p) at which the spatia probability density P(x,T) takes its maximum value: $P(x,T_p) \equiv \max [P(x,T)]$. Inside the barrier the wave function is certainly not spatially localized as it always takes its spatia maximum at the left edge x=-w/2, but it is *temporally localized* such that the peak time $T_p(x)$ can be unambiguously defined and calculated under the barrier. Before the wave packet arrives at the boundary, the function $T_p(x)$ is given by the linear dependence $T_p(x)=(x-x_0)/v_0$ for $x<<-w/2$, if the wave packet spreading is not significant. [11,12]

In Figure 3 we display the location x as a function of this temporal peak time T_p around and inside the barrier. For comparison the straight line was computed from the same wave packet without any tunneling barrier, $x^{(0)}(T_p)=v_0\,T_p+x_0$. There are two striking observations. First, due to the details of the scattering process at the left edge of the barrier and the resulting interference of the incoming and reflected wave packet, the peak time T_p for the tunneling case is actually different from that of the corresponding force-free wave packet at x=-w/2. This arrival time is actually *larger* than that associated with the free particle. Secondly, the continuous graph $x=x(T_p)$ permits us to define a

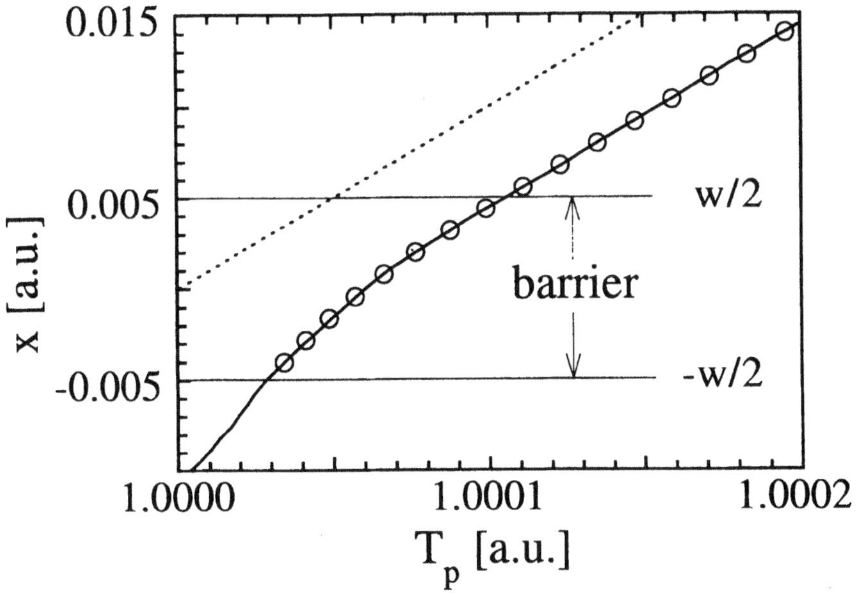

Figure 3. The location x as a function of the temporal-peak time T_p. The dashed line corresponds to the graph associated with a free wave packet in the absence of any scattering potential. The open circles denote the prediction according to the analytical formula based on the stationary phase approximation [Eq. 4]. [The parameters were x_0=-100 a.u., Δx=20 a.u., v_0=100 a.u., w=0.01, W_0=1.5E_0]

tunneling speed *inside* the potential, $v_T(x) \equiv [\, dT_p(x) / dx \,]^{-1}$ The decreasing slope of the graph in Fig. 3 suggests a significant slow-down of the particle across the barrier.

For the special case of the square well potential, it is possible within the framework of a generalized stationary phase approximation to derive analytical formulas for the instantaneous tunneling velocity from the peak time:

$$T_p(x) \;=\; \frac{W_0(\Gamma^{-1}+\Gamma)B_1\tanh(\kappa w)+2w(\Gamma^{-1}-\Gamma)B_2\,\mathrm{sech}^2(\kappa w)}{4\left[1+\dfrac{(1-\Gamma^2)^2}{4\Gamma^2}\tanh^2(\kappa w)\right]} \;+$$

$$+\;\frac{W_0\Gamma\kappa B_1\tanh[\kappa(x-w/2)]+2\Gamma B_2(x-w/2)\,\mathrm{sech}^2[\kappa(x-w/2)]}{2\kappa\{1+\Gamma^2\tanh^2[\kappa(x-w/2)]\}} \tag{4a}$$

where

$$B_1=\left[\frac{1}{E_0(W_0-E_0)}+\frac{1}{(E_0-W_0+2c^2)(E_0+2c^2)}\right], \quad B_2=\left[-1+\frac{(W_0-E_0)}{c^2}\right],$$

$$\kappa \;\equiv\; \frac{1}{c}\sqrt{c^4-(E_0+c^2-W_0)^2} \tag{4b}$$

and

$$\Gamma \;\equiv\; \sqrt{\frac{E_0(E_0+2c^2-W_0)}{(E_0+2c^2)(W_0-E_0)}}. \tag{4c}$$

The circles in Fig. 3 represent the predictions of this formula and the agreement with the exact data is superb. We should note that formula (4) is based only on the first spinor component of the wave function. The estimate of Eq. (4) produces a relative error, in this case, of less than $10^{-3}\%$.

The new microscopic tunneling picture predicts significantly different tunneling speeds than the traditional definition (3). Let us analyze the source of this discrepancy and comment on the role of v_e in the microscopic picture for the specific numerical example introduced above. We found that for x_0=-100 a.u., v_0=100 a.u., w=0.01 a.u. the spatial delay was $\langle x(T)\rangle_t - \langle x^{(0)}(T)\rangle = -5.5 \times 10^{-3}$ a.u. for a Schrödinger wave packet that was associated with a total temporal delay $\Delta T = -5.5 \times 10^{-5}$ a.u. Using Eq. (3) one would associate this delay with an effective average tunneling speed of $v_e \approx 64$ a.u. [see Fig. 1] For these parameters, Fig. 3 shows that the center of the wave packet formed from the incoming and reflected waves reaches the left edge at time T_L=1.0000305 a.u., which due to the interference is already delayed compared to the arrival time of a force-free wave packet [$T_L^{(0)}$=0.9999500 a.u.]. Fig. 3 also shows that the tunneled portion reaches the right edge x=w/2 at a time T_R= 1.0001068 a.u.. As a result, the exact time spent by the particle under the barrier is only T_R-T_L =0.0000763 a.u. which is shorter than the corresponding time $T_R^{(0)}$-$T_L^{(0)}$=0.0001 a.u. of the force-free electron. This time would amount to an average tunneling speed of $\overline{v}_T$=w/(T_R-T_L) = 131 a.u., which is more than *twice* the effective tunneling speed v_e calculated above. This huge difference (between 64 and 131 a.u.) is due to the delay of the wave function already present on the left edge of the barrier before the electron enters the potential, $T_L > T_L^{(0)}$. The speed-up of the particle under the barrier (associated with $\overline{v}_T$=131 a.u.) is compensated by the slow-down before entering barrier so that the effective speed which takes both mechanisms into account amounts to a net value of w/(T_R-$T_L^{(0)}$)$\approx$64 a.u. consistent with v_e.

In summary, we have shown numerically and analytically that the Dirac theory can predict superluminal speeds that exceed those predicted by the Schrödinger equation. We have provided a time resolved picture of the tunneling process and proposed a novel tunneling speed v_T. In contrast to all speeds discussed to date, v_T is a function of the position under the barrier and may take a significantly different numerical value. This new concept may be useful in reexamining experimental data obtained recently in tunneling of optical pulses in frustrated total internal reflection [13], in photonic band gaps [7] or in dielectric media [14].

4. Acknowledgement

We acknowledge discussions with W. Becker, M. Dörr, S.D. Hassani, G.H. Rutherford and H. Wanare, and numerical assistance by R.E. Wagner and P.J. Peverly. This work has been supported by the NSF under grant PHY-9970490. We also acknowledge support from the Research Corporation for Cottrell Science Awards and ISU for URGs. The numerical work has been performed at NCSA.

5. References

1. MacColl, L.A. (1932) *Phys. Rev.* 40, 621.
2. Wigner, E.P. (1955) *Phys. Rev.* 98, 145.
3. For reviews on the controversial issue of tunneling times, see, e.g., Hauge, E.H. and Stovneng, J.A. (1989) *Rev. Mod. Phys.* 61, 917; Landauer, R. and Martin, T. (1994) *Rev. Mod. Phys.* 66, 217; Chiao, R.Y. and Steinberg, A.M. (1997) *Prog. in Optics*, ed. E. Wolf (Elsevier, Amsterdam, p.345. and references therein.
4. Leavens, C.R. and Aers, G.C. (1989) *Phys. Rev.* B 40, 5387 and Leavens, C.R. and Mayato, R.S. (1998) *Annalen der Physik* 7, 662.
5. Martin, T. and Landauer, R. (1992) *Phys. Rev.* A 45, 2611.
6. Steinberg, A.M., Kwiat, P.G. and Chiao, R.Y. (1993) *Phys. Rev. Lett.* 71, 708
7. Spielmann, C., Szipöcs, R., Stingl, A. and Krausz, F. (1994) *Phys. Rev. Lett.* 73, 2308.
8. Gasparian, V., Ortuno, M., Ruiz,.J. and Cuevas, E. (1995) *Phys. Rev. Lett.* 75, 2312.
9. Braun, J.W., Su, Q. and Grobe, R. (1999) *Phys. Rev.* A 59, 604 .
10. Thaller, B. (1992) *The Dirac Equation* (Springer, Berlin).
11. Su, Q., Smetanko, B.A. and Grobe, R. (1998) *Opt. Express* 2, 277.
12. Su, Q., Smetanko, B.A. and Grobe, R. (1998) *Las. Phys.* 8, 93.
13. Balcou, Ph. and Dutriaux, L. (1997) *Phys. Rev. Lett.* 78, 851.
14. Wang, L.J., Kuzmich, A. and Dogariu, A. (2000) *Nature* 406, 277.

RELATIVISTIC ENERGY AND ANGULAR DISTRIBUTIONS OF ELECTRONS IN THE IONIZATION OF ATOMS BY SUPER-INTENSE LASER RADIATION

V.P. KRAINOV
Moscow Institute of Physics and Technology
141700 Dolgoprudny, Moscow Region, Russia
e-mail: krainov@cyberax.ru

1. Introduction

The general simple theory of tunnelling ionization of atoms by low-frequency intense laser field is based on *Landau-Dykhne adiabatic approach*. This approach is applicable when the laser frequency is small compared to the ionization potential of the considered atom. The ionization rate is given by the simple expression

$$W_{iE} = W_0 \exp\left\{ -2\,\mathrm{Im} \int_0^{\tau} \left[E_f(F,\omega,t) + E_i \right] dt \right\}, \qquad E_f(F,\omega,\tau) = -E_i. \tag{1}$$

Here E_i is the unperturbed binding energy of the initial atomic state, and $E_f(F, \omega, t)$ is the energy of the final continuum state of an electron in the field of the laser radiation. F and ω are the radiation field strength and field frequency, respectively. The quantity W_0 is ionization rate at the atomic field strength. The atomic system of units is used throughout this paper. The upper limit in Eq. (1) corresponds to the some imaginary time when the energies of the initial and the final states coincide each with other. The energy of the initial state is supposed to be unperturbed by the laser field since this field distorts the final continuum state much stronger than the initial bound state. It is valid with exponential accuracy.

The well-known non-relativistic energy spectra in the tunnelling ionization are obtained using non-relativistic simple expression for the energy of a free electron in the field of the electromagnetic wave.

a) *linear polarization* (at $\theta = 0$):

$$W_{iE} = W_0 \exp\left\{ -\frac{2E\gamma^3}{3\omega} \right\}, \qquad \gamma \equiv \frac{\omega\sqrt{2E_i}}{F} << 1 \tag{2}$$

Here E is electron kinetic energy, γ is the Keldysh parameter, θ is the angle between the laser polarization and the electron momentum in the final continuum state. Maximum of the energy distribution occurs, obviously, at $E = 0$. The quantity W_0 is the tunneling ionization rate at $E = 0$. We conclude that the most of electrons are ejected with the small velocities in the tunnelling regime. This process takes place when the laser field strength is nearly of the maximum value. In the classical consideration of

B. Piraux and K. Rzążewski (eds.), Super-Intense Laser-Atom Physics, 379–384.

this process there is cut off for the energy spectrum at the electron kinetic energy which is equal to twice ponderomotive energy: $E = 2U_p$. In the quantum consideration there is no such cut off, but the probability decreases exponentially at the large kinetic energies.

b) *circular polarization* (at $\vartheta = 0$):

$$W_{iE} = W_0 \exp\left\{ -\frac{\gamma}{2\omega U_p}\left(E - U_p\right)^2 \right\}, \qquad U_p = \frac{F^2}{2\omega^2} \tag{3}$$

Here U_p is the averaged ponderomotive energy in the circularly polarized field. Maximum of the energy distribution occurs at $E = U_p$. The quantity W_0 is the ionization rate at $E = U_p$. The quantity ϑ is the angle between the polarization plane of the laser radiation and the direction of the electron momentum. We give here the energy distribution only when this angle is equal to zero, since the probability is strongly decreased at the deflection from this value of the angle. Thus, it is seen the important difference with the case of the linearly polarized field.

The well-known non-relativistic angular distributions in the tunnelling ionization are of the form:

a) *linear polarization*:

$$W_{iE}(\theta) = W_0 \exp\left\{ -\frac{2\sqrt{2E_i}E}{F}\theta^2 \right\}. \tag{4}$$

$(\theta \ll 1)$.

Here quantity W_0 is the ionization rate at the angle $\theta = 0$. The quantity E is the kinetic energy of the ejected electron. We conclude that the angular distribution depends on the kinetic energy of the ejected electron: when this energy increases the width of the angular distribution diminishes. Besides of this, the width of the angular distribution increases with the increase of the laser intensity and does not depend on the laser frequency.

b) *circular polarization* (in the maximum of the energy spectrum where the angular distribution is the most interesting, from the practical point of view):

$$W_{iE}(\vartheta) = W_0 \exp\left\{ -\frac{\sqrt{2E_i}F}{\omega^2}\vartheta^2 \right\}, \qquad E = U_p. \tag{5}$$

($\vartheta \ll 1$). Here quantity W_0 is the ionization rate at the angle $\vartheta = 0$. In contrast to the case of the linearly polarized field, the width of the angular distribution diminishes with the increasing of the laser field strength and with the decreasing of the laser frequency.

2. Energy Spectra for Relativistic Electrons in the Tunnelling Ionization of Atoms by Linearly Polarized Short Laser Pulses

Results of Ref. [1] are described below. Now the known relativistic expression is used for the electron energy in the field of the electromagnetic wave. We did not give this expression here due to its cumbersome form and restrict ourselves by the final results only.

The energy spectrum of relativistic electrons is of the form:

$$W_{iE} = W_0 \exp\left\{-\frac{2E\gamma^3}{3\omega} - \frac{E^2\gamma}{\omega c^2}\right\} \quad \text{at } \theta = 0. \tag{6}$$

Here the quantity W_0 is the ionization rate at $E = 0$ as before. The first term in the exponent reproduces non-relativistic Eq. (2) as it should be.

The second term in exponent of (6) is responsible for relativistic effects. We conclude from Eq. (6) that the relativistic effects are important at the moderate (non-relativistic!) values of the electron kinetic energy $E > \gamma^2 mc^2$, but not at the relativistic values $E > mc^2$. Hence, they become to be important already at the relatively small laser intensities.

Let us consider the typical example of the tunnelling ionization by a strong laser field: $\gamma = 0{,}1$; $E > 5$ keV. Then according to Eq. (6) the CO_2 – laser with the high intensity $I = 10^{15}$ W/cm^2 produces relativistic energy spectrum of ejected electrons in keV range.

In Fig. 1 the dependence (6) of the ionization rate on the electron kinetic energy E is presented qualitatively for the above example. Arrow shows the begin of the region where relativistic effects become observable.

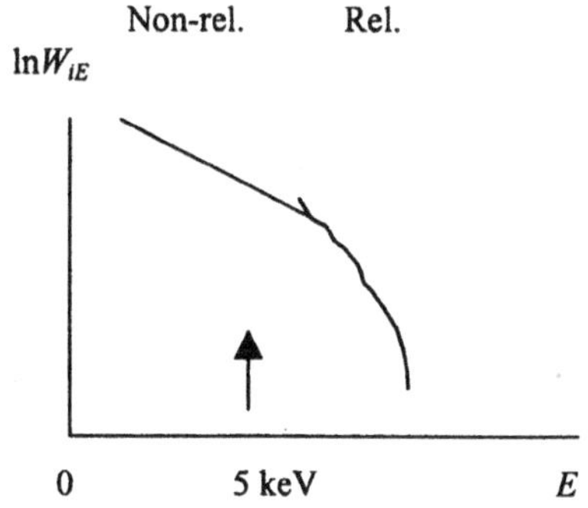

Fig. 1

3. Energy Spectra for Relativistic Electrons in the Tunnelling Ionization of Atoms by Circularly Polarized Short Laser Pulses

The results of Ref. [2] are described below in this Section. The maximum of the angular distribution in the relativistic case is found at the nonzero angle, unlike the above considered non-relativistic case (see Eq. (4)) [3]:

$$\vartheta_0 = \arctan \frac{F}{2\omega c} \tag{7}$$

This angle increases with the field strength and diminished with the laser frequency. The relativistic energy distribution at this angle is of the form

$$W_{iE} = W_0 \exp\left\{ -\frac{\gamma}{2\omega U_p^{rel}} \left(E - U_p^{rel} \right)^2 \right\}. \tag{8}$$

According to this relation, the relativistic "mass" shift of the ponderomotive energy corresponds to the increasing of the ponderomotive energy in comparison with the non-relativistic case:

$$U_p = \frac{F^2}{2\omega^2} \to U_p^{rel} = \frac{F^2}{2\omega^2}\left[1 + \left(\frac{F}{2\omega c} \right)^2 \right] > U_p. \tag{9}$$

According to (8) the relativistic energy spectrum is broader than the non-relativistic spectrum, since its width is proportional to the relativistic ponderomotive energy. In the non-relativistic case the energy distribution (8) reduces to Eq. (3) as it should be.

Unlike the case of linear polarization, the most of electrons are ejected with the ponderomotive energy which can be of the same order as the rest energy of the electron at relatively low laser intensity and low laser frequency, i.e. to be essentially relativistic quantity.

It follows from Eq. (9) that the relativistic mass of an electron is less than the rest electron mass, and the relativistic mass decreases with the increase of the laser intensity. Of course, this result is specific for a considered case. In other cases the value and the sign of the relativistic mass shift can be quite different.

4. Angular Distributions for Relativistic Electrons in the Tunneling Ionization of Atoms by Circularly Polarized Short Laser Pulses

In the maximum of the energy distribution ($E = U_p^{\rm rel}$) the angular distribution is the most interesting. It is found to be

$$W_{iE}(\vartheta) = W_0 \exp\left\{-\left[1+\left(\frac{F}{2\omega c}\right)^2\right]\frac{F\sqrt{2E_i}}{\omega^2}(\vartheta-\vartheta_0)^2\right\}. \tag{10}$$

The relativistic angular distribution is narrower than the non-relativistic distribution (5). It has a maximum at the nonzero angle given by Eq. (7).

The case of elliptically polarized light is considered within the same approach [4]. The angular distribution is of the form nearly the maximum (compare with (10)):

$$W_{iE}(\vartheta) = W_0 \exp\left\{-\left[1+\left(\frac{gF}{2\omega c}\right)^2\right]\frac{gF\sqrt{2E_i}}{\omega^2}(\vartheta-\vartheta_g)^2\right\}. \tag{11}$$

Here $0 < g < 1$ is the ellipticity of laser radiation (except of the limit $g \ll 1$, since in the limit of a linearly polarized field the angular distribution is determined by the small amgle between the field strength and the direction of the ejected electron, instead of the small angle between the polarization plane and the direction of the ejected electron). Now the angle where the distribution has the maximum is (compare with (7))

$$\vartheta_g = \arctan\frac{gF}{2\omega c}. \tag{12}$$

We do not consider here the relativistic angular distribution in the case of the linearly polarized radiation, since according to Eq. (4) the corresponding non-relativistic angular distribution is the most prominent for low (non-relativistic) kinetic electron energies, unlike the case of circularly polarized radiation.

5. Conclusion

It follows from above consideration that the circular and elliptical laser polarizations are more preferable for the experimental investigation of the relativistic effects in the energy and angular spectra of ejected electrons at the tunneling ionization than the linear laser polarization. However, in the case of the linearly polarized field the relativistic effects begin to be important at the relatively small kinetic energies of the ejected electron compared to its rest energy.

All obtained results are valid both for tunneling ionization and for barrier-suppression ionization by super-intense laser field, but in the latter case exponents should be substituted everywhere by the Airy functions according to the simple rule (see Ref. [5] for details):

$$\exp(-x) \rightarrow \mathrm{Ai}[(3x/2)^{2/3}]. \tag{13}$$

The main qualitative relativistic effects for circular and elliptical polarization are:

1) The relativistic increasing of the peak in the energy spectrum of ejected electrons;
2) The preferential ejection of electrons out of the polarization plane of laser radiation.

Our discussion restricts only by the case of direct ionization of relativistic electrons. This is the relatively low-energy part of the electron energy spectrum. We did not consider the rescattering processes which produce plateau of hot electrons. This problem is unsolved up to now for relativistic electrons.

This work was supported by the research grant N 99-02-17810 of the Russian Foundation for Fundamental Research.

6. References

1. Krainov, V.P. (1998) *Optics Express* **2**, 268.
2. Krainov, V.P. (1999) *J. Phys. B: At. Mol. Opt. Phys.* **32**, 1607.
3. Reiss, H.R. (1990) *J. Opt. Soc. Am.* B **7**, 574.
4. Ortner, J. (2000) *J. Phys. B: At. Mol. Opt. Phys.* **33**, 383.
5. Krainov, V.P. (1997) *J. Opt. Soc. Am.* B **14**, 425.

PART VI

RYDBERG ATOMS AND STRONG FIELD COHERENT CONTROL

ATOMIC WAVE PACKETS CREATED BY SHORT ELECTRIC PULSES

I.BERSONS, A.KULSH, AND R.VEILANDE
Institute of Atomic Physics and Spectroscopy,
University of Latvia,
Raina boulevard 19, Riga, LV-1586, Latvia

1. Introduction

A charged particle emits (absorbs) electromagnetic radiation efficiently in the region where its acceleration is at a maximum. In an atom this region is close to the nucleus, where the electron motion can be approximated by a parabola. Based on this approximation, Kramers (see Ref. [1]) has derived very simple and accurate expressions for the transition probabilities between hydrogenic bound-bound and bound-continuum states. This law has a more expressed manifestation in the multiphoton absorption process [2,3].

The interaction of atoms with short electric pulses differs sharply from that for laser pulses. In recent years the ionisation of atoms in high-lying Rydberg states by unipolar so-called half-cycle pulses (HCP's) has been studied extensively [4-6]. Here we discuss only the limit of a very short HCP whose duration τ is much shorter than the classical electron orbital period $T_K = 2\pi n^3$, where $E_n = -1/2n^2$ is the electron initial energy (atomic units are used throughout). In this limit, the HCP with the peak value $\mathbf{F}_0$ delivers a momentum $\mathbf{q} = \mathbf{F}_0\tau$ to the electron. As shown in Ref. [7], the large distances from the atomic core play an important role in the interaction of the HCP with a Rydberg atom, and states with a large angular momentum are involved in the process of HCP absorption.

In this work, we consider the classical theory of the Rydberg atom kicked by one and two HCP's. On the basis of the semiclassical approximation, we derive also the expression for the atomic wave packet created by a short unipolar electric pulse.

2. 1D kicked atom. Classical theory

In the 1D model, the electron moves along the x axis ($x > 0$) and its coordinate as a function of time t is given by

$$x = n^2 (1 - \cos\xi), \quad t = n^3 (\xi - \sin\xi). \tag{1}$$

The electron is kicked by impulse $\mathbf{q}$ at some point of its trajectory, after which it moves along a new trajectory. The parameters of the new trajectory (prime signed ones) are defined by two equations

$$x = x', \tag{2}$$

$$p + q = p'. \tag{3}$$

B. Piraux and K. Rzążewski (eds.), Super-Intense Laser-Atom Physics, 387–394.

Equation (2) can be interpreted as a condition that the initial and final trajectories have a common point at the moment of momentum transfer. Equation (3) is the momentum conservation equation.

Using Eqs. (1)-(3), we can find the electron energy after a kick:

$$E_{n'} = E_n + q^2/2 + \frac{q \sin \xi}{n(1 - \cos \xi)}. \tag{4}$$

The last term in Eq. (4) is a product of q and the kinetic momentum $\pm (2E_n + 2/x)^{1/2}$ of the electron. If the electron is kicked at the outer turning point of its trajectory, the electron final energy is equal to $E_n + q^2/2$. The electron final energy is large positive (negative), when it is kicked near the nucleus and the directions of the pulse and electron motion coincide (are opposite). The value of the coordinate x, at which the electron changes its energy from E_n to $E_{n'}$, is given by

$$x_n^{n'} = \frac{2q^2}{Q_{nn'}}, \tag{5}$$

where

$$Q_{nn'} = \left(E_{n'} - E_n - q^2/2\right)^2 + q^2/n^2. \tag{6}$$

In order to find the final energy distribution of the electron and the ionisation probability, the following identity

$$\frac{1}{T_k} \int_0^{T_k} dt = 1 \tag{7}$$

is exploited. Using Eqs.(1) and (4), we obtain

$$\frac{2\,|q|^3}{\pi n^3} \int_{-\infty}^{\infty} \frac{dE_{n'}}{Q_{nn'}^2} = 1. \tag{8}$$

So the final energy distribution of the electron is given by

$$\frac{dP_{nn'}}{dE_{n'}} = \frac{2\,|q|^3}{\pi n^3 Q_{nn'}^2}. \tag{9}$$

This distribution has a sharp peak, especially for small q, at the energy $E_{n'}$ equal to $E_n + q^2/2$. This means that, on the average, the electron is kicked when it is far from the origin. The electron spends most of its time at large x. Near the nucleus it moves very fast.

The integration in Eq. (8) between the limits 0 and infinity yields the ionisation probability

$$W_{ion} = \frac{1}{\pi}\left[\frac{\pi}{2} - \arctan u - \frac{u}{1+u^2}\right], \tag{10}$$

where

$$u = \frac{1 - q^2 n^2}{2\,|q|\,n}. \tag{11}$$

Let us consider now application of two pulses with a time delay t between them. Recently the ionisation of the 1D atom by two short unipolar pulses has been studied experimentally [8]. All parameters of the electron new trajectory after the first kick are defined by Eqs. (2) and (3). The first kick occurs at the instant of time $t' = n'^3 \left(\xi' - \sin \xi'\right)$. The second kick occurs at the instant of time

$$n'^3 \left(\eta - \sin \eta\right) = n'^3 \left(\xi' - \sin \xi'\right) + t, \tag{12}$$

after a time interval t. The electron trajectory after the second momentum q_2 transfer is defined by the equations

$$x' \left(\eta\right) = x'', \tag{13}$$

$$p' \left(\eta\right) + q_2 = p''. \tag{14}$$

To find the ionisation probability, the variables η and ξ' in Eq. (12) should be expressed as functions of E_n, $E_{n'}$ and $E_{n''}$. This can be realised on the basis of Eqs. (2), (3), (13) and (14). Then Eq. (12) is transformed to a three-term recurence relation for energies before, between and after the kicks. Calculation of the ionisation probability as a function of a time delay t is in progress. The classical theory developed above can be generalized when many kicks to the atom are applied. Recently [9,10] classical and quantum mechanical calculations for the 1D atom subject to a train of kicks have been performed, and the interesting chaotic phenomena have been observed.

3. 3D kicked atom. Classical theory

In the 3D case, it is convenient to define the orientation of the initial trajectory in the space by the Euler angles, as shown in Fig. 1.

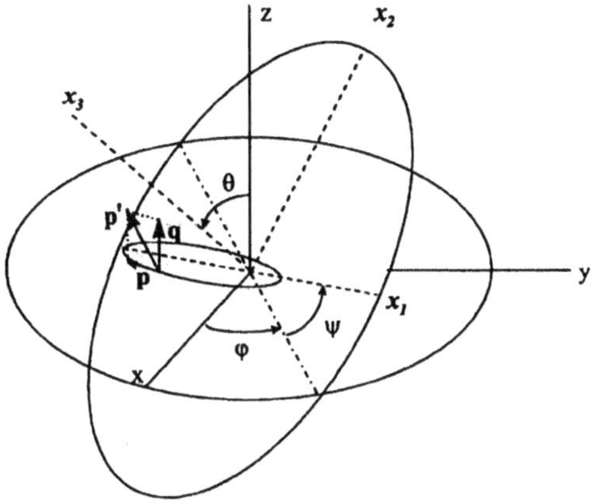

Fig. 1. Orientation of the electron trajectory in the space is defined by the Euler angles θ, ψ and φ. The sudden external impulse q changes the electron momentum from p to p'.

We are choosing the direction of $\mathbf{q}$ along the z axis, the direction of the projection m of the electron angular momentum. Then this projection is conserved and all expressions do not depend from the Euler angle φ. Instead of scalar Eqs. (2) and (3), we have now two vectorial equations. The vectorial product of them yields the angular momentum conservation equation

$$\mathbf{l} + \mathbf{r} \times \mathbf{q} = \mathbf{l}'. \tag{15}$$

The expression for the electron energy after the kick looks like the 1D Eq. (4). Only difference is that the last term of Eq. (4) depends also on the Euler angles θ and ψ [11]. In much the same way, the square of the angular momentum of the electron after the kick can be found [11]. After averaging over the Euler angles and over the period of the orbital motion, we obtain

$$\overline{l'^2} = l^2 + \frac{q^2 n^2}{3}\left(5n^2 - 3l^2\right). \tag{16}$$

When $l = 0$, the mean final angular momentum $\overline{l'} = 1.29qn^2$. In the experiment by Reinhold et al. [12], $n = 417$, $pn = 0.53$ and $\overline{l'} = 285$. Therefore, a very large angular momentum is transferred to the electron. Again, the reason is that the electron is kicked when it is far from the nucleus. But the electron angular momentum is proportional to r.

In order to find the transition probabilities, the 3D generalization of Eq. (7) is used:

$$\frac{1}{2\pi T_K}\int_0^{T_K} dt \int_0^{2\pi} d\psi = 1. \tag{17}$$

The integration over t and ψ may be replaced by n' and l' integrations. The Jacobian then determines the transition probability $W_{nlm}^{n'l'm}$, derived in [11]. Averaging $W_{nlm}^{n'l'm}$ over m, we have get the same expression for the transition probability $W_{nl}^{n'l'}$ as derived in [13] by a different theoretical method.

When $l = 0$, the probability of $n00 \to n'l'0$ transition is given by

$$W_{n00}^{n'l'0} = \frac{l'}{2\pi\left(nn'q\right)^3}\left|1 - \frac{l'^2}{q^2r^2}\left(3 - \frac{r}{n^2}\right)\right|^{-1}, \tag{18}$$

where the equation

$$\left(\frac{1}{n^2} - \frac{1}{n'^2} - q^2\right)^2 = 4\left(\frac{2}{r} - \frac{1}{n^2}\right)\left(q^2 - \frac{l'^2}{r^2}\right), \tag{19}$$
$$\frac{l'}{q} < r < 2n^2,$$

determines the distance r (the real and positive roots of Eq. (19)) at which this transition occurs.

In Fig. 2 the electron final-state energy and angular momentum distribution, equal to $n'^3 W_{n00}^{n'l'0}$, is plotted as a function of l' and scaled final energy $E_0 = E_{n'}/|E_n|$ for initial $n = 417$ and $qn = 0.53$. The electron energy distribution is symmetric with respect to the energy $E_n + q^2/2$. Figure 2 confirms the fact that the classical transition is singular at some combinations of parameters.

It also shows that the transition probabilities to states with the large angular momentum l' are dominant.

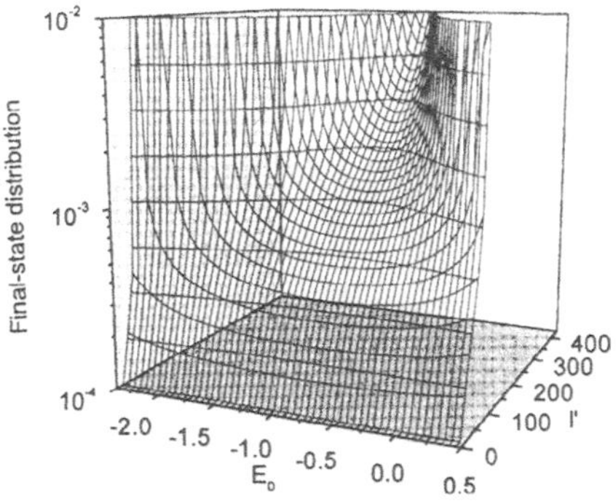

Fig. 2. The electron final-state energy and angular momentum distribution for the initial state with $n = 417$ and $l = m = 0$. The energy axis is scaled to the energy of the initial state, $E_0 = E_{n'}/|E_n|$. The parameter qn is equal to 0.53.

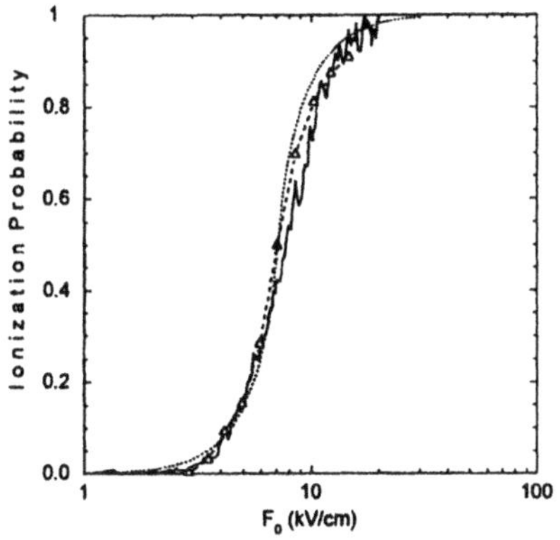

FIG. 3. The ionization probability from the $35d$ state vs. the peak field $\mathbf{F}_0$ at pulse duration $\tau = 0.5$psec. Solid line: data from experiment of Ref. [4]. Open triangles: numerical calculation from Ref. [4]. Dashed line: the classical theory, Eq. (20).

Integrations of $W_{n00}^{n'l'0}$ over l' and positive final energies yield the ionisation probability

$$W_{ion}(n,0) = \frac{2}{\pi}\left[\arctan(qn) + \frac{q^2n^2-1}{2qn}\ln\left|\frac{q^2n^2+1}{q^2n^2-1}\right|\right], \tag{20}$$

deduced previously on the basis of the binary encounter and semiclassical theories [14]. In Fig. 3 Eq. (20) is compared with the experimental data and numerical integration of the classical equations of the electron motion [4]. The agreement between all the data is very good.

4. Semiclassical theory of 1D kicked atom

If the atom is initially in some stationary state $\psi_n(x)$, the wave function of the electron immediately after application of a short pulse is $\exp(iqx)\psi_n(x)$ (see Ref. [12]). The final electronic wave function can be written as

$$\psi(t,x) = \sum_{n'} e^{-iE_{n'}t}\psi_{n'}(x)\,T_{nn'}, \tag{21}$$

where

$$T_{nn'} = \int \psi^*_{n'}(x)\,e^{iqx}\psi_n(x)\,dx. \tag{22}$$

In the semiclassical approximation, the matrix element $T_{nn'}$ is reduced to

$$T_{nn'} = \frac{1}{2\pi(nn')^{\frac{3}{2}}}\sum_{l_1,l_2=\pm1}\int\frac{\exp[iS(x)]}{(k_nk_{n'})^{\frac{1}{2}}}dx, \tag{23}$$

where

$$S(x) = l_1S_n(x) + l_2S_{n'}(x) + qx - \frac{\pi}{2}(l_1+l_2), \tag{24}$$

$$S_n(x) = \int_0^x k_n(x')\,dx', \tag{25}$$

$$k_n(x) = (2E_n + 2/x)^{1/2}, \tag{26}$$

and similarly are defined $S_{n'}(x)$ and $k_{n'}(x)$ for the final state.

The integral (23) can be estimated by the method of stationary phase. The stationary point is defined by the equation

$$l_1k_n(x) + l_2k_{n'}(x) + q = 0 \tag{27}$$

and is equal to $x_n^{n'}$ given by Eq. (5) in the classical mechanics. The method of stationary phase yields ($q > 0$)

$$T_{nn'} = \frac{x_n^{n'}}{(2\pi qn^3n'^3)^{1/2}}\exp\left[iS\left(x_n^{n'}\right) + i\frac{\pi}{4}l_1l_2\right]. \tag{28}$$

Only one term is remaining in Eq. (23). As it follows from Eq. (27), the phase of the matrix element (28) is different in different energy regions:

$$\begin{array}{lll} l_1 = -1, & l_2 = 1, & E_{n'} < E_n - q^2/2, \\ l_1 = l_2 = -1, & & E_n - q^2/2 < E_{n'} < E_n + q^2/2, \\ l_1 = 1, & l_2 = -1, & E_{n'} > E_n + q^2/2. \end{array} \tag{29}$$

The transition probability from the state n to n' is equal to $|T_{nn'}|^2$ and, multiplied by n'^3, coincides with the electron energy distribution given in the classical mechanics (Eq. (9)).

Replacing summation over n' by $E_{n'}$ integration, the wave packet (21) in the semiclassical approximation can be presented as

$$\psi(t,x) = \frac{q^{\frac{3}{2}}}{\pi n^{\frac{3}{2}}} \int dE_{n'} \frac{1}{Q_{nn'} k_{n'}^{\frac{1}{2}}(x)} \sum_{l=\pm 1} \exp\left[i S_n^{n'}\left(t, x, x_n^{n'}\right)\right], \tag{30}$$

where

$$S_n^{n'}\left(t, x, x_n^{n'}\right) = -tE_{n'} + l\left[S_{n'}(x) - \frac{\pi}{4}\right] + S\left(x_n^{n'}\right) + \frac{\pi}{4} l_1 l_2. \tag{31}$$

Again using the method of stationary phase for estimation of the integral (30), we obtain the following condition on time

$$\left| t(x) \pm t\left(x_n^{n'}\right)\right| = t, \tag{32}$$

where $t(x)$ is the time of electron classical motion with energy $E_{n'}$:

$$t(x) = \int_0^x \frac{dx'}{k_{n'}(x')}. \tag{33}$$

Equation (32) determines the stationary point of the energy $E_{n'}$.

If the second pulse is applied with a time delay t between the two pulses, the amplitude of transition from the initial state n to the final n'' is given by

$$T_{nn''} = \sum_{n'} T_{n'n''}(q_2)\, e^{-itE_{n'}} T_{nn'}(q_1). \tag{34}$$

Then instead of Eq. (32), we have the following equation

$$\left| t\left(x_{n'}^{n''}\right) \pm t\left(x_n^{n'}\right)\right| = t, \tag{35}$$

which coincides with Eq. (12) of the classical mechanics. However, the transition probabilities in both mechanics are different, because the semiclassical transition probabilities depend on the phase difference of the matrix elements $T_{nn'}$ and $T_{n'n''}$.

5. References

1. Bethe, H. A. and Salpeter, E. E. (1977) *Quantum Mechanics of One- and Two- Electron Atoms*, Plenum/Rosetta, New York.

2. Bersons, I. (1981) *Multiphoton ionization of high Rydberg states*, Phys. Lett. **84A**, 364-366.
3. Bersons, I. (1982) *Multiphoton ionization of highly excited atomic states*, Zh. Eksp. Teor. Fiz. **83**, 1276-1286 [Sov. Phys. JETP **56**, 731-831].
4. Jones, R. R., You, D., and Bucksbaum, P. H. (1993) *Ionization of Rydberg atoms by subpicosecond half-cycle electromagnetic pulses*, Phys. Rev. Lett. **70**, 1236-1239.
5. Jones, R. R., Tielking, N. E., You, D., Raman, C., and Bucksbaum, P. H. (1995) *Ionization of oriented Rydberg states by subpicosecond half-cycle electromagnetic pulses*, Phys. Rev. **A51**, R2687-R2690.
6. Frey, M. T., Dunning, F. B., Reinhold, C. O., and Burgdörfer, J. (1996) *Ionization of very-high-n Rydberg atoms by half-cycle pulses in the short-pulse regime*, Phys. Rev. **A53**, R2929-R2932.
7. Bersons, I., Kulsh, A. (1999) *Large angular-momentum changing in a short half-cycle pulse interaction with a Rydberg atom*, Phys. Rev. **A60**, 3144-3147.
8. Bromage, J. and Stroud, C. R. (1999) *Excitation of a three-dimensionally localized atomic electron wave packet*, Phys. Rev. Lett. **83**, 4963-4966.
9. Yoshida, S. et. al. (1999) *Floquet analisys of the dynamical stabilization of the kicked hydrogen atom*, Phys. Rev. **A59**, R4121-R4124.
10. Yoshida, S., Reinhold, Ç. O., and Burgdörfer, J. (2000) *Quantum localization of the kicked Rydberg Atom*, Phys. Rev. Lett. **84**, 2602-2605.
11. Bersons, I. and Kulsh, A. (2000) *Classical form factor for* $nlm \rightarrow n'l'm$ *transitions in the hydrogen atom*, Phys. Lett. (accepted).
12. Reinhold, C. O., Burgdörfer, J., Frey, M. T., and Dunning, F. D. (1996) *Dynamics of Rydberg wave packets generated by half-cycle pulses*, Phys. Rev. **A54**, R33-R36.
13. Vrinceanu, D. and Flannery, M. R. (1999) *Classical and quantal atomic form factors for arbitrary transitions*, Phys. Rev. **A60**, 1053-1069.
14. Bersons, I. and Kulsh, A. (1999) *Excitation and ionization of Rydberg atoms by short half-cycle pulses*, Phys. Rev. **A59**, 1399-1403.

STRONG FIELD COHERENT CONTROL

P.H. BUCKSBAUM
Physics Department, University of Michigan
Ann Arbor, MI 48109-1120 U.S.A.

1. Introduction

This paper reviews recent progress in experiments to control quantum and classical dynamics in condensed phase and gas phase systems, using shaped ultrafast radiation. Many of the same techniques that have led to laser pulses in the 10-100 fsec range can also be applied to the control of quantum systems with similar dynamical time scales. Systems under study include electron wave packets in Rydberg atoms, chemical dynamics in molecular liquids and lattice dynamics in crystalline solids. The feature common to each of these systems is ultrafast response. Shaped intense ultrafast radiation initiates the dynamics, which can then be studied using several new techniques.

Coherent control aims to go beyond observations of atomic and molecular properties, to manipulate the dynamics of quantum systems using strong external fields. One of our goals has been the construction of a "learning machine," that is, an automated experimental apparatus that can use feedback signals from a quantum system to help optimize control strategies for a particular application. The learning machine consists of a learning algorithm, together with programmable experimental inputs and readouts. Learning machines could be particularly useful for controlling systems where the Hamiltonian is not known completely.

On a different level, wave packets are forms of information stored in the quantum system. We have studied some simple examples where information is stored as quantum phase in a Rydberg wave packet, and then efficiently extracted by converting the phase information to quantum amplitudes.

I will describe some basic aspects wave packet shaping and control, and then summarize our work on learning algorithms and quantum information storage.

2. Sculpting quantum wave packets

2.1 OPTICAL PULSE SHAPING

Strong field control of matter is based on ultrafast amplified laser pulses, which have been developed over the past ten years. The "front-end" of an ultrafast laser system is a Kerr-Lens mode-locked titanium sapphire laser. This laser produces light pulses that are approximately 50 to 100 fsec in duration through a nonlinear mechanism

B. Piraux and K. Rzążewski (eds.), Super-Intense Laser-Atom Physics, 395–410.

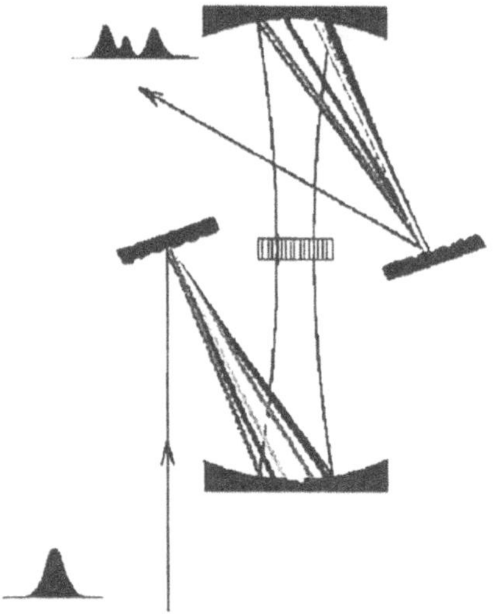

Figure 1: Fourier filter optical pulse shaper

called Kerr-lens mode locking, which ensures that only the highest peak power, and hence the shortest pulse, gets all the gain in the laser. Much has been made of the short duration of these pulses, which makes possible new technologies and precise timing for dynamics experiments.[1] We are more interested in the coherence of the spectrum, which is an equivalent property: The only way to make a short pulse is to coherently superpose a broad spectrum with great precision. All of the Fourier coefficients must be relatively real so that their corresponding spectral components add in phase.

The amplifiers for ultrafast lasers take advantage of this spectral precision through a technique called chirped pulse amplification (CPA). The idea of CPA is to spectrally disperse the frequency components in the pulse, so that they travel through the amplifier at different times. This minimizes the peak power in the amplifier, which can then be run at higher total amplified pulse power. Here's how chirped pulse amplification works. The light passes through a grating and a telescope, which disperses the spectrum laterally. It also delays the various colors with respect to each other, just because the different exit angle off the grating makes their path through the telescope different. If a second telescope and grating is inserted in the mirror-reversed configuration, the result is to disperse the spectrum a second time by the same amount, while putting all the colors back into overlap. Thus the net effect is to stretch, or "chirp" the pulse. Following amplification, a second grating pair can be inserted to reverse the chirp and restore the short pulse. Chirped pulse amplifiers are responsible for nearly all of the very high intensity laser-atom physics reported in this NATO volume.

Pulse shapers take this technology to the next logical step. Rather than simply stretch the pulse, a filter is inserted into the laser beam at the midpoint of the pulse stretcher, just between the two telescopes, as shown in figure 1. The filter can attenuate selected colors in the laser beam or even add phase delays to them. The result is not a chirped pulse, but rather a chirped and shaped pulse. If the stretcher is adjusted so that the net stretching is nullified, then the device becomes a simple Fourier filter. Such filters can reshape the pulse to any new shape subject to only two limitations: First, no feature in the shaped pulse can rise or fall faster than the limit imposed by the total spectral bandwidth, which is the same limit that determines the minimum duration of the input pulse. Second, the total length of the shaped pulse is limited to the Fourier transform of the narrowest spectral feature that was altered by the filter: in other words, the size of a single "pixel" on the filter is inversely proportional to the maximum pulse length.

2.2 RYDBERG WAVE PACKET SCULPTING

The shaped optical field creates a shaped wave packet in an atom through photoexcitation from a pure atomic state. In lowest order perturbation theory, the shaped optical pulse $\mathbf{F}(t) = \hat{z}F(t)$ with complex spectrum

$$F(\omega) = \int_{-\infty}^{\infty} F(t)e^{i\omega t}dt \tag{1}$$

excites a wave packet

$$\Psi(t) = \sum_n c_n \phi_n e^{-i\omega_n t} \tag{2}$$

where each c_n is a complex constant given (in atomic units) by

$$c_n = \int_{-\infty}^{\infty} F(t)e^{i(\omega_0-\omega_n)t}\langle\phi_n|z|\phi_0\rangle \tag{3}$$

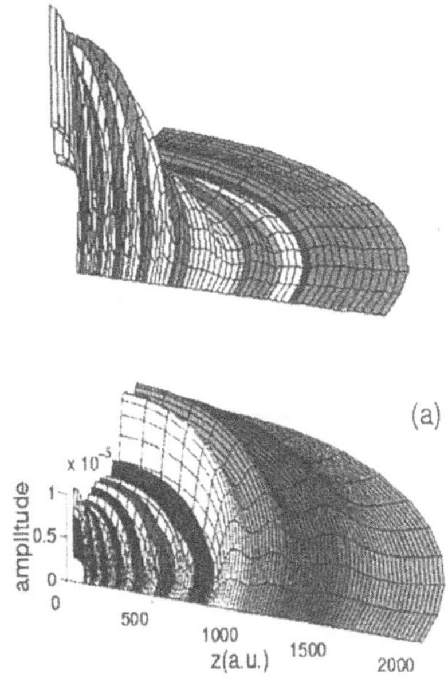

Figure 2: Rydberg wave packets Top: near the inner turning point. Bottom: near the outer turning point.

If the orbitals that make up the wave packet differ only by their principal quantum number n, then the matrix elements are all relatively real. Furthermore, since the "launch state" ϕ_0 is at or near the ground state of the atom, the matrix elements generally scale according to a constant energy coulomb scaling rule, $\langle\phi_n|z|\phi_0\rangle \propto n^{-3/2}$. If this can be taken as approximately constant for the range of n states in the wave packet, then the similarity of equation (1) and equation (3) means that $c_n \propto F(\omega_n - \omega_0)$. Thus the wave packet coefficients can be engineered from the pulse spectrum.

Our wave packet sculpting experiments have used atomic cesium in an effusive beam. Atoms of ground state cesium are first excited to the 7s state via two photons from a Raman shifted pulsed dye laser at a wavelength of 1.08 μm. Then the 7s atoms are excited to Rydberg wave packets using 100 fsec pulses from an amplified Ti:Sapphire laser operating at a central wavelength of 795 nm. Approximately 5-10% of the atoms in the laser focal volume reach Rydberg states on every pulse of the laser. This number is limited by the need to avoid multiphoton absorption, and by the pulse shaping itself, which filters the light as required to produce the desired shape. The use of cesium is dictated only by the convenient location of the 7s level, which lies below the Rydberg series by the energy of an 800 nm Ti:Sapphire photon. Other sources of broadband coherent radiation, such as an optical parametric amplifier, can extend the range of sculpted radiation throughout the visible and near-visible spectrum.

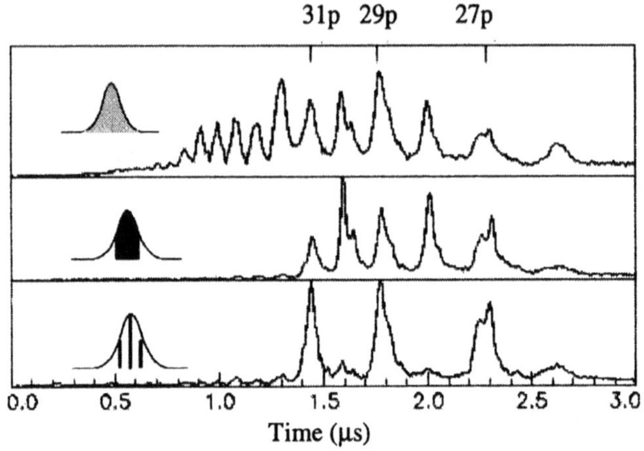

Figure 3 Ramped Field ionization of three sculpted cesium wave packets. The inset figures show the general shape of the excitation spectra.

2.3 WAVE PACKET SHAPES

The wave packet excited in a single Rydberg electron covers a radial range out to approximately n^2 Bohr radii. If the wave packet is created with relatively real coefficients c_n it is more compact, and concentrated near the ion core. Figure 2a shows a wedge-shaped section of such a packet. The ion is located at the point of the wedge, where the wave function is piled up. The alternating light and dark rings show the quantum phase evolution along the radial direction. Rapid phase changes indicate high momentum, where the wave packet is moving rapidly away from the ion. Figure 2 bottom shows the situation after one-half cycle of motion, when the wave packet is localized at the outer turning point of its orbit.

2.4 VIEWING WAVE PACKET SCULPTURES

The images shown in figure 2 represent data from cesium wave packet sculpture experiments. The basic technique used to obtain these pictures is state-selective ramped field ionization, which works as follows: Rydberg atoms are unstable against field ionization above a critical field. Each Rydberg energy eigenstate ionizes at the field where the combined static plus coulomb potential surface develops a saddle point that dips below the binding energy of the state. For a state with energy E_n, the critical field is $F_{crit} = E_n^2 / 4$ in atomic units.

When a Rydberg wave packet is subjected to a slowly varying ramped field, the critical field for ionization is not well defined; rather, there is a probability for ionization at any critical field corresponding to the constituent energy levels in the wave packet. Each level ionizes with probability $|c_n|^2$, and an ensemble of identically prepared wave packets displays a series of ionization peaks during the ramp, as shown in figure 3. If the amplitudes of different frequencies are sculpted away by the pulse shaper, the atomic state that can be excited at that frequency is sculpted as well.

With care to avoid problems like saturation in the charged particle detector, the ramped field ionization distribution can be a faithful measure of the amplitudes of each state in the packet. Unfortunately, though, this signal is totally insensitive to the relative phase of the different states in the wave packet. Although the *absolute* phase of a Schrödinger wave function is not an observable, the relative phases are necessary to determine the shape of the wave function. If the phases are relatively real and positive, then the wave function is piled up around the ion, as in the top image in figure 2. This phase information is obtained through wave packet cross correlation interference, a technique which is similar to optical holography. A reference wave packet is created in superposition with the shaped wave packet at the moment when the shape measurement is made. Ramped field ionization of the superposed wave packets then can reveal the relative phase of the shaped orbitals with respect to the reference. In practice, we can detect and control the individual orbitals to within a few degrees of phase.

The full description of a wave packet sculpture can be contained in a simple table of the amplitude and phase of every orbital participating in the wave packet. More insight, though, comes from viewing the sculptures as reconstructed images in configuration space, momentum space, or a mixture. One particularly useful view is a Wigner distribution. This is a six-dimensional phase space representation of the wave packet (three space plus three momentum axes), given formally by the following transformation from the Schrödinger wave function:

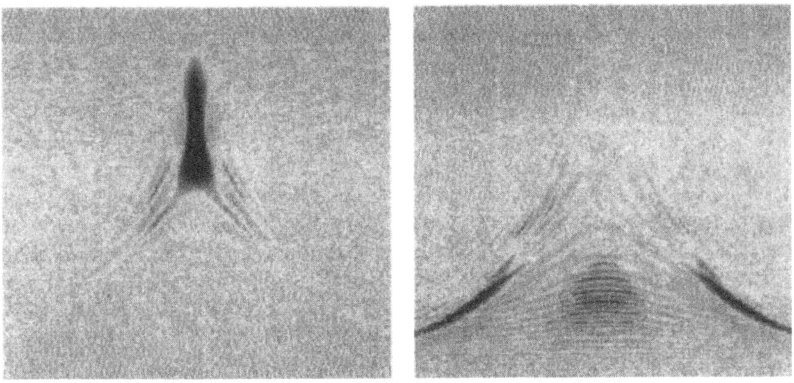

Figure 4 Wigner distributions for the wave packets in figure 2. The horizontal direction indicates the radial momentum with zero in the middle; the vertical direction plots radial position, with the atomic nucleus located at the bottom. Left: wave packet at the outer turning point. Right: wave packet at the inner turning point.

$$\rho(r, p_r) = (1/\pi)\int_{-\infty}^{\infty} ds \exp(2ip_r s)\Psi^*(r-s)\Psi(r+s) \tag{4}$$

The radial Wigner plots for our radial Rydberg wave packet sculptures display the motion more clearly than the Schrödinger view, as shown in figure 4.

Rydberg wave packets can also be used to store and manipulate information in the form of quantum amplitudes and phase. A simple example of this kind of quantum information storage and retrieval is to simply encode a binary number into the phase of a Rydberg wave packet. Figure 5 displays data from a recent experimental test of the retrieval of the binary number 0000010000, which was encoded into a Rydberg wave packet by assigning each orbital to represent one bit. If the value of the bit is 0, the state is real in positive; if the value is binary 1, the corresponding state is real and negative.[2]

Ramped field ionization will not reveal which bits are 1's and 0's, since it only reads the amplitude of each state and not its phase. However, a simple form of wave packet holography can decode the information, if we superpose the shaped wave packet with a reference "universal decoder" wave packet, whose states are all real and negative. Such a superposition amplifies the binary 1 states, and eliminates the binary 0 states, so that ramped field ionization can detect the location of the binary 1 bits. This form of data retrieval has been discussed in the context of quantum algorithms, that may be useful computational tools for some problems.[3]

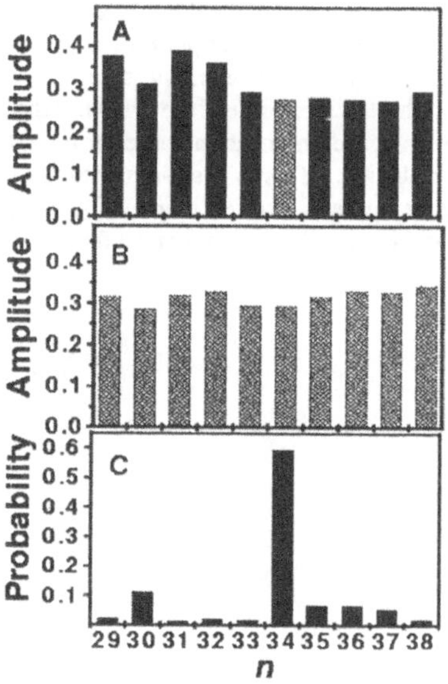

Figure 5 A: 10-state Rydberg wave packet encoded so that the relative phases store the binary number 0000010000. B: Universal decoding wave packet. C: Decoded wave packet. (From Reference 2.)

3. Strong Field Applications of Sculpted Wave Packets

The Rydberg wave packet sculpting techniques described above require atoms excited in the perturbative limit. However, strong fields can come into play in various ways that we have studied. One important application is the excitation of the sculpted wave packets themselves with strong fields, leading to wave packet trapping. The theory of this process is discussed elsewhere in this volume[4]. Here we will summarize our recent efforts to control the shape of the wave function with strong far-infrared fields following their manufacture using the wave packet sculpting apparatus that was described in section 2.

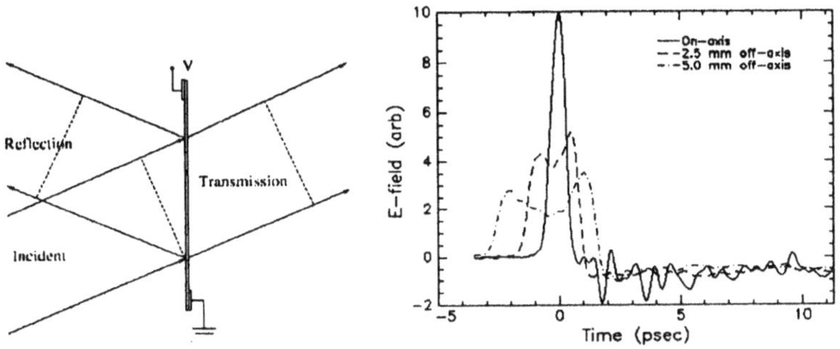

Figure 6 Left: Half-cycle pulses are made by exciting a biased photoconductive antenna with a short optical pulse. Right: The measured shape of a half cycle pulse in and near a focus. (From Reference 5.)

3.1 HALF-CYCLE PULSES AND RYDBERG STATES

Strong far-infrared broadband coherent radiation pulses are produced in the laboratory by shorting out a biased large area semiconductor switch, as shown schematically in figure 6.[5] The radiation, when focused, appears as a half-cycle pulse. Peak fields up to 10^5 *V/cm* can be focused onto the Rydberg atoms in an atomic beam. This leads to rapid ionization and/or redistribution of coherent population.[6]

Photoionization of Rydberg states by half-cycle pulses cannot be described by single-order perturbation theory, because a large number of photons and many intermediate states are involved in the process. One simple strong-field approach to the problem is the impulse approximation. Here the HCP is modeled as a delta function in time, which transfers momentum to the atomic electron. In the impulse limit, the initial wave function of the electron in momentum space $\phi(\mathbf{p})$ is advanced by the impulse

$$\mathbf{Q} = \int \mathbf{F}(t)dt \tag{5}$$

$$\phi(\mathbf{p}) \rightarrow \phi(\mathbf{p}+\mathbf{Q}). \tag{6}$$

The corresponding effect on the configuration space representation of the same wave function is to multiply the state by a spatially varying phase:

$$\Psi(\mathbf{x}) \rightarrow \Psi(\mathbf{x})e^{i\mathbf{Q}\bullet\mathbf{x}} \tag{7}$$

The simple form of this transformation makes it possible to use HCP ionization to analyze the spatial structure of a bound wave function. The ionization rate for a given **Q** depends on the probability that the wave function is transported into the continuum following the impulse. This depends, in turn, on the local WKB momentum of the state as a function of position. For a state with bound energy *E<0*, the WKB momentum has a magnitude related to the local kinetic energy:

$$|\mathbf{p}(\mathbf{x})| = \sqrt{2(E - V(x))}\,. \tag{8}$$

After the impulse, the wave function energy depends on position:

$$E'(\mathbf{x}) = \frac{|\mathbf{p}(\mathbf{x})+\mathbf{Q}|^2}{2} + V(\mathbf{x}) \qquad (9)$$

The ionization probability is equal to the integral of $\psi(\mathbf{x})$ over the region where $E'>0$. Thus more deeply bound parts of the wave function contribute to ionization at lower values of **Q**. This has led to the development of wave packet tomography,[7] and also simplifies the analysis of wave packet ionization experiments.[8]

The impulse model has reasonable success in describing photoionization, so long as the natural time scale for evolution of the phase of the quantum state is much longer than the duration of the HCP. For Rydberg atoms, this time scale is the Kepler orbit time $t_K=2\pi n^3$, which is much faster than the pulse shown in figure 6 for $n>20$. Some important features of HCP ionization, such as the asymmetry in the in the ionization of oriented states, depend critically on dynamics in the part of the wave function close to the core. Here the phase is evolving much faster, and the impulse approximation is not valid. Reliable calculations of these processes require a full integration of Schrödinger's equation.[9]

3.2 STRONG FIELD CONTROL OF WAVE PACKETS WITH HCP'S

The effect of strong half cycle pulses on sculpted wave packets can be illustrated by returning to the data base search problem described in section 2.1. Suppose we construct a wave packet to store binary information as phase, just as illustrated in the top section of figure 5. An n-state wave packet with one flipped phase in state m can also be described as a wave packet with equal phases for all the states, superposed with an out-of-phase eigenstate with twice the amplitude:

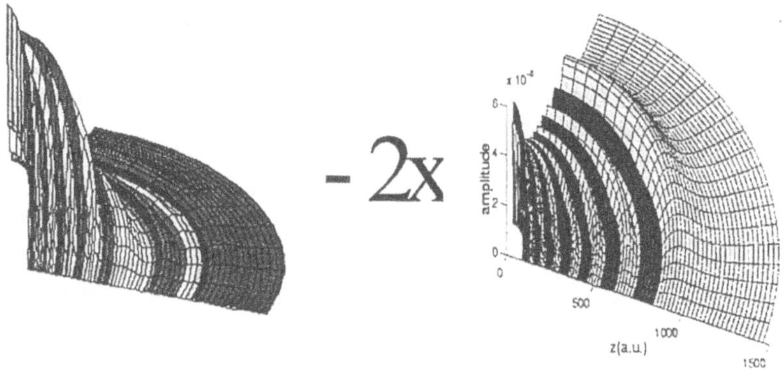

Figure 7: A wave packet with one flipped phase can be represented as the superposition of a compact n-state wave packet, minus 2x an eigenstate.

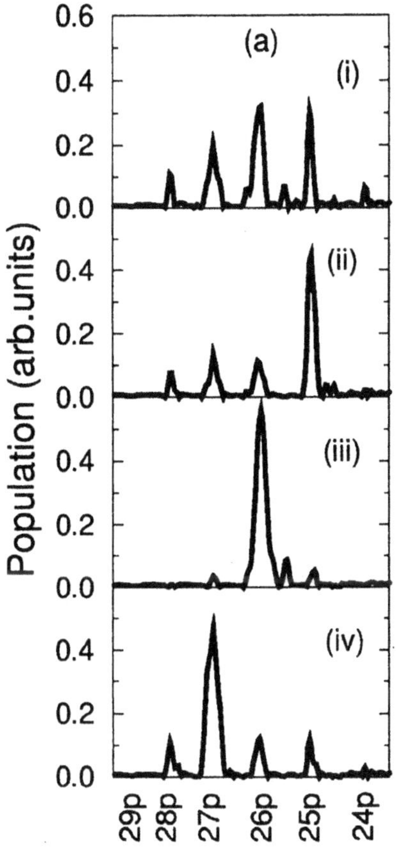

Figure 8. (i): Ramped field ionization of a Rydberg wave packet. One of the states is phase-reversed, but the ramped field signal cannot detect it. (ii)-(iv): Ramped field ionization following redistribution with a half-cycle pulse, for three different phase-flipped states. The phase reversal is converted to a large amplitude enhancement by the HCP. (Ref. 10)

$$\Psi(\mathbf{x}) = \frac{1}{\sqrt{n}}\sum_i \phi_i(\mathbf{x}) - \frac{2}{\sqrt{n}}\phi_m(\mathbf{x}) \qquad (10)$$

This superposition is shown in figure 7. The first term corresponds to a compact wave packet located around the core, where the local WKB momentum is very high. The second term is an eigenstate, which is not localized near the core. Therefore, a value of Q can be found that will ionize the wave packet in the first term, while leaving the second term intact. Thus a single half-cycle pulse will reveal the location of the flipped bit. Impulse calculations on model Rydberg wave packets confirm this idea, as do full integrations of Schrödinger's equation.

We have recently performed an experiment to test this strong field search mechanism, and the results are shown in figure 8. A simple 5-state radial wave packet was constructed in cesium, with one of the states phase-flipped with respect to the others. At values of momentum transfer predicted by the impulse model, we see a suppression of the unflipped states. Thus the half cycle pulse, followed by ordinary ramped field ionization, is able to successfully find the flipped state in the data base.[10]

One largely unexplored area of strong field terahertz physics is the interaction of shaped intense far-infrared pulses with atoms. It may be possible to shape far-infrared pulses by sculpting the optical radiation incident on the semiconductor switch. Following the early work by Weling and Auston on this topic,[11] we have modulated the optical excitation in order to produce narrow-band terahertz pulses and observed strong-field effects in Rydberg state redistribution.[12] This should be a fertile area for control in the future.

4. Strong Field Control in Crystals

4.1 CONTROL OF LARGE AMPLITUDE COHERENT LATTICE MOTION

Coherent motion of a crystalline lattice can be induced via impulsive excitation, just as we have been discussing for Rydberg atoms. Observations of high amplitude acoustic or optical phonon modes is straightforward, because the motion produces a time-varying linear index.[13] Two examples are shown in figure 9. The left figure shows a very high-Q optical phonon mode excited in a transparent crystal BGO by impulsive Raman scattering by 100 fsec optical pulses. The probe is optical transmission. The figure on the right shows a series of Raman excitations by two time-delayed optical pulses incident on crystalline bismuth, probed in this case by optical reflectivity. The second pulse can either enhance or suppress the motion, depending on the relative phase of the two impulses. This, then, is a form of coherent control.

4.2 X-RAY OBSERVATIONS OF COHERENT ACOUSTIC MOTION

X-ray diffraction is very sensitive to lattice motion, so x-rays are important probes of coherent dynamics in crystals. There has even been a recent suggestion that this connection could be turned around, so that coherent crystalline motion could be used to modulate x-ray beams for a number of dynamic probe applications.[14] Since the modulation can be as fast as an optical phonon period such as the ones shown in figure 9, this may lead to sub-picosecond x-ray switches.

We have been studying this problem in a new laser laboratory that we constructed with the Michigan-Lucent-Howard collaborative access team on an undulator beam line at the Advanced Photon Source at Argonne National Laboratory. The laser produces 50 fsec high intensity optical pulses that are synchronized to the 50 psec x-ray pulses. Similar laboratories now also exist at Berkeley's Advanced Light Source and at the European Synchrotron Radiation Facility in Geneva.

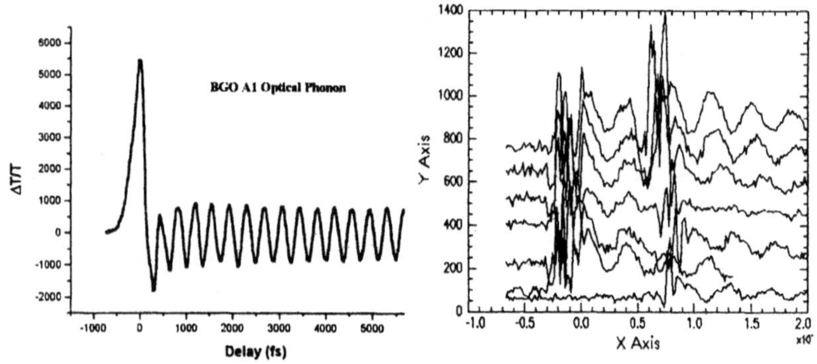

Figure 9: Left: Impulsive excitation of coherent optical phonons in BGO. Right: Two-pulse excitation of optical phonons in bismuth. The oscillations can be enhanced or suppressed depending on the relative time delay.

Early results of this work have concentrated the large coherent acoustic motion that the laser imparts to InSb crystals.[15] An example is shown in figure 10. This shows the motion induced by a pulse of 10-12 mJ/cm^2 incident on an InSb crystal with [111] orientation. Bragg diffraction by 10-keV x-rays was measured as a function of crystal angle, around the principal (111) Bragg peak. The peak moves by an amount roughly equal to its width following laser excitation. Analysis of the comparison with a simulation, also shown, reveals that approximately equal amounts of energy are deposited as heat, and as coherent motion.

The time resolution of these images is limited by the duration of the synchrotron pulse, which is 1000 times longer than the laser pulse. Future work will concentrate on new methods to improve this time resolution, so that sub-picosecond motion can be detected. Ultimately, we hope to use this knowledge to build x-ray modulators for measurement applications at synchrotrons.

5. Feedback with Strong Field Control: Towards Bond-Selective Chemistry

5.1 FEEDBACK LEARNING ALGORITHMS AND QUANTUM DYNAMICS

In principle, shaped broadband coherent radiation pulses should be able to transfer population between any two arbitrary states of a quantum system. The practical limitations to this are divided in two categories:

5.1.1 Technical limitations

Some pulse shapes are not possible because of technical limitations. For example, the laser system may have a bandwidth of only a few percent, centered on a particular central frequency, and this may not include enough frequency components for the job at hand. Another limitation in this first category is the technical operation of the pulse shaping filter. Its resolution is ultimately limited by the rules governing propaga-

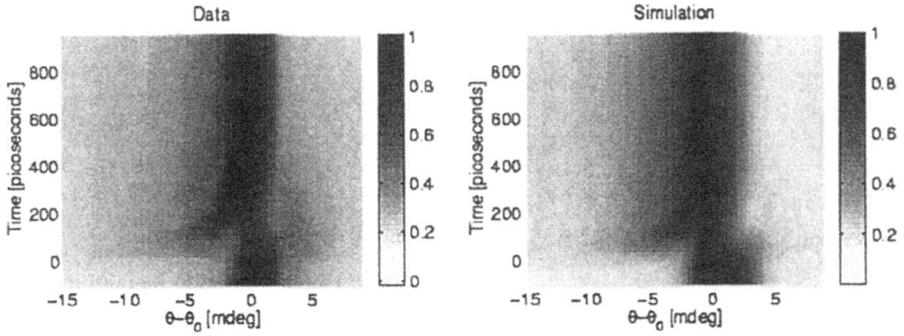

Figure 10: X-ray diffraction data (left) and simulation (right), displaying the coherent acoustic motion of an InSb crystal lattice following impulsive excitation with a strong ultrafast laser pulse. Diffraction angle is plotted on the x-axis, and time delay between the impulsive excitation and the x-ray probe is shown on the y-axis. (From Reference 15)

tion of any electromagnetic radiation. The only way to overcome this first category of limitations is through technological innovations. Other contributions to this NATO volume describe some of these, such as the new strong field techniques that are used to generate coherent radiation in the vacuum ultraviolet, and the prospects for generating coherent light at x-ray frequencies.

5.1.2 System knowledge

A second category of limitations on coherent control is imperfect knowledge of the material system that we wish to alter. For example, the initial state of the system may not be pure, but rather could be broadened by finite temperature to a Boltzman distribution among many possible states. There is then no wave function that can describe this situation. Instead, density matrices and master equations can describe the average dynamical properties of the incoherent mixture. Recently there have been theoretical efforts to describe how coherent control might be useful to aid in laser cooling of model quantum systems at finite temperature, but the general problem remains very difficult.[16]

Also in this second category are pure systems whose evolution cannot be calculated because the system Hamiltonian or its energy eigenstates are not completely known. Most few-body quantum systems fall into this category, such as polyatomic molecules or multi-electron atoms with configuration interactions.

5.1.3 Learning Control

A learning algorithms is an approach to solve coherent control problems when the limitations are caused by imperfect knowledge of the system.[17] A learning algorithm is a search for the best control path to reach a particular goal, or target state of the system. The search proceeds by comparing the relative fitness of different control patterns, and using this fitness to correct the control. Learning algorithms are widely used to help solve control-engineering problems of all sorts, but they are fairly new to optical control problems for quantum systems. We have begun to test these ideas to control some specific problems in strong field quantum dynamics.

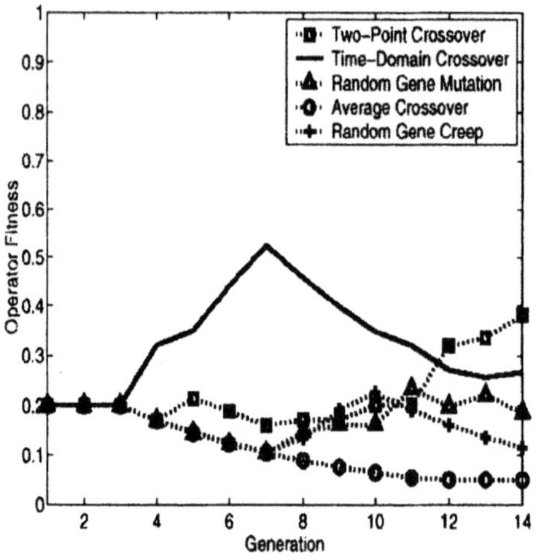

Figure 11: Fitness of five different operators used to learn how to optimize the yield of non-dissociative ionization of Na_2. (From Ref. 19)

The flow diagram for a learning algorithm proceeds as follows: The algorithm begins with the random selection of approximately 20 pulse shapes from the programmable pulse shaper shown in figure 1. Each of these shapes can be encoded into a string of numbers describing the pulse. For example, the numbers might record the amplitude and phase of each of 100 different Fourier components of the pulse. This string is called the *genome* for that pulse

Each of these pulses is then used in the experiment. For example, if we wish to control the final state in dissociative ionization of Na_2, then each pulse is focused into a molecular beam of sodium, and the results of multiphoton dissociative ionization are recorded. If the signal is weak, the experiment might be integrated over several trials, but the aim is to measure the *fitness* of this particular pulse for the control goal.

Following the trial, the fitness of each trial pulse in the initial population is compared. The pulses that are relatively more successful are then used to generate a new set of second-generation pulses, and the process is iterated. In this way, the pulses evolved towards better solutions to the problem.

The key to this evolutionary approach is the means by which the new generation of pulses is derived from the most successful individual pulses in the parent generation. There are many different approaches to this problem of procreation. For example, the characteristics of two successful parents can be traded to produce several offspring with some traits from each parent. This kind of mixing is called the "genetic algorithm." Alternatively, one parent pulse shape can be altered through random mutation to try to improve its performance in the next round. Other generating operators include averaging between successful parents, Fourier filtering the parent pulses, polynomial fits, and so on.

We have found that no single generation operator is optimal for our learning algorithms, so we use several different methods to produce each generation. The algorithm program then keeps track of the operator that produced the fittest offspring, by compiling an operator *genealogy* for each pulse. Operators that produce more successful pulses are given credit, and become more likely to be employed in producing the next generation.

An example of the use of genealogy is shown in figure 11. The problem was to maximize the yield of non-dissociative ionization of Na_2 while minimizing dissociative ionization. The best pulse was found by a combination of conventional cross-over operators, which mix genetic material from two parents; and *time-domain crossover*, where the genome is rewritten with an inverse Fourier transform so that it records the pulse shape in the time domain, rather than the frequency domain. We believe that this form

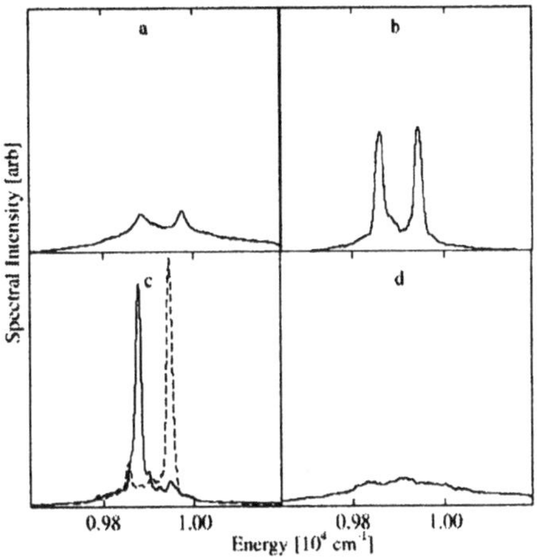

Figure 12: Forward scattered spectrum in methanol. a: for an unshaped 150 fsec pulse. b: Pulse shape to optimize both C-H stretch modes. c: Optimizing the asymmetric (solid) or symmetric (dotted) mode. d: suppressing both modes. (From Reference 18)

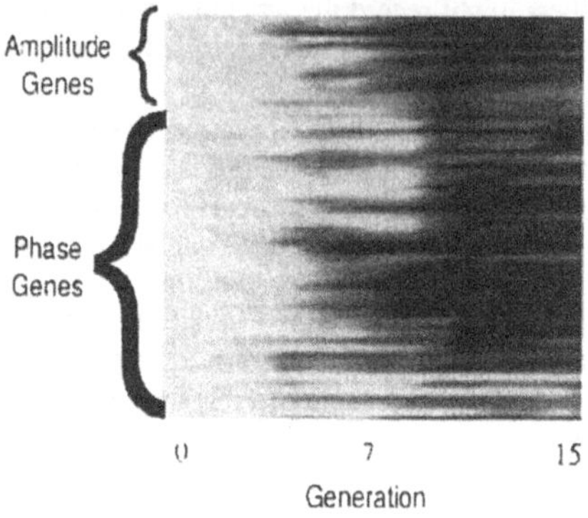

Figure 13: The genetic variation across the individuals in a generation is plotted as shading. Darker shades indicate less variation. Genes that affect the outcome show less variation. (From Ref. 19)

of operator worked better because, following the initial optimization steps, the best solution became more sensitive to the pulse width than to its color.

5.2 LEARNING CONTROL OF MODE EXCITATION IN METHYL ALCOHOL

Strong fields may be necessary to employ coherent control in molecular liquids, because of the problem of rapid dephasing that destroys the interference between different excitation paths. We have studied the selective excitation of C-H stretch modes in liquid room temperature methyl alcohol.[18] An intense 800 nm 150 fsec laser focused into the liquid undergoes self-phase modulation and stimulated Raman scattering, during which many modes of the molecule are excited. We used the SRS spectrum to optimize the laser energy deposited in the symmetric C-H stretch mode, and found we were able to shape the optical field to enhance the C-H symmetric stretch while suppressing the nearby asymmetric stretch, and the self-phase modulation. We were also able to run the algorithm to optimize the asymmetric stretch and suppress the symmetric stretch, or to enhance them both, or suppress them both. This initial SRS spectrum, along with several optimized spectra following completion of learning algorithms for different goals, are shown in figure 12.

The evolution of the pulse shapes toward the final goal can be viewed in several different ways. Figure 13 shows a plot of the variation in the different genes in the genome as a function of generation for one of the learning algorithms designed to optimize the symmetric stretch.[19] The genes all begin with wide variation, shown as a light color. As they evolve, some more successful characteristics become general features of all of the individuals in the generations, and the genes that code for these characteristics show less variation, and appear as darker shades on the map.

The pulse shape that corresponds to one of the successful individuals in the last generation is shown in Figure 14. This is a Husimi representation of the color spectrogram. Husimi plots are related to Wigner distributions. They display two conjugate variables at the same time, in this case the pulse field intensity as a function of both spectrum and time. To construct a Husimi plot, one begins with the Wigner distribution for the field:

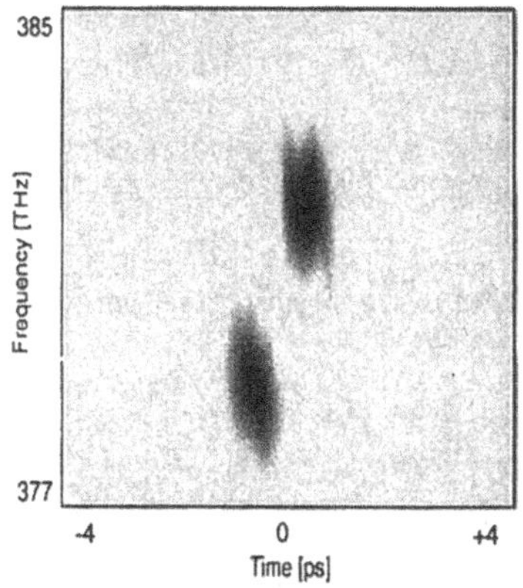

Figure 14 Husimi distribution of a shaped laser pulse that has been optimized to excited the C-H symmetric stretch in methyl alcohol. (From Reference 18)

$$S(t,\upsilon)=\int E(\upsilon)E^*(\upsilon-\upsilon')e^{2i\upsilon' t}d\upsilon' \quad (11)$$

This is a complete description of the field. The integral of S over time yields the intensity as a function of frequency, while the integral over frequency yields the intensity vs. time. Though complete, this function is difficult to interpret, and this can be improved if the function is convolved with a two dimensional Gaussian function. This is the definition of the Husimi distribution $Q(\nu,\nu')$:

$$Q(t,\upsilon)=\iint dt'd\upsilon' S(t',\upsilon')e^{-(\upsilon-\upsilon')^2(t-t')^2} \quad (12)$$

The Husimi distribution is essentially an optical spectrogram. Integration over one variable no longer yields the intensity vs. the other variable, because of the convolution; but the Husimi density displays the pulse as intensity vs. color and time simultaneously. The Husimi display of the solution to the symmetric stretch learning algorithm is shown in figure 14.

6. Conclusion and Acknowledgements

Shaping of ultrafast high intensity laser pulses has led to new tools and new insights in several different areas of optical physics. Coherent control may someday be an important tool in research in chemical physics, quantum information, and materials science.

Many researchers contributed to the science reviewed in this article. I especially acknowledge the contributions of Tom Weinacht, Matt Decamp, Jae Ahn, Brett Pearson, and David Reis, for providing the data used in the experimental examples in this review. This work was supported by the National Science Foundation.

7. References

1. Elsässer, Th., Fujimoto, J.G., Wiersma, D.A., and Zinth, W., editors, (1999) Ultrafast Phenomena XI, *Springer Series in Chemical Physics* **63**, Springer-Verlag Publishers, Berlin; and previous volumes in this series.
2. Ahn, J., Weinacht, T.C., and Bucksbaum, P.H. (2000) Quantum information storage and retrieval in Rydberg wave packets, *Science* **287**, 463-466.
3. Grover, L.K. (1997) *Physical Review Letters* **79**, 325-328; Grover, L.K., (1997) *Physical Review Letters* **79**, 4709-4912.
4. Federov, M., (2000) elsewhere in this volume.
5. Budiarto, E., Pu, N.-W., Jeong, S., and Bokor, J., (1998) *Optics Letters* **23**, 213-215.

6. Bucksbaum, P.H. (1999) Wave packets and Half-cycle Pulses, in *The Physics and Chemistry of Wave Packets*, ed. by J. Yaezell, John Wiley Publishers, New York, 159-192.
7. Campbell, M.B., Bensky, T.J., and Jones, R.R. (1998) Probing electronic radial wave packets using impulsive momentum retrieval, *Physical Review A* **58** 514-520.
8. Raman, C.S. Conover, C.W.S., Sukenik, C.I., and Bucksbaum, P.H. (1996) Ionization of Rydberg Wavepackets by Sub-Picosecond, Half-Cycle Electromagnetic Pulses, *Physical Review Letters* **76**, 2436-2439.
9. Raman, C.S., Weinacht, T.C., and Bucksbaum, P.H., (1997) Stark Wavepackets Viewed with Half-Cycle Pulses, *Physical Review A* **55**, R3995-R3998.
10. Ahn, J., Huchinson, D.N., Rangan, C., and Bucksbaum, P.H. (2000) Quantum phase retrieval of a Rydberg wave packet using half cycle pulses, LANL preprint archive quant-ph/0009064.
11. Weling, A.S., Hu, B.B., Froberg, N.M., and Auston, D.H. (1994) Generation of Tunable Narrow-Band THz Radiation from Large-Aperture Photoconducting Antennas, *Applied Physics Letters* **64** 137-139.
12. Raman, C.S., DeCamp, M.F., and Bucksbaum, P.H., (1997) Redistribution of atomic Rydberg states by tunable narrow band picosecond far-infrared pulses, *Optics Express* **1**, 186-196.
13. Cheng, T.K., Brorson, S.D., Kazeroonian, A.S., Moodera, J.S., Dresselhaus, G., Dresselhaus, M.S., and Ippen, E.P., (1990) Impulsive Excitation of Coherent Phonons Observed in Reflection in Busmuth and Antimony, *Applied Physics Letters* **57**, 1004-1006; Hase, M., Mizoguchi, K., Harima, H., et al. (1996) Optical control of coherent optical phonons in bismuth films, *Applied Physics Letters* **69**, 2474-2476.
14. Bucksbaum, P.H., and Merlin, R.D. (1999) The phonon Bragg switch: A proposal to generate sub-picosecond x-ray pulses, *Solid State Communications*, **111**, 535-539.
15. Reis, D.A., DeCamp, M.F., Bucksbaum, P.H., Clarke, R., Hertlein, M., Merlin, R., Dufresne, E., Williams, E., Falcone, R., Kapteyn, H., Murnane, M., Larsson, J., Missalla, Th., and Wark, J., (2000) Xe-ray probe of ultrafast strain propagation, submitted for publication.
16. Tannor, D.J., Bartana, A. (1999) On the interplay of control fields and spontaneous emission in laser cooling, *Journal of Physical Chemistry A* **103**, 10359-10363.
17. Judson, R.S., and Rabitz, H., (1992) Teaching Lasers to Control Molecules, *Physical Review Letters* **68**, 1500-1503.
18. Weinacht, T.C., White, J.L., and Bucksbaum, P.H. (1999) Toward Strong Field Mode-Selective Chemistry, *Journal of Physical Chemistry A* **103**, 10166-10168.
19. Pearson, B.J., White, J.L., Weinacht, T.C., and Bucksbaum, P.H., (2000) Adaptive learning algorithms for coherent control, submitted for publication.

Subject Index

Author Index

GPSR Compliance

The European Union's (EU) General Product Safety Regulation (GPSR) is a set of rules that requires consumer products to be safe and our obligations to ensure this.

If you have any concerns about our products, you can contact us on ProductSafety@springernature.com

In case Publisher is established outside the EU, the EU authorized representative is:

Springer Nature Customer Service Center GmbH
Europaplatz 3
69115 Heidelberg, Germany

Batch number: 10873696

Printed by Printforce, the Netherlands